S0-BMG-663

Biochemistry of Cell Differentiation II

Publisher's Note

The *International Review of Biochemistry* remains a major force in the education of established scientists and advanced students of biochemistry throughout the world. It continues to present accurate, timely, and thorough reviews of key topics by distinguished authors charged with the responsibility of selecting and critically analyzing new facts and concepts important to the progress of biochemistry from the mass of information in their respective fields.

Following the successful format established by the earlier volumes in this series, new volumes of the *International Review of Biochemistry* will concentrate on current developments in the major areas of biochemical research and study. New volumes on a given subject generally appear at two-year intervals, or according to the demand created by new developments in the field. The scope of the series is flexible, however, so that future volumes may cover areas not included earier.

University Park Press is honored to continue publication of the *International Review of Biochemistry* under its sole sponsorship beginning with Volume 13. The following is a list of volumes published and currently in preparation for the series:

Volume 1: **CHEMISTRY OF MACROMOLECULES** (H. Gutfreund)
Volume 2: **BIOCHEMISTRY OF CELL WALLS AND MEMBRANES** (C.F. Fox)
Volume 3: **ENERGY TRANSDUCING MECHANISMS** (E. Racker)
Volume 4: **BIOCHEMISTRY OF LIPIDS** (T.W. Goodwin)
Volume 5: **BIOCHEMISTRY OF CARBOHYDRATES** (W.J. Whelan)
Volume 6: **BIOCHEMISTRY OF NUCLEIC ACIDS** (K. Burton)
Volume 7: **SYNTHESIS OF AMINO ACIDS AND PROTEINS** (H.R.V. Arnstein)
Volume 8: **BIOCHEMISTRY OF HORMONES** (H.V. Rickenberg)
Volume 9: **BIOCHEMISTRY OF CELL DIFFERENTIATION** (J. Paul)
Volume 10: **DEFENSE AND RECOGNITION** (R.R. Porter)
Volume 11: **PLANT BIOCHEMISTRY** (D.H. Northcote)
Volume 12: **PHYSIOLOGICAL AND PHARMACOLOGICAL BIOCHEMISTRY** (H. Blaschko)
Volume 13: **PLANT BIOCHEMISTRY II** (D.H. Northcote)
Volume 14: **BIOCHEMISTRY OF LIPIDS II** (T.W. Goodwin)
Volume 15: **BIOCHEMISTRY OF CELL DIFFERENTIATION II** (J. Paul)
Volume 16: **BIOCHEMISTRY OF CARBOHYDRATES II** (D.J. Manners)
Volume 17: **BIOCHEMISTRY OF NUCLEIC ACIDS II** (B.F.C. Clark)

(Series numbers for the following volumes will be assigned in order of publication)
AMINO ACID AND PROTEIN BIOSYNTHESIS II (H.R.V. Arnstein)
BIOCHEMISTRY AND MODE OF HORMONES II (H.V. Rickenberg)
DEFENSE AND RECOGNITION II (E.S. Lennox)
BIOCHEMISTRY OF NUTRITION (A. Neuberger)
BIOCHEMISTRY OF CELL WALLS AND MEMBRANES II (J.C. Metcalfe)
MICROBIAL BIOCHEMISTRY (J.R. Quayle)

Consultant Editors: H.L. Kornberg, Sc.D., F.R.S., Department of Biochemistry, University of Cambridge; and D.C. Phillips, Ph.D., F.R.S., Laboratory of Molecular Biophysics, Department of Zoology, University of Oxford

INTERNATIONAL REVIEW OF BIOCHEMISTRY

Volume 15

Biochemistry of Cell Differentiation II

Edited by

J. Paul, Ph.D.,
F.R.C.P., F.R.C. Path., F.R.S.E.

The Beatson Institute for Cancer Research
Glasgow, Scotland

UNIVERSITY PARK PRESS

Baltimore • London • Tokyo

UNIVERSITY PARK PRESS
International Publishers in Science and Medicine
Chamber of Commerce Building
Baltimore, Maryland 21202

Typeset by The Composing Room of Michigan, Inc.
Manufactured in the United States of America by Universal Lithographers, Inc., and The Optic Bindery Incorporated

Library of Congress Cataloging in Publication Data

Paul, John, 1922–
Biochemistry of cell differentiation II.

(International review of biochemistry ; v. 15)
Includes bibliographical references and index.
1. Cell differentiation. 2. Cytochemistry.
I. Title. II. Series. [DNLM: 1. Biochemistry.
2. Cell differentiation. W1 IN8296 v. 15 / QH607
B615]
QP501.B527 vol. 15 [QH607] 574.1'92'08s
ISBN 0-8391-1079-0 [574.8'761] 77-12003

Consultant Editors' Note

The MTP *International Review of Biochemistry* was launched to provide a critical and continuing survey of progress in biochemical research. In order to embrace even barely adequately so vast a subject as "progress in biochemical research," twelve volumes were prepared. They range in subject matter from the classical preserves of biochemistry—the structure and function of macromolecules and energy transduction—through topics such as defense and recognition and cell differentiation, in which biochemical work is still a relatively new factor, to those territories that are shared by physiology and biochemistry. In dividing up so pervasive a discipline, we realized that biochemistry cannot be confined to twelve neat slices of biology, even if those slices are cut generously: every scientist who attempts to discern the molecular events that underlie the phenomena of life can legitimately parody the cry of *Le Bourgeois Gentilhomme,* "Par ma foi! il y a plus de quarante ans que je dis de la Biochimie sans que j'en susse rien!" We therefore make no apologies for encroaching even further, in this second series, on areas in which the biochemical component has, until recently, not predominated.

However, we repeat our apology for being forced to omit again in the present collection of articles many important matters, and we also echo our hope that the authority and distinction of the contributions will compensate for our shortcomings of thematic selection. We certainly welcome criticism—we thank the many readers and reviewers who have so helpfully criticized our first series of volumes—and we solicit suggestions for future reviews.

It is a particular pleasure to thank the volume editors, the chapter authors, and the publishers for their ready cooperation in this venture. If it succeeds, the credit must go to them.

H. L. Kornberg
D. C. Phillips

Contents

Preface

In the preface to *Biochemistry of Cell Differentiation I,* I briefly mentioned the brilliant constellation of scientists who, in the thirties, vainly sought to explain phenomena of cell differentiation in biochemical terms but in the end had to retire baffled because nothing was known of DNA, much less the genetic code and protein synthesis. The problem itself had challenged biologists for decades before that; however, it is only now that, quite suddenly, a succession of dramatic technical advances has transformed the prospects.

This field has advanced incredibly rapidly in the past four years. In 1973, globin messenger RNA had been definitively demonstrated only two years earlier, and complementary DNA had been made, using reverse transcriptase, the year before. These were crucial events in the history of developmental biology. The words "genetic engineering" had not yet appeared in their current context, but the demonstrated value of analysis by nucleic acid hybridization and restriction enzymes presaged the present intense activity and optimism in this area. It would no longer be rash to claim that we are within sight of a detailed understanding of the problems pursued by that earlier generation. Moreover, we know that we will obtain answers with a precision our predecessors could never have envisaged.

The impact of genetic recombinant techniques in developmental biology research is so recent that many of the most exciting results have appeared in press since the manuscripts in this volume were completed at the beginning of this year. Dr. Glover's review provides a succinct account of the achievements of the two or three years during which the field has been emerging. The application of the methods in individual studies is mentioned in several other chapters, notably those by Dr. Weinberg and Drs. Firtel and Jacobson.

A remarkable feature of the present upsurge of interest in this area has been the greatly improved communications among scientists all across the field, which has led to the marriage of these new techniques to the study of classical systems. This is very well exemplified in the chapter by Dr. Sommerville, which is on the application of molecular methods to the study of lampbrush chromosome behavior, and in the chapter by Drs. Case and Daneholt, which details a biochemical analysis of the products of puffs in giant chromosomes. Most of the other chapters in this volume are reports of the analysis of individual developing systems by sophisticated biochemical techniques. With relation to invertebrates, information handling in sea urchins is discussed in detail by Dr. Weinberg, and Drs. Firtel and Jacobson discuss it in slime molds. Many investigators believe that mouse teratocarcinoma cell lines may provide a paradigm of overall early vertebrate development, and Dr. Hogan has prepared a review of present knowledge in this area. The three remaining chapters deal with specific vertebrate systems, the lens crystallins discussed by Prof. Bloemendal, muscle development by Dr. Buckingham, and erythroid maturation by Dr. Harrison. In the latter two contributions the use of differentiating cell lines is important. Again, these may serve as paradigms for normal development and it is perhaps worth noting the increasing importance of systems of this kind in the biochemical analysis of cell differentiation.

J. Paul

Biochemistry of Cell Differentiation II

International Review of Biochemistry
Biochemistry of Cell Differentiation II, Volume 15
Edited by J. Paul

1
Gene Cloning: A New Approach to Understanding Relationships Between DNA Sequences

D. M. GLOVER
Department of Biochemistry,
Imperial College of Science and Technology,
London, England

The ability to clone segments of DNA from any organism within *Escherichia coli* has been heralded as a major breakthrough in molecular genetics. Indeed, the potential applications of this approach are only beginning to be realized. The success of molecular cloning hinges upon the relative ease with which it is now possible to join DNA molecules in vitro. DNA from eukaryotes as diverse as slime molds and frogs has been covalently joined to bacterial plasmids to form hybrid molecules that retain the replication properties and some marker function of the plasmid. The heterogeneous population of in vitro recombinant DNA molecules can then be introduced into a bacterial strain lacking the marker function of the plasmid vector. Transformed colonies, selected by their acquisition of this marker function, arise from the infection of one bacterium by a single DNA molecule. A homogeneous population of hybrid plasmid DNA can, therefore, be prepared from a culture grown from such a colony. Similarly, foreign DNA can be covalently inserted into the genome of bacteriophage λ, and homogeneous isolates of segments of the foreign DNA are then obtained from plaques produced by the hybrid bacteriophage. A system has, therefore, been established whereby DNA from virtually any source can be propagated and cloned within a prokaryotic organism. Recently, the converse situation has been attained, and experiments have been reported whereby prokaryotic DNA has been covalently linked to DNA from the eukaryotic virus SV40 and the resulting hybrids propagated in monkey cells.

These basic techniques have wide-ranging implications. Already an impact has been made on our understanding of the organization of eukaryotic chromosomes resulting from the characterization of cloned segments of the chromosomal DNA. It is also now possible to isolate prokaryotic genes for which there is no transducing phage. There are obvious industrial applications in terms of introducing functional genes for hormones, antibodies, or specific antigens into organisms which can readily be grown in bulk. Hershfield et al. (1) have demonstrated an elevated level of expression of the *E. coli* enzymes involved in the biosynthesis of tryptophan in cells carrying the *trp* genes cloned in a plasmid. This is a most encouraging result in terms of using hybrid plasmids to produce large quantities of commercially useful prokaryotic gene products.

Although it brings such potential benefits, genetic manipulation of this kind could inadvertantly have some harmful effects. Some of the first in vitro DNA recombinants made were between SV40 DNA and a segment of DNA containing the galactose operon of *E. coli* (2). The aim was to develop a method for transducing foreign sequences into mammalian cells. The potential hazards of bacterial plasmids covalently linked to tumor virus DNA (2) were immediately realized, and therefore such molecules were never introduced into *E. coli*. Concomitant with the developments in techniques that facilitated in vitro DNA recombination came the realization among the scientific community of our ignorance of the hazardous potential of *E. coli* carrying eukaryotic DNA. An early assessment of the problems is to be found in the Ashby Committee Report (3) and in the summary statement of the Asilomar Conference on recombinant DNA molecules (4). As is often the case when hard evidence is lacking, viewpoints became sharply polarized. The whole issue has since provoked much debate, upon which this chapter does not dwell. The main fear seems to be that a potentially hazardous gene from a eukaryote might inadvertently be cloned within *E. coli;* that such bacteria might successfully colonize the intestinal tracts of animals; and that synthesis of the eukaryotic gene product by these bacteria might precipitate some disastrous pandemic (5, 6). The types of gene products that might be dangerous in this way range from mammalian hormones to the genomes of oncogenic viruses. On the other extreme is the view that prokaryotic organisms within nature are frequently in contact and must often take up eukaryotic DNA from, for example, decaying plant and animal matter. Then, given the enormity of the bacterial population of the earth, it is likely that recombinational events such as the ones that can now be carried out in vitro have already occurred, and that these recombinant organisms have no selective advantage and, therefore, die out (7). In drafting guidelines for the precautions to be followed in research of this type, a middle-of-the-road approach has been taken. Experiments have been categorized with respect to the best estimate of their potential hazard. Containment of the potentially hazardous organism is to be attained through a) the safe design and operation of the laboratory and b) the use of a vector DNA and host cell that are genetically enfeebled and so have little chance of survival outside the laboratory. Suggested levels of physical and biological containment are described in the *Guidelines for Research Involving Recombinant DNA Molecules* of the National Institutes of Health, Bethesda, Maryland (8), and in the report of the Williams committee in Great Britain (9). Some aspects of biological containment are discussed below.

ENZYMOLOGY OF IN VITRO RECOMBINATION

The ease with which DNA molecules can now be joined in vitro is a consequence of the recent availability of many restriction endonucleases, enzymes that recognize specific sequences in DNA and then cleave both strands of the duplex. These enzymes have been found in many prokaryotes and are likely to be responsible for the degradation of DNA from alien species, indigenous DNA being protected from degradation by the action of a modification enzyme, usually a methylase. The properties and utilization of these enzymes have recently been reviewed (10). There are two classes of restriction enzyme: class I enzymes, which recognize DNA at specific sites, but do not cleave the DNA at this site and class II enzymes, which cleave DNA at, or near, specific sequences that are usually several nucleotides long and rotationally symmetrical about the central nucleotide pairs. It is this latter class that has been invaluable in the development of procedures for cloning DNA.

The two methods by which DNA molecules can be joined are discussed below.

"Tailing"

The foundations for in vitro DNA recombination were laid in an approach developed by Jackson et al. (2) and Lobban and Kaiser (11), illustrated in Figure 1. The linear molecules to be joined are treated with λ-exonuclease, which successively removes deoxymononucleotides from the 5′-phosphoryl termini of double-stranded DNA, rendering the 3′-OH termini single-stranded. A single-stranded 3′-OH terminus is a good primer for terminal transferase, which is used to add deoxyadenylate and deoxythymidylate residues to the 3′-OH termini on the respective molecules. The two different molecules having dA or dT tails were annealed and then covalently sealed by the action of exonuclease III, DNA polymerase I, and DNA ligase. In this reaction, the exonuclease III removes 3′-phosphate groups from 3′-phosphoryl, 5′-OH nicks introduced by an endonuclease contaminating the terminal transferase. DNA polymerase I would be inhibited by 3′-phosphoryl groups. The 3′→5′-exonuclease of DNA polymerase I

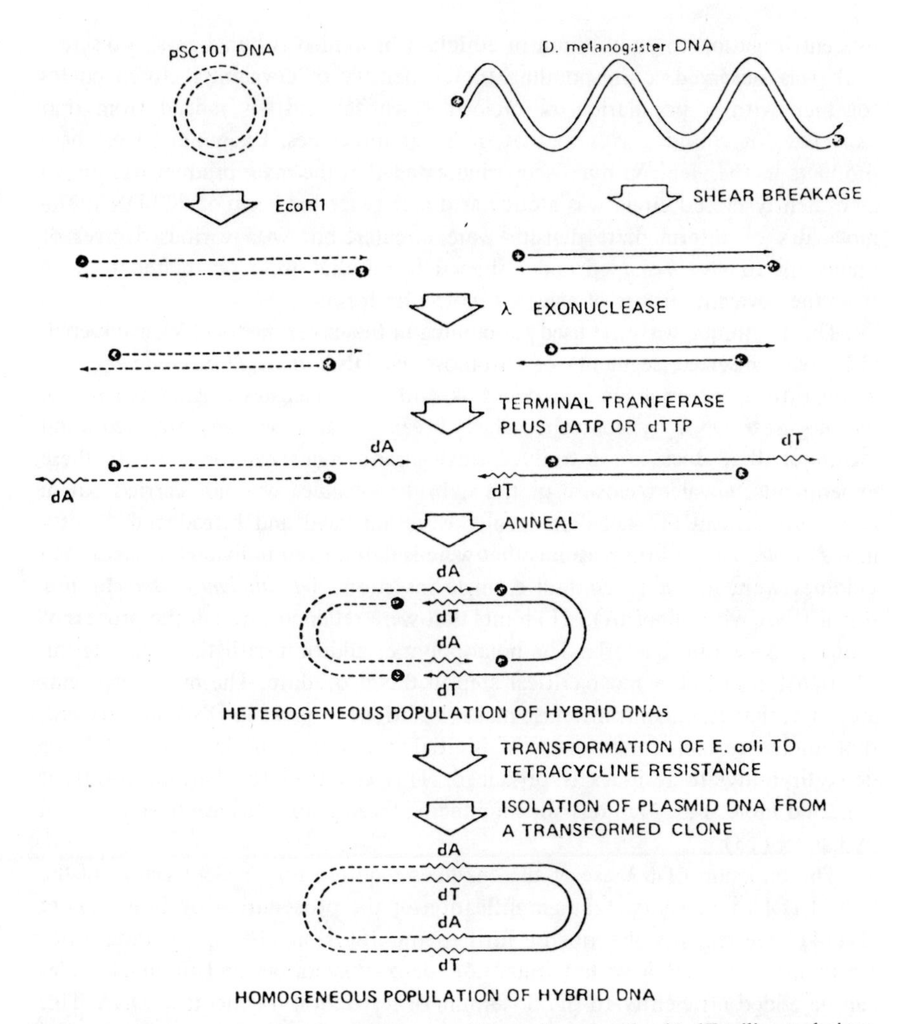

Figure 1. Construction of in vitro DNA recombinants by using the dA-dT tailing technique (2, 11), first used to construct viable hybrid molecules between *Drosophila* DNA and a bacterial plasmid. Reproduced from Wensink et al. (12) with permission of MIT Press.

should remove any "overhanging" single-stranded oligo(dA) or oligo(dT) tails on the annealed molecules. The 5′→3′-exonuclease can remove the 5′-nucleotide from nicks containing 5′-OH groups or other unusual end groups, leaving 5′-phosphoryl termini. The polymerizing activity fills in single-stranded gaps by the template-directed addition of deoxymononucleotides to the 3′-OH termini, finally to be displaced by ligase that can covalently seal nicks with 3′-OH and 5′-phosphoryl termini.

Lobban and Kaiser (11) joined two populations of P22 DNA molecules tailed with oligo(dT) and oligo(dA) and separated covalently joined molecules by centrifugation to equilibrium in ethidium bromide-CsCl gradients. A discrete peak was observed, corresponding to the density of covalently closed circles together with a population of molecules whose densities ranged from that expected for closed circles to that of linear molecules. Examination of these products in the electron microscope indicated that the main product banding as a covalently closed circle was a circular dimer twice the lenth of P22 DNA. The molecules of intermediate density were circular, but with various degrees of branching to give θ-shaped and σ-shaped forms that were postulated to arise from the covalent closure of aberrant molecular forms.

This technique was first used for cloning in the experiments of Wensink et al. (12), who inserted segments of chromosomal DNA from *Drosophila melanogaster* into a bacterial plasmid. Wensink and his colleagues fragmented the *D. melanogaster* DNA by controlled shear breakage, although any fragmentation technique that does not introduce single-strand nicks can be used. In these experiments, covalent closure of the hybrid molecules was not carried out in vitro, but instead the tailed molecules were annealed and introduced directly into *E. coli*. The hybrid plasmids that were isolated from individual transformed colonies were found to contain a single insertion of *D. melanogaster* chromosomal DNA with oligo(dA):(dT) joints that were retained through the process of propagation within bacteria. The homopolymer addition with the use of terminal transferase is the most critical step in this procedure. The major problems are, first, that terminal transferase makes additions to nicks in DNA and, second, that under certain conditions a self-initiated polymer of deoxyadenylate or deoxythymidylate residues is produced. This can lead to aberrant forms of annealed molecules. Methods for minimizing these problems have been reviewed by Glover (13).

The omission of the use of the enzymes required for covalent sealing of the hybrid DNA represents great simplification of the procedure. Roychoudhury et al. (14) have recently described a further simplification whereby, without prior treatment of the DNA with exonuclease, deoxynucleotides and ribonucleotides can be added efficiently to the 3′ termini of restriction fragments of DNA. This technique should prove particularly advantageous for experiments in which the amount of DNA is limitingly small.

Joining at Cohesive Ends Created by Certain Restriction Endonucleases

The class II restriction endonucleases can themselves be divided into two groups: those which when they cleave DNA generate fully base-paired ends and those which make staggered single-stranded cleavages generating mutually cohesive termini. This latter type of scission was first recognized for the enzyme Eco RI by Mertz and Davis (15), who were able to observe circular forms of Eco RI-cleaved DNA if the DNA was spread for electron microscopy at low temperature. From the temperature stability of this hydrogen bonding, they were able to deduce that the length of the single-stranded terminal nucleotides was on the order of 4 residues. Moreover, they were able to covalently seal molecules cleaved by Eco RI with the use of *E. coli* DNA ligase. This technique has been extensively applied in the construction of hybrid DNA molecules (Figure 2). In principle, the same method may be applied to the cleavage products of any of the restriction enzymes listed in Table 1. Plasmids are now characterized that will serve as vectors for DNA fragments produced by most of these enzymes; these are described below.

The physical location of restriction sites is often disadvantageous when the interest lies in cloning a large polypeptide coding sequence. In addition, the size limitation of restriction fragments places a restriction upon the length of a contiguous DNA sequence that one can clone in an individual hybrid plasmid. This limitation need not arise if the "tailing" approach is used, as the chromosomal DNA to be cloned may be sheared to any size before tailing. Glover et al. (26) have inserted partial Eco RI digests of *D. melanogaster* into a bacterial plasmid. This, however, is quite a laborious technique because the partial digestion products and the ligased Eco RI fragments had to be carefully sized in order to discriminate against oligomeric structures containing, for example, two *D. melanogaster* DNA segments and one bacterial plasmid or vice versa. Oligomeric structures of this type inevitably arise in ligation reactions with plasmid vectors. In the experiments of Morrow et al. (27), for example, on the order of 25% of the transformed bacterial colonies contained hybrid plasmids, and of these about 25% contained more than one inserted Eco RI fragment of *Xenopus* DNA. If the interest is in the study of a contiguous DNA segment, then these plasmids must be discarded. Another disadvantage of joining at cohesive ends generated by restriction endonucleases is the frequency of self-cyclization of the vector plasmid. This results in a background of colonies in the bacterial trans-

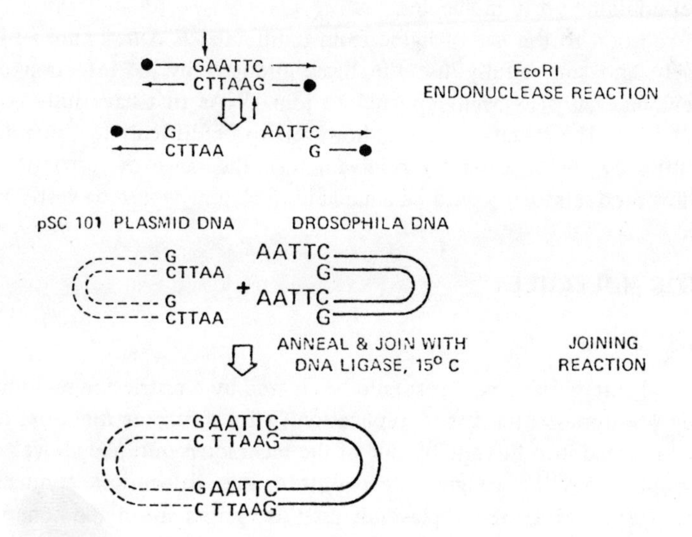

Figure 2. Construction of hybrid plasmids for *Drosophila* DNA and pSC101 by the Eco RI ligase method.

Table 1. Some of the restriction endonucleases that generate sticky ends

Microorganism	Enzyme	Sequence (5′ → 3′)	Reference
Bacillus amyloliquefaciens	Bam I	↓GGATCC	16, 17
Bacillus globigii	Bgl II	↓AGATCT	17
E. coli (RY13)	Eco RI	↓GAATTC	18, 19
Haemophilus influenzae d	Hind III	↓AAGCTT	20, 21
Providencia stuartii	Pst I	CTGCA↓G	22
Streptomyces albus	Sal I	?	23
Streptomyces stanford	Sst I	?	24
Xanthomonas malvacearum	Xma I	↓CCCGGG	25

formation (75% in the experiment quoted above) which do not contain hybrid plasmids. On the other hand, in the "tailing" approach, vector plasmid DNAs that have homopolymeric single-strand extensions cannot self-cyclize. Consequently, almost all of the bacterial colonies transformed by DNA joined in this manner contain hybrid molecules.

In addition to the use of ligase from uninfected *E. coli*, a number of groups routinely and successfully use the ligase induced by T4 infection of *E coli.* This enzyme has also been reported to join DNAs that have fully base-paired termini (28). This reaction seems, however, to be difficult to control. If these difficulties can be successfully resolved, then the range of restriction enzyme digestion products that would be amenable to cloning would be vastly increased.

VECTOR MOLECULES

Plasmid Vectors

A plasmid suitable as a vector should be cleaved by a restriction endonuclease at a single site nonessential for its replication or for a marker function. DNA may then be inserted into this site by one of the techniques outlined above.

Plasmids may be roughly divided into two categories: conjugative and nonconjugative. Most of the plasmids used as vectors are of the nonconjugative type, but there are now moves toward the use of derivatives of conjugative plasmids as vectors. Conjugative plasmids such as the F plasmids or R_p plasmids can bring about the transfer of DNA as a result of conjugation. They are capable of integrating into the bacterial chromosome and consequently can transfer bacterial DNA to which they are covalently joined. In addition, they are capable of mobilizing DNA to which they are not covalently linked and, therefore, can promote the transfer of coexisting plasmids from one cell to another. The conjugal transfer of DNA is under the control of the *tra* operon, which contains about 30 kb of DNA and amounts to about 30% of the genome of the F plasmid, for example. Some of the other genetic markers on F plasmids are discussed below, as functional fragments of the genome have now been cloned. This group of plasmids is said to be under stringent replicative control because a 1:1 ratio is maintained between the copy number of the plasmid and bacterial chromosome within the cell. This is in contrast with the smaller nonconjugative plasmids, which are more desirable for cloning not only because they are under relaxed replicative control and are present as 10–15 copies per cell, but also because they present less of a potential biohazard since their DNA cannot be conjugally transferred as readily from one cell to another.

Both conjugative and nonconjugative plasmids are frequently associated with either genetically characterized insertion sequence (IS) segments or other inverted repeats, which form a hairpin loop type of structure detectable in heteroduplex studies by electron microscopy. The IS sequences are also present in the *E. coli* chromosome and cause strongly polar mutations when, for example, they are inserted into the *gal* or *lac* operons (29, 30). They have a number of probable functions. IS2 is asymmetric, in that when it is inserted in one direction it carries stop signals for gene expression and when inserted in the other direction it carries a strong promoter (31). The inverted sequence repeats seem to have a role in *recA* independent recombinational and translocational processes. These sequences may, therefore, play a considerable role in the evolution of plasmids. They will be mentioned in the following discussion of the plasmids used as cloning vectors. The physical maps of some of these plasmids are shown in Figure 3.

pSC101 The pSC101 plasmid specifies tetracycline resistance and was isolated from a sheared DNA preparation obtained from cells carrying the R plasmid, R6-5. It is possible that it arose from the rejoining of the sheared R factor DNA within the recipient *E. coli*. Alternatively, it is possible that the molecule was formed in vivo and was coisolated along with the R factor DNA. Because of its small size, the supercoiled form would have been resistant to the shear breakage. pSC101 carries genes for its own replication and resistance to tetracycline, but none of the other drug-resistant determinants carried by R6-5. Insertion of a DNA segment at the single Eco RI site of pSC101 does not interfere with the expression of tetracycline resistance or with the replication functions of the plasmid (27, 32). pSC101 contains a single site for Hind III, which is just 30 base pairs removed from the RI site. Carroll and Brown (33) have been able to clone Hind III fragments of the 5 S DNA of *Xenopus laevis* within this site and retain the expression of tetracycline resistance. Other

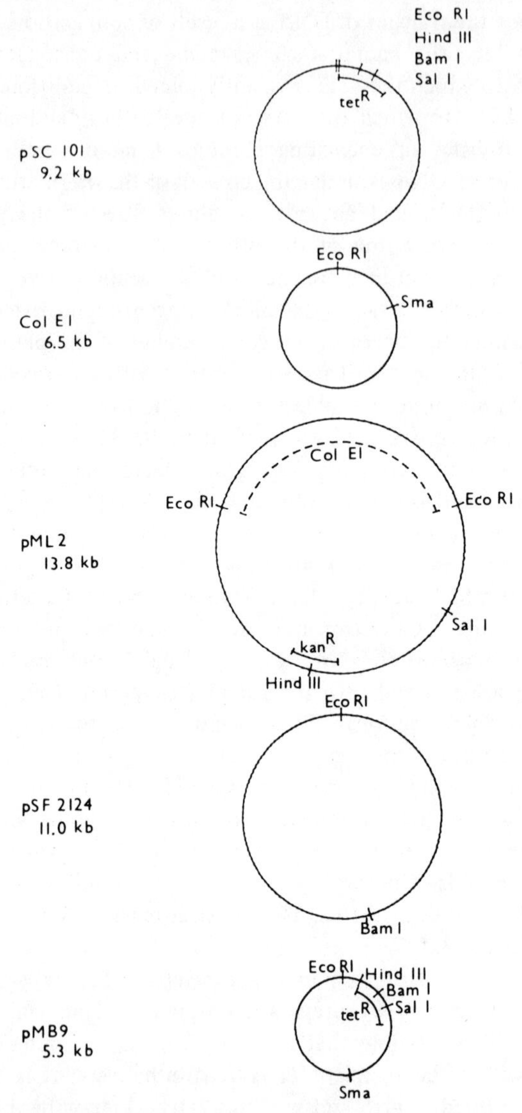

Figure 3. (At *left* and *right*.) Physical maps of some of the bacterial plasmids that are currently being used as cloning vectors. The isolation and characterization of these plasmids are described in the text.

workers have found that the insertion of foreign DNA at this site inactivates the gene for tetracycline resistance. Evidence has been presented (34) that the segment of DNA between the Hind III and Eco RI site of pSC101 contains a promoter for *tet*^{R} gene transcription. The expression of tetracycline resistance then depends upon whether the added DNA can provide a substitute promoter

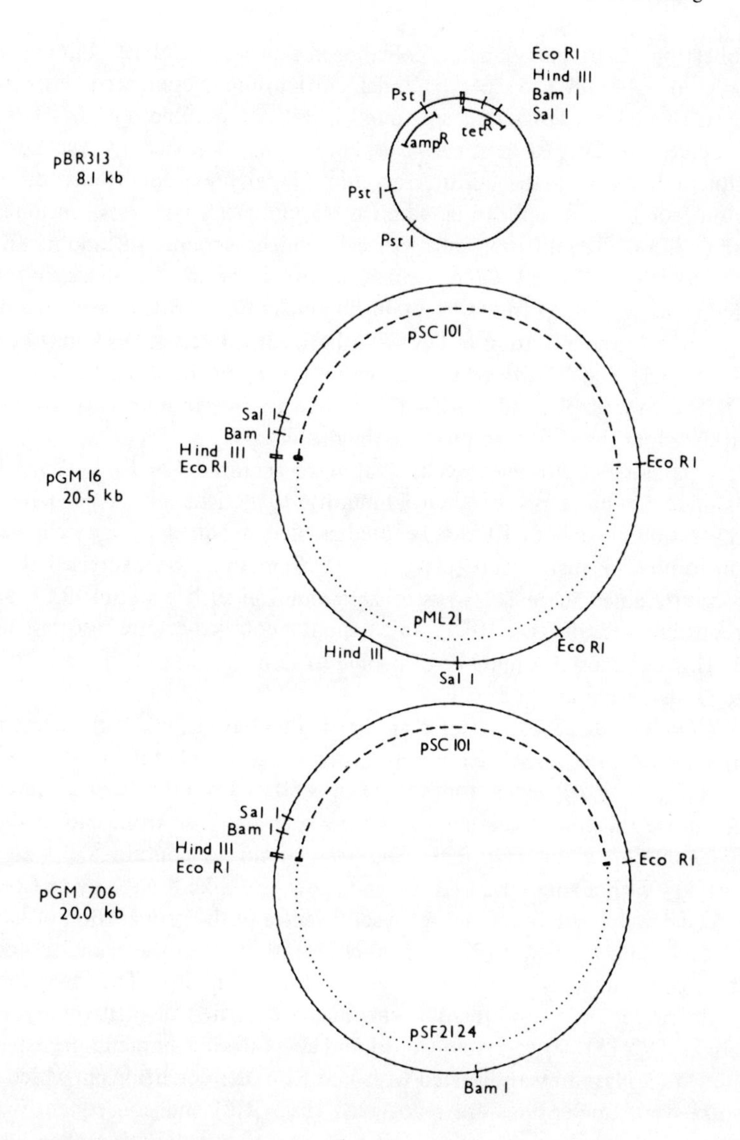

for this transcription. It has been found, for example, that sequences within the Col EI-kan plasmid (see below) can provide this function. This promoter-like function has been narrowed down to one of the two Hind III cleavage fragments of the Eco RI fragment that determines kanamycin resistance (see the map of Col EI kanamycin in Figure 3).

Col EI The colicinogenic plasmid Col EI determines the production of the antibiotic protein colicin EI. Strains of *E. coli* that carry that Col EI plasmid are immune to colicin EI, whereas other strains are sensitive. Col EI, unlike pSC101, requires the product of the *polA* gene, DNA polymerase I, for its replication. Cells carrying the colicinogenic plasmid Col EI contain about 24 copies of plasmid (35) per bacterial chromosome, compared with only 1–2 copies of plasmid DNA that are found in cells containing pSC101. The number of copies of Col EI per cell can be further increased by the addition of chloramphenicol to the culture medium. Under these conditions, the bacterial chromosome stops replicating, whereas plasmid DNA synthesis continues to give 1,000–3,000 plasmid molecules per cell, which accounts for about 50% of the cellular DNA. Col EI DNA, which accumulates in the presence of chloramphenicol, has been found to retain ribonucleotides, which have been proposed as primers in the initiation of DNA synthesis (36). Foreign DNA may be inserted at the single Eco RI site of Col EI without altering these replication properties (1). The hybrid plasmids retain the ability to confer immunity to colicin EI, but they lose the ability to produce the protein.

It is possible to select cells that have acquired Col EI or Col EI hybrid plasmids on the basis of their immunity to colicin EI. A partially purified preparation of colicin EI can be used, which is added to the cells during the transformation just before plating (13). Care must be exercised during this procedure since there is always a background in such experiments from a high proportion of cells (ca. 10^{-5}) which spontaneously become resistant to colicin EI. This has caused a number of people to turn toward Col EI plasmids carrying drug resistance markers.

Mini-Col EIs Two plasmids are in use that have acquired the name mini-Col EI. One of these arose from the experiments of Hershfield et al. (1), who inserted *E. coli trp* genes from the phage ϕ 80 pt 190 into the plasmid Col EI. In the course of these experiments, a *trp* segregant arose from one of the Col EI *trp*-containing clones (1, 37). This was found to contain a 3.4-kb plasmid (pVH51) with a single Eco RI site and showing a 3-kb homology to Col EI. This molecule was itself found to be a useful vector in the propagation of kanamycin resistance and *trp* genes (37) and more recently part of the gene for globin (38). The second mini-Col EI plasmid was constructed in vitro. The genes for the Eco RI restriction and modification enzymes are carried on a naturally occurring plasmid (RY13) derived from Col EI and also carrying ampicillin resistance (39). The RY13 plasmid was digested with Eco RI under conditions in which only the central four nucleotides are recognized (Eco RI*) and consequently a greater number of cleavages are made. The products of this digestion were ligated and introduced into *E. coli*, and clones were selected that were immune to colicine EI, but that had lost ampicillin resistance. One of these clones contained a 2.7-kb plasmid pMB8 with a single Eco RI site (40). When heteroduplexed with Col EI, pMB8 showed a 3.75-kb deletion loop, corresponding to Col EI sequences that are deleted from pMB8 (41). pMB8 is one of a series of plasmids constructed in vitro and is discussed below.

Mini-Col EIs and mini-Col EI-derived plasmids are less readily mobilized by self-transmissible plasmids than are wild type Col EI. The use of mini-Col EIs, therefore, represents a significant step toward biosafety. Further steps in this direction have been taken by Charney and Berg (42), who have constructed a mini-Col EI derivative of pJC307, a Col EI mutant which is temperature-sensitive for its own DNA replication. The approach was, first, to protect the single Eco RI site of pJC307 by in vitro methylation with the use of the Eco RI modification enzyme and, second, to carry out an Eco RI* digestion followed by direct transformation of bacteria with the DNA fragments. A colicine-immune transformant was selected which carried a 2.7-kb plasmid, pDC1. An additional safety feature was incorporated into this molecule in the form of an Eco RI fragment of λ-DNA which carries a temperature-sensitive CI gene and the *kil* gene. This plasmid can replicate at 25°C, although if the temperature is raised to 37°C, then, in addition to the block in plasmid DNA replication, the thermosensitive repressor made by CI is inactivated and transcription is initiated from the P_L promoter, resulting in the expression of *kil* and death to the bacterium.

Col EI-kan The Eco RI fragment that determined resistance to kanamycin originated in the R factor R6-5. It was transferred into pSC101 by Cohen et al. (32) and from pSC101 into Col EI by Hershfield et al. (1) to form a plasmid designated pML2. pML2 has two Eco RI sites, although Covey et al. (43) have constructed deletion mutants which lack one of the Eco RI sites. pML2 was first partially digested with Eco RI and the partial digest given a short treatment with λ-exonuclease. This removed the single-stranded 5′ ends created by Eco RI and left protruding 3′ single-strand ends. These full length linear molecules were purified away from residual circles by gel electrophoresis and then transferred back into *E. coli,* in which transient pairing presumably occurred between fortuitous homology in the single-stranded 3′ termini from either end of the molecule. This hydrogen-bonded circle was then repaired and replicated by the bacterium.

Two derivatives of pML2 were characterized, one in which there was a deletion of about 25–30 nucleotide pairs (pCR1) and in the other a deletion of about 1,500 nucleotide pairs (pCR11). pCR1 has already found use in cloning a fragment of DNA synthesized in vitro from a globin messenger RNA (mRNA) template (44).

Col EI-kan has one cleavage site for the restriction endonuclease Hind III, located in the kanamycin resistance-determining Eco RI fragment. Hind III cleavage products have been excised from cloned segments of the genome of *D. melanogaster* and have themselves been cloned following their insertion into the Hind III site of pML2 (34).

Inverted repeat sequences of 1 kb in length form boundaries to the kanamycin resistance gene, and these have been used as characteristic markers in the examination of these in vitro recombinants in the electron microscope (see below).

A hybrid molecule has been constructed between mini-Col EI and the kanamycin resistance Eco RI fragment (37). This molecule (pML21) has similar properties to those of pML2, but is not as readily mobilized by self-transferrable plasmids.

Col EI-ampicillin (pSF2124) The gene for ampicillin resistance is located on a 4.5-kb piece of DNA, known as TnA, which can be translocated from one replicon to another (45). It is similar to the kanamycin resistance determinant in that it is bounded by inverted repeat sequences which resemble the IS sequences in their structure. The sequence can inactivate the gene into which it inserts. So et al. (41) were able to translocate TnA from an R plasmid (RI*drd*19) to Col EI by cocultivation of the two plasmids within the same bacterial cells. A number of recombinants were isolated, and the position of insertion of TnA was determined by heteroduplex analysis with pMB8 (40). When pMB8 is cleaved with Eco RI and heteroduplexed with Eco RI-cleaved Col EI, a deletion loop is formed which serves as a positional marker on the molecule. Insertion of TnA close to one of the Eco RI ends results in inactivation of the gene affecting colicins biosynthesis, as might have been predicted knowing that insertion of DNA into the Eco RI site has this effect. One recombinant, pSF2124, still capable of specifying the biosynthesis of colicin and coding for ampicillin resistance, was found to be suitable for cloning DNA. This plasmid has since been used to clone *D. melanogaster* DNA (46).

pGM706 and pGM16 Hamer and Thomas (47) have inserted DNA segments carrying Col EI replication functions and the gene for kanamycin resistance into both the Bam HI and Sal I sites of pSC101. In both cases, kanamycin-resistant, colicine-immune clones were obtained in *polA*I cells, indicating that the hybrid plasmid was using the pSC101 replication cistons. These clones were, however, sensitive to tetracycline. These authors then went on to construct vehicles, coding for resistance to tetracycline plus a second drug which permits easy discrimination between recyclized vector molecules and the in vitro recombinants. pGM706 is a hybrid constructed between pSC101 and pSF2124 joined at their RI sites. Insertion of DNA into its single Sal I site inactivates tetracycline resistance but not ampicillin resistance. pGM16 is a hybrid constructed between Eco RI-cleaved pSC101 and the partial Eco RI digestion product of pML21. Insertion of DNA into its single Bam HI site again inactivates tetracycline resistance functions but not kanamycin resistance. These plasmids are somewhat large and contain genes which are clearly redundant in the role of a cloning vector. They have been used as vectors for *Drosophila* DNA (45) and also for a segment of the sea urchin histone genes (48).

pMB9 This plasmid results from an experiment designed to introduce a tetracycline resistance marker into pMB8 (40). pMB8 was cleaved with Eco RI, ligated to the fragments from an Eco RI* digest of pSC101, introduced into *E. coli,* and tetracycline-resistant, colicin EI-immune cells were selected. One of these colonies contained a plasmid with two Eco RI sites. One of these sites was deleted by digesting again with Eco RI* and religating. This resulted in a 5-kb plasmid which combines the advantages of pSC101 and Col EI. It has a single Eco RI site into which foreign DNA can be inserted; the tetracycline resistance marker facilitates the selection of transformants and retains the replication properties of Col EI. It has been used successfully to clone globin genes synthesized in vitro by Maniatis et al. (49).

pBR313 The TnA transposon has been translocated into pMB9 by cocultivation of this plasmid with pSF2124 (40). The resulting recombinant plasmids which determine resistance to tetracycline and ampicillin were subjected to Eco RI* digestion and religation in order to regenerate ampicillin-resistant, tetracycline-resistant plasmids with certain restriction sites deleted. pBR313 contains single sites for the enzymes Eco RI, Hind III, Bam I, and Sal I. These latter three sites are in the gene for tetracycline resistance, and insertion of DNA into these sites inactivates this gene. Ampicillin-resistant clones can, therefore, be screened for whether they contain hybrid plasmids which are sensitive to tetracycline. Rodriguez et al. (40) have enriched for hybrid plasmids in a mass culture. Cells were first transformed to ampicillin resistance, the ampicillin was removed, and the culture was propagated in tetracycline-containing media. The growing cells were killed by the addition of D. cycloserine. When the D. cycloserine was removed, a high proportion of the viable cells were found to contain hybrid plasmids. The disadvantage of this approach is that single transformed bacteria have the opportunity to replicate in liquid culture so that when the cells are eventually plated a particular hybrid molecule is represented in more than one colony. It is likely that the Hind III site is in a promoter for the tetracycline resistance gene, as seems to be the case with pSC101. In this case, therefore, one would still retain expression of tetracycline resistance if the inserted DNA contained a sequence recognizable as a promoter. This plasmid combines in a much smaller molecule the advantages of the Bam I and Sal I vectors constructed by Hamer and Thomas (47).

Bacteriophage λ

Initially, phage were developed which would serve as vectors for Eco RI fragments. Wild type phage λ-DNA has five Eco RI cleavage sites; therefore, derivatives of phage λ were constructed with a reduced number of Eco RI sites (50–52). The Eco RI sites which remain in these phage are in the nonessential region of the genome (sites 1, 2, and 3) so that the phage DNA can be completely cleaved with Eco RI and foreign DNA inserted into this region. The three groups of workers (50–52) initially made phage which completely lacked Eco RI sites. Rambach and Tiollais (51) and Murray and Murray (50) used deletion phages that had lost the two left-most sites (sites 1 and 2) (Figure 4). Thomas et al. (52) used a phage which had in addition lost site 3 as a result of a *bio 69* substitution. This phage also contained a duplication in the left arm (*a 200*) in order to restore DNA to a suitable length to assist propagation. In all cases, mutants lacking the remaining sites were selected by making lysates alternately on an *E. coli* strain containing no Eco RI restriction system and on a strain harboring the Eco RI restriction-modification system. This was continued for several cycles until the efficiency of plating indicated that the DNA was no longer restricted. The resulting phage, which completely lacked Eco RI sites, were crossed with phage containing all the Eco RI sites, and recombinants were selected which contained only sites 1 and 2 or 1, 2, and 3. Two of the groups (50, 52) carried out a third step to create a deletion between sites 1 and 2 or sites 1 and 3 by the in vitro joining of the lefthand and righthand fragments

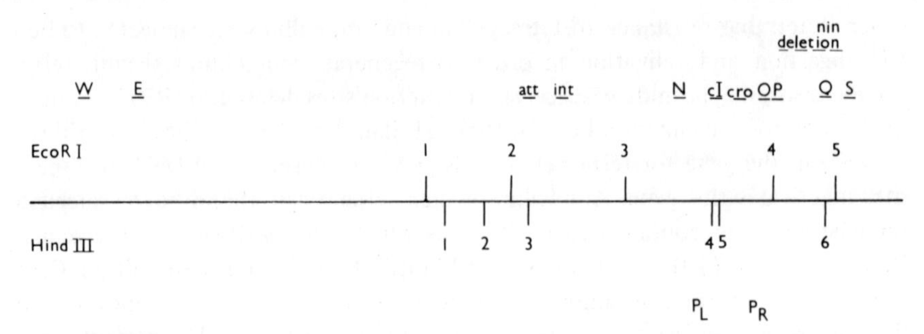

Figure 4. The Eco RI and Hind III sites of bacteriophage λ-DNA and the relative positions of some of the genetic markers that are discussed in the text.

following Eco RI digestion. These phage can be cleaved with Eco RI and Eco RI fragments of foreign DNA inserted between the left and right fragments.

Derivatives of λ have also been obtained which serve as vectors for Hind III fragments (53). The wild type phage DNA has six sites for this enzyme. Deletion of the fragment between Eco RI sites 1 and 2 removes two Hind III sites. The substitution of imm^{21} for imm^{λ} also removes two sites. A cross between *Qam* and a *φ 80* hybrid phage has resulted in the substitution of *φ 80* DNA and the loss of site 6, which is very close to *Q*. The resulting phage has a single remaining site into which foreign DNA can be inserted.

Further derivatives have been constructed which contain imm^{434}, the CI gene which contains single cleavage sites for Hind III and Eco RI. Insertion of foreign DNA into such a position results in clear plaque morphology (54). This enables hybrid phage to be distinguished from the phage without inserted DNA which give turbid plaque morphology.

The formation of plaques by hybrid bacteriophage DNA molecules can be made to depend on the presence of foreign DNA within the molecule. The phage of Thomas et al. (52) lose the ability to form plaques if the Eco RI fragments between sites 1 and 3 are deleted, since the phage has then lost 27% of its DNA and is too small to be packaged. Plaques resulting from the joining of foreign DNA to the purified Eco RI left fragment and right fragment must contain additional foreign DNA. Alternatively, the unfractionated Eco RI fragments from the phage λgt-λC may be used in such experiments. This phage contains the Eco RI fragment between sites 2 and 3 (fragment C). Some of the products of the joining reaction contain this fragment. These phage can be distinguished because this fragment contains the attachment site and the *int* and *xis* genes. These phage form stable lysogens which give light-colored colonies when spotted onto an eosin-methylene blue plate with no sugar and overlaid with λb2cI⁻ phage.

This latter principle corresponds to the theory behind the "replacement" vectors of Murray et al. (54). One of these vectors contains a replaceable Eco RI fragment which carries the gene for a mutant transfer RNA (tRNA) gene from *E. coli, supE*. These phage are recognized by the suppression of an amber mutation in the *lacZ* gene of the bacterial recipient, detected by plating on color indicator agar. A similar replacement vector for Hind III fragments is described which contains a Hind III fragment for the *supF* gene. This was derived from a λ*trpE* transducing phage following a series of crosses which introduced deletions from *att* through to *red,* together with imm^{λ} or imm^{21}.

In order to introduce biological safety into these vectors, amber mutations have been introduced into the *E, W,* and *S* genes (55). Phage λ, carrying an *Eam,* cannot assemble heads or cut concatomeric DNA into infectious monomeric molecules. The *W* gene product allows the joining of heads to tails. These phage are, therefore, unlikely to cause multiple infections of wild type *E. coli*. The *S* mutation prevents host lysis and can be suppressed in strains carrying *supF* but not by strains carrying *supE.* A double suppressor strain is, therefore, needed for phage propagation. An additional advantage of the *S* mutation is that for preparative purposes it permits a high phage yield from smaller cultures. Most of the phage developed for cloning DNA contain extensive deletions which usually include the *nin 5* deletion. This ensures that late functions are expressed, making it improbable that such phage could be propagated in prophage or plasmid states. Autonomously replicating plasmids have been constructed in vitro from Hind III, Bam I, or Sal I restriction fragments of λ-DNA which include the replication origin together with other replication functions (56). Several fragments of chloroplast DNA have been cloned within such "λdv-like" plasmids.

Eukaryotic Vectors

The genome of SV40 has been subject to extensive physical mapping. It is only natural, therefore, that several groups have used this molecule as a vector to propagate DNA within monkey cells (57–60). SV40 contains 5 kb of circular DNA, which can be divided into early and late regions, each occupying about one-half of the genome. Passage of the virus at high multiplicities of infection can generate defective viral genomes; several such defectives have been isolated which contain reiterations of the initiation site for viral DNA replication. One such defective, which contains five tandem repeats of an 0.88-kb sequence, has been used as a vector for fragments of λ-DNA (58). Digestion of this defective SV40 DNA with Hind III yields 80 base pair fragments to which has been joined a 520 base pair Hind III fragment of λ-DNA. Although the SV40 DNA fragment contains the origin of replication, it contains no functional genes and, therefore, the infection of monkey cells by recombinant molecules has to be helped by wild type DNA. This results in the production of a mixed virus population. A similar "triplication" mutant of SV40 has also been used as a vector (57). In this case, digestion of the monomer segment (1.67 kb) with Bam I and Hind III yields a fragment which contains the replication origin. This has been covalently joined to a fragment of DNA which has one Bam I cohesive end and one Hind III cohesive end and which contains the lefthand operator (O_L) of the phage. After propagation in monkey cells, this fragment of λ-DNA was found to retain its affinity for λ repressor.

This cloning approach has problems: many of the plaques in these experiments contain just the wild type helper virus, and even in plaques with mixed virus populations the helper can outgrow the recombinant viruses. In order to eliminate this problem, experiments have been designed whereby a helper virus is used which carries a temperature-sensitive mutation which is complemented by a segment of DNA within the vector. Hamer et al. (59), for example, have joined a 3.7-kb fragment of wild type SV40 DNA, produced by cleavage with Eco RI and Hpa II, to an 0.88-kb fragment of a bacterial plasmid DNA which carries a single copy of the $tRNA^{Tyr}$ $su^{+}III$ gene of *E. coli*. The vector DNA carries the early region and the replication origin but lacks late functions. These late functions are supplied by a helper which is temperature-sensitive in the single early gene. At high temperature, therefore, virus is produced only in cells infected with both the recombinant and the helper. Hamer et al. (59) have detected transcription products from the bacterial DNA within the monkey cells, but these are either long molecules which probably result from read-through of transcripts initiated at the viral promoter or short heterogeneous molecules. Mature suppressor tRNAs were not detected.

Goff and Berg (60) have used a similar approach with a fragment generated by cleavage with Bam I and Hpa II, which also carries the early genes and the replication origin. They have cloned a fragment of λ-DNA generated by cleavage with Hind III and RI which contains the *cro* and *CII* genes along with four λ promoters. Unlike the other groups, these workers used dA-dT "tailing" to join the two DNA species. Again, a helper virus was used with a temperature-sensitive mutation in the *A* gene.

BACTERIAL RECIPIENT AND INTEGRITY OF HYBRID MOLECULES

Transformation of the recipient bacteria with respect to a marker function carried by a plasmid or to the infectious process of phage is achieved by pretreating the bacteria with calcium ions and then incubating the cells in the presence of DNA (61). At its best, this technique is an ineffective process, resulting in the transformation or infection of 1 cell in 10^5. In the case of the phage λ, however, an in vitro packaging system has now been developed which will increase efficiency of this process by an order of magnitude (62). The choice of a suitable strain of *E. coli* is important, not only because the experimenter wishes his cloned DNA molecules to be replicated with high fidelity, but also because it is desirable that potentially hazardous recombinant DNAs are not allowed to wander promiscuously throughout the biosphere.

In general, the bacterial strains that are used as recipients for the hybrid molecules lack restriction functions in order that foreign DNA is not degraded; they are, in addition, *recA*$^-$ in order to reduce the possibility of loss of foreign

DNA by recombinational events between tandemly repeated sequences frequently found in eukaryotic DNA. Some of the cloned hybrid plasmids containing *D. melanogaster* DNA (for example, pDm103 (26)) appear to be stable for many generations in culture since no gross changes can be detected in molecular length of the plasmid or its restriction enzyme cleavage products. A similar analysis has been carried out with mouse mitochondrial DNA cloned in pSC101 (63). After 125 generations, there is no detectable change in melting profile, contour length, or restriction pattern of this plasmid. Furthermore, no changes have been detected by heteroduplex analysis. The only change that is observed in this case is the replacement of a small block of ribonucleotide sequences normally found in mouse mitochondrial DNA with the corresponding deoxyribonucleotides in the DNA propagated in *E. coli.* The effect of the *recA* mutation upon the fidelity of eukaryotic sequences has been examined by Carroll and Brown (33). The 5 S DNA from *X. laevis* is known to contain a repeating DNA sequence of approximately 15 nucleotides (see below). This DNA has been cloned in pSC101 and the hybrid plasmids grown in both *recA*$^+$ and *recA*$^-$ cells. No differences in length were observed in plasmid DNA obtained from either bacterial strain with the use of a gel system that is capable of detecting as little as a 15 base pair deletion.

Deletions can also occur in hybrid plasmids containing Eco RI fragments of prokaryotic DNA. The *trp*$^-$ segregant of a Col EI *trp*$^+$ clone isolated by Hershfield et al. (1, 37) (discussed above) is an example of such an event.

Difficulties in replication may conceivably impose limits on the length of foreign DNA that can be inserted into plasmids, although Col EI plasmids have been propagated which carry *D. melanogaster* DNA inserts as large as 40 kb (64). It is certainly true, however, that certain hybrid plasmids are able to replicate more easily than others (see below).

In the case of bacteriophage λ, there is a limit on the amount of foreign DNA that can be inserted since phage containing greater than 108% of the wild type amount of DNA cannot be packaged. It might be imagined that this fact itself could impose considerable selection pressure favoring the deletion of foreign DNA sequences. Indeed, deletions have been detected in in vitro recombinants of bacteriophage λ and part of the *E. coli trp* operon (50). The λ *trp* phage had 3% more DNA than wild type, whereas, had the whole fragment been incorporated into the vector, a phage with at least 10% more DNA than wild type would have been expected.

One potential cause for deletions has already been discussed. If DNA for the terminal transferase reaction is heavily nicked, then the enzyme is capable of adding homopolymer blocks onto the 3′-OH termini of such nicks. The in vitro recombinational event with such DNA can lead to loss of either plasmid vector or donor DNA sequences. The desirable features of strains of *E. coli* which make them suitable for cloning eukaryotic DNA also endow both the bacteria and the recombinant DNAs which they contain with decreased ability to survive in nature. Unmodified phage DNA, for example, plates with an efficiency of 10^{-2}–10^{-5} on host strains with restriction systems. Similarly, *recA*⁻ cells grow slowly and are very sensitive to ultraviolet irradiation.

Curtiss et al. (65) have developed a recipient strain suitable as a recipient for plasmid vectors (but unfortunately resistant to phage λ) which requires the metabolite diaminopimelic acid, which is not normally present in the human gut. In addition, the strain exhibits increased sensitivity to a number of antibiotics, detergents, and bile salts. It is a weak strain and is readily outgrown by contaminants, although this can be minimized by including in the medium nalidixic acid, cycloserine, or trimethoprim, to which the strain is resistant. As the strain is extremely sensitive to detergents, it is very important to use well rinsed glassware and to transform the strain with DNA solutions which have been extensively dialyzed in order to remove any detergent remaining from the preparative procedures.

SELECTION OF RECOMBINANTS COMPLEMENTARY TO RNA

In the absence of a selection system for a genetic marker, as is almost invariably the case when eukaryotic DNA is being cloned, the approach of Grunstein and Hogness (66) is invaluable for detecting a colony containing a hybrid plasmid complementary to a particular RNA. A heterogeneous population of hybrid colonies is replica-plated onto a nitrocellulose filter on an agar plate. The bacterial colonies are allowed to grow on the filter and are then lysed under conditions which allow the DNA to denature and fix to the filter in the position of the bacterial colony. The colonies containing the hybrid plasmid of interest can then be localized by autoradiography after hybridization with radioactive RNA. This technique has found extensive use in the isolation of colonies with plasmids containing ribosomal DNA (rDNA) (34, 66), sequences complementary to polyadenylated RNAs (49), and a variety of complementary RNAs (cRNAs) (34, 66, 67) synthesized in vitro from other hybrid plasmids. A similar technique for use with phage λ vectors has also been developed (68, 69).

CLONED SEGMENTS OF PROKARYOTIC DNA

Drug Resistance Markers

Much of the groundwork in the development of cloning technology was covered in work on derivatives of R plasmids. pSC101 was itself possibly derived from the R plasmd R6-5 (70), although not by in vitro recombination. pSC101 does not, therefore, correspond in size to any of the Eco RI fragments of R6-5 which were successfully "reassorted" and propagated by Cohen et al. (32). It proved possible to transform *E. coli* with the Eco RI fragments of R6-5 and to obtain clones which carried some of the drug resistance markers of the intact parent plasmid. In this case, ligation of the Eco RI fragments presumably occurred in vivo. One clone (pSC102), for example, had 3 of the 12 Eco RI fragments from

R6-5 and coded for resistance to kanamycin, neomycin, and sulfonamide. One of these RI fragments determined kanamycin resistance and was transferred onto pSC101 (to form the plasmid pSC105). The efficiency of hybrid plasmid formation could be increased substantially by treating the Eco RI fragments with *E. coli* ligase. In addition, DNA from a non-self-transmissable plasmid isolated from *Salmonella typhimurium* (RSF1010) was ligated to pSC101 and propagated in *E. coli* (32). This plasmid, which has a single Eco RI site, was later inserted into Col EI DNA (71) to form a hybrid plasmid which confers resistance to streptomycin and immunity to colicin EI. Eco RI-cleaved DNA of *D. melanogaster* (72) was cloned within this hybrid plasmid, although it is not a convenient Eco RI fragment vector because it has two susceptible cleavage sites necessitating the use of partially digested vector molecules.

Plasmid Replicons

The mini-Col EI derivatives (37, 40, 42) which have been constructed in vitro would seem to contain very little information other than that required for DNA replication. Similarly, the in vitro construction of plasmid derivatives of λ represents another way of isolating clustered genes involved in DNA replication (56). The Eco RI fragments which contain replication regions of the conjugative plasmids, F^1lac (73) and R6-5 (74), have been isolated as in vitro recombinants between single Eco RI fragments carrying drug resistance markers. The 9-kb Eco RI fragment from the F prime plasmid was found to carry incompatibility functions in addition to replication functions.

Timmis et al. (75) have shown that an in vitro recombinant between pSC101 and Col EI can use either of its two replicons. Replication is specified by pSC101 if the Col EI replication is prevented in a Pol I mutant; the replication mechanism of Col EI operates in the presence of chloramphenicol when the pSC101 function is inoperative.

F Plasmid Genes

Restriction enzyme analysis and molecular cloning are now proving to be powerful tools for dissecting the larger plasmids. The F sex factor of *E. coli*, for example, is 94.5 kb. The *tra* operon constitutes about 30 kb of the plasmid and is responsible for the transfer proficiency of DNA. The operon contains 12 identified genes, 9 of which are responsible for the formation of F pili and hence the absorption of male-specific phage (76). These, together with two other genes, are necessary for transfer, whereas the remaining gene serves to reduce transfer from one cell which carries F to another (surface exclusion). The operon is under the positive control of *traJ*, which is itself under the negative control of the *traO* and *finP* genes (fertility inhibition). These, together with other markers on the plasmid (the origin of transfer, replication genes, immunity to lethal zygosis (Ilz), and inhibition of replication of female-specific phage such as T7) have been identified within in vitro recombinants between pSC101 and Eco RI fragments of F (77). A number of partial digestion products of F were cloned and characterized. These DNA segments are shown aligned against the Eco RI cleavage map of F in Figure 5.

The plasmids pRS5, pRS21, and pRS30 were all deficient in surface exclusion (*Sfx*⁻). Only pRS5 contains fragment 7, and only strains carrying this plasmid were resistant to T7 infection, indicating that genes for female-specific phage inhibition are on this fragment. Strains carrying pRS30, which contains the contiguous segment of F with Eco RI fragments 6, 15, and 1, are sensitive to *fd*, unlike pRS5, which contains fragment 6 but lacks fragments 15 and 1 or pRS26, which contains fragments 15 and 1 but lacks fragment 6. This locates the genes responsible for F pilus formation, in agreement with previous heteroduplex and complementation analysis. Strains carrying pRS26, pRS15, and pRS31, which have fragments 17, 19, and 2 in common, are all proficient in surface exclusion. This location is in agreement with the map position of *traS*, the gene responsible for surface exclusion. An anomolous result here is that none of these plasmids contain *traJ*, which in F is responsible for the control of *traS*, and it is supposed that the inserted Eco RI fragments are under the control of some element in pSC101. The three plasmids pRS26, pRS15, and pRS31 all express immunity to lethal zygosis.

Escherichia coli DNA

Fragments of the *E. coli* genome have been cloned with the use of both phage and plasmid vectors. Cameron et al. (78), for example, have used the phage λgt-λB as a vector for the *E. coli* gene for DNA ligase. This phage is *red*⁻ and will not, therefore, grow on *E. coli lig*ts7 unless it carries the gene for DNA ligase. The hybrid phage used in this experiment and also hybrid phage containing yeast DNA (79) have been maintained as several pools. When the DNA from these pools is cleaved with Eco RI and examined on a gel, it does not have as high a

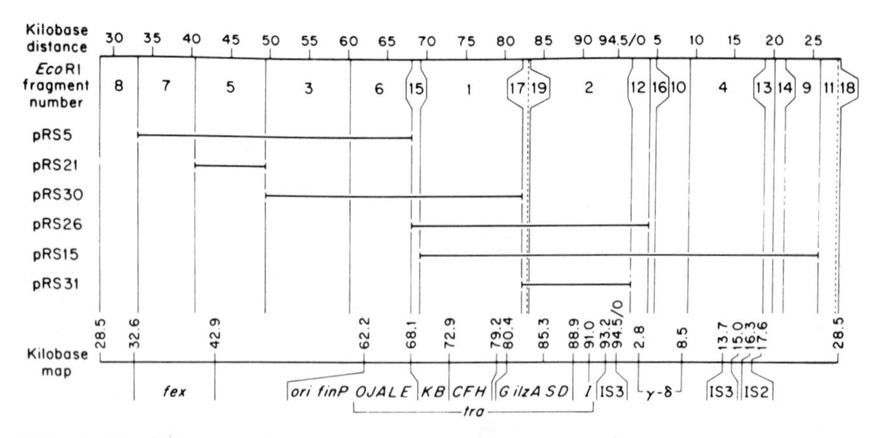

Figure 5. Physical map of F DNA indicating the relative positions of Eco RI cleavage sites and several genetic markers, together with the segments of the genome present in the pRS plasmids. Reproduced from Skurray et al. (77) with permission of The National Academy of Sciences, U.S.A.

complexity as either whole *E. coli* DNA or whole yeast DNA. This is partly to be expected from the packaging limits of phage λ, but may also be a reflection of different growth rates among these hybrid phages, resulting in the under-representation of certain Eco RI fragments within the pool. This differential replication of bacteria containing different hybrid molecules within the pool is also a major problem of working with heterogeneous populations of hybrid plasmids maintained within one culture.

Pools of hybrid phage containing DNA from *Klebsiella aerogenes* and *Klebsiella pneumoniae* have also been constructed, and the genes for DNA polymerase I isolated from these organisms (79).

Selections for certain genes carried by hybrid phage have been performed with the phage in the lysogenic state. Because hybrid phage formed with λgt-λB do not have the gene for integration or the phage attachment site, the hybrids were integrated into the bacterial chromosome as double lysogens by using an int^+ att^+ helper phage. The *Klebsiella* genes *aroA* (3-enolpyruvylshikimate-5-P synthetase) and *aroC* (chorismate synthetase) have been isolated with the use of such lysogens (77).

The effectiveness of Col EI as a cloning vector was first demonstrated by Hershfield et al. (1), who successfully cloned the kanamycin resistance-determining Eco RI fragment from pSC105, and also a segment of the *trp* operon. If tryptophan repression is relieved, then cells carrying the Col EI *trp* plasmid can contain 20–25% of their protein as *trp* enzymes. The donor DNA used in these experiments was a transducing phage carrying part of the *trp* operon. Other groups have since cloned fragments of *E. coli* DNA carried by transducing phages or by F^1 plasmids. Collins et al. (80) have, for example, made in vitro recombinants between DNA from a number of transducing phage and Col EI. These hybrid molecules contain ribosomal RNA genes or the gene for RNA polymerase (*rif*). Similarly, the *capR* gene of *E. coli* has been transferred from an F^1 plasmid into pSC101 (81). A mutation in *capR* of *E. coli* causes cells to overproduce capsular polysaccharide and the enzymes involved in its synthesis. In addition, these strains are highly sensitive to ultraviolet irradiation and become filamentous and die when grown in rich media. This gene seems to play an important part in the control of cell division. The in vitro recombinants were introduced into a $capR^-$ strain, which normally gives mucoid colonies. Two transformants were isolated which were nonmucoid and which contained a recombinant plasmid. Cells containing the recombinant plasmid show wild type levels of the enzymes UDP-glucose pyrophosphorylase and UDP-glucose dehydrogenase, although the cells remain sensitive to ultraviolet irradiation. It seems, therefore, as if cloning has permitted the separation of different markers in the *capR* gene.

Clarke and Carbon (82) made full use of the advantages of the dA-dT tailing techniques, with Col EI as the plasmid vector for fragments of whole *E. coli* DNA. They successfully isolated portions of the *ara* and *trp* operons by selections of hybrid plasmid pools on auxotropic strains. The hybrid plasmids carrying the *ara* region transform a leucine-requiring strain to Leu^+. They can estimate, therefore, from the size of the plasmid that there are no more than 15 kb between the *ara* and *leu* operons.

Rather than construct fresh batches of hybrids for testing on auxotropic strains, Clarke and Carbon (83) have gone on to make a bank of hybrid plasmids each of which contains a sheared segment of *E. coli* DNA (mean molecular length 13 kb) inserted into Col EI by the dA-dT technique. The hybrid molecules were introduced into a strain of *E. coli* harboring an F plasmid and approximately 2,000 individual colonies picked and stored. The genes carried by the hybrid plasmid can then be identified by F-mediated transfer into an F^- auxotroph in a replica mating experiment. Clarke and Carbon calculate that there is a 99% probability that any given *E. coli* gene is present in their bank and, indeed, plasmids carrying about 40 known genes have already been identified. In some cases neighboring genes have been identified in clones screened for their ability to complement an auxotroph. Two Col EI-*xyl* plasmids, for example, were found to overproduce glycine-tRNA synthetase (product of the *glyS* gene). The great potential of this procedure is clear in the establishment of gene banks for prokaryotic and lower eukaryotic organisms. On the other hand, the establishment of such a colony bank with genes of higher eukaryotes is not to be recommended because of the ease with which F plasmids promote plasmid transfer into other bacterial hosts.

CLONING EUKARYOTIC DNA

The DNA from most eukaryotes may be divided into three classes on the basis of the kinetics of its reassociation following denaturation. The proportions of these different classes vary. In *D. melanogaster*, for example, the rapidly annealing fraction, the highly reiterated simple sequence "satellite" DNA, amounts to about 12% of the genome; a further 12% are the moderately repetitive sequence (mrs) DNA, which are sequences represented on average about 70 times in the genome; the remaining DNA reassociates at a rate which indicates that these sequences are represented about once in the genome.

It has been possible, largely by the use of density gradients, to achieve considerable purification of highly repetitive and moderately repetitive sequences which are tandemly arranged. The characterization of this material has served to delineate the essential features of the organization of this DNA. The advent of restriction endonuclease and DNA cloning technology has, however, permitted a much finer analysis of tandemly repeated genes.

5 S and Ribosomal DNA

Xenopus laevis The genes for 5 S RNA and ribosomal RNA occur as tandemly repeated clusters within eukaryotic genomes. These genes are separated by so-called spacer sequences which are not transcribed. In *X. laevis* the units of 5 S DNA, each approximately 700 base pairs long, are present in thousands of copies and the 12-kb units of rDNA are present in 500 copies.

Earlier characterization of density gradient-purified material indicated that the genes were tandemly arranged and that each unit of transcription was separated from the adjacent unit by nontranscribed spacer sequences (84, 85). In the case of rDNA, the 18 S and 28 S genes are cotranscribed as precursor molecules which are subsequently processed into 28 S and 18 S rRNA.

The elucidation of the main features of the organization of these genes raised the question as to the mechanism by which tandemly arranged sequences are kept relatively homogeneous and how the genes and their spacers have evolved together. Two main theories have been proposed. On the one extreme are the sudden correction mechanims such as Callan's "master-slave" hypothesis in which many genes are simultaneously corrected against a master template. On the other extreme, gradual correction mechanisms have been proposed which involve unequal crossing over between homologous chromatids in such a way that variants could be spread or eliminated from tandem genes (86). Individual genes for 5 S DNA and ribosomal DNA in *X. laevis* have now been cloned (27, 33, 87) and subjected to elegant molecular analysis, which has cast some light on these problems.

The 5 S DNA of *X. laevis* has an AT-rich spacer which consists largely of repeated units of simple sequence DNA. Carroll and Brown (88) have recently described evidence which indicates that there is considerable length heterogeneity in this spacer sequence. Cleavage with the restriction endonuclease Hind III occurs at one site in each 5 S gene and generates a series of closely spaced bands, each of approximately 700 base pairs. These bands differ in length by regular increments of 15 base pairs, which is the length of the simple sequence unit described by Brownlee et al. (89). Carroll and Brown (33) have made partial Hind III digests of gradient-purified 5 S DNA and have cloned these fragments; each cloned segment contains from one to five of the 5 S genes. Denaturation mapping of these cloned 5 S genes indicates that the spacer regions in adjacent genes can be heterogeneous in length. It is unlikely that such arrangements could be found if sudden correction mechanisms were operating during evolution, and Carroll and Brown are strongly in favor of a mechanism of unequal crossing over between the AT-rich spacers in which the subrepeats are essentially homologous. The result of such a recombination would be units which varied in length by integral multiples of the subrepeat length, as is observed.

Wellauer et al. (87) have made a similar analysis of the ribosomal genes from *X. laevis*. There are two Eco RI fragments which can be excised from the gene unit. One of these is homogeneous in length and contains most of the 28 S gene plus some transcribed spacer and part of the 18 S gene. The other class of fragment is heterogeneous in length and contains nontranscribed spacer sequences. Four representatives of this latter class have been cloned within the Eco RI site of pSC101 and further characterized. When denatured and allowed to self-reassociate, these cloned rDNA segments were found to form a varying proportion of "imperfect" homoduplexes when examined in the electron microscope. These had loops which varied in size and position along the molecule, which could be explained if a region within each molecule consisted of multiple repeats of a simple sequence (see Figure 6). The nontranscribed spacer sequences within these cloned segments were also analyzed in experiments in which the shorter cloned DNA segment was examined in the electron microscope heteroduplexed with longer cloned Eco RI fragments, all containing the spacer sequences (CD42, XIr4, and XIr5, which are 6.3, 8.1, and 9.9 kb, respectively). The position of the deletion loop within these structures was used to demarcate the repetitious region. The interpretation of these experiments is shown in Figure 6. The nontranscribed spacer is divided into four areas: a region devoid of loops (*A*), two regions which have variable loops indicating internally repetitious sequences (*B* and *D*), and a region separating *B* and *D* in which loops are sparse (*C*). The smallest loops seen were about 50 base pairs long, which is, therefore, the maximal size of the subrepeat. Most of the length heterogeneity is due to variation in length of region *D;* again, the speculation is that the heterogeneity arises from out-of-register recombinational events within the rDNA.

The pattern of spacer lengths is found to be significantly different from one animal to another; therefore, it can be followed as a genetic marker (90). The patterns of spacer heterogeneity are stably inherited from one generation to the next. A total of 50 progeny have been examined from three matings. Out of

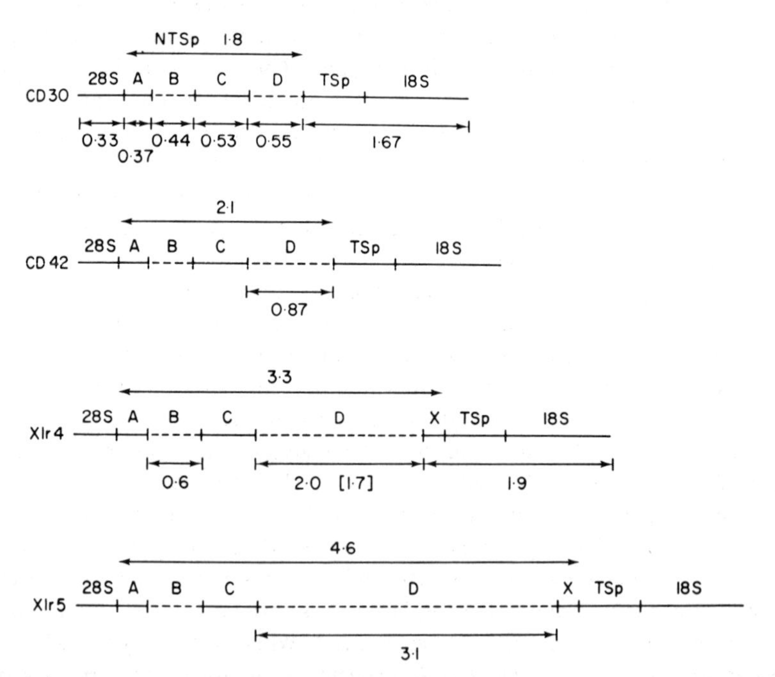

Figure 6. Maps of four cloned rDNA segments from *X. laevis* showing the locations of the two repetitions, regions B and D, as determined by heteroduplex mapping. Reproduced from Wellauer et al. (87) with permission of Academic Press.

these animals, only two showed changes in their spacer patterns. The mechanism by which these two changes arose remains an open question.

The amplification of the ribosomal genes in the oocytes of *X. laevis* is thought to occur by a rolling circle mechanism of DNA replication. If this is so, then one would predict that adjacent rDNA genes from oocytes would be more homogeneous in length than adjacent rDNA genes from somatic tissue. Wellauer et al. (91) have examined heteroduplex structures formed between either long single-strand chromosomal or amplified rDNA and a cloned segment of spacer-containing rDNA (CD30) and have found that this is indeed the case.

Drosophila melanogaster The nucleolus organizers of *D. melanogaster* are found on both the X and Y chromosomes. In wild type males there is apparently no meiotic recombination and so the two nucleolus organizers are genetically isolated on nonhomologous chromosomes, although the rRNA transcribed from either chromosome seems to be identical. A cloned segment of *Drosophila* rDNA has been described by Glover et al. (26). This is a 17-kb Eco RI fragment (Dm103) with the coding capacity for one molecule of 28 S RNA and one molecule of 18 S RNA. The kinetics of self-reassociation of the single strands of Dm103 indicates that it contains about 3 kb of internally repetitious "simple" sequence. Digestion of total DNA with Eco RI indicates that there are two major size classes of rDNA which will hybridize to complementary RNA (cRNA) transcribed from Dm103—one in which the repeating gene unit is 17 kb in length and the other in which the repeating gene unit is 11–12 kb in length. Tartof and Dawid (92) have found that, although the 11–12-kb unit is present on both the X and Y chromosomes, the 17-kb unit is found only on the X. This suggests that indeed recombination is not occurring between the X and Y rDNAs, unless there is some specific mechanism for eliminating the 17-kb units from the Y chromosome.

Dm103 was found to be a typical representative of this 17-kb class. In order to obtain a physical map of this plasmid, two approaches were taken. In one case, the Hind III fragments produced by digestion of pDm103 were themselves cloned in either pSC101 or Col EI-kan (34). These subclones were then tested for their degree of complementarity toward 28 S and 18 S rRNA. The physical map produced by these experiments is in good agreement with the more direct "R loop" mapping approach (93). An R loop is the structure formed when RNA hybridizes to its complementary strand of DNA to form a region of DNA-RNA duplex and to displace a single strand of DNA. The structure of R loops formed by pDm103 or a 12-kb unit of *D. melanogaster* rDNA is shown in Figure 7. It can be seen that the 17-kb unit has a most unusual feature in that the 28 S gene contains a 5-kb insertion. The function of this DNA is a matter for speculation.

Histone Genes

The histone genes from sea urchins, like rDNA and 5 S DNA, are tandemly arranged and can be purified by equilibrium density gradient centrifugation. The sea urchin 9 S polysomal histone mRNA has been used as a probe in order to

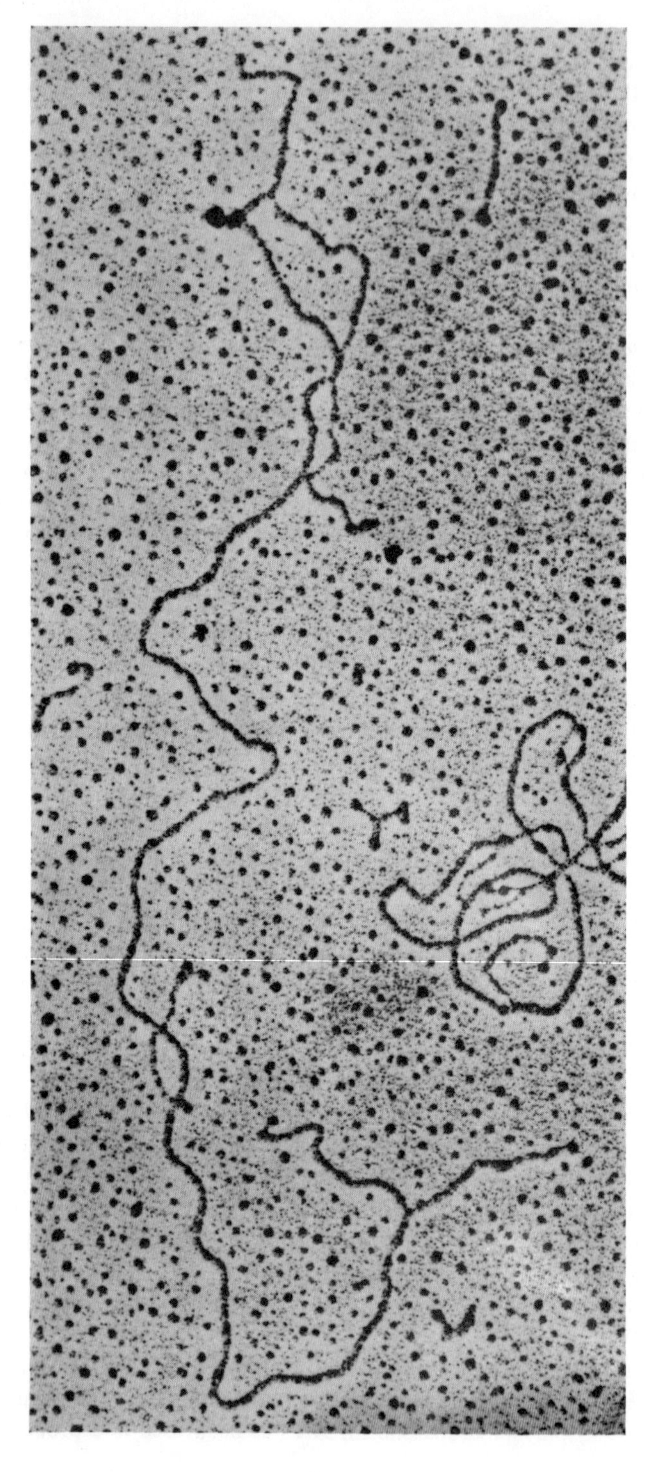

isolate cloned fragments of histone genes from several species of sea urchin and also from *D. melanogaster*. The properties of histone mRNAs and genes have been recently reviewed (94). Kedes et al. (95) have cloned the 2-kb and 4.6-kb Eco RI fragments of the histone DNA of *S. purpuratus* within pSC101. These two plasmids, pSP17 and pSP2, have themselves been subcloned (48). pSP17 contains a Hind III site within the histone DNA and a Hind III site within the pSC101 genome. The two Hind III fragments which result from the digestion of pSP17 have been cloned within pML21. pSP2 has two Bam I sites, one of which is in pSC101. A 1.3-kb Bam I fragment from this plasmid was cloned in pGM16. The 6-kb Hind III unit of *P. miliaris* has been cloned within a λ vector (96).

The major features of the gene unit have been established from hybridization studies with restriction fragments from either cloned DNAs or gradient-purified DNA (Figure 8). The 6-kb Hind III fragment of *P. miliaris* DNA can, for example, be cleaved into five fragments with the enzymes Eco RI and Hind II (97), each of which hybridizes to one of the five individual histone mRNAs. Digestion of the histone DNA with λ-exonuclease followed by DNA-RNA hybridization with the individual mRNAs has been used to establish the polarity of transcription. If the 6-kb Eco RI unit of *P. miliaris* is digested with λ-exonuclease (96), for example, then the hybridization efficiency of H4 mRNA is reduced, indicating that in this fragment the H4 gene sequence is near the 5′ terminus. If the 6-kb Hind III unit is digested with λ-exonuclease, however, then the H4 mRNA hybridizes to the digested fragment without denaturing the DNA, indicating that in this fragment the H4 sequence is near the 3′ terminus. These experiments establish the polarity of transcription of the H4 gene. It is possible to separate the DNA strands of the λ phage carrying the *P. miliaris* unit. All of the mRNAs are then found to hybridize to one of the strands, and since the five coding sequences within the repeat unit are arranged in tandem they must all have the same polarity. Similar experiments with the cloned segments of *S. purpuratus* DNA have established the gene order and its polarity (48).

The exact physical locations of the cloned *P. miliaris* genes have been determined from the examination of partially denatured molecules in the electron microscope (98). The AT-rich spacer sequences are now being subject to DNA sequence analysis, which indicates that unlike the spacer sequences of rDNA and 5 S DNA they are not a regular subrepeat of a simple sequence (99). The locations of the genes and spacers within pSP2 and pSP17 have also been determined with the use of the electron microscope (100). In this case, single-stranded plasmid DNA was annealed to mRNA, and the structures were visual-

---

Figure 7. R loops formed between Dm103 and rRNA of *D. melanogaster*. The fork at one end of the molecule results from the hybridization of 18 S rRNA and the three internal loops from 28 S rRNA. Insect 28 S RNA contains a hidden break about halfway down the molecule. The largest of the three 28 S R loops is due to the hybridization of one of the 28 S half-molecules. The insertion into the 28 S gene occurs into the sequences coding for the other half-molecule, which consequently forms the smaller separated R loops. This insertion is not present in the 12-kb unit where one sees two 28 S R loops of equal size (93).

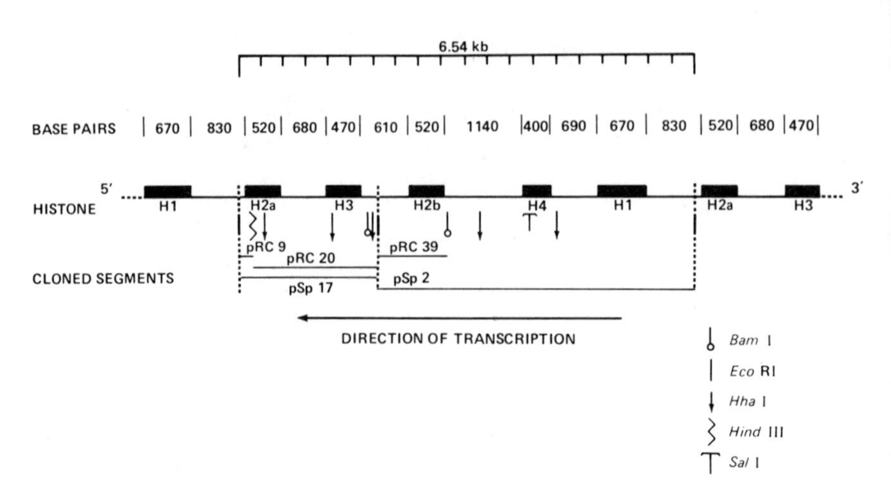

Figure 8. Physical map of the histone gene repeat unit of *S. purpuratus* showing the segments of the repeat unit which have been cloned within bacterial plasmids. Reproduced from Cohn et al. (48) with permission of MIT Press.

ized by the gene 32 protein-ethidium bromide method, which permits easy distinction between double- and single-stranded regions.

As the histones are a highly conserved group of polypeptides, it has proved possible to carry out cross-hybridization experiments between sea urchin histone mRNA and *D. melanogaster* DNA. This mRNA hybridizes in situ to a single region on the polytene chromosomes 39DE (101), comprising 12 chromomeres, and it has proved possible to screen hybrid plasmids containing *D. melanogaster* DNA in order to isolate a plasmid (cDm500) which will hybridize to the sea urchin histone mRNA and which will also hybridize in situ to region 39DE. This plasmid has been covalently linked to cellulose and used to isolate five *D. melanogaster* mRNAs coding for histones. This cloned segment of *D. melanogaster* DNA contains sequences present in the genome as 4.8-kb tandem repeats. It has been mapped with several restriction enzymes and shows transcriptional organization that differs from the sea urchin's in that genes are found on each DNA strand (102).

Analysis of *Drosophila* Genome

Polytene State To date, the most significant advances in the use of the DNA cloning technology as a means to analyze other eukaryotic genes have been with one organism, *D. melanogaster. D. melanogaster* has several advantages which make it a popular model organism for the study of the organization of the eukaryotic chromosome. It not only has cells which contain the diploid chromosome complement, but also cells in which the DNA has undergone as many as 10 replication events to produce the giant banded chromosomes. Not all the DNA sequences of the genome acquire the same degree of polyteny. The highly repetitive DNA sequences can be scarcely detected in polytene tissue. Highly radioactive copies of satellite DNA hybridize in situ to the centromeric regions of mitotic chromosomes and to the chromocenter of the polytene chromosomes (103). This region, the heterochromatin, has different histochemical staining properties and is proportionally very much reduced in the polytene state. A wealth of cytogenetic evidence in *D. melanogaster* has shown that the majority of genes are confined to the other part of the mitotic chromosomes, the euchromatic region. This euchromatic DNA undergoes extensive replication to form the banded arms of the polytene chromosomes, which may be imagined to consist of coordinate aggregates of identical chromatin fibers.

The genetic significance of the polytene chromosomes was realized in the 1930s by several groups of workers who observed that the banding pattern was specific and that by examining flies with chromosomal rearrangements it was possible to correlate cytological and genetic maps. The view quickly developed that each of the faint bands corresponded to a single genetic locus. The central paradox in our understanding of chromosomal organization lies in attempts to correlate the amount of DNA with the number of genes. In prokaryotes, there seems to be just sufficient DNA to account for the expected number of genes, whereas in most eukaryotes there is at least one order of magnitude more DNA than one would deem necessary. This is well illustrated in *D. melanogaster*, in which attempts have been made to quantitate the total number of genes by attempting to saturate well defined chromosomal regions with mutations (104–106). The main outcome of such experiments is that there seems to be an approximate correspondence between the number of complementation groups and the number of bands observed on the polytene chromosomes. If this is indeed true, then the total number of recognizable complementation groups within the organism would be about 5,000, which enables one to estimate, knowing the haploid complement of DNA, a mean gene length of about 26 kb. This is sufficient DNA to code for 20–30 average-sized polypeptides!

Randomly Cloned Fragments This problem can at last be approached. The ability to clone segments of chromosomal DNA in plasmids permits detailed physical mapping such as has been successfully applied to many viral genomes in recent years. At first, random isolates of cloned *D. melanogaster* DNA were characterized (12, 26). This proved a useful stepping stone to the isolation and characterization of selected segments of DNA. Of the nine randomly isolated segments characterized in detail, six contained long inserts of *D. melanogaster* DNA (about 15 kb) which was single copy by the criterion of reassociation kinetics, with no detectable mrsDNA. When highly radioactive copies of these hybrid DNA molecules were hybridized to polytene chromosomes in situ and the position of annealing was detected by autoradiography, each cloned single copy segment was found to hybridize to a single euchromatic region.

The other segments each contained mrsDNA and showed different patterns of in situ hybridization. One plasmid (pDm101) showed hybridization to the chromocenter. It was speculated that this chromosomal segment originated from the Y chromosome because it contained not only sequences represented about

30 times in the genome, but also sequences represented as 0.3 copy per genome equivalent of DNA. Y chromosomal DNA would be under-represented to about this extent in the mixed male-female DNA used to drive the reassociation reactions.

Interspersed Moderately Repetitive Sequence DNA One other plasmid was characterized (pDm1) (12) which contains a 3-kb *D. melanogaster* DNA segment which is present in 90 copies in the haploid genome. When hybridized to the polytene chromosomes in situ, pDm1 shows homology to 15 regions of the polytene chromosomes and also to the heterochromatin of the chromocenter. Smith et al. (21) have studied in detail a similar segment (Dm27) which hybridizes to about 60 chromosomal locations (Figure 9). These cloned segments have the properties of interspersed mrsDNA. There are two extreme interpretations of these in situ hybridization results. All the sequences within Dm27 could, for example, be present at each of the 60 chromosomal locations or there could be a subset of sequences within Dm27, all of which are present at a single location, with only one member of this subset being present at the other locations.

Rubin et al. (67) have used the colony hybridization technique (66) to screen for hybrid plasmids containing *D. melanogaster* DNA complementary to cRNA synthesized in vitro from Dm27 DNA. In this way, several plasmids have been isolated which show sequence homology to Dm27, but which originate from different chromosomal locations. The sequence homologies between electrophoretically separated restriction fragments from this group of plasmids have been analyzed in hybridization experiments with cRNA made against individual hybrid plasmids of this group with the use of the gel transfer hybridization system developed by Southern (107). This analysis, together with the results of in situ hybridization of radioactive copies of these plasmids to polytene chromosomes, indicates that Dm27 contains a cluster of different repetitive sequences, each of which is represented at many chromosomal sites.

DNA Coding for Abundant Messages One of the plasmids (cDm412) containing Dm27 sequences also contains sequences complementary to an abundant messenger RNA from *Drosophila* tissue culture cells. Six adjacent restriction fragments resulting from Hind III and Eco RI digestion of cDm412 show homology to an mRNA, 7 kb in length. Radiolabeled RNA complementary to cDm412 hybridizes in situ to about 70 euchromatic regions, as might be expected because the plasmid contains mrsDNA. A surprising result in found, however, when in situ hybridization is performed with radiolabeled restriction fragments from within the mRNA coding sequence. Thirty-three well separated chromosomal sites show homology to this mRNA sequence, suggesting that this coding sequence is found at many chromosomal locations. This conclusion is borne out by two further lines of evidence.

First, if total *D. melanogaster* DNA is digested with Hind III and Eco RI and the resulting fragments are fractionated on an Agarose gel, then ^{32}P-labeled restriction fragments internal to the mRNA sequence hybridize to a single band

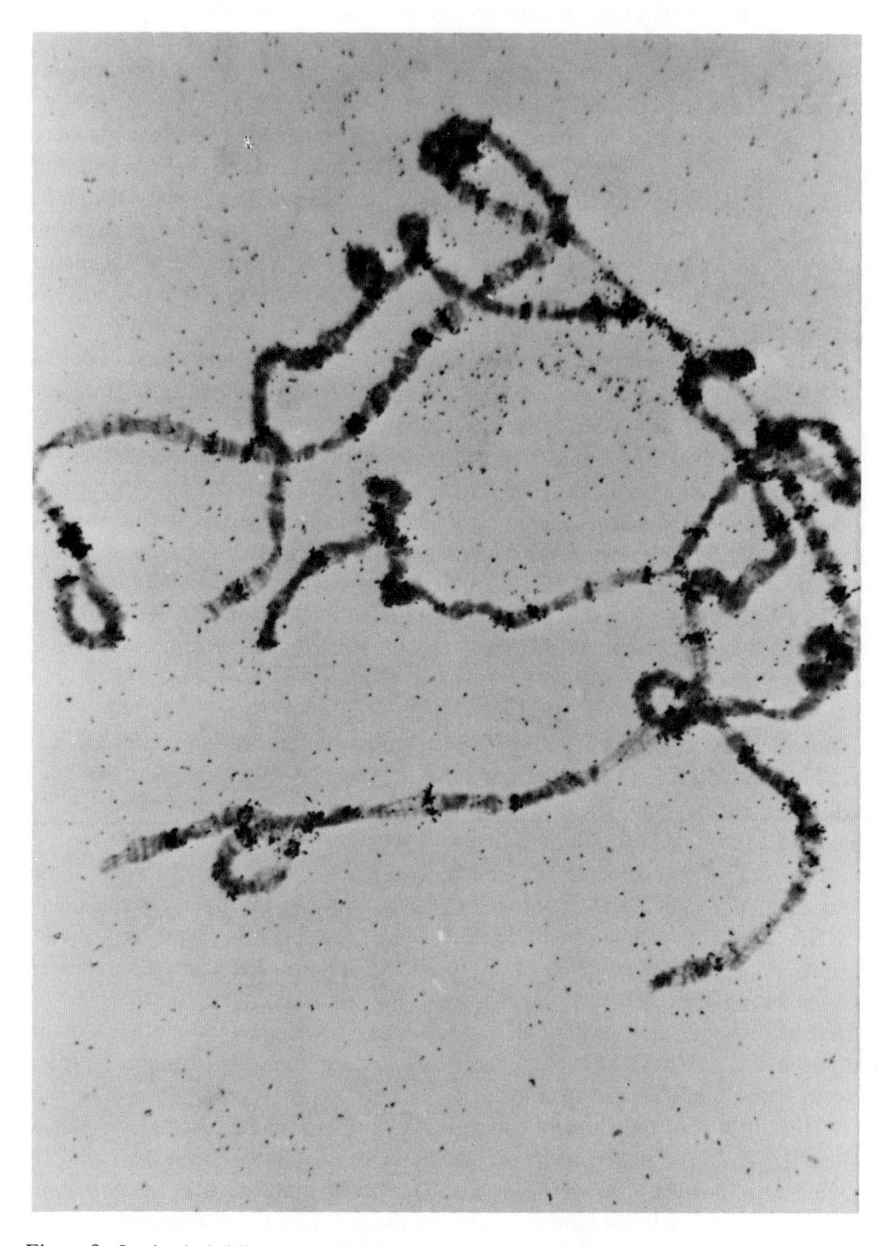

Figure 9. In situ hybridization of Dm27 to the salivary gland polytene chromosomes. Dm27 sequences labeled to a high specific activity with ^3H hybridize to 60 euchromatic regions. The silver grains deposited during autoradiography indicate the site of the homology. (Dr. David. J. Finnegan, unpublished photograph.)

from the total DNA digest which has the same size as the labeled fragment. If ^{32}P-labeled fragments containing the 3′ and 5′ termini of the message sequence are used, then these show hybridization to a number of restriction fragments of different sizes within the digest of total *D. melanogaster* DNA.

Second, several thousand independently cloned hybrid plasmids were screened in a colony hybridization test for complementarity to a restriction fragment within the mRNA sequence. Two of these (cDm454 and cDm468) were mapped with restriction enzymes and gave identical fragment patterns in the internal regions complementary to the mRNA, whereas the fragments complementary to the ends of the mRNA, as well as to the adjacent non-mRNA sequences, were different in each case.

Heat Shock Genes Other segments of chromosomal DNA which code for mRNA have now been cloned and have given equally surprising results. These are segments of DNA which are transcribed when *Drosophila* cells are subjected to heat shock. This response is a phenomenon which is particularly suited for analysis at the molecular level. If larvae are exposed to elevated temperatures, the polytene chromosomes undergo intensive RNA synthesis at several specific sites to produce a small number of new proteins. Similarly, when tissue culture cells are shifted from 25°C to 37°C, pre-existing polysomes are rapidly degraded and new polysomes are made. The polyadenylated RNA made at 37°C hybridizes in situ to the locations on the polytene chromosomes which undergo the intense RNA synthesis following heat shock. Lis et al. (108) have used ^{32}P-labeled polyadenylated RNA synthesized at 37°C to screen for plasmids containing the complementary DNA with the use of the colony hybridization technique. Lis and his co-workers have isolated a number of plasmids which contain DNA that hybridizes to the major heat shock locus. Two of these plasmids hybridize to about 15 additional chromosomal sites and presumably contain mrsDNA.

Restriction enzyme mapping and heteroduplex experiments between these four segments indicate that the 87 C locus contains multiple repeats of an mRNA coding sequence. In at least one of these plasmids there is a 7.4-kb spacer sequence between two of these coding regions. The tandemly repeated sequences do not, however, always have the same arrangment. One plasmid contains sequence arrangement of the type AAAAAA, whereas another has the arrangement ABABABAB. The spacer regions are thought to separate these blocks of tandemly repeating sequences.

This analysis of the *Drosophila* genome by using cloned segments of chromosomal DNA is certainly producing its share of surprises. These two mRNA coding chromosomal segments which have been studied in detail are both complex genes, each containing coding sequences which are repeated but distributed quite differently throughout the chromosomes. Single copy coding sequences have not yet been subjected to this scrutiny. The chromosomal arrangements of the cDm412 class of segments and the heat shock genes and the ribosomal DNA are all somewhat bizarre. Genes of the Dm412 type would almost certainly never be detected by the classic genetic approaches. One wonders how many more genes will be of this type and whether the genome will be much more plastic than the classic recombination maps have led us to expect.

Enzymic Synthesis of Single Copy Genes for Cloning

One recent approach to cloning eukaryotic sequences complementary to mRNAs has been to synthesize DNA complements with reverse transcriptase. Several groups have attempted this latter approach with the use of globin mRNA as a model system (38, 44, 49, 109). The hybrid plasmids cloned by Maniatis et al. (49) according to the scheme given in Figure 10 have been most extensively characterized. Globin mRNA was annealed with an oligo(dT) primer which paired with the stretch of poly(A) sequences at the 3′ end of the message and full length complementary DNA synthesized by the action of the reverse transcriptase from avian myeloblastosis virions. The RNA template was then destroyed by alkali to leave the complementary DNA, of which 2–8% was resistant to the single-strand-specific endonuclease *SI*. These *SI*-resistant fragments proved to be hairpin structures at the 3′ terminus of the cDNA which made excellent primers for DNA polymerase I, which was used to synthesize the second DNA strand. The major product of this reaction was 580 base pairs in

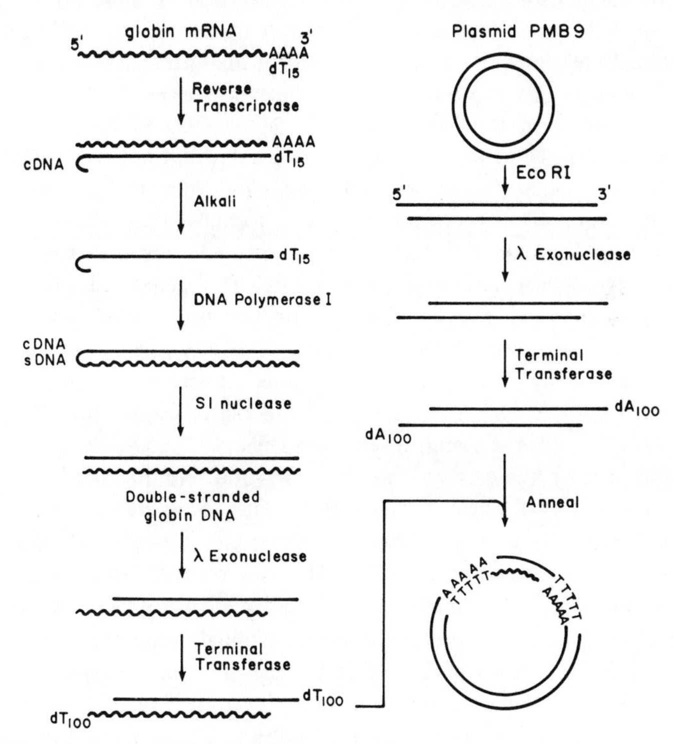

Figure 10. The scheme by which Maniatis et al. (49) constructed in vitro DNA recombinants between globin cDNA and the bacterial plasmid pMB9. Reproduced from Maniatis et al. (49) with permission of MIT Press.

length and was purified by preparative gel electrophoresis. The terminal hairpin was cleaved by using endonuclease *SI* to generate a denaturable double-stranded DNA molecule. This was treated with λ-exonuclease, followed by terminal transferase to effect its tailing with homopolymer blocks of deoxythymidylate residues. Maniatis et al. (49) then annealed these tailed molecules to linear forms of the plasmid pMB9 which had been tailed with deoxyadenylate residues. About 80% of the bacterial colonies transformed by this DNA mixture showed complementarity to globin mRNA in the colony hybridization screening test of Grunstein and Hogness (66). It is clearly important for many experiments that the sequences within the mRNA are copied with high fidelity. One clone was arbitrarily selected for further analysis by digestion with a number of restriction endonucleases. A physical restriction map was generated which could be aligned with the nucleotide sequence established either from the RNA sequence or by inference from the amino acid sequence of globin. Two restriction sites out of the 18 mapped require one nucleotide charge to get alignment. This is best explained by assuming that the source of the mRNA was a genetic variant since these nucleotide charges would result in a conservative amino acid substitution similar to polymorphisms of globin that have been previously described. Maniatis and co-workers estimate that the cloned insertion is some 60–80 base pairs shorter than the mRNA sequence, but that it contains all the coding region plus a substantial position (150 nucleotide pairs) of the untranslated region.

The other groups which have cloned globin DNA have used slightly different procedures, but have not cloned the full length of the globin coding sequence. Rougeon et al. (44) devised their experiments so that the Eco RI site within the plasmid DNA might be reconstructed on the sides of the inserted globin sequence, although this does not appear to have been achieved. The four nucleotide single-stranded 5′ termini of the Eco RI-cleaved vector were repaired by avian myeloblastosis reverse transcriptase. This generated a linear molecule with full base-paired termini. The vector was then tailed with oligodeoxy-cytidylate and annealed to the synthetic globin DNA tailed with deoxyguanidyl-ate residues. One bacterial colony transformed by a plasmid showing complementarity to cDNA was selected. This plasmid was estimated to contain 100–140 nucleotides of globin-specific sequence.

Rabbitts (38) has used yet another variation. Additions of oligo(dT) tails were made to Eco RI-cleaved mini-Col EI. The adenylated mRNA annealed to these tails, permitting cDNA to be synthesized directly onto the plasmid at both termini, thereby extending the oligo(dT) tails with globin cDNA. The globin cDNA sequences were then further extended with oligo(dT) tails resulting from a second incubation with terminal transferase. Finally, the molecules were mixed with mini-Col EI tailed with oligo(dA), and the mixture was denatured and allowed to reanneal before introducing the hybrid molecules into *E. coli*. This approach relies on the bacteria to repair the single-stranded cDNA within the hybrid molecules. About 7% of the bacterial transformants were estimated to contain a plasmid with globin sequences, and once again it seems likely that only a part of the globin sequence was successfully inserted.

Another approach has been to attempt to insert covalently a cDNA-mRNA hybrid structure into the plasmid vector Col EI (109). Since the replicative intermediate of Col EI is known to contain oligoribonucleotides, it was hoped that an extensive region of DNA-RNA duplex could be successfully replicated with *E. coli*. The cDNA-mRNA heteroduplex resulted from the action of reverse transcriptase and was tailed with oligo(dA) by using terminal transferase. This material was then ligated to Eco RI-cleaved Col EI DNA with the use of DNA polymerase I and T4 ligase. Once again, cloned plasmids containing globin sequences were isolated which contained 45–62% of the complete globin sequence.

In order to begin an analysis of a family of eukaryotic genes, it may be useful initially to clone synthetic DNA with the use of an approach such as that described by Maniatis et al. (49). It would then be possible to use this cloned sequence to identify sequences flanking the gene, for example, by hybridization to restriction fragments separated by gel electrophoresis. These flanking sequences could then in turn be cloned within plasmids. In preparing DNA synthesized in vitro from a population of mRNAs, one may have to be wary about not obtaining a true representation of the population within the synthesized DNA sequences. Efstratiadis et al. (110) report that, although they used a mixture of α- and β-mRNA and the cDNA contained substantial copies of both sequences, the β sequence was predominantly synthesized by DNA polymerase I. Indeed, Maniatis et al. (49) only detected β-globin sequences in their clones, although 520 base pair DNA from α-mRNA may have been selected against in their size fractionation step. Reverse transcription of a given RNA results in the synthesis of a population of cDNAs with a well defined electrophoretic banding pattern. This presumably is a consequence of sequences or secondary structure within the RNA which act as stop signals and may be an additional problem with this type of approach.

EXPRESSION OF EUKARYOTIC DNA WITHIN *ESCHERICHIA COLI*

Mini-cell Analysis

The expression of cloned segments of eukaryotic DNA has been analyzed at the molecular level in mini-cell-producing strains of *E. coli*. Mini-cells are cellular "buds" which contain no chromosomal DNA, but do contain plasmid DNA together with the cellular machinery for transcription and translation. When purified away from the parental *E. coli* they represent a means of studying plasmid transcription and translation with no host cell background. The main outcome of experiments with mini-cells containing hybrid plasmids is that in some cases the eukaryotic DNA is both transcribed and translated, but there is little evidence to suggest whether or not the final polypeptide bears any relationship to that expected in the homologous system. In the case of pSC101 carrying mouse mitochondrial DNA (111), for example, initiation of RNA synthesis seems to occur predominantly on the light strand of the mitochondrial DNA, irrespective of its orientation within the plasmid vector. The molecular weight distribution of the polypeptides which are ultimately produced differs from that observed in mouse L cell mitochondria.

Rambach and Hogness (112) have carried out similar experiments on cloned segments of *D. melanogaster* DNA. Novel polypeptides coded by *D. melanogaster* DNA have been frequently found in mini-cells containing hybrid plasmids. In one case the new polypeptide has been shown to be encoded by a particular restriction fragment of *D. melanogaster* DNA, and the same polypeptide is synthesized irrespective of the orientation of this restriction fragment or of the plasmid vector in which it is being carried. Similar experiments have been performed with a cloned extrachromosomal element from yeast, which codes for a number of high molecular weight polypeptides irrespective of the orientation of the cloned DNA within the vector plasmid (111).

Complementation of *Escherichia coli* Auxotrophs

The most convincing evidence for functional expression of eukaryotic DNA within *E. coli* so far comes from the complementation of *E. coli* auxotrophs with cloned segments of yeast DNA. Struhl et al. (79) have cloned a 10-kb Eco RI fragment of yeast DNA which complements a nonrevertible histidine auxotroph lacking the enzyme imidazole glycerol phosphate (IGP) dehydratase. The hybrid molecules were integrated into the bacterial chromosome with the use of integration helper phage. Transcription was probably initiated within the yeast DNA since with this lysogen the λ promoters were either strongly repressed or deleted. In addition, reversal of the orientation of the inserted yeast DNA had no effect upon expression. Complementation with this hybrid phage, λgtSc*his*, was observed for both revertible and nonrevertible mutants in IGP dehydratase and only for this gene of the histidine operon.

The same segment of yeast DNA has been independently cloned by Ratzkin and Carbon (114, 115). These authors have made hybrid plasmids between poly(dT)-tailed Col EI DNA and poly(dA)-tailed yeast DNA and selected those capable of complementing deletion mutants in the *leuB* gene (β-isopropylmalate dehydrogenase) or the *hisB* gene (IGP dehydratase) of *E. coli.* The rate of reassociation of these plasmid DNAs following denaturation can be increased by the addition of total yeast DNA, indicating that the plasmids contain yeast DNA sequences. Furthermore, the rate of reassociation of the plasmid which complements the *hisB* mutation can be increased by λgtSc*his* DNA, indicating that both the phage and the plasmid contain overlapping yeast DNA segments.

A total of 16 transformants were isolated in which the *leu 6* mutation in *E. coli* was either complemented or suppressed by Col EI plasmids carrying yeast DNA. These divide into four groups on the basis of their Eco RI restriction patterns. A representative of one of these groups, pY*eleu 10* DNA, is capable of transforming three different mutations in the *leuB* gene. This plasmid has also been trasnferred to mutants of *S. typhimurium* carrying deletions in the *leuA, leuB, leuC*, and *leuD* genes. Colonies were obtained from this F-mediated cross only with the *leuB* deletion recipients. It seems highly likely, therefore, that the yeast DNA in pY*eleu 10* is producing some polypeptide which is capable of complementing this deletion mutation. The three other groups of hybrid plasmids, on the other hand, would only complement one of three mutations in the *leuB* gene. It seems possible in this case that missense suppression has been observed by yeast tRNA or aminoacyl tRNA synthetase genes carried on the plasmid.

Here then is substantial evidence that some meaningful expression of eukaryotic DNA can occur within *E. coli*. It remains to be seen whether the same is possible for DNA from the higher eukaryotes.

Promoting Expression

One problem that is recognized even with yeast DNA plasmids is the possibility of inefficient levels of expression. The enzyme activities can approach wild type levels in *E. coli* carrying the yeast *hisB* plasmid, but are only 1% in the case of the yeast *leuB* plasmid. The detailed understanding of gene expression in λ phage is a useful asset in devising vectors in which inserted DNA fragments can be efficiently transcribed. The rationale is to put the inserted DNA under the control of the phage promoter P_L, which is on the order of 10-fold more efficient than the average *E. coli* promoter (116). The presence of the *N* gene product is also important because this protein interacts with RNA polymerase and permits transcription through sequences which would normally act as termination signals (117). The phage should ideally be *cro*⁻ in order to relieve the repressive effect of the *cro* gene product on transcription from the P_L and P_R promoters. Since *cro*⁻ phage grow poorly, it is possible to use a hybrid immunity phage that contains cro^{434} and $P_R{}^{434}$, along with the promoter P_L of λ which is not repressed by the cro^{434} (118).

These control systems can be incorporated into plasmid genomes. Helinski et al. (119) have constructed an in vitro recombinant of Col EI and a λ*trp* phage which carries the *trpB, C,* and *D* genes and the *N*, P_L, CI*ts*, P_R, *cro*⁻ genes. At 40°C the thermosensitive repressor of the *CI* gene is inactivated and transcription is then initiated off one strand from P_L and the other strand from P_R. The *trp* genes have been found to be temperature-inducible within cells carrying this plasmid, although it remains to be determined whether or not enhanced levels of expression of inserted foreign DNA can be promoted.

Another encouraging approach has been to incorporate the promoter of the *lac* operon into the plasmid vector. Polisky et al. (120) have inserted an Eco RI fragment containing the *lac* operator, promoter, and most of the structural gene for β-galactosidase into pSF2124 and then subsequently deleted one Eco RI site from this recombinant by using the technique of Covey et al. (43). The transcription of an Eco RI fragment containing 28 S rDNA of *X. laevis* inserted into this plasmid was found to increase about 9-fold upon induction of the *lac* operon. The *lac* promoter is carried on a 203 base pair fragment generated by Hae III, an enzyme which produces "flush" ends with 3′-terminal (dG) residues. An intriguing approach was used by Backman et al. (121) to convert these Hae III termini into termini identical with those produced by Eco RI. A plasmid vector was digested with Eco RI, and the single-stranded termini were converted to double-stranded termini with the use of DNA polymerase I. The Hae III fragment was then joined to these flush Eco RI termini with T4 ligase to produce a molecule from which the *lac* promoter fragment could then be cleaved with Eco RI. These authors have also used a similar approach to replace the small DNA fragment between the Hind III and Eco RI sites of pMB9 with a DNA fragment carrying the λ *CI* gene but with Hind III and Hae III termini. This recombinant plasmid retains its Eco RI site at which it was possible to insert the portable *lac* promoter. Strains carrying plasmids containing the *CI* gene but not the promoter were found to produce 5 times more λ repressor than a single lysogen, whereas strains with plasmids containing both the *CI* gene and two copies of the *lac* promoter produced 35 times more repressor than did a single lysogen.

CONCLUDING REMARKS

When the problems of enhancing the transcription of foreign DNA have been overcome, it is probable that appropriate translational controls will have to be incorporated into the novel genetic element. The introduction of such controlling elements into the vector molecules is important if the full potential of the applications of in vitro DNA recombination is to be achieved. At the same time, a parallel exploitation of cloning technology is needed for plasmids which have a broader host range, in order that prokaryotic genes of commercial importance might be usefully propagated within a variety of microorganisms. The technology of in vitro DNA recombination marks a new era in molecular biology which may enable much of the knowledge gained in the last 20 years to find useful application. The technology is, however, still in its infancy, and the development of this prodigy will be watched with great interest.

ACKNOWLEDGMENTS

I would like to thank Peter Rigby for his criticism of this chapter, Di Barnett and Peter Burrows for their help with the diagrams, and Val Monrabal for typing the manuscript.

REFERENCES

1. Hershfield, V., Boyer, H. W., Yanofsky, C., Lovett, M. A., and Helinksi, D. R. (1974). Proc. Natl. Acad. Sci. U.S.A. 71:3455.

2. Jackson, D. A., Symons, R. M., and Berg, P. (1972). Proc. Natl. Acad. Sci. U.S.A. 69:2904.
3. The Ashby Committee Report (1975). Her Majesty's Stationery Office Publication, London.
4. Berg, P., Baltimore, D., Brenner, S., Roblin, R. O., and Singer, M. F. (1975). Science 188:991.
5. Sinsheimer, R. (1976). Trends Biochem. Sci. 1:N178.
6. Chargaff (1976). Science 192:938.
7. Davis, B. (1976). Trends Biochem. Sci. 1:N178.
8. NIH Guidelines, National Institutes of Health, Bethesda, Maryland.
9. The Williams Committee Report (1976). Her Majesty's Stationery Office Publication, London.
10. Nathans, D., and Smith, H. O. (1975). Ann. Rev. Biochem. 44:273.
11. Lobban, P. E., and Kaiser, A. D. (1973). J. Mol. Biol. 78:453.
12. Wensink, P. C., Finnegan, D. J., Donelson, J. E., and Hogness, D. S. (1974). Cell 3:315.
13. Glover, D. M. (1976). *In* B. Smith and P. Pain (eds.), New Techniques in Biophysics and Cell Biology, Vol III, pp. 126–143. John Wiley International.
14. Roychoudhury, R., Jay, E., and Wu, R. (1976). Nucleic Acid Res. 3:101.
15. Mertz, J. E., and Davis, R. W. (1972). Proc. Natl. Acad. Sci. U.S.A. 69:3370.
16. Wilson, G. A., and Young, F. E. (1975). J. Mol. Biol. 97:123.
17. Roberts, R. J., Wilson, G. A., and Young, F. E., unpublished observations.
18. Yoshimori, R. N. (1971). Ph.D. thesis, University of California, San Francisco.
19. Hedgepeth, J., Goodman, H. M., and Boyer, H. W. (1972). Proc. Natl. Acad. Sci. U.S.A. 69:3448.
20. Old, R., Murray, K., and Roizes, G. (1975). J. Mol. Biol. 92:331.
21. Smith, H. O., and Wilcox, K. W. (1970). J. Mol. Biol. 51:379.
22. Smith, E. I., Blattner, F. R., and Davies, J. (1976). Nucleic Acid Res. 3:343.
23. Arrand, J. R., Myers, P. E., and Roberts, R. J., unpublished observations.
24. Goff, S., and Rambach, A., unpublished observations.
25. Endow, S. A., and Roberts, R. J., unpublished observations.
26. Glover, D. M., White, R. L., Finnegan, D. J., and Hogness, D. S. (1975). Cell 5:149.
27. Morrow, J. F., Cohen, S. N., Chang, A. C. Y., Boyer, H. W., Goodman, H. M., and Helling, R. B. (1974). Proc. Natl. Acad. Sci. U.S.A. 71:1743.
28. Sgaramella, V. (1972). Proc. Natl. Acad. Sci. U.S.A. 69:3389.
29. Hirsch, H. J., Starlinger, P., and Brachet, P. (1972). Mol. Gen. Genet. 119:191.
30. Hirsch, H. J., Saedler, H., and Starlinger, P. (1972). Mol. Gen. Genet. 115:266.
31. Saedler, H., Reif, H. J., Hu, S., and Davidson, N. (1974). Mol. Gen. Genet. 132:265.
32. Cohen, S. N., Chang, A. C. Y., Boyer, H. W., and Helling, R. B. (1973). Proc. Natl. Acad. Sci. U.S.A. 70:3240.
33. Carroll, D., and Brown, D. D. (1976). Cell 7:477.
34. Glover, D. M., and Hogness, D. S. (1977). Cell 10:167.
35. Clewell, D. B., and Helinski, D. R. (1972). J. Bacteriol. 110:1135.
36. Blair, D. G., Sherratt, D. J., Clewell, D. B., and Helinski, D. R. (1972). Proc. Natl. Acad. Sci. U.S.A. 69:2518.

37. Hershfield, V., Boyer, H. W., Chow, L., and Helinski, D. R. (1976). J. Bacteriol. 126:447.
38. Rabbitts, T. H. (1976). Nature (Lond.) 260:221.
39. Smith, H. R., Humphreys, G. O., Willshaw, G. A., and Anderson, E. S. (1976). Mol. Gen. Genet. 143:319.
40. Rodriguez, R. L., Bolivar, F., Goodman, H. W., Boyer, H. W., and Betlach, M. *In* D. Nierlich and W. Rutter (eds.), ICN-UCLA Symposium on Molecular and Cellular Biology, Vol. 5. Academic Press, New York, in press.
41. So, M., Gill, R., and Falkow, S. (1975). Mol. Gen. Genet. 142:239.
42. Charney, D., and Berg, P., personal communication.
43. Covey, E., Richardson, D., and Carbon, J. (1976). Mol. Gen. Genet. 145:155.
44. Rougeon, F., Kourilsky, P., and Mach, B. (1976). Nucleic Acid Res. 2:234.
45. Heffron, F., Rubens, C., and Falkow, S. (1975). Proc. Natl. Acad. Sci. U.S.A. 72:3623.
46. Schedl, P., Artovams-Tsakonis, S., and Gehring, W. Personal communication.
47. Hamer, D. H., and Thomas, C. A. (1976). Proc. Natl. Acad. Sci. U.S.A. 73:1537.
48. Cohn, R. H., Lowry, J. C., and Kedes, L. M. (1976). Cell 9:147.
49. Maniatis, T., Kee, S. E., Efstratiadis, A., and Kafatos, F. (1976). Cell 8:163.
50. Murray, N. E., and Murray, E. (1974). Nature (Lond.) 251:476.
51. Rambach, A., and Tiollais, P. (1974). Proc. Natl. Acad. Sci. U.S.A. 71:3927.
52. Thomas, M., Cameron, J. R., and Davis, R. W. (1974). Proc. Natl. Acad. Sci. U.S.A. 71:4579.
53. Murray, K., and Murray, N. E. (1975). J. Mol. Biol. 98:551.
54. Murray, N. E., Brammar, W. J., and Murray, K. Mol. Gen. Genet., in press.
55. Equist, L., Tiemeier, D., Leder, P., Weisberg, R., and Steinberg, N. (1976). Nature (Lond.) 259:596.
56. Hobom, G., personal communication.
57. Nussbaum, A. L., Davoli, E., Ganem, D., and Fareed, G. C. (1976). Proc. Natl. Acad. Sci. U.S.A. 73:1068.
58. Ganem, D., Nussbaum, A. L., Davoli, D., and Fareed, G. C. (1976). Cell 7:349.
59. Hamer, D. H., Davoli, D., Thomas, C. A., and Fareed, G. C., personal communication.
60. Goff, S., and Berg, P. (1976). Cell 9:695.
61. Mandel, M., and Higa, A. (1970). J. Mol. Biol. 53:159.
62. Hohn, B., and Murray, K., personal communication.
63. Brown, W. M., Watson, R. M., Vinograd, J., Tait, K. M., Boyer, H. N., and Goodman, H. M. (1976). Cell 7:517.
64. Hogness, D. S., personal communication.
65. Curtiss, R., Pereira, D. A., Hsu, J. C., Hull, S. C., Clark, J. E., Maturin, L. J., Inone, M., Goldschmidt, R., Moddy, R., and Alexander, L., personal communication.
66. Grunstein, M., and Hogness, D. S. (1975). Proc. Natl. Acad. Sci. U.S.A. 72:3961.
67. Rubin, G. M., Finnegan, D. J., and Hogness, D. S. (1976). Prog. Nucleic Acid Res. Mol. Biol. 19:221.

68. Jones, K. W., and Murray, K. (1975). J. Mol. Biol. 96:455.
69. Kramer, R. A., Cameron, J. R., and Davis, R. L. (1976). Cell 6:227.
70. Cohen, S. N., and Chang, A. C. Y. (1973). Proc. Natl. Acad. Sci. U.S.A. 70:3240.
71. Tanaka, T., and Weisblum, B. J. Bacteriol., in press.
72. Tanaka, T., Weisblum, B., Schnoss, M., and Inman, R. Biochemistry, in press.
73. Lovett, M. A., and Helinski, D. R. J. Bacteriol., in press.
74. Timmis, K., Cabello, F., and Cohen, S. N. (1975). Proc. Natl. Acad. Sci. U.S.A. 72:2242.
75. Timmis, K., Cabello, F., and Cohen, S. N. (1974). Proc. Natl. Acad. Sci. U.S.A. 71:4556.
76. Achtman, M., and Helmuth, R. (1975). *In* D. Schlessinger (ed.), Microbiology 1974, pp. 95–103. American Society for Microbiology, Washington, D.C.
77. Skurray, R. A., Nagaishi, H., and Clark, A. J. (1976). Proc. Natl. Acad. Sci. U.S.A. 73:64.
78. Cameron, J. R., Panasenko, S. M., Lehman, I. R., and Davis, R. W. (1975). Proc. Natl. Acad. Sci. U.S.A. 72:3416.
79. Struhl, K., Cameron, J. R., and Davis, R. W. (1976). Proc. Natl. Acad. Sci. U.S.A. 73:1471.
80. Collins, J., Fiil, N. P., Jorgensen, P., and Friesen, J. D. (1976). Control of Ribosome Synthesis: Alfred Benzon Symposium IX, Munksgaard, Copenhagen.
81. Berg, P. E., Gaydan, R., Avni, H., Zehnbauer, B., and Markovitz, A. (1976). Proc. Natl. Acad. Sci. U.S.A. 73:697.
82. Clarke, L., and Carbon, J. (1975). Proc. Natl. Acad. Sci. U.S.A. 72:4361.
83. Clarke, L., and Carbon, J. (1976). Cell 9:91.
84. Brown, D. D., Wensink, P. C., and Jordan, E. (1971). Proc. Natl. Acad. Sci. U.S.A. 68:3175.
85. Wensink, P. C., and Brown, D. D. (1971). J. Mol. Biol. 60:235.
86. Smith, G. (1973). Cold Spring Harbor Symp. Quant. Biol. 38:507.
87. Wellauer, P. K., Dawid, I. B., Brown, D. D., and Reeder, R. H. (1976). J. Mol. Biol. 105:461.
88. Carroll, D., and Brown, D. D. (1976). Cell 7:467.
89. Brownlee, G. G., Cartwright, E. M., and Brown, D. D. (1974). J. Mol. Biol. 89:703.
90. Reeder, R. H., Brown, D. D., Wellauer, P. K., and Dawid, I. B. (1976). J. Mol. Biol. 105:507.
91. Wellauer, P. K., Reeder, R. H., Dawid, I. B., and Brown, D. D. (1976). J. Mol. Biol. 105:487.
92. Tartof, K. D., and Dawid, I. B. (1976). Nature (Lond.) 263:27.
93. White, R. L., and Hogness, D. S. (1977). Cell 10:177.
94. Kedes, L. H. (1976). Cell 8:321.
95. Kedes, L. H., Cohn, R. H., Lowry, J. C., Chang, A. C. Y., and Cohen, S. N. (1975). Cell 6:359.
96. Gross, K., Schaffner, W., Telford, J., and Birnsteil, M. (1976). Cell 8:479.
97. Schaffner, W., Gross, K., Telford, J., and Birnsteil, M. (1976). Cell 8:471.
98. Birnsteil, M., personal communication.
99. Schaffner, W., personal communication.

100. Wu, M., Holmes, D. S., Davidson, N., Cohn, R. H., and Kedes, L. H. (1976). Cell 9:163.
101. Pardue, M. L., Weinberg, E., Kedes, L., and Birnsteil, M. (1972). J. Cell Biol. 55:199.
102. Karp, R., Goldberg, M., and Hogness, D. S., personal communication.
103. Gall, J. G., Cohen, E. H., and Polan, M. L. (1971). Chromosoma 33:319.
104. Judd, B. M., and Young, M. W. (1973). Cold Spring Harbor Symp. Quant. Biol. 38:573.
105. Hochman, B. (1973). Cold Spring Harbor Symp. Quant. Biol. 38:581.
106. Lefevre, G. (1973). Cold Spring Harbor Symp. Quant. Biol. 38:591.
107. Southern, E. M. (1975). J. Mol. Biol. 98:503.
108. Lis, J., Prestidge, L., and Hogness, D. S., personal communication.
109. Wood and Lee. (1976). Nucleic Acid Res. 3:1961.
110. Efstratiadis, A., Kafatos, F., Maxam, A., and Maniatis, T. (1976). Cell 7:279.
111. Chang, A. C. Y., Lansman, R. A., Clayton, D. A., and Cohen, S. N. (1975). Cell 6:231.
112. Rambach, A., and Hogness, D. S., personal communication.
113. Hollenberg, C. P., Kustermann-Kuhn, B., and Royer, H.-D. Gene, in press.
114. Ratzkin, B., and Carbon, J. (1977). Proc. Natl. Acad. Sci. USA 74:487.
115. Carbon, J., Clarke, L., Ilgen, C., and Ratzkin, B. Tenth Miles Symposium on the Impact of Recombinant Molecules on Science and Society, in press.
116. Davidson, J., Brammar, W. J., and Bruce, F. (1974). Mol. Gen. Genet. 130:9.
117. Franklin, N. C. (1974). J. Mol. Biol. 89:33.
118. Murray, N. E., and Murray, K., personal communication.
119. Helinski, D. R., Hershfield, V., Figurski, D., and Meyer, R. J. (1976). Tenth Miles Symposium on the Impact of Recombinant Molecules on Science and Society, in press.
120. Polisky, B., Bishop, R. J., Gelfand, D. H. (1976). Proc. Natl. Acad. Sci. U.S.A. 73:3900.
121. Backman, K., Ptashne, M., and Gilbert, W. (1976). Proc. Natl. Acad. Sci. USA 73:4174.

International Review of Biochemistry
Biochemistry of Cell Differentiation II, Volume 15
Edited by J. Paul

2
Cellular and Molecular Aspects of Genetic Expression in *Chironomus* Salivary Glands

S. T. CASE[a] AND B. DANEHOLT

Department of Histology,
Karolinska Institutet,
Stockholm,
Sweden

The authors' research was supported by grants from the Swedish Cancer Society, Magnus Bergvalls Stiftelse, and Karolinska Institutet (Reservationsanslaget).

[a]Recipient of a National Institutes of Health National Research Service Award from the National Institute of General Medical Sciences.

Our present ability to discuss the expression of genetic information in differentiated cells of Diptera is founded on a series of genetic and cytological investigations which began in the early part of this century. Basic genetic concepts, such as the chromosome theory of heredity (1), were established by experimentation with *Drosophila.* Later on, it was discovered that certain Dipteran tissues, in particular the salivary glands, contained polytene chromosomes of giant size (2–4). Because of their large dimensions and characteristic banded morphology, these chromosomes proved useful in mapping the chromosomal location of defined genes (3, 5). During subsequent years, experiments employing both *Drosophila* and *Chironomus* provided a wealth of information as to the distribution and synthesis of macromolecules along the giant chromosomes (6–10). Numerous studies made it possible to consider the organization of these chromosomes in terms of functional units, such as genetic (11), replication (12), and transcription (13, 14) units. Therefore, it is not surprising that these chromosomes have now served for several decades as models for eukaryotic chromosomes in general. Finally, studies in *Chironomus* have recently permitted a detailed morphological and biochemical examination of the transfer of genetic information from discrete portions of giant chromosomes to the cytoplasm (10, 15) and polysomal structures (16).

Although *Drosophila* has been the Dipteran of choice in many studies, *Chironomus* offers certain unique opportunities to comprehend gene expression at the cellular and biochemical levels. This is mainly a result of the large chromosomal puffs, called Balbiani rings, found on certain salivary gland polytene chromosomes. Beermann (17) proposed, and this review attempts to document, that a chromosomal puff represents an active genetic locus in a highly differentiated cell. This chapter provides an overview of available data on the function of the salivary gland cells and discusses the synthesis, transport, and function of puff RNA in some detail, concentrating on the products of Balbiani ring 2.

SALIVARY GLANDS IN *CHIRONOMUS*

The primary function of the salivary glands is to produce a proteinaceous secretion used by *Chironomus* larvae to spin gelatinous, tubelike burrows (18,

19). It has also been observed that a larva uses this secretion to close one end of the burrow with a funnel-shaped net (20). By means of undulating movements inside the tube, the larva creates a stream of water through the tube, and various particles, including plankton, are then caught in the net. The larva consumes the net as well as the attached particles, and immediately afterward it builds a new net. The larval salivary glands are, therefore, to be regarded as spinning glands rather than as digestive ones. Furthermore, the behavior of larvae suggests that the salivary glands are more or less continuously producing the proteinaceous secretion, not only for the spinning of new nets but also for entirely new tubes.

The salivary glands are paired and located on either side of the esophagus in the second and third thoracic segments (18). The glands are immersed in hemolymph and are suspended by ligaments (21). Each gland delivers its secretion anteriorly through a secretory duct which eventually fuses with the duct from the opposite gland just prior to entry into the hypopharynx (18, 22). The salivary glands are actively secreting throughout the larval period, but degenerate and disappear during later developmental stages (21, 23).

Structure of Secretory Cell

Each salivary gland harbors 35–40 cells, which are located in one peripheral layer surrounding a central lumen (Figure 1*A*). The most characteristic feature of the salivary gland is the unusual size of its cells and their nuclei. The cell nucleus is conspicuous with a diameter of about 75 μm. Inside the nucleus there are four giant chromosomes, two of which carry a nucleolus (Figure 1*B*). These chromosomes are discussed in more detail below.

The cytoplasm has several ultrastructural properties characteristic of a secretory cell (23–25). It has an abundant granular endoplasmic reticulum, mainly of the tubular type. A great number of Golgi complexes and secretory granules are recognized in various parts of the cytoplasm. It is, furthermore, characteristic that secretory granules are particularly abundant in the apical region of the cell, i.e., those cell areas close to the lumen. The surface facing the glandular lumen is provided with a brush border, indicating the presence of transport processes. These various structural features are evident in Figure 2, in which one can see the endoplasmic reticulum as well as the secretory products in various stages of condensation within the cisternae and vacuoles of the Golgi complex. The secretory granules are presumably flowing toward the brush border, where their contents will be delivered into the lumen of the gland. Biochemical analysis of the secretion has revealed that it is composed of several different secretory polypeptides, the nature of which is discussed below. The flow of the secretory products has been followed in autoradiographic (25, 26) and biochemical (27, 28) studies. It was established that the secretion entered the salivary gland lumen in less than 30 min after its synthesis (27, 28) and was expelled from the larva after about 15 hr (26). It should also be added that the production of secretion is quite impressive: in 1 day the gland produces an amount of secretion equivalent to its own weight (the secretory proteins excluded) (29).

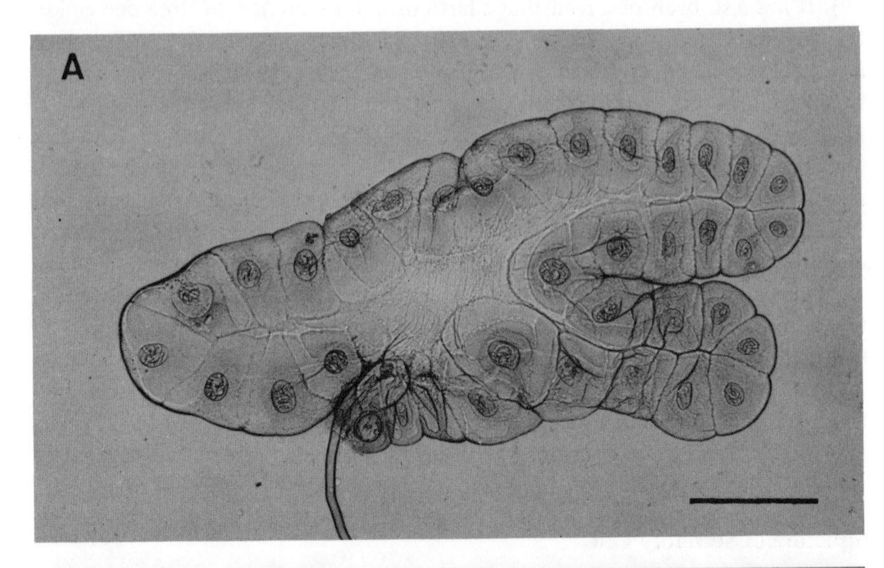

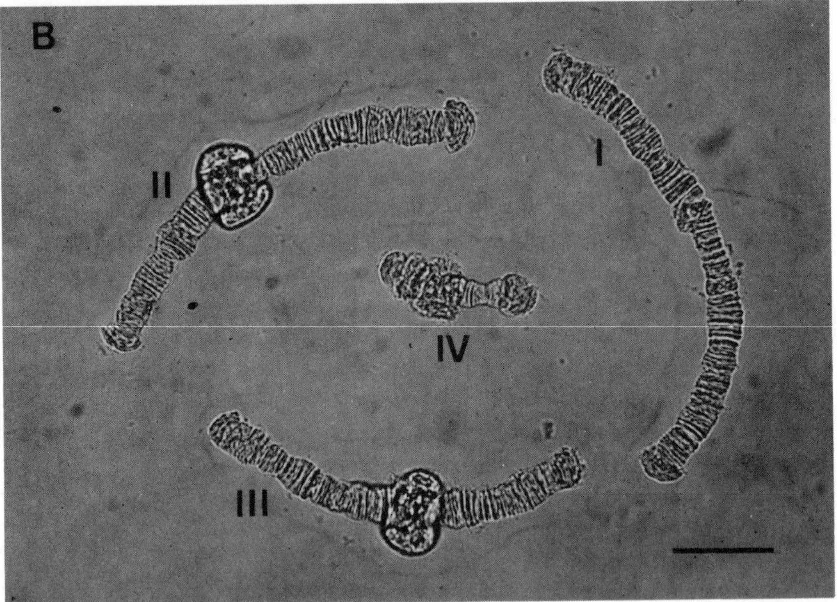

Figure 1. *A*, an isolated salivary gland from *C. tentans*. Nuclei are visible within the large secretory cells which surround the central lumen. The secretory duct exits from the lumen toward the bottom of the photograph. The *bar* represents 400 μm. *B*, *C. tentans* polytene chromosomes I–IV microdissected from a salivary gland secretory cell. The chromosomes display their characteristic banding patterns. One nucleolus is visible on chromosome II and one on chromosome III. The *bar* represents 50 μm.

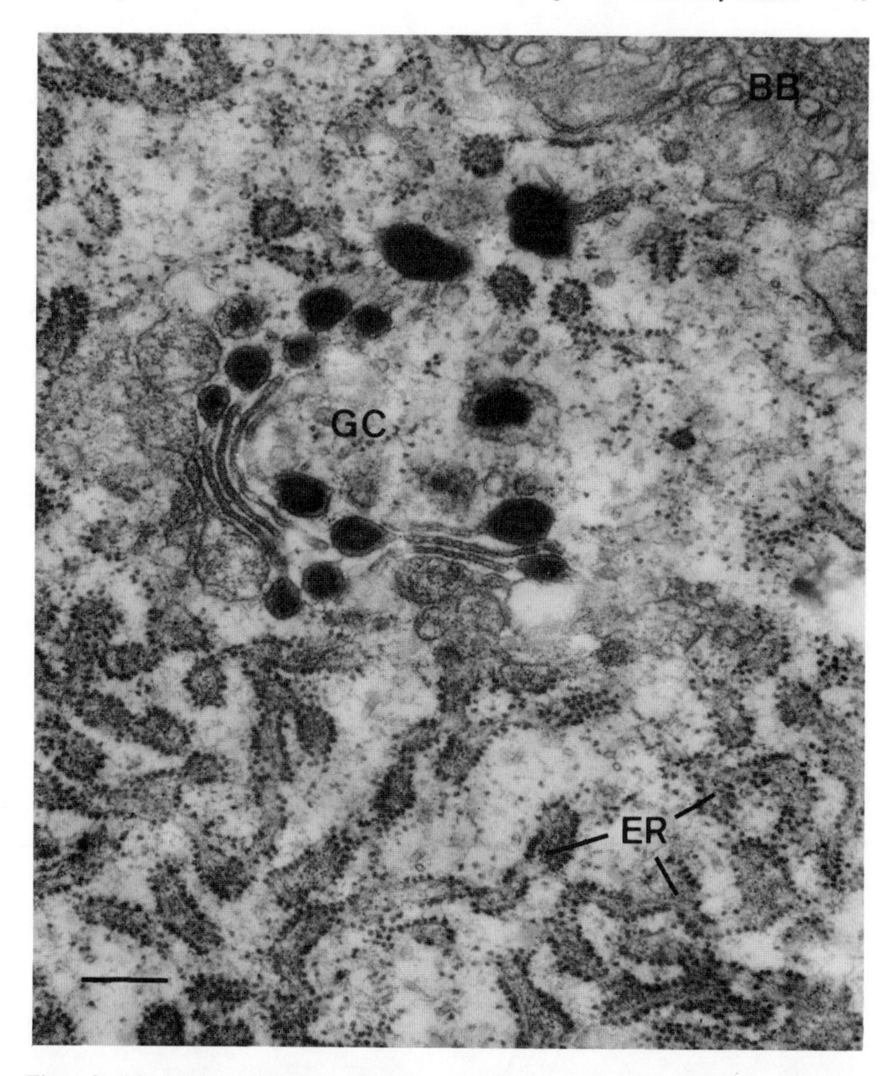

Figure 2. An electron micrograph showing a portion of the cytoplasm in a salivary gland secretory cell. Secretory products are synthesized in the glandular endoplasmic reticulum (ER), condensed within the darkly staining cisternae and vacuoles of the Golgi complex (GC), and finally transferred to the brush border (BB) of the cell where they will be delivered into the lumen of the gland. The *bar* represents 0.2 μm.

Although the functional significance is less clear, some additional ultrastructural features of the cytoplasm should be mentioned. Mitochondria are mainly observed in a well demarcated zone at the basal region of the cell facing the hemolymph. The cell membrane at the hemolymph surface is heavily invaginated, suggesting that absorption or secretion or both takes place there. Biochemical data (26) are also available, indicating an exchange of material between the hemolymph and the salivary gland cells.

Polytene Chromosomes

The structure of polytene chromosomes in *Drosophila* and *Chironomus* has been investigated in considerable detail (8, 30, 31). The chromosomes attain their giant size due to a polytenization process taking place during the growth of a larva. Each chromosome is duplicated many times without intervening nuclear or cell divisions. The resulting homologous daughter chromosomes remain closely associated side by side in such a way that multistranded, cable-like structures are obtained. Each daughter chromosome, usually designated a chromatid, is likely to consist of a single deoxyribonucleoprotein (DNP) fibril (32), which in certain regions is heavily coiled into condensed structures (33) called chromomeres (Figure 3). In the interchromomeric regions this DNP fibril is extended (34, 35). Along a given chromatid the chromomeres are located at certain defined positions, and the chromomere at each position is of a specific size. Chromatids are lined up side by side in such a perfect manner that homologous chromomeres form a transverse band. As a result of this organiza-

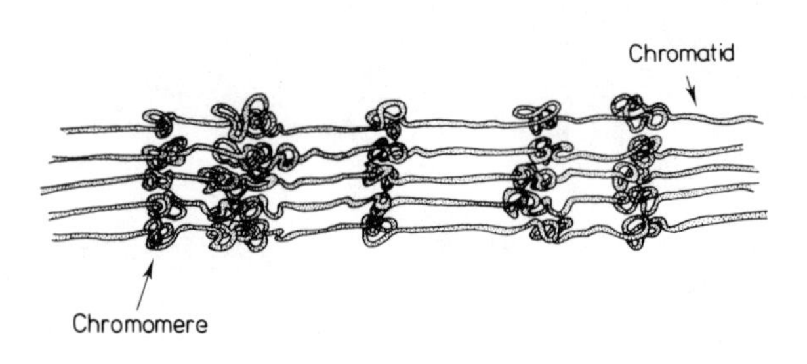

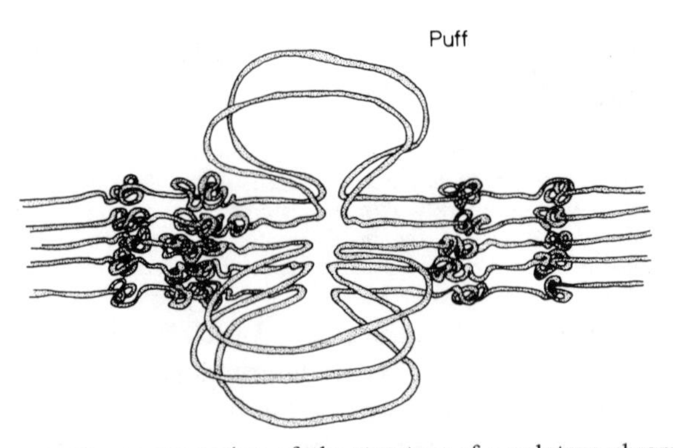

Figure 3. A diagrammatic representation of the structure of a polytene chromosome (*upper*) and the change in this structure which takes place when a puff is formed (*lower*). It can be observed how the heavily coiled chromatid within each chromomere of one chromosomal band extends into a loop.

tion, each polytene chromosome displays a specific pattern of bands and interbands. Clearly, the banding patterns are useful morphological landmarks in the analysis of these chromosomes.

It is consistent with the course of the DNP fibril that most of the DNA, by far, is located in the chromomeres, whereas only small amounts (less than 5%) reside in the interchromomeric regions (6, 31). There are, however, considerable differences in DNA content among the bands. Ultramicrospectrophotometric measurements (36) show that the DNA content of individual bands varies over two orders of magnitude. The distribution of histones follows closely that of DNA (6), whereas nonhistone proteins are largely restricted to the RNA-containing puffs (see below) (6, 37).

Chromosomal Puffs

Although the chromosome banding pattern is the same in all tissues, the pattern of the morphological modifications, called puffs, differs (8, 17). Puffs represent more or less swollen chromosomal segments, and at least some of them are tissue- or developmental stage-specific or both (for discussion, see 8). Certain puffs can be induced or repressed experimentally by various physical and chemical factors (for review, see 8, 38, 39). Of special interest to developmental biologists are the effects caused by the insect hormone ecdysone. By injection of this hormone into the hemolymph of *Chironomus tentans* larvae it was possible to rapidly achieve the formation of puffs at two particular chromosomal sites (I-18C and IV-2B) (37, 40). This result mimics the events occurring during normal development, just before and during the larval/larval (41) and larval/pupal (42) molts, which are controlled by the ecdysone titer in the hemolymph (37).

Upon close examination in the electron microscope, it is apparent that the heavily coiled DNP fibrils of the chromomeres in a band uncoil and form large loops during the formation of a puff (Figure 3) (43, 44). It has been established that puffed chromosomal regions are rich in RNA (6, 45) and nonhistone proteins (6, 37). The RNA is rapidly labeled as shown by autoradiography (45, 46). If a puff is induced experimentally, there is a concomitant increase in the uptake of [^{3}H]uridine in this chromosome region (47, 48). Furthermore, the synthesis of puff RNA is sensitive to drugs that interfere with the transcription process. Actinomycin D (48–50) and α-amanitin (51–53), known to inhibit elongation, block the labeling of puff RNA. The initiation inhibitor DRB (5,6-dichloro-1-β-D-ribofuranosylbenzimidazole) also prevents labeling, but only after a certain delay (54) (RNA polymerases already on the template continue elongation, whereas no new polymerases start synthesizing RNA). Therefore, several lines of evidence (including section "Balbiani Ring 2: An Active Gene") suggest that puffs are likely to be the sites for RNA synthesis, which is in good agreement with the idea that puffs are active genes.

Nonhistone proteins accumulate within puffs concurrently with the production of RNA (37, 48, 55). However, the nonhistone proteins do not seem to be synthesized within the puffs, since puff proteins remain essentially unlabeled even during prolonged labeling times (56–58). Moreover, when a puff is induced, nonhistone proteins appear in that chromosome region in the absence of protein synthesis (37). The function of nonhistone proteins in the puffs is not known, but proteins bound to RNA in ribonucleoprotein (RNP) particles (44, 59) probably constitute the major part of the nonhistone proteins in puffs (6). No changes in the histone proteins have been observed when puffs are formed (6, 60).

The salivary gland cells in *Chironomus* are highly differentiated secretory cells. It has been claimed that more than 50% of the proteins made are secretory polypeptides (26–28). It might be anticipated that such a highly specialized cell is suitable for an analysis of the pathway(s) for expression of genetic information coding for the major products, the salivary polypeptides. Such a view was earlier supported from cytogenetic data indicating that certain large cell-specific puffs, the so-called Balbiani rings (BRs), correlated with the production of the salivary polypeptides (27, 29, 61).

BALBIANI RING 2: AN ACTIVE GENE

This section discusses the proposed (27, 29, 61) functional relationship between synthetic activities in BRs and salivary polypeptide synthesis in light of present understanding of the transfer of genetic information. In other words, the information encoded in this gene must be transcribed into RNA, and these transcriptional products must be transferred to the cytoplasm, where they will function in polysomes to direct the synthesis of specified proteins. The concept of a puff representing an active genetic locus (13, 17, 61) receives considerable support from studies of BR 2 on chromosome IV in *C. tentans* salivary glands.

Transcription at BR 2

BR 2 as Site of RNA Synthesis BR 2 is the most conspicuous puff in the salivary glands of *C. tentans;* it also contains more RNA than any other puff. The amount of RNA in BR 2 (20 pg) (62) corresponds to that of 50 or more average-sized puffs (45). BR 2 RNA is rapidly labeled with radioactive precursors at a linear rate until saturation is reached, usually within 20–30 min (63). This is somewhat less time than it takes to saturate the majority of other puffs (see under "Synthesis of Chromosomal RNA"). The lack of delay and early saturation of labeling suggest that BR 2 RNA is not synthesized at other chromosomal sites and subsequently transported to and stored at BR 2. Although both autoradiographic (45) and biochemical (62) data strongly favored a local synthesis, they were not conclusive. However, it was possible to prove this point by RNA/DNA hybridization experiments. When radioactive BR 2 RNA was hybridized in situ to squashed chromosomes, it was observed that the hybrids were almost exclusively localized in the BR 2 region of chromosome IV (64, 65). It could, therefore, be concluded that the DNA template for BR 2 RNA is, in fact, located within BR 2.

Morphological Nature of BR 2 Products The structure of the BRs was first studied in the electron microscope by Beermann and Bahr (43). They established that the BRs in *C. tentans* contained, as the major structural component, loops that had a lampbrush-like appearance due to an array of stalked granules projecting outwardly from an ill-defined axis (43). Examples of these loop structures can be seen in Figures 4 and 5.

One of the most detailed morphological studies of stalked granules has been performed on the BR granules of *Chironomus thummi.* Stevens and Swift (44) observed that the largest BR granules were about 400–500 Å in diameter and

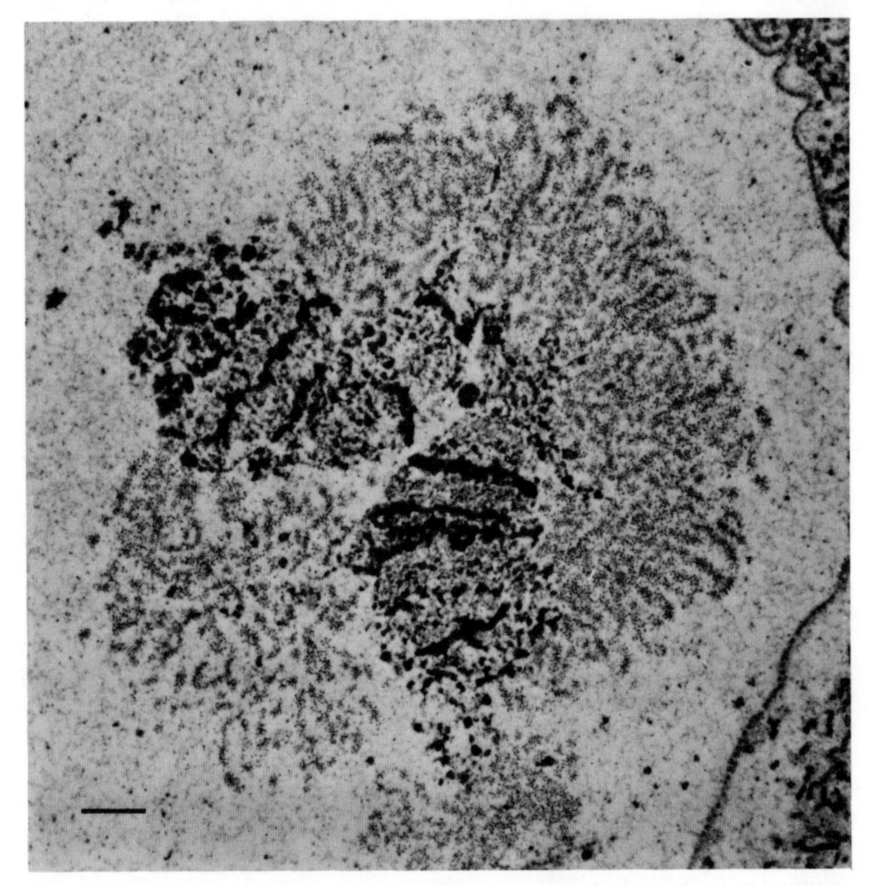

Figure 4. Large BR on chromosome IV from a *C. tentans* salivary gland cell. A great number of lampbrush-like loops can be observed within the BR. On both sides of the BR, one can recognize the banded structure of the polytene chromosome. The nculear envelope and cytoplasm can be seen along the right side of the electron micrograph. The *bar* represents 1 μm.

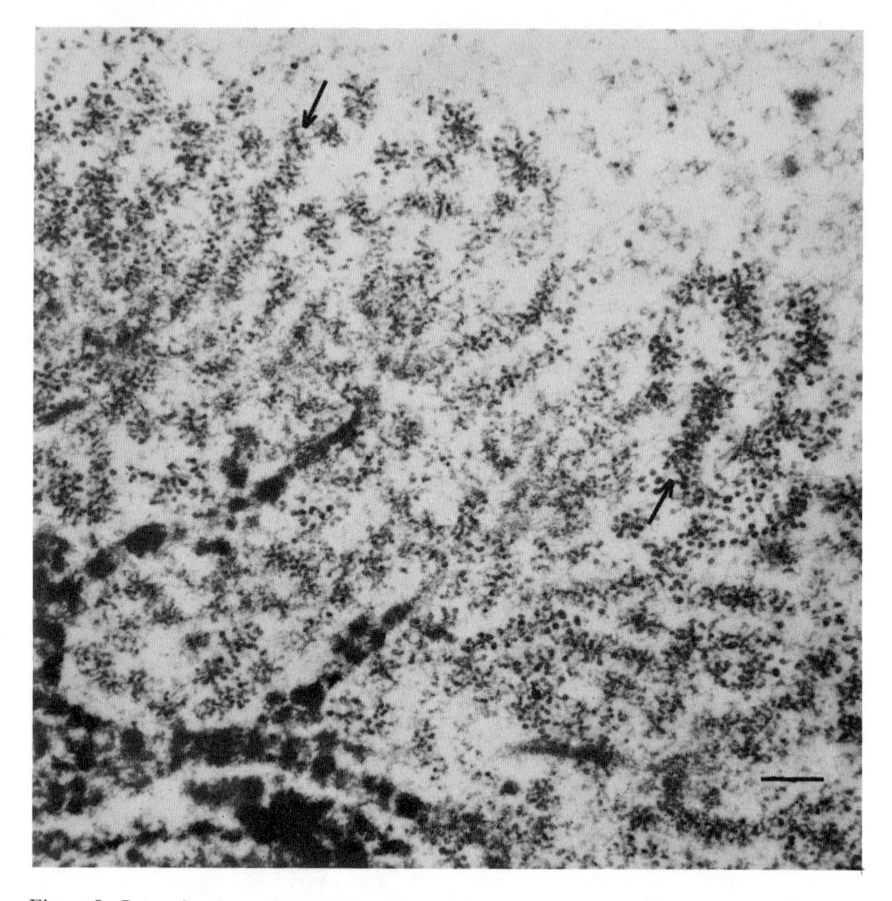

Figure 5. Part of a large BR with loop structures which are composed of stalked granules projecting from an ill-defined axis. *Arrows* point to two loops which, when compared to each other, contain different sizes of BR granules. The *bar* represents 0.5 μm.

were situated at the end of a short stalk which was 180–220 Å in diameter. Such structures have also been recognized in *C. tentans* (16) and are shown in Figure 6*A*. By selective staining and enzymatic treatments, Stevens and Swift demonstrated that the above structures were RNP, closely associated with and projecting from a 50–100-Å DNP fibril.

Existing evidence suggests that the stalked RNP granules are composed of a heavily coiled RNP fibril. First, upon examining glands from larvae exposed for various times to actinomycin D, Stevens (66) found that BR granules could not be found in the puff. Instead, only fibrillar structures were observed, which were postulated to represent the unraveling of partially formed or degraded BR granules. Such partially uncoiled granules can also be found in untreated larvae. Granules with ordinary stalks are shown in Figure 6*A*, whereas partially uncoiled, fibrillar structures are presented in Figure 6*B*. After a high resolution

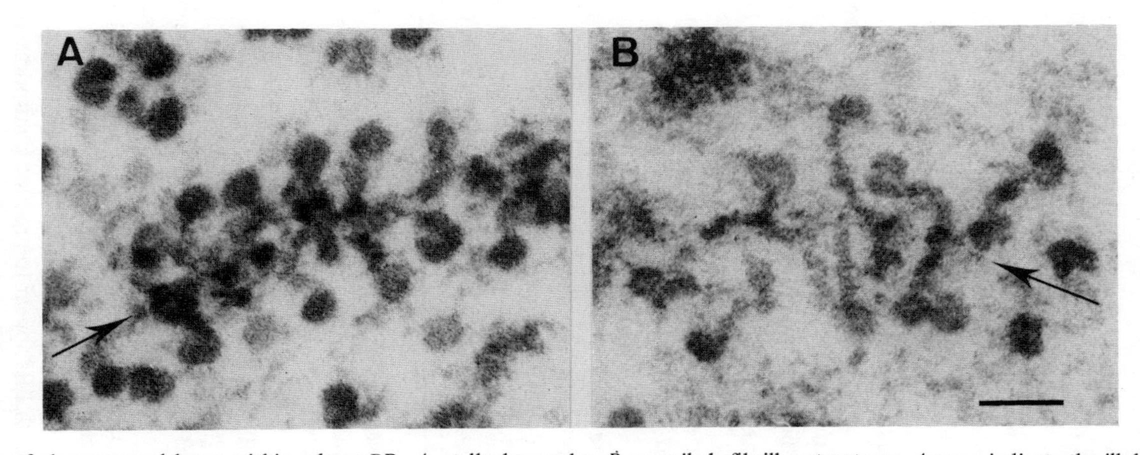

Figure 6. Details of chromosomal loops within a large BR. *A*, stalked granules. *B*, uncoiled, fibrillar structures. *Arrows* indicate the ill-defined axis of the loops. The *bar* represents 0.1 μm.

electron microscopic examination of the granules, Vazquez-Nin and Bernhard (59) concluded that a BR granule consists of heavily coiled RNP fibrils of 15 Å diameter.

Only granules within the BRs have stalks. Granules of the BR type released into the nuclear sap are more or less spherical (see under "Morphological Evidence for Transport"), and free stalks are never observed (44). The connection of BR granules to the DNP axis suggests that the stalked granules represent growing RNP fibrils or recently completed but not released products. The brushlike loops would then represent the arrangement of transcription complexes as they occur in situ. If this is true, one might expect that the granules would exhibit a size gradient along the transcription complex. Such a gradient has not been observed, probably due to the circumstance that only a minor segment of a given transcription complex can be studied in a single section of the electron microscopic preparation. Nevertheless, it should be feasible to detect various loops within the same BR which, when compared to each other, contain granules of different sizes. Two such loops with different granule sizes are pointed out in Figure 5. The largest stalked BR granules attain a diameter of 400–500 Å, i.e., the size of the largest and most frequently found granules in the nuclear sap. Finally, if the stalked granules do represent transcription products in statu nascendi, each one should consist of a single, heavily coiled RNP fibril.

Biochemical Nature of BR 2 Products Perhaps the most important feature of *Chironomus* salivary gland chromosomes is the tremendous size of their BRs. This led Edström and Beermann (62) to apply microdissection techniques to this material in order to obtain defined chromosomal segments or other cellular fractions. Extraction and analysis of their constituent RNA molecules have permitted some of the first detailed investigations of eukaryotic RNA synthesis at a nonribosomal locus.

It was initially observed (67, 68) that both BR 2 and BR 1 contained large RNA transcripts. It was found that the relative content of newly synthesized radioactive puff RNA was directly proportional to the size of that puff (68). After improving the extraction technique, Daneholt (14) was able to show that the major product of BR 2 was an RNA molecule with a sedimentation coefficient of about 75 S (Figure 7). The molecular weight of 75 S RNA was calculated to be $15–35 \times 10^6$, as determined by its sedimentation value in sucrose gradients and electrophoretic mobility in Agarose gels, respectively (14). BR 2 RNA retains its relatively slow mobility in Agarose gels after denaturation in formaldehyde (69) and has, therefore, been considered to be a single strand of RNA. It has been recently shown that BR 1 RNA also migrates as a 75 S RNA molecule (70). Thus, both BRs appear to produce giant primary transcripts.

A particular feature of the BR RNA profile is its asymmetry with regard to the gradual slope on the low molecular weight side of the 75 S peak (14, 70) (Figure 7). Various lines of evidence suggest that this is a nascent profile representing the full spectrum of growing RNA molecules (for discussion, see 14), but direct support for this hypothesis has been obtained only recently.

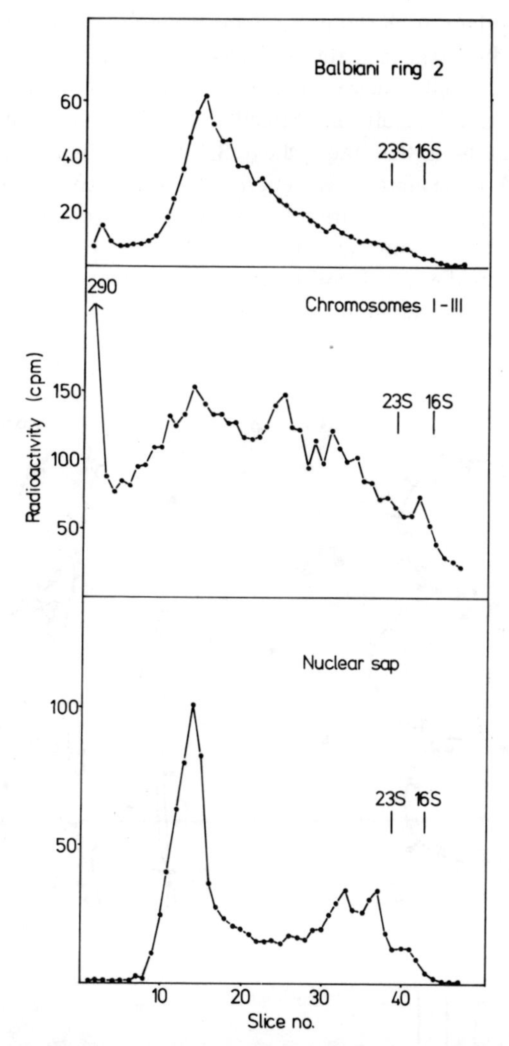

Figure 7. Radioactive RNA profiles obtained by Agarose gel electrophoresis of RNA extracted from microdissected nuclear components. Explanted glands were incubated in the presence of isotopes for 90 min in vitro. *Escherichia coli* RNA was used as markers. Modified from Daneholt (14) with permission of *Nature*.

Egyházi (70) has shown that by DRB treatment, which inhibits the initiation of RNA synthesis (54), labeling was diminished in the low molecular weight portion of the BR RNA profile prior to a reduction of labeling in the 75 S region. This experiment negates the possibility that smaller molecules in the BR profile are degradation products of 75 S RNA. Therefore, the molecules forming the gradual slope on the low molecular weight side of the 75 S RNA peak are likely to be molecules in statu nascendi.

It seems appropriate at this time to draw attention to similarities between morphological and biochemical data available on the BR products. Results of electron microscopic studies suggest that stalked granules in BRs are growing RNP fibrils. Biochemical results indicate that BR RNA is mainly nascent. A fundamental hypothesis is that the stalked BR granules in *C. tentans* contain nascent 75 S RNA molecules. This hypothesis remains to be directly tested. Figure 8 shows the proposed relationship between BR granules and BR RNA synthesis as interpreted from morphological (Figures 4–6) and biochemical (Figure 7) data. When the transfer of BR products to the cytoplasm is discussed below, the reader should keep in mind the events depicted in Figure 8.

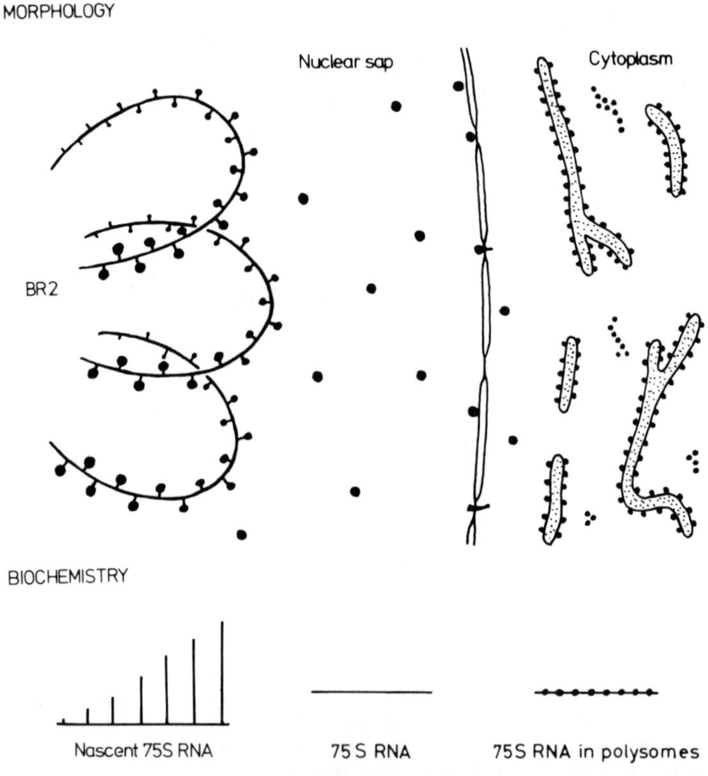

Figure 8. The transfer of genetic information from BR 2. This diagram shows the authors' interpretation of available morphological and biochemical data pertaining to the transfer of products from BR 2 in *C. tentans.* It is proposed that the lampbrush-like loops within BR 2 represent transcription complexes containing nascent 75 S RNA and associated proteins. The completed products are released as ribonucleoprotein granules containing 75 S RNA. These granules undergo configurational (and perhaps compositional) modifications upon passage through the nuclear pore complexes. Once they are in the cytoplasm, the granules rapidly disappear, and the 75 S RNA component becomes incorporated into membrane-bound polysomes where it acts as a messenger to direct the synthesis of salivary polypeptides.

Potential Study of Transcriptional Regulation The technique of Miller and Beatty (71), whereby genes actively transcribing RNA can be visualized in the electron microscope, has now been applied to a number of biological systems (72, 73). Although transcription of ribosomal genes has been most extensively examined (73–76), the initial characterization of nonribosomal transcription units has been recently reported in *Drosophila melanogaster* (73, 75, 76) and *Oncopeltus fasciatus* (75, 77). Due to the observed density of RNP fibrils (Figures 5 and 6) and potential size of a 75 S RNA transcription unit, BR 2 should be amenable to a similar study. If the technology involved in visualizing transcription complexes can be adapted to BR 2, there will be a new dimension of analysis possible in this system at the transcriptional level. From such a study, one could determine the size of the transcription unit, the RNP fibril density, initiation and termination sites, and, by knowing the transcription time, calculate initiation and elongation rates.

When studies of transcriptional regulation are considered, it is important to point out that the BR 2-containing region of chromosome IV can be obtained in three biologically distinct states. From salivary glands, BR 2 occurs puffed and active in RNA synthesis. By exposing larvae to galactose in their culture media, BR 2 can be made to progressively regress; removal of the sugar results in puff reformation (78). Finally, chromosome IV can be obtained from other tissues (e.g., Malpighian tubules) in which the region containing BR 2 sequences is never puffed (17). It would be interesting if these morphological states could be translated into terms of RNA synthesis via transcription complex analyses.

Another problem that might be approached by the visualization of transcription complexes in BR 2 is whether there is more than one BR 2 transcription unit. In the case of tandem genes it would be possible to determine their polarity with respect to each other. The existence of multiple transcription units is discussed in the following section.

BR 2 Transcript Related to BR 2 Gene Some information is available as to the number and structure of the transcription units in BR 2. The amount of DNA complementary to BR 2 RNA has been estimated in two studies. By a microadaptation (79, 80) of the conventional RNA/DNA filter hybridization technique, Lambert (65) was able to demonstrate that 0.17% of the *C. tentans* haploid genome was complementary to microdissected, ^{3}H-labeled BR 2 RNA. Recently, Hollenberg (81) determined the saturation level with the use of ^{125}I-labeled poly(A)-containing salivary gland RNA. In spite of the complex population of RNA used (it hybridizes in situ to at least BR 1, BR 2, BR 6, and puff I-17B), he obtained a saturation level of only 0.04% of the haploid genome. Based on a haploid DNA content of 1.2×10^{11} daltons (82) and the assumption that the primary transcript is about $15–35 \times 10^{6}$ daltons (14) and all of its sequences have hybridized in these two studies, one would calculate that there are at least 1–2 copies (81), but perhaps as many as 5–15 copies (65) of the BR 2 gene.

From the saturation values (65, 81) and haploid DNA content (82), the possibility remains that copies of the BR 2 gene could occupy 0.5–3.5 average chromomeres of 60×10^6 daltons (32). Although this does not rule out the possibility that a chromomere is a single transcription unit (13, 14), it raises the possibility that homologous units may be located in more than one chromomere. In fact, it has been demonstrated that BR 2 RNA hybridizes to a 3–5-μm region of an unpuffed chromosome IV from rectum cells, which could be indicative of the presence of complementary sequences in three to five chromosomal bands (83).

In order to further analyze the chromomere-transcription unit relationship in BR 2, the problem must be settled as to whether the BR 2 sequences are distributed in one or several chromomeres. Moreover, it must be determined whether any single chromomere contains more than one transcription unit. Finally, it must be known whether a transcription unit is confined only to the chromomere or whether it extends into interchromomeric or even the adjacent chromomeric DNA. As to the number and possible distribution of BR 2 genes within or among different chromomeres, more precise localization techniques are required. This might be accomplished by electron microscopically mapping (84, 85) BR 2 RNA hybridized to microdissected DNA molecules from restricted regions of unpuffed chromosome IV. Regardless of the potential number of copies of the BR 2 gene, the number functioning at any one time may only be assessed via analysis of transcription complexes in the electron microscope (see under "Biochemical Nature of BR 2 Products").

The internal structure of the BR 2 transcription unit can be considered from experiments carried out with the transcriptional product. BR 2 RNA has an unusual base composition compared to other chromosomal RNA, particularly with respect to its high adenine and low uracil content (62). There was a similar BR 2-specific CMP to UMP ratio in each size class of molecules examined from the nascent BR 2 RNA profile, suggesting that there is a similar average base composition along the entire length of the BR 75 S RNA molecule (86).

Subsequently, it was demonstrated that BR 2 RNA rapidly and selectively hybridized in situ to BR 2, suggesting that this molecule is at least partially composed of internally repeated nucleotide sequences (64). Furthermore, the specific hybridization rate of BR 2 RNA relative to nucleolar RNA indicates that there are 200 binding sites, possibly represented as internal repetitions of $1-2 \times 10^6$ daltons within the BR 2 gene(s) (65).

One technique which is applicable to study the distribution as well as the internal organization of the BR 2 gene is that of recombinant DNA. Recent studies by Wensink et al. (87) and Glover et al. (88) with cloned random DNA segments from *D. melanogaster* have shown that such fragments can be characterized with respect to their internal sequence repetition and chromosomal distribution. The colony hybridization technique of Grunstein and Hogness (89) now permits the in situ detection of bacterial clones containing specific DNA fragments which are complementary to any purified radioactive RNA species. Such fragments would not only be valuable in determining the intra- and interchromomeric arrangement of the BR 2 gene, but would also aid in the detection of potential regulatory sequences, such as the type proposed by Britten and Davidson (90, 91).

Transport of BR 2 Primary Transcripts

Morphological Evidence for Transport As soon as the initial observations of stalked BR granules in BRs were made, the presence of morphologically similar granules which lacked stalks was noted throughout the nuclear sap (43). The characteristic size and staining of these sap granules initially suggested their chromosomal origin at the BRs (43). A rather uniform distribution of BR granules is found throughout the sap (16, 44), suggesting a random distribution of the RNP particles upon their release from the chromosome.

An additional and more dramatic change takes place at the nuclear envelope. Granules at the inner edge of the nuclear envelope appear unaltered (Figure 9*A*). However, as they become associated with and pass into nuclear pore complexes, striking configurational changes are noted (Figure 9, *B–D*). The granule first becomes elliptical in shape and begins to assume an elongated, rodlike appearance (16, 44). These penetration structures usually end up with a diameter of 200 Å and a length of up to 1,700 Å. After 2-hr exposure to actinomycin D no penetration complexes are observed in the nuclear pores (66).

Granules of the BR type have also been observed in the cytoplasm, close to the nuclear envelope (16). It was estimated that these cytoplasmic granules constitute less than 5% of all BR granules observed at a given moment outside the BRs. Their distribution is a sharply decreasing gradient that occupies a zone outside the nuclear envelope of about 0.1 μm.

Biochemical Evidence for Transport The appearance of labeled RNA in various microdissected cellular components was studied, and it was shown that a prominent radioactive 75 S RNA species is present in the nuclear sap (Figure 7), which after a distinct delay appears in the cytoplasm (92). It was concluded that the primary BR transcripts are delivered to the cytoplasm without a substantial reduction in size (92). The fact that BR sequences are actually present in nuclear sap (64) and cytoplasmic (93) RNA was convincingly demonstrated by their selective in situ hybridization to BRs. Furthermore, when total cytoplasmic RNA was electrophoretically fractionated, only RNA molecules from the 75 S region were shown to hybridize to BR 1 and BR 2 (94). Although this latter experiment does not conclusively rule out the possibility that smaller RNA molecules which are complementary to the BRs exist in the cytoplasm, it does provide strong support to the conclusion that BR 75 S RNA is delivered from the nuclear sap to the cytoplasm without a concomitant size reduction (92).

Egyházi (63) has recently demonstrated a remarkable aspect of BR RNA transport in *C. tentans*. By measuring the kinetics of appearance and quantitating the amount of ^{3}H-labeled 75 S RNA in different cellular compartments during in vitro incubation of the glands, he concluded that, of all the BR 75 S

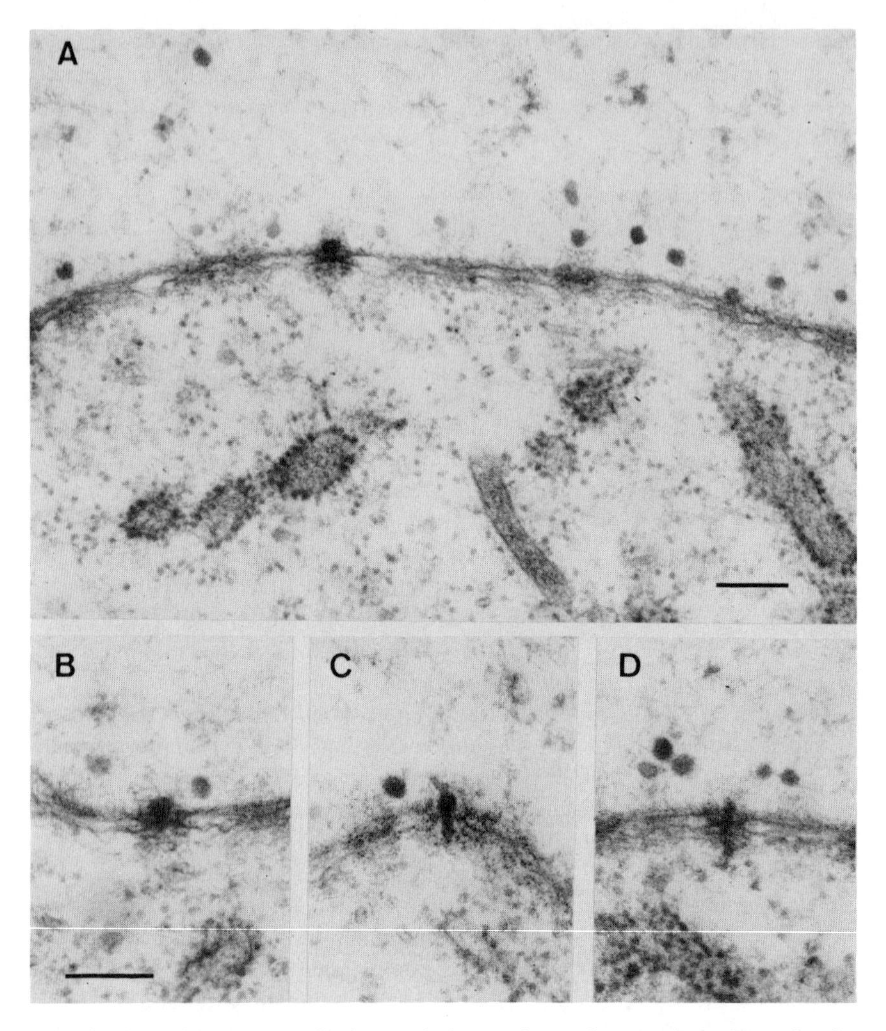

Figure 9. *A*, nuclear sap granules in proximity to the nuclear envelope. Note the size difference between these granules and both free and endoplasmic reticulum-bound ribosomes (lower portion of the electron micrograph). The *bar* represents 0.2 μm. *B–D*, granules in various stages of transfer through nuclear pore complexes. The magnification is the same in *B, C,* and *D.* The *bar* represents 0.2 μm.

RNA transcripts synthesized, only 14–17% can be located in the nuclear sap and 4–7% recovered as cytoplasmic 75 S RNA. These kinetic experiments were further supported by actinomycin D or DRB chase experiments, which showed that a similarly minor amount of 75 S RNA transcripts can be recovered in the cytoplasm. Although a majority of the BR transcripts synthesized in explanted glands are not destined for cytoplasmic transport, it would be desirable to know whether an equivalent amount is transported in vivo.

Several characteristics of the 75 S RNA which does enter the cytoplasm have been established thus far. Although only 5% of the number of transcripts synthesized enter the cytoplasm (63), they represent 95% of the cellular 75 S RNA, which is about 1.5% of the total RNA in the cell (92). These numbers imply that 75 S RNA has a high metabolic stability. Values for the half-life of 75 S RNA range from 1.5 days (92) to as much as 6 days (95). A major portion of the cytoplasmic 75 S RNA molecules binds to Millipore filters or poly(U)-Sepharose under conditions used for the binding of polyadenylic acid sequences, suggesting that 75 S RNA is a poly(A)-containing messenger-like RNA molecule (69, 95).

It has been shown recently that 75 S RNA can be isolated from Agarose gels by a microelectrophoretic elution technique (96) which yields an essentially pure and undegraded preparation (Figure 10). This should greatly facilitate the structural analysis of this molecule. For example, it has been possible to expose purified, predominantly cytoplasmic 75 S RNA to extreme denaturing conditions and to electrophoretically demonstrate its structural integrity. Furthermore, 75 S RNA has now been visualized in the electron microscope (S. T. Case and B. Daneholt, manuscript in preparation). Such characteristics as recognized regions of intramolecular hydrogen bonding (secondary structure), the presence and location of poly(A) sequences, and determination of the size of this molecule should all be considered fundamental tasks for an electron microscopic study. In the case of 75 S RNA, such studies are of particular interest because this cytoplasmic molecule may reflect the entire structure of the corresponding transcription unit.

Post-transcriptional Regulation From the studies of BR transcript turnover and transport, two important functional parameters are apparent. Intranuclear metabolism of the BR transcripts destined for nucleocytoplasmic transport does not include a considerable size reduction (63, 92), as is found for most mammalian heterogeneous nuclear RNA (hnRNA) (97). On the other hand, those transcripts which are not transported enter a pathway of in toto intranuclear degradation (63), an unprecedented finding in that this would then include the breakdown of putative amino acid coding sequences (see under "BR 2 RNA and Its Relationship to Salivary Polypeptides"). The significance of this finding may be in the indication of an interesting post-transcriptional level of potential regulation of BR gene expression within these cells.

The ability to isolate discrete cellular components may also be of further value in pinpointing the sites of various post-transcriptional modifications (98). For example, if 75 S RNA is a poly(A)-containing (69, 95) messenger RNA (mRNA) molecule in the sense demonstrated for other poly(A)-containing mRNAs, certain predictions can be made. Evidence for transcriptional versus post-transcriptional polyadenylation could be ascertained and, in the case of the latter, the chromosomal or nuclear sap location of such an event determined. The determination and localization of other post-transcriptional modifications, such as methylation (99), are equally amenable to this approach. What is most

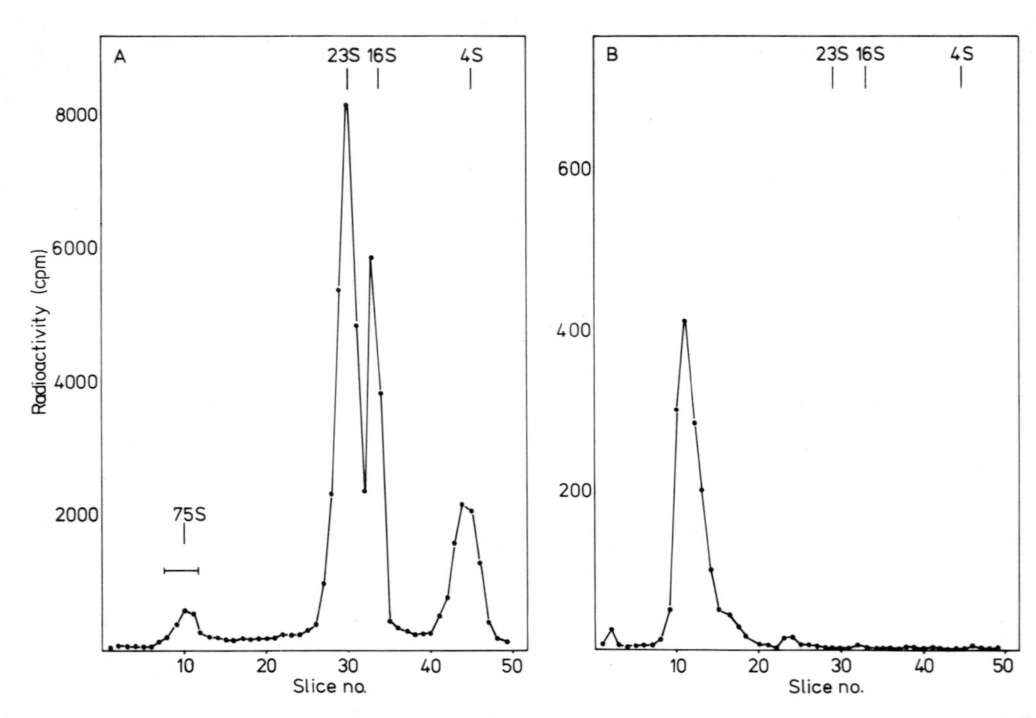

Figure 10. *A*, electrophoretic profile of total salivary gland RNA labeled for 3 days in vivo. *B*, re-electrophoresis of 75 S RNA obtained by microelectrophoretic elution of the 75 S region indicated in profile *A*. *E. coli* RNA was used as markers.

frequently lacking at this time is adaptation of the techniques applied to other systems in which the quantity of RNA is not as limited.

BR 2 RNA and Its Relationship to Salivary Polypeptides

Up to this point, this chapter has developed the puffing concept for BR 2 in so far as a size- and sequence-specific transcript is synthesized there (see under "Transcription at BR 2") and this RNA is transported to the cytoplasm (see under "Transport of BR 2 Primary Transcripts"). If this RNA is to serve as the vehicle for BR 2 gene expression by coding for the tissue-specific salivary polypeptides, it must be demonstrated that BR transcripts reside in polysomes and that these transcripts actually contain the specific amino acid coding sequences.

BR 2 RNA in Polysomes Polysomes were first examined in *C. tentans* by Clever and Storbeck (100). With the use of a conventional homogenization procedure they obtained radioactive profiles of polysomes with their nascent polypeptide chains labeled with [^{3}H]leucine. These polysome profiles were shown to be sensitive to RNase, EDTA, puromycin, and actinomycin (depending upon the developmental stage at which they were isolated), yet stable after exposure of the glands to cyclohexamide. With the use of a modified homogenization medium, Hosick and Daneholt (101) also examined *C. tentans* salivary gland polysomes. By examining profiles in which either the RNA or nascent polypeptides were radioactively labeled, they determined that the polysomes had an average sedimentation value of 300–400 S and were also sensitive to EDTA, puromycin, and RNase. When the RNA in the polysome extract was analyzed by Agarose gel electrophoresis, it exhibited a heterogeneous size profile with respect to high molecular weight RNA up to and including 75 S RNA.

A more recent study (102) shows an improved solubilization procedure for *C. tentans* salivary glands (i.e., no homogenization) which results in polysomes with an average sedimentation coefficient of 700 S, the range being 200–2,000 S (Figure 11). Fractions obtained from slowly (400 S) and rapidly (1,500 S) sedimenting portions of the polysome profile were examined in the electron microscope. EDTA-sensitive polysomal structures were observed (Figure 12*A*) which contained an average of 12–14 and 55–65 ribosomes for the slowly and rapidly sedimenting fractions, respectively, although some of the structures in the latter sample contained more than 100 ribosomes (102). On the basis of radioactive RNA profiles obtained from polysomal or supernatant regions of sucrose gradients, in which control or EDTA-treated samples were sedimented, it was calculated that these polysomes contained at least 60% of the cellular high molecular weight RNA (30–75 S) and that a minimum of 40% of cytoplasmic 75 S RNA was in polysomes. In addition, there was a heavy enrichment of polysomal 75 S RNA in the most rapidly sedimenting polysomes of about 1,000 S or greater (102) (Figure 13). Finally, the EDTA-sensitive polysomal RNA was shown to hybridize in situ to the sites known to contain sequences for ribosomal RNAs (rRNAs) and BRs RNAs (Figure 12*B*), perhaps showing some enrichment

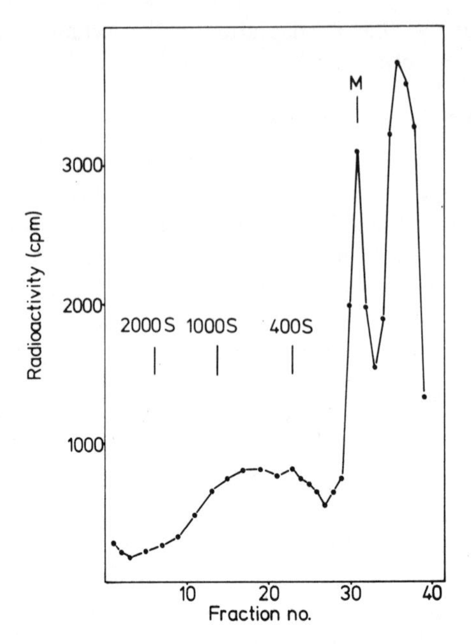

Figure 11. Sucrose gradient sedimentation profile of *C. tentans* salivary gland polysomes labeled for 3 days with radioactive RNA precursors. The monosome (M) peak is indicated.

for sequences complementary to BR 2 in the most rapidly sedimenting (greater than 1,000 S) polysomes (16, 103).

Although it was never shown that it was exclusively cytoplasmic 75 S RNA (rather than some other size) contained within the largest polysomes which hybridized to the BRs, the following indirect evidence supports this assumption.

1) 75 S RNA is predominantly synthesized at the BRs (14, 70).
2) A percentage (63) of BR-specific (64, 93) 75 S RNA molecules undergoes nucleocytoplasmic transport without a substantial size reduction (14, 63, 92).
3) Cytoplasmic 75 S RNA accumulates, due to its long half-life (92, 95), to become the predominant, nonribosomal, nontransfer RNA species (92).
4) At least 40% of cytoplasmic 75 S RNA is in polysomes (102).
5) Thus far, 75 S RNA is the only size of cytoplasmic RNA shown to hybridize in situ to the BRs (94).
6. Polysomal RNA (102), especially from the most rapidly sedimenting polysomes (103), hybridizes to the BRs.

Therefore, at this time it seems quite reasonable to conclude that BR 2, and perhaps BR 1, transcripts at least enter and possibly function as mRNA in salivary gland polysomes as 75 S RNA.

Secretory Polypeptides Cytogenetic data exist to support the idea that BRs in salivary gland cells are sites coding for salivary polypeptides (27, 29, 61).

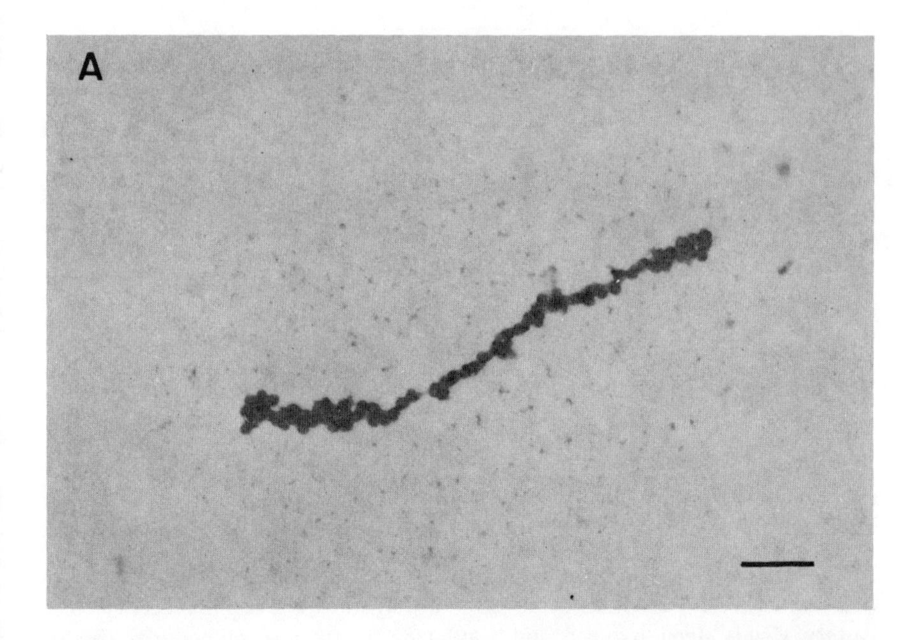

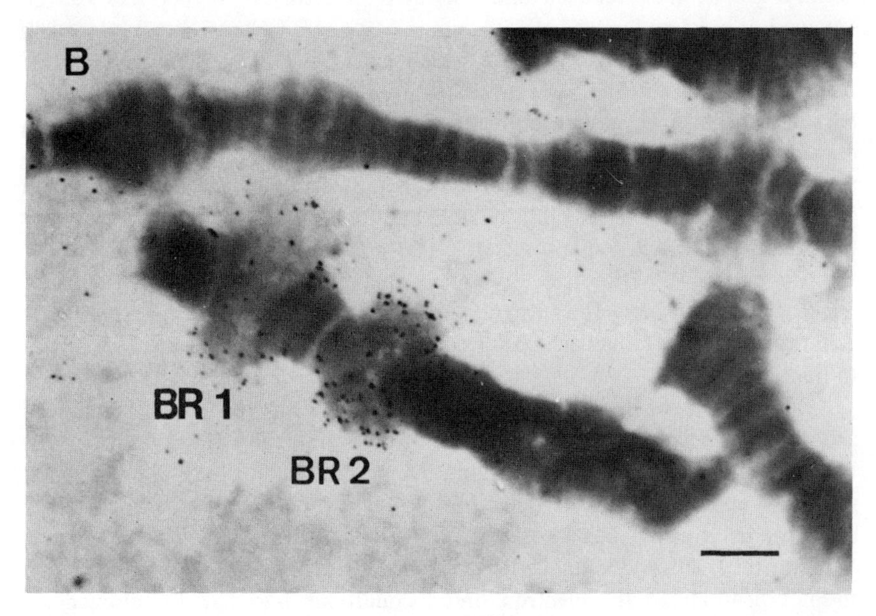

Figure 12. *A*, an electron micrograph of a polysomal structure obtained from a sucrose gradient fraction corresponding to about 1,500 S (see Figure 11). On this polysome there are more than 60 ribosomes. The *bar* represents 0.1 μm. *B*, radioactive polysomal RNA hybridized in situ to salivary gland chromosomes. Notice the grains over BR 1 and BR 2, whereas the remaining visible portions of the chromosomes are unlabeled. The *bar* represents 10 μm. Reproduced with permission of L. Wieslander.

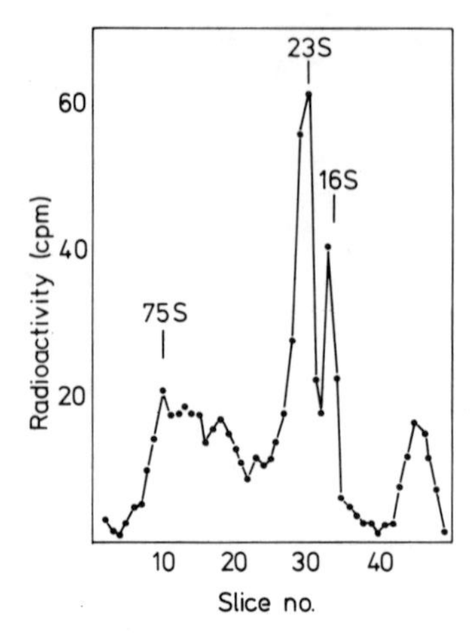

Figure 13. Electrophoretic profile of radioactive RNA extracted from polysomes sedimenting at greater than 1,000 S (see Figure 11).

Grossbach (27, 29, 104) has demonstrated that when secretion is isolated from salivary glands, then reduced and alkylated, five major polypeptides can be obtained in polyacrylamide gels containing 8 M urea. Of most direct interest to this discussion are correlative data which exist for two of these polypeptide fractions. Grossbach (27) has cytogenetically demonstrated a direct correlation between the size or presence of BR 1 and protein fraction 3 (see 27 for nomenclature). Then under culturing conditions for galactose-induced modification of puff size (78), Grossbach (105) has confirmed the correlation of protein fraction 3 to BR 1 and, additionally, has indicated a relationship between fraction 2 and BR 2. Pankow et al. (106) now report a developmental stage-specific pattern of BR sizes, to which they correlate relative synthesis of secretory polypeptides. They also conclude that (under the same electrophoretic conditions used by Grossbach) BR 1 correlates to fraction 3 and BR 2 correlates to fraction 2.

The most recent data from Grossbach (107) indicate that when the reduced and alkylated secretion is first separated in urea-containing gels and subsequent fractions reanalyzed in SDS-containing 7% polyacrylamide gels, the molecular weight of fraction 2 is in the range of $1.4–1.6 \times 10^5$. Fraction 3 is reported to be somewhat smaller, but its size is not yet definitive (107). This recent size estimate for fraction 2 is smaller than original estimates (27, 104), probably due to modifications in the electrophoretic conditions (see 107 for discussion). Concurrently, feel that the present value is not in disagreement with the data of Rydlander and Edström (108), who report that most of the secretory polypeptides eluted from a Sepharose column as $3–6 \times 10^4$ daltons when isolated in the presence of glycine ethyl ester. In fact, the distribution of their eluate ranges from $2 \times 10^4 - 2 \times 10^5$ daltons, which could readily contain a polypeptide fraction of the molecular weight presently reported for the BR 2 polypeptide.

From the previous discussion, it is apparent that a detailed analysis of the size and chemical composition of individual secretory polypeptide fractions is still needed. However, for the purpose of this discussion, it seems reasonable to conclude that BR 1 and BR 2 are correlated to two distinct polypeptides, the latter of which has a molecular weight of about $1.4-1.6 \times 10^5$. The possible occurrence of amino acid coding sequences within 75 S RNA with regard to a polypeptide of this size is discussed in the next section.

Protein Coding Potential of 75 S RNA Unequivocal evidence for an mRNA sequence organization in BR 2 or 75 S RNA or both can only come from the translation of these molecules in heterologous cells or cell-free systems. However, certain predictions can be made at this time based upon indirect evidence which is available. Since there is a preferential enrichment of 75 S RNA in the most rapidly sedimenting (greater than 1,000 S) polysomes and polysomes sedimenting at about 1,500 S were shown to contain an average of 55–65 ribosomes (102), the size of the polypeptide which might be synthesized on structures of this size can be gauged by making comparisons with other differentiated cells (Table 1). It would not seem unreasonable to expect that the more rapidly sedimenting 75 S RNA-containing polysomes described above could be synthesizing a polypeptide of about 2×10^5 daltons. This is in reasonable agreement with the estimates that the BR 2 polypeptide is of $1.4-1.6 \times$ daltons (107).

Although the above agreement on the size of the BR 2 polypeptide seems reasonable, it results in an enigmatic situation in predicting the organization of coding sequences within BR 2 polysomal RNA. 75 S RNA has been reported to have a molecular weight between $15-35 \times 10^6$ (14), although initial electron microscopic measurements tend to favor the smaller of these two values (S. T. Case and B. Daneholt, manuscript in preparation). If totally translated, this RNA could contain sufficient sequences to code for a protein of approximately 1.5×10^6 daltons. On the other hand, as has just been pointed out, the size of salivary gland polysomes (102) and the size of the BR 2 polypeptide (107) would indicate a protein of one-tenth of the potential coding length of 75 S RNA. If there is only one translational unit of this size in BR 2 75 S RNA, there

Table 1. Examples of polysomes on which large polypeptides are synthesized

Polypeptide	Molecular weight	Ribosomes per polysome
Vitellogenin	170,000 (109)[a]	30–40 (110)
Myosin	200,000 (111)	50–60 (112)
Fibroin	350,000 (113)	50–80 (114)

[a]Numbers in parentheses refer to references.

must be a vast excess of nucleotides in this giant molecule, the function of which is virtually unknown. Alternatively, 75 S RNA may yet prove to be polycistronic.

CHROMOSOMAL RNA

There are a great number of puffs on polytene chromosomes (6–8). In *C. tentans* salivary glands, Pelling (45) mapped 272 RNA-containing puffed regions (BRs not included). Since these puffs were identified by staining, this amount should be regarded as a minimal estimate due to limitations in sensitivity and resolution of the method used. Although individual puffs are amenable to morphological studies, it would seem more difficult to carry out biochemical studies on single puffs of ordinary size, as has been done for the giant BRs. For example, an average puff on chromosome I in *C. tentans* contains about 0.4 pg of RNA (45). These amounts are so minute that at the present time it is not feasible to analyze individual puffs with conventional biochemical techniques. Therefore, as a rule workers are left to study a complex mixture of RNA molecules originating from numerous chromosomal loci. Although it is evident from autoradiographs (45) that most chromosomal RNA is transcribed at puffs, it is also conceivable that RNA is synthesized at nonpuffed regions. During the following discussion it should be remembered that chromosomal RNA refers to molecules attached to or in intimate association with the polytene chromosomes, whereas the term heterogeneous nuclear RNA represents these chromosome-associated molecules, as well as those already released from the chromosomes but still contained within the nucleus.

Synthesis of Chromosomal RNA

It has been shown by autoradiography (45) that the chromosomal puffs become rapidly labeled with RNA precursors, indicating that RNA is synthesized within the puffs. Considering the large number of sites active in RNA synthesis (45), it is not surprising that newly synthesized chromosomal RNA is heterogeneous and covers a broad size range (10–100 S) with the peak at about 35 S (14, 115, 116).

Biochemical evidence suggests that in *C. tentans* at least some of the chromosomal RNA molecules remain associated with their template after the completion of their synthesis (117). The transcription time for chromosomal RNA has been determined to be less than 30 min in experiments in which DRB was added at various times before addition of isotope (117). Since the time required to reach plateau of chromosomal labeling with RNA precursors (45 min (49) or even up to 90 min (54)) is considerably longer than the transcription time, it was concluded that at least some chromosomal RNA is stored on the chromosomes for a certain period of time prior to its release into the nuclear sap (or, alternatively, is degraded) (117). In *Drosophila hydei,* Bisseling et al. (118) have shown that a 40 S RNA species accumulates at its site of synthesis, the subterminal puff 2–48 BC.

Transport of Chromosomal RNA

It is, of course, essential to investigate the fate of chromosomal RNA in order to establish its function. RNA extracted from the nuclear sap (14, 116), cytoplasm (69, 92, 116), and polysomes (B. Daneholt, unpublished data) has been characterized by electrophoresis. All three RNA samples contain molecules in a broad size range (10–75 S). Moreover, poly(A)-containing RNA molecules extend over an equally broad size range (95). Thus, it can be concluded that RNA of many different sizes reaches the cytoplasm and enters polysomes. In other words, it would seem that a number of puffs other than the BRs are synthesizing mRNA sequences. To draw more precise conclusions, it will be necessary to perform experiments with molecules synthesized at defined loci.

Although the BR primary transcripts appear not to be cleaved during their transport to the cytoplasm (92), it is important to remember that this does not exclude the possibility that some, or even most, other puff products can be cleaved. In fact, there is evidence for such a process in the salivary glands of *C. tentans.* During labeling experiments, a 35 S RNA species is rapidly detected in the nuclear sap (116), and, after a distinct delay, 35 S RNA can also be found in the cytoplasm (63, 116) and in polysomes (B. Daneholt, unpublished data). An interesting feature of 35 S RNA metabolism was noted in the following experiment (116). Salivary gland RNA was labeled in the presence of an initiation inhibitor (DRB) for 40 min, after which time mainly the larger labeled transcripts remained on the chromosomes. (Most of the smaller ones had been completed and released from the chromosomes, and DRB had prevented new ones from being initiated.) Although no distinct 35 S RNA could be observed on the chromosomes, it was found in the nuclear sap. Subsequently, as the DRB chase was prolonged, labeled 25–100 S RNA was released from the chromosomes, whereas in the nuclear sap mainly the 35 S RNA peak increased. The small amount of 35 S RNA delivered from the chromosomes could not alone account for the increase of 35 S RNA in the nuclear sap. These results, therefore, suggest that the 35 S RNA species arises from cleavage of a larger precursor molecule.

Stability of RNA in Nucleus and Cytoplasm

Most of the newly synthesized chromosomal RNA is degraded into acid-soluble products within 3 hr (49). In addition, little or no RNA has a lifetime longer than a few hours within the nucleus (95). It is, therefore, possible that most chromosomal RNA is degraded rapidly within the nucleus. If so, then the turnover of nuclear RNA sequences may be comparable to that found in other eukaryotic cells (97). Whereas nuclear RNA is remarkably labile, cytoplasmic RNA is rather stable. Edström and Tanguay (95) pointed out that there are different stabilities of cytoplasmic RNA. Although 75 S RNA is quite stable (a half-life between 1.5 days (92) and 6 days (95)), most RNA of lower molecular weights (10–40 S) has a half-life well below 24 hr (95). Future experimentation aimed at an understanding of the differential metabolic stability of various cytoplasmic as well as nuclear RNA species will undoubtedly be important.

NUCLEOLI AND RIBOSOMAL RNA SYNTHESIS

In the salivary glands of *C. tentans* there are ordinarily two active nucleolar organizers, one located on chromosome II and one on chromosome III (17, 30, 119). As a rule, there are also two separate nucleoli in a nucleus (Figure 1*B*), but sometimes they are fused into one. In the electron microscope, it has been shown that the nucleoli in *Chironomus* salivary gland cells consist of a fibrillar core (pars fibrosa) surrounded by a granular component (pars granulosa) (66, 120, 121). Chromosomal fibers are closely associated with the central fibrillar component (121). Autoradiographic evidence (121) suggests that RNA is synthesized in the fibrillar portion and then moves into the granular one.

When nucleolar RNA synthesis in *C. tentans* was further characterized, it was shown that a 38 S RNA species is first formed (67, 122). Later on, this molecule is split within the nucleolus and two RNA species appear, 30 S and 23 S RNA (67, 122). Further conversions (30 → 28 S and 23 → 18 S RNA) also take place within the nucleus (123, 124), but the exact intranuclear locations have not been established. Moreover, 28 S RNA has an internal "hidden" nick approximately in the middle of the molecule (125, 126). The end products, 28 S and 18 S RNA, are delivered into the cytoplasm (124), 18 S RNA somewhat earlier than 28 S RNA (95). Edström and Lönn (127, 128) have also been able to follow the flow of ribosomal RNA within the cytoplasm by investigating three cytoplasmic zones at increasing distances from the nuclear envelope. During pulse labeling experiments (127) they established that the three ribosomal RNA components, 28 S, 18 S, and 5 S RNA, appeared in steep, puromycin-sensitive radioactivity gradients. After long labeling times (6 days), no gradients could be detected (128). The authors concluded that after the entry into cytoplasm the ribosomal RNA species becomes rapidly incorporated into polysomal structures which slow down the peripheral spread of the ribosomal RNA. Finally, it is known that cytoplasmic ribosomal RNA has a half-life of about 15 days (95).

When discussing nucleolar RNA, one should also mention that one ribosomal RNA species, 5 S RNA, is found in the nucleoli (129) but is probably not synthesized there. The corresponding genes are located in a distinct region comprising two bands on chromosome II (130, 131). It is therefore likely that a transfer of 5 S RNA occurs from these bands on chromosome II into the nucleoli. Whether this transport takes place via nuclear sap or directly by a physical association (maybe of a transient nature) of the two chromosomal regions must still be decided. Perhaps it is relevant that Wieslander (131) found that a frequently occurring heterozygous inversion on chromosome II resulted in a physical association of the bands containing the 5 S genes on one chromosome with the nucleolar organizer of the homologous chromosome.

CONCLUDING REMARKS

Although there is much to learn in understanding the mechanisms by which differential gene expression is regulated, insight has been gained into the processes involved in the transfer of genetic information from those genes which are active. For example, the pathway by which a transcriptional product is transported from BR 2 to polysomes within *C. tentans* salivary gland cells has been outlined. If the model for the transfer of this product (Figure 8) is correct, it will form the basis for an integrated biochemical morphological analysis which is rarely available in other experimental systems. However, it should be pointed out that this pathway may not be the only type of pathway operational within the secretory cell. Indeed, if 35 S RNA does result from the cleavage of a larger precursor molecule, then within these cells there are at least two different pathways concurrently available to primary transcripts: transport with cleavage or transport without cleavage. Once workers are able to identify defined pathways for genetic transcripts, it can be determined whether they are primarily regulated at the transcriptional or post-transcriptional levels and what specific processes are involved. Such information will be fundamental in answering more specific questions about gene expression in differentiated cells, e.g., whether there are divergent pathways for "household" and "luxury" cellular products. Perhaps at this point it will be possible to outline the molecular processes which bring about the differentiated state of a cell.

ACKNOWLEDGMENTS

During various aspects of our work, we have received excellent technical assistance from Mrs Sigrid Sahlén, Miss Eva Mårtensson, and Miss Jeanette Nilsson. We wish to especially thank Mrs Birgit Frideen for typing this manuscript and Mrs Louise Nelson for drawing the figures. For offering helpful comments and ideas, we are grateful to our colleagues L. Wieslander, L. Nelson, and M. M. Lamb.

REFERENCES

1. Morgan, T. H. (1926). The Theory of the Gene. Yale University Press, New Haven, Connecticut.
2. Heitz, E., and Bauer, H. (1933). Z. Zellforsch. Mikrosk. Anat. 17:67.
3. Painter, T. S. (1933). Science 78:585.
4. King, R. L., and Beams, H. W. (1934). J. Morphol. 56: 577.
5. Lindsley, D. L., and Grell, E. H. (1968). Carnegie Inst. Wash. Publ. 627.
6. Swift, H. (1962). *In* J. M. Allen (ed.), The Molecular Control of Cellular Activity, pp. 73–125. McGraw-Hill, New York.
7. Beermann, W. (1967). *In* A. Brink (ed.), Heritage from Mendel, pp. 179–201. University of Wisconsin Press, Madison, Wisconsin.
8. Ashburner, M. (1970). *In* J. W. L. Beament, J. E. Treherne, and V. B.

Wigglesworth (eds.), Advances in Insect Physiology, pp. 1–95. Academic Press, London.
9. Berendes, H. D. (1973). Int. Rev. Cytol. 35:61.
10. Daneholt, B. (1974). Int. Rev. Cytol. (Suppl.) 4:417.
11. Judd, B. H., Shen, M. W., and Kaufman, T. C. (1972). Genetics 71:139.
12. Keyl, H. -G., and Pelling, C. (1963). Chromosoma 14:347.
13. Beermann, W. (1964). *In* Genetics Today, pp. 375–384. Pergamon Press, Oxford.
14. Daneholt, B. (1972). Nature (New Biol.) 240:229.
15. Daneholt, B. (1975). Cell 4:1.
16. Daneholt, B., Case, S. T., Hyde, J., Nelson, L., and Wieslander, L. Prog. Nucleic Acid Res. Mol. Biol. 19:319.
17. Beermann, W. (1952). Chromosoma 5:139.
18. Miall, L. C., and Hammond, A. R. (1900). The Structure and Life History of the Harlequin Fly (*Chironomus*). Clarendon Press, Oxford.
19. Sadler, W. O. (1935). Cornell University (Agricultural Experiment Station), Memoir 173.
20. Walshe, B. M. (1947). Nature (Lond.) 160:474.
21. Kroeger, H. (1973). Z. Morphol. Tiere 74:65.
22. Churney, L. (1940). J. Morphol. 66:391.
23. Schin, K. S., and Clever, U. (1968). Z. Zellforsch. Mikrosk. Anat. 86:262.
24. Kloetzel, J. A., and Laufer, H. (1969). J. Ultrastruct. Res. 29:15.
25. Kloetzel, J. A., and Laufer, H. (1970). Exp. Cell Res. 60:327.
26. Doyle, D., and Laufer, H. (1969). J. Cell Biol. 40:61.
27. Grossbach, U. (1973). Cold Spring Harbor Symp. Quant. Biol. 38:619.
28. Wobus, U., Popp, S., Serfling, E., and Panitz, R. (1972). Mol. Gen. Genet. 116:309.
29. Grossbach, U. (1969). Chromosoma 28:136.
30. Beermann, W. (1962). *In* Protoplasmatologia, Vol. 6D, pp. 1–165. Springer-Verlag, Vienna.
31. Beermann, W. (1972). *In* W. Beermann, J. Reinert, and H. Ursprung (eds.), Results and Problems in Cell Differentiation, Vol. 4, pp. 1–33. Springer-Verlag, Berlin.
32. Kavenoff, R., and Zimm, B. H. (1973). Chromosoma 41:1.
33. Sorsa, M., and Sorsa, V. (1968). Ann. Acad. Sci. Fenn. (Biol.) 127:3.
34. Berendes, H. D. (1969). Ann. Embryol. Morphol. (Suppl.) 1:153.
35. Sorsa, M. (1969). Ann. Acad. Sci. Fenn. (Biol.) 150:1.
36. Rudkin, G. T. (1965). Genetics 52:665.
37. Clever, U. (1967). *In* L. Goldstein (ed.), The Control of Nuclear Activity, pp. 161–186. Prentice-Hall, Englewood Cliffs, New Jersey.
38. Berendes, H. D. (1972). *In* W. Beermann, J. Reinert, and H. Ursprung (eds.), Results and Problems in Cell Differentiation, Vol. 4, pp. 181–207. Springer-Verlag, Berlin.
39. Kroeger, H., and Lezzi, M. (1966). Annu. Rev. Entomol. 11:1.
40. Clever, U. (1961). Chromosoma 12:607.
41. Clever, U. (1963). Chromosoma 14:651.
42. Clever, U. (1962). Chromosoma 13:385.
43. Beermann, W., and Bahr. G. F. (1954). Exp. Cell Res. 6:195.
44. Stevens, B. J., and Swift, H. (1966). J. Cell Biol. 31:55.
45. Pelling, C. (1964). Chromosoma 15:71.
46. Rudkin, G. T., and Woods, P. S. (1959). Proc. Natl. Acad. Sci. U.S.A. 45:997.

47. Berendes, H. D. (1967). Chromosoma 22:274.
48. Berendes, H. D. (1968). Chromosoma 24:418.
49. Daneholt, B., Edström, J. -E., Egyházi, E., Lambert, B., and Ringborg, U. (1969). Chromosoma 28:399.
50. Egyházi, E. (1974). Nature (Lond.) 250:221.
51. Beermann, W. (1971). Chromosoma 34:152.
52. Egyházi, E., D'Monte, B., and Edström, J.-E. (1972). J. Cell Biol. 53: 523.
53. Wobus, U., Panitz, R., and Serfling, E. (1971). Experientia 27:1202.
54. Egyházi, E. (1974). J. Mol. Biol. 84:173.
55. Holt, T. K. H. (1971). Chromosoma 32:428.
56. Pelling, C. (1959). Nature (Lond.) 184:655.
57. Pettit, B. J., and Rasch, R. W. (1966). Cell Comp. Physiol. 68:325.
58. Holt, T. K. H. (1970). Chromosoma 32:64.
59. Vazquez-Nin, G., and Bernhard, W. (1971). J. Ultrastruct. Res. 36:842.
60. Gorovsky, M. A., and Woodword, J. (1967). J. Cell Biol. 33:723.
61. Beermann, W. (1961). Chromosoma 12:1.
62. Edström, J.-E., and Beermann, W. (1962). J. Cell Biol. 14:371.
63. Egyházi, E. (1976). Cell 7:507.
64. Lambert, B., Wieslander, L., Daneholt, B., Egyházi, E., and Ringborg, U. (1972). J. Cell Biol. 53:407.
65. Lambert, B. (1972). J. Mol. Biol. 72:65.
66. Stevens, B. J. (1964). J. Ultrastruct. Res. 11:329.
67. Edström, J. -E., and Daneholt, B. (1967). J. Mol. Biol. 28:331.
68. Daneholt, B., Edström, J. -E., Egyházi, E., Lambert, B., and Ringborg, U. (1969). Chromosoma 28:418.
69. Edström, J. -E., and Tanguay, R. (1973). Cold Spring Harbor Symp. Quant. Biol. 38:693.
70. Egyházi, E. (1975). Proc. Natl. Acad. Sci. U.S.A. 72:947.
71. Miller, O. L., Jr., and Beatty, B. R. (1969). Science 164:955.
72. Hamkalo, B. A., and Miller, O. L., Jr. (1973). Annu. Rev. Biochem. 42:379.
73. McKnight, S. L., and Miller, O. L., Jr. (1976). Cell 8:305.
74. Reeder, R. H., Higashinakagawa, T., and Miller, O. L., Jr. (1976). Cell 8:449.
75. Laird, C. D., Wilkinson, L. E., Foe, V. E., and Chooi, W. Y. (1976). Chromosoma 58:169.
76. Laird, C., and Chooi, W. Y. (1976). Chromosoma 58:193.
77. Foe, V. E., Wilkinson, L. E., and Laird, C. D. (1976). Cell 9:131.
78. Beermann, W. (1973). Chromosoma 41:297.
79. Lambert, B., Egyházi, E., Daneholt, B., and Ringborg, U. (1973). Exp. Cell Res. 76:369.
80. Lambert, B., and Daneholt, B. (1975). *In* D. M. Prescott (ed.), Methods in Cell Biology, Vol. 10, pp. 17–47. Academic Press, New York.
81. Hollenberg, C. P. (1976). Chromosoma 57:185.
82. Daneholt, B., and Edström, J. -E. (1967). Cytogenetics 6:350.
83. Lambert, B. (1975). Chromosoma 50:193.
84. Wu, M., and Davidson, N. (1975). Proc. Natl. Acad. Sci. U.S.A. 72:4506.
85. Thomas, M., White, R. L., and Davis, R. W. (1976). Proc. Natl. Acad. Sci. U.S.A. 73:2294.
86. Daneholt, B. (1970). J. Mol. Biol. 49:381.

87. Wensink, P. C., Finnegan, D. J., Donelson, J. E., and Hogness, D. S. (1974). Cell 3:315.
88. Glover, D. M., White, R. L., Finnegan, D. J., and Hogness, D. S. (1975). Cell 5:149.
89. Grunstein, M., and Hogness, D. S. (1975). Proc. Natl. Acad. Sci. U.S.A. 72:3961.
90. Britten, R. J., and Davidson, E. H. (1969). Science 165:349.
91. Davidson, E. H., and Britten, R. J. (1973). Q. Rev. Biol. 48:565.
92. Daneholt, B., and Hosick, H. (1973). Proc. Natl. Acad. Sci. U.S.A. 70:442.
93. Lambert, B. (1973). Nature (Lond.) 242:51.
94. Lambert, B., and Edström, J. -E. (1974). Mol. Biol. Reports 1:457.
95. Edström, J. -E., and Tanguay, R. (1974). J. Mol. Biol. 84:569.
96. Case, S. T., and Daneholt, B. (1976). Anal. Biochem. 74:198.
97. Lewin, B. (1975). Cell 4:11.
98. Perry, R. P. (1976). Annu. Rev. Biochem. 45:605.
99. Perry, R. P., and Kelly, D. E. (1974). Cell 1:37.
100. Clever, U., and Storbeck, I. (1970). Biochim. Biophys. Acta 217:108.
101. Hosick, H., and Daneholt, B. (1974). Cell Diff. 3:273.
102. Daneholt, B., Andersson, K., and Fagerlind, M. (1977). J. Cell Biol. 73:149.
103. Wieslander, L., and Daneholt, B. (1977). J. Cell Biol. 73:260.
104. Grossbach, U. (1971). Habilitationsschrift, Universität Stuttgart-Hohen-
105. Grossbach, U. (1975). *In* Nachrichten der Akademie der Wissenschaften in Göttingen, Vol. 2, pp. 44–47.
106. Pankow, W., Lezzi, M., and Holderegger-Mähling, I. (1976). Chromosoma 58:137.
107. Grossbach, U. *In* W. Beermann (ed.), Biochemical Differentiation in Insect Glands. Springer-Verlag, Berlin, in press.
108. Rydlander, L., and Edström, J. -E. (1975). *In* Proceedings of the Tenth FEBS Meeting, pp. 149–156. Societé de Chimie Biologique, Paris.
109. Roskam, W. G., Tichelaar, W., Schrim, J., Gruber, M., and Ab, G. (1976). Biochim. Biophys. Acta 435:82.
110. Roskam, W. G., Gruber, M., and Ab, G. (1976). Biochim. Biophys. Acta 435:91.
111. Dreizen, P., Hartshorne, D. J., and Stracher, A. (1966). J. Biol. Chem. 241:443.
112. Heywood, S. M., Dowben, R. M., and Rich, A. (1967). Proc. Natl. Acad. Sci. U.S.A. 57:1002.
113. Sprague, K. U. (1975). Biochemistry 14:925.
114. McKnight, S. L., Sullivan, N., and Miller, O. L., Jr. Progr. Nucleic Acid Res. Mol. Biol. 19:313.
115. Daneholt, B., Edström, J. -E., Egyházi, E., Lambert, B., and Ringborg, U. (1969). Chromosoma 28:379.
116. Egyházi, E. (1974). *In* Proceedings of Ninth FEBS Meeting, Vol. 33, pp. 57–62. Akademiai Kiadó, Budapest.
117. Egyházi, E. (1976). Nature (Lond.) 262:319.
118. Bisseling, T., Berendes, H. D., and Lubsen, H. H. (1976). Cell 8:299.
119. Pelling, C., and Beermann, W. (1966). Natl. Cancer Inst. Monogr. 23:393.
120. Kalnins, V. J., Stich, H. F., and Bencosme, S. A. (1964). Can. J. Zool. 42:1147.
121. von Gaudecker, B. (1967). Z. Zellforsch. Mikrosk. Anat. 82:536.

122. Ringborg, U., Daneholt, B., Edström, J. -E., Egyházi, E., and Lambert, B. (1970). J. Mol. Biol. 51:327.
123. Ringborg, U., Daneholt, B., Edström, J. -E., Egházi, E., and Rydlander, L. (1970). J. Mol. Biol. 51:679.
124. Ringborg, U., and Rydlander, L. (1971). J. Cell Biol. 51:355.
125. Rubinstein, L., and Clever, U. (1971). Biochim. Biophys. Acta 246:517.
126. Pelling, C. (1970). Cold Spring Harbor Symp. Quant. Biol. 35:521.
127. Edström, J. -E., and Lönn, U. (1976). J. Cell Biol. 70:562.
128. Lönn, U., and Edström, J. -E. (1976). J. Cell Biol. 70:573.
129. Egyházi, E., Daneholt, B., Edström, J.-E., Lambert, B., and Ringborg, U. (1969). J. Mol. Biol. 44:517.
130. Wieslander, L., Lambert, B., and Egyházi, E. (1975). Chromosoma 51:49.
131. Wieslander, L. (1975). Mol. Biol. Reports 2:189.

International Review of Biochemistry
Biochemistry of Cell Differentiation II, Volume 15
Edited by J. Paul

3
Gene Activity in the Lampbrush Chromosomes of Amphibian Oocytes

J. SOMMERVILLE

Department of Zoology, The University, St. Andrews, Scotland

In general, gene activity in eukaryotes relates to transcriptional and post-transcriptional events occurring in the interphase nucleus and as such is most conveniently studied in cells that contain, during this period, well defined chromosomes. The classic example in which morphological, cytochemical, and autoradiographical studies have been related to a biochemical analysis of the products of RNA transcription is to be found in the puffing phenomenon in the polytene chromosomes of insect nuclei (1). However, there is a period in the life history of many eukaryotes when there is a special type of genetic activity that in certain favorable organisms is associated with elaborately organized giant chromosomes. This period is the initial developmental process of oogenesis, and the chromosomes are referred to as lampbrush chromosomes. At a certain stage of meiosis, during an extended first meiotic prophase, synthetic events are geared mainly to the production of massive amounts of materials such as ribosomes (2–4), transfer RNA (5–7), messenger RNA (8–11), histones (12), DNA polymerase (13), and RNA polymerases (14), all of which are required for the rapid and uninterrupted early development of the embryo from the fertilized holoblastic egg. There also is accumulation of ancillary components required for high metabolic activity, such as mitochondria (15–17), glycogen (18), and yolk proteins (18–21). In addition to its synthetic role, it has been postulated that RNA synthesis during meiosis may have some influence in the subsequent gene expression of the organism (22).

This chapter is concerned with recent work on the genetic organization of lampbrush chromosomes and with the transcriptional and translational activity that occurs during oogenesis. Although lampbrush chromosomes have been observed in the oocytes of a wide variety of organisms (23), and also in spermatocytes (24–26), this discussion has been restricted to the class of animals in which they have been studied in most detail—the Amphibia. In doing so, it is at times necessary to relate the chromosomal studies that have been performed with the use of urodeles, such as the newt *Triturus*, to the biochemical and developmental studies that are often more conveniently studied with the use of anurans, such as the toad *Xenopus* and to a lesser extent the frog *Rana*. It is a basic assumption that most of the processes that are discussed are a general feature of all Amphibia, especially in view of their highly probable monophyletic origin (27).

In considering gene activity in lampbrush chromosomes, any detailed description of extrachromosomal transcription has been omitted, in particular, nucleolar activity. Although the amplification of ribosomal genes and the control of synthesis and processing of the precursors of 28 S and 18 S ribosomal RNA are important events in the oogenic process, these aspects have been dealt with in detail in earlier reviews (28–30). Other relevant articles on such topics as nucleic acid synthesis during oogenesis and early embryogenesis (31), RNA synthesis during oogenesis (32), and molecular events during oocyte maturation (33) should be referred to for a comprehensive coverage of these various aspects.

GENERAL FEATURES OF OOGENESIS

Oogenesis involves a series of complex processes that brings about the formation, growth, development, and differentiation of the oocyte, a cell that provides the primary cellular link between one generation and the next. Although a complicated series of interactions at the cellular, physiological, and biochemical levels are continually effective in the ovarian tissue, with the eventual production of mature oocytes, the main features of oogenesis can be listed individually. Such considerations help to put lampbrush chromosome activity in perspective as only a part, albeit an essential part, of the oogenic process and also emphasize that lampbrush chromosome activity is not an isolated intranuclear event but is subject to many external influences.

Chromosomal Reorganization

The meiotic process involves not only the eventual reduction by half of the diploid chromosome number but also the generation of new combinations of genes. This is accomplished by two major events: crossing over between the chromatids of homologous chromosome pairs while they are synapsed during the early stages of the first meiotic prophase and the reassortment of homologous chromosomes to give a haploid complement at the first meiotic division. The period of genetic activity that is manifested by the chromosomes adopting a lampbrush organization is slotted between these two events, yet is subject to little temporal restriction, for the later stage of the first meiotic prophase (diplotene) is extended over periods of up to several months. Eventually, as the oocyte reaches its maximal size, the chromosomes condense and gather in the center of the nucleus in preparation for metaphase.

Macromolecular Synthesis and Organelle Development Within Oocyte

Two aspects can be considered: first, the formation of macromolecules required for the metabolism and growth of the oocyte from a cell of but several micrometers which lacks many of the microstructures typical of somatic cells such as ribosomes, membranous organelles, and glycogen (18, 34) to a highly differentiated mature oocyte of up to a few millimeters in diameter; second, the accumulation of molecules and organelles to prepare the oocyte for the rapid progression of cell division and differentiation that characterizes early embryogenesis. Both of these interrelated aspects rely upon the high transcriptional activity that occurs during the lampbrush stage, although it must be emphasized that a large part of oocyte growth is due to the uptake and accumulation of yolk proteins that occur during the later vitellogenic period of oogenesis (see next section). However, it is with respect to the postulated accumulation of stable messenger RNA (mRNA) molecules that special attention has been paid to the study of transcription in oocytes. On the other hand, it would be an oversimplification to accept lampbrush chromosome activity as representing solely the transcription of those genes whose coding sequences are required for protein synthesis at specified times during early embryogenesis. Although maternal inheritance of mRNA has been inferred from the apparent successful early development of actinomycin D-treated eggs (35) and of enucleated eggs (36, 37), a formal demonstration of the presence of mRNA molecules which were synthesized during first meiotic diplotene and activated during early embryogenesis has yet to be made. Recent data on the stability of RNA molecules in oocytes (38, 39) make the concept of a stable pool of stored mRNA somewhat equivocal. Nevertheless, the fact remains that the informational content of oocytes and embryos is similar (9) (see under "Sequence Complexity and Abundance Classes of Messenger RNA").

Influences of Ovarian Tissue and Other External Agents

During its development, the oocyte is an integral part of the ovarian tissue, being completely surrounded by an inner layer of follicle cells, an intermediate connective tissue layer containing blood vessels and an outer surface epithelium. Cell contacts are made by means of macrovilli extending from the follicle cells to meet microvilli on the adjacent surface of the oocyte (18, 40). Therefore, the influence of materials external to the oocyte must be mediated through the follicle cells or derived from the follicle cells themselves. Autoradiographic studies have shown that in *Triturus* the follicle cells incorporate little [^{3}H]uridine compared with the oocytes (41). In addition, *Xenopus* follicle cell RNA appears to be rather unstable (32). The major exogenously synthesized components to influence oogenesis are proteins, particularly yolk and hormones.

Although protein synthesis undoubtedly occurs within oocytes, as demonstrated by autoradiographic studies (42–44) and biochemical analysis (21, 45–46), the majority of oocyte protein, in the form of yolk proteins, is synthesized within the liver and transported to the developing oocyte (19). The endogenous contribution toward labeled protein in vitellogenic *Xenopus* oocytes is no more than a few percent of the newly synthesized protein. Moreover, the endogenously synthesized proteins are more unstable than the proteins that are incorporated into yolk platelets (26). Therefore, lampbrush chromosome activity can be discounted as a major contributory factor in the formation of protein food reserves. The synthetic machinery of the oocyte is used for other important functions.

The final stages of oocyte development are also controlled by external influences. Pituitary gonadotropins, and possibly also ovarian steroid hormone (progesterone), are involved in oocyte maturation and ovulation. The ways in which these hormones might direct the series of events culminating in the dissolution of the oocyte nucleus (germinal vesicle) and the second meiotic division have been recently reviewed (33). The maturation events can be considered as demarcating the possibility of further lampbrush chromosome activity.

An important consequence of the breakdown of the germinal vesicle is that its contents are released into the ooplasm. If the nucleus is considered to act as a repository for proteins synthesized during the lampbrush stage of oogenesis, then at maturation these proteins are distributed to the appropriate regions of the egg. Proteins of this type tend to have some potential regulatory function, such as the histones that are accumulated in *Xenopus* oocyte nuclei (12) and the o^+ substance that is synthesized during the lampbrush stage in *Ambystoma* oocytes. (When released from the nucleus at maturation, the o^+ substance ensures development through gastrulation by inducing a stable state of nuclear activation in the embryo (47–50)).

PHYLOGENY, C VALUE, AND KARYOTYPES OF AMPHIBIA

There is a large range of DNA content in the genomes of different amphibian taxa. The haploid chromosome amount in picograms is referred to as the C value (51), which ranges from about 1–100 pg within the class Amphibia (Table 1). Obviously, this 100-fold difference does not relate to any differences in the organizational complexity of similarly structured and phylogenetically related organisms. Presumably, there is a similar number of diverse informational genes in all Amphibia, irrespective of genome size, but this should be formally demonstrated. Several questions can be asked about the significance of variable and often apparently excessive amounts of DNA. These questions relate in the first instance to the evolutionary implications of C value differences and in the second to functional activity in terms of the extent of genome transcription and the number of informational RNA sequences formed.

It should be emphasized at the outset that large differences in C value in Amphibia are not accompanied by a corresponding increase in chromosome number (Table 1), nor are they due to a polyneme series of increasing, laterally identical strands of DNA double helix per chromatid. There are several lines of evidence for this last statement, particularly where it concerns the structural organization of amphibian chromosomes in their lampbrush configuration.

a) The kinetics of deoxyribonuclease digestion of the lampbrush chromosomes of *Notophthalmus* are compatible with the scission of only a single double helix per lateral loop, i.e., chromatid (62).
b) The trypsin-resistant, deoxyribonuclease-sensitive chromatid fibers, when examined by electron microscopy, can be seen to have a width of 2–3 nm, dimensions that could accommodate no more than one DNA double helix (63).
c) More recently, Straus (54) has studied the reassociation kinetics of DNA extracted from a variety of Amphibia including the anurans *Scaphiopus couchi,*

Table 1. Genome characteristics of some Amphibia

Order	Species	Haploid chromosome number	C value (pg)[a]	Genome represented by nonrepetitive sequences[b] (%)
Apoda				
	Gymnopis multiplatica		3.7 (52)[c]	
Anura			1–15	
	Scaphiopus couchi	13	0.93 (53)	60.4 (54)
	Xenopus laevis	18	3.1 (15, 55)	55 (10)
	Rana pipiens	13	5.0 (53)–7.4 (56)	22 (54)
Urodela			10–100	
	Pleurodeles waltlii	12	19.5 (57)	
	Triturus cristatus	12	23 (53)	<45 (58)
	Notophthalmus viridescens	11	35 (57)	
	Plethodon cinereus	14	20 (60)	40 (60)
	Plethodon vehiculatum	14	38 (60)	20 (60)
	Ambystome maculatum	14	26 (57)–44 (56)	<20 (61)
	Amphiuma means	14	65 (53)–96 (56)	
	Necturus maculosus	19	78 (53)–102 (56)	<10 (61)

[a]Values are variable due to procedural differences and use of different standards. In some instances a range of values is given.

[b]In high C value Amphibia, it is difficult to discriminate a distinct nonrepetitive fraction from reassociation kinetics (59); therefore, in some instances low frequency values (<100 copies) are quoted.

[c]Numbers in parentheses refer to source of data.

Bufo marinus, and *Rana clamitans*, whose C values are in the ratio 2:7:10. In each instance it was found that some sequences were represented only once per haploid genome, a finding that indicates a unineme rather than a polyneme organization.

Single copy DNA has also been demonstrated in *Xenopus* (10, 64), but in the generally higher C value genomes of urodeles it is more difficult to demonstrate the presence of uniquely occurring DNA sequences. This is particularly evident in examining the reassociation kinetics of DNA extracted from pedogenetic genera such as *Ambystoma, Amphiuma*, and *Necturus* (54, 58, 61). Although part of the difficulty may be of a technical nature, due to the very long times of incubation required to reassociate high complexity DNA sequences, those sequences detected as being nonrepetitive tend to constitute a proportionately smaller fraction of the genome in higher C value Amphibia (Table 1). Therefore, any final single copy transition may be difficult to detect in a reassociation reaction. Nonrepetitive sequences may be more easily detected by means of molecular hybridization. For instance, Rosbash et al. (65) have shown that the mRNA sequences found in *Triturus* oocytes are preferentially transcribed from nonrepetitive DNA sequences although no clear-cut single copy transition is apparent from the DNA reassociation kinetics.

It is to be expected then that the larger C value genomes of Amphibia are contained in correspondingly larger chromosomes. This view is substantiated by karyotype analysis of the pedogenetic urodeles (66), species of the genus *Bufo* (67), and species of the genus *Plethodon* (60, 68, 69). In all instances of C value variation within a phylogenetically related group, not only the chromosome number, but also their relative dimensions, remain similar; only their absolute sizes differ.

Several consequences of the evolutionary increase in C value have been postulated; these should be examined in the light of what is now known about the organization of the amphibian genome and lampbrush chromosome activity.

Extensive Tandem Duplication and Conservation of Repeats

Because increase in genome size is generally regarded as resulting from internal duplication of DNA, the karyological observations mentioned above would support the view that genetic sequences have been duplicated in a symmetrical fashion throughout the genome. This could arise by a localized duplication of nucleotide sequences about the centromere. The sequences in this region are highly repetitive and constitute what is known as the pericentromeric heterochromatin. In fact, most of the highly repetitive (satellite) DNA of *Notophthalmus* (70) and *Plethodon* (71) has this particular chromosomal location, and where there is an increase in C value of *Plethodon* species, and consequently in chromosome size, there is some correlation with the amount of pericentromeric heterochromatin (72). On the other hand, reassociation experiments show no correlation between C value and the percentage of rapidly reassociating DNA in a range of amphibian genera (54, 58). The satellite sequences are not transcribed, and their function, as in mammals, probably relates to chromosome structural organization and movement and possibly to speciation (73).

The occurrence of satellite DNA, or very rapidly reassociating sequences, cannot alone account for the major increases in C value. It is likely that tandem duplication is more extensive and involves sequences at various interstitial sites, although still maintaining the relative dimensions of the chromosomal arms. The most extreme case is that of tandem duplication of all sequences more or less in proportion. Callan (74) has proposed that informational sequences transcribed from lampbrush chromosomes are present in the genome as tandem repeats and that the homogeneity of each gene cluster is maintained by correction against a master template during the course of lampbrush activity. The fact that oocyte mRNA has been shown to be preferentially transcribed from nonrepetitive DNA sequences of *Triturus*, which has a fairly high C value, apparently discounts this notion as representing a general phenomenon (65). However, this conclusion is based on the interpretation of reassociation kinetic data and might exclude the possibility of detecting low level repeated sequences (genes) that have diverged to an extent sufficient to exclude stable hybridization but that may well exist as transcriptionally active sequences (75). A general model that would accommodate these considerations has been proposed by Paul (76). Certain sequences, including the genes for 5 S RNA and 28 S and 18 S ribosomal RNA, are maintained as clusters of tandem repeats which do not deviate, although variability in the constitution of spacer sequences makes it difficult to explain the nature of the correction mechanism (see under "DNA Base Sequence Organization"). Furthermore, it has been shown that the kinetics of hybridization of ribosomal RNA to excess amounts of *Xenopus* DNA and *Triturus* DNA is similar, in spite of a 7-fold difference in C value (65). This implies that *Triturus* has 7 times as many ribosomal genes as *Xenopus*. Estimations of ribosomal gene number by filter hybridization have shown that there is a tendency for higher C value Amphibia to have a larger number of ribosomal genes, although an upper limit is reached in urodeles (77). To extend the argument, *Notophthalmus*, which has 14 times as much DNA as *Xenopus*, is estimated to have about 12 times as many 5 S genes (78). These values support the concept of duplication of all nucleotide sequences in proportion to increase in C value, although this effect might be detected only in highly conserved gene clusters. Birnstiel et al. (29) has suggested that the increase in number of ribosomal genes is roughly correlated with the evolutionary increase in DNA content.

Increased Numbers of Different Informational Sequences

The percentage of the genome that is detected as nonrepetitive sequences, or at least low frequency repeats, varies from about 55% in *Xenopus* (10) to 22% in *Rana* (54), through 40–20% in different species of *Plethodon* (60) to less than 10% in *Necturus* (61). In terms of DNA amount, these values correspond to about 1.5 pg for the anuran genera and about 8 pg for the urodeles.

If all nonrepetitive sequences are potentially coding sequences, the corresponding number of genes could amount to 10^6 in anurans and more than 5×10^6 in urodeles. These values are greatly excessive in view of gene number estimates for eukaryotes (in the range of 5×10^3–5×10^4) based on genetic considerations, cytological observations, and molecular hybridization data (79, 80). Furthermore, the number of different informational sequences is quite independent of C value. Rosbash et al. (65) demonstrated by molecular hybridization that the number of different mRNA sequences in *Triturus* oocytes is the same as the number of different mRNA sequences in *Xenopus* oocytes, despite the marked difference in C value. The sequence complexity of informational sequences in amphibian oocytes has been estimated by a number of workers to be equivalent to 1–2×10^4 genes (10, 65, 75). Nevertheless, Sommerville and Malcolm (75) have suggested that in higher C value Amphibia, such as *Triturus*, the informational sequences exist as low frequency repeats which have diverged during evolution. Preliminary results indicate that this is a general phenomenon linked to C value increase (81).

Increased Number of Control Sequences

Genetic control sequences have been postulated to be derived from the intermediate repetitive fraction of the genome, and it has been suggested that increase in C value results in sequences available for a finer control of gene activity (82, 83). Furthermore, it has been demonstrated by Davidson and co-workers (84–86) that there is a general interspersion of nonrepetitive sequences with repetitive sequences in the genome of *Xenopus* and a variety of other eukaryotes. In *Xenopus* up to 80% of the genome consists of regions of sequence interspersion, and, in at least 50% of the genome, repetitive sequences of 300 ± 100 nucleotides average length are interspersed with nonrepetitive sequences of 800 ± 200 nucleotides average length (84). The basic assumption is that some of the nonrepetitive sequences are equivalent to coding genes. However, the portion of the *Xenopus* genome transcribed in oocytes is but a few per cent (10), and the general sequence organization of the total genome may have little relevance to the control of transcription of this small fraction. In addition, it is difficult to see how this model relates to Amphibia of higher C value. Although it has been shown that sequence interspersion is also characteristic of the genome of *Triturus* (75), this was related to the observation that *Triturus* has a distribution of sequence frequency classes somewhat similar to that of *Xenopus*. In general, Amphibia of larger genome size have higher frequencies of sequence repetition than Amphibia with smaller genomes (54, 58, 60, 87). Straus (54) concludes that the larger genomes have been constructed, at least in part, by repeating small fractions of the genome a large number of times. However, it would be unwise to place too much weight on reassociation data alone, due to restrictions in obtaining uniform DNA fragment lengths, in standardizing incubation conditions, and in interpreting hydroxylapatite binding assays, particularly when the DNA from very high C value Amphibia is being studied. Certainly in many instances it is difficult to discriminate distinct frequency classes. In *Ambystoma, Amphiuma,* and *Necturus* there appears to be a complete spectrum of reiteration frequencies between $C_0t\ 10^{-1}$ and $C_0t\ 10^5$ (54, 58), whereas in *Xenopus* two distinct transitions are apparent (10), at $C_0t\frac{1}{2} \sim 10^{\circ}$, representing intermediate repetitive sequences, and at $C_0t\frac{1}{2} \sim 4 \times 10^3$, representing nonrepetitive sequences. It is, therefore, difficult to draw any definite conclusion about the function of the majority of intermediate repetitive sequences. Whether or not increased amounts of nuclear DNA require a corresponding increase in the number of control sequences for stoichiometric reasons remains an open question.

General Increase in DNA Serving as Physiological Adaptation

To account for the exceedingly high C values in pedogenetic Amphibia, it has been suggested that the excess DNA might have some selective advantage in being correlated with factors such as length of mitotic or meiotic cycle, basic metabolic rate, cell size, and environmental conditions (61).

Accumulation of Nonfunctional DNA

Most of the DNA of high C value organisms may be simply difficult to eliminate (87, 88). This DNA may serve as a partition between genes, the nontranscribed spacer being of selective advantage in confining the deleterious effect of nonsense or frameshift mutations to the single, adjacent genetic locus. In fact, the regions of transcription in lampbrush chromosomes are organized as well spaced units which are separated by regions of nontranscribed (condensed) chromatin (see under "Visualization of RNA Transcription").

STRUCTURE OF LAMPBRUSH CHROMOSOMES

Macromolecular Composition of Oocyte Nuclei and Chromosomes

DNA Lampbrush chromosomes, because they are manifested as such during the first meiotic prophase of the oocyte, contain a 4-C amount of chromosomal DNA. This DNA is contained in two homologous sets of chromosomes, each chromosome consisting of two sister chromatids which are paired in an intimate and complexly organized manner (see under "Spatial Distribution of Chromatin in Lampbrush Chromosomes"). In spite of early evidence suggesting a substantial increase in the DNA content of the chromatids of *Notophthalmus* and *Xenopus* when in the lampbrush state (89, 90), this excess is most probably due to contamination of the oocyte nuclear preparations with mitochondrial DNA. Therefore, there is no good reason to suspect any deviation from the general rule of constancy in C value, irrespective of cell type. The amphibian oocyte is, therefore, truly tetraploid. The considerable variation in C value between amphibian taxa has already been mentioned. Estimations of chromosomal DNA can be affected to a variable extent by contamination from two major sources: extrachromosomal nucleolar DNA and cytoplasmic DNA. For instance, in *Xenopus* oocytes, the 12 pg of chromosomal DNA (15) are in fact less than the 30 pg of DNA found in the nucleoli (91, 92) and are eventually vastly exceeded by the 1–2 μg of DNA which accumulate in the cytoplasm (15). Most of the cytoplasmic DNA is derived from the enormous numbers of mitochondria that are formed during oogenesis (16); a mature *Xenopus* oocyte contains about 10^5 times as many mitochondria as does a liver cell (17). Another possible source of cytoplasmic DNA is from yolk platelets (93). As a result of the insufficient amounts of chromosomal DNA that can be extracted from oocytes in purified form, the general properties of amphibian chromosomal DNA are derived from erythrocyte or liver preparations.

RNA Early measurements of the RNA content of lampbrush chromosomes and the oocyte nucleus were indicative of both a high transcriptional activity and a considerable accumulation of RNA in the nucleoplasm (89, 94). It must be kept in mind that values for total RNA content and the relative proportions of different types of RNA are highly dependent upon the stage of oogenesis (see under "Temporal Sequence of RNA Synthesis During Oogenesis") which itself is difficult to standardize, even among oocytes from the same species. Therefore, quantitative comparisons between results obtained by different workers, often using different amphibian taxa, are difficult to make. Nevertheless, the general interpretation is similar in each instance.

In *Triturus* it has been estimated that there is about 1–2 ng of RNA associated with the lampbrush chromosomes, 2.5 ng in the nucleoli and 20–30 ng in the nucleoplasm of 1 mm diameter (midvitellogenic) oocytes (94, 95). This compares with 7 ng of RNA, out of an average nuclear content of 26 ng, which is extracted with the lampbrush chromosomes of the related genus *Notophthalmus* (89). Such results are found to be more consistent and more meaningful when expressed in terms of RNA to DNA ratio. In the lampbrush chromosomes of both *Triturus* and *Notophthalmus*, the RNA to DNA ratio is about 10, which can be compared with values of 0.02–0.1 for the chromatin of somatic cells from a variety of sources (96). It appears, therefore, that transcriptional activity in lampbrush chromosomes is at least 100 times more extensive than in the chromosomes of somatic tissues.

As a check on purity of extraction, the base composition of *Triturus* oocyte chromosomal RNA (G + C = 45.8%) is similar to that of chromosomal DNA (G + C = 44.4%), but quite different from that of nucleoli (G + C = 60.4%) (94).

The smaller chromosomes of anuran Amphibia are more difficult to isolate and assay for nucleic acid content. However, the 100 ng of RNA found in the nuclei of mature oocytes of *Rana* (97), when compared with the measurements at the earlier lampbrush stage of urodeles quoted above, suggest that there is considerable accumulation of RNA in the oocyte nucleus prior to maturation. A quantitative relationship between chromosomal RNA content and the amount of RNA that accumulates in the oocyte nucleus is difficult to ascertain because 95% or more of the RNA synthesis occurring during the lampbrush stage is ribosomal (4, 98) and, as such, is mostly extrachromosomal in origin. This RNA may, in fact, accumulate in the nucleus as ribosomal precursors (99), making the content of informational RNA sequences in the oocyte nucleus difficult to establish.

Protein Measurements of the protein content of lampbrush chromosomes and concentration of protein in the nucleoplasm reinforce the view that high levels of metabolic activity occur in oocytes. The protein content of the vitellogenic oocyte nucleus of *Notophthalmus* is 1.7 μg, of which 0.43 μg is associated with the lampbrush chromosomes (89). This estimation gives a protein to DNA ratio for lampbrush chromosomes of 550, a value very much greater than values of 1–2 which are quoted for somatic cell chromatins (96). Again, there appears to be accumulation of protein, as well as RNA, in the nucleus as oogenesis progresses, there being 5 μg of protein in the mature oocyte nucleus of *Rana* (97). Undoubtedly, the vast majority of the protein of lampbrush chromosomes is associated with the RNA transcripts (100–103) and most of the nucleoplasmic protein is in the form of chromosome-derived ribonucleoproteins

(104, 105). However, there may well be significant amounts of soluble protein in the oocyte nucleus. For instance, it has been proposed that about 0.1 μg (>2%) of the protein present in the nuclei of mature oocytes of *Xenopus* consists of stored histone molecules (12).

Thus, compared with most other types of cell, the 100-fold greater chromosomal transcriptional activity and the 500-fold greater chromosomal protein content single out oocytes as being very special. That some of the special features are reflected in the structural organization of lampbrush chromosomes is discussed in the next few sections.

Development of Lampbrush Chromosomes

Prophase of the first meiotic division in oocytes is much more extended than the corresponding stage of mitosis, and during this time the chromosomes undergo a complex series of activities. As a result of DNA replication in premeiotic S phase, each chromosome entering meiosis consists of two identical chromatids. Little is known of the morphology of the chromosomes during early prophase, but it is generally assumed that they are short and compact and that, in this state, homologous chromosomes pair and crossing over occurs between nonsister and sister chromatids. (In somatic cells of *Triturus* cultured in 1–100 μg/ml of BUdR, the frequency of sister chromatid exchange is found to range from 20–50 per cell (106), although the naturally occurring frequency during, or prior to, first meiotic prophase is probably much lower.) An additional complex mechanism involves the tendency of nonhomologous chromosomes to cluster during early stages of meiosis. In *Xenopus* spermatocytes, different regions of the chromosomes, which are the sites of particular types of repetitive sequence, become associated at different stages. For instance, interstitial AT-rich satellite sequences associate at preleptotene, centromeric heterochromatin associate during early leptotene, and telomeric 5 S DNA associate during zygotene and pachytene (107). In early diplotene the paired chromosomes separate along most of their length, but homologs remain attached to one another at one or more points (chiasmata) where chromatid exchange is presumed to have taken place. As diplotene progresses in oocytes, the chromosomes lengthen considerably and decondense, each chromatid forming lateral looplike projections at many points along its length. These extensions are referred to as lampbrush loops. The decondensation and extension progress and culminate at the peak of lampbrush chromosome development when the chromosomes and their lateral loops reach their maximal lengths. This state of maximal dispersion of chromatin may last for several months, although localized differences may occur in the state of individual loop pairs in terms of extension and regression. In relation to other oogenic events, maximal loop extension occurs just prior to vitellogenesis and during early vitellogenesis—all this while the homologs remain attached at sites on the chromosome axes where condensed chromatin is located, but never at sites on the lateral loops (41). Whether this restriction is due to noninvolvement of lateral loop DNA in crossing over or to breakage of such connections during lampbrush chromosome development, or in the isolation of the chromosomes, is not known. Eventually the chromosomes shorten, with increasing amalgamation of the axial condensed chromatin, and their loops retract; by late diplotene, when considerable amounts of yolk have accumulated and the oocyte is reaching maturity, the chromosomes revert to their loopless and contracted state. In the mature oocyte, the chromosomes are in the form of condensed metaphase bivalents.

Spatial Distribution of Chromatin in Lampbrush Chromosomes

Descriptions of the structure of lampbrush chromosomes have been derived mainly from studies on urodele genera such as *Triturus, Notophthalmus, Pleurodeles*, and *Ambystoma*, which have reasonably high C values and, consequently, large, well developed chromosomes. In fact, the lampbrush chromosomes of species of each of these genera have been mapped with respect to relative chromosome lengths, position of the centromeres, and position of morphologically distinct loops and other distinguishing features (108). The chromosomes of anurans, such as *Xenopus* and *Rana*, are generally too small to reveal much detail by light microscopy, although the lampbrush chromosomes of *Rana esculenta* have been mapped (109). Nevertheless, the general features of lampbrush chromosome organization can be seen to pertain to all Amphibia.

Isolated lampbrush chromosomes have been extensively studied by phase contrast microscopy (110), and whole mount preparation (105) and sectioned chromosomes (111) have been examined by electron microscopy. The chromosomal isolation procedures, no matter how carefully controlled, may produce states of organization somewhat different from those existing within the oocyte nucleus. Especially important may be condensation effects stimulated by the rupture of the nuclear membrane and release of the chromosomes into an isotonic saline solution. Although structural observations on isolated chromosomes have to some extent been corroborated by analysis of sectioned nuclei (105, 111), even here isolation artifacts may arise–for instance, from the rapid exit of ions and small molecules from the nucleus (112, 113). What follows is a brief account of the structural features of lampbrush chromosomes, which are judged to be as near as possible to their native state of organization.

The chromatin of lampbrush chromosomes is organized in such a way that a clear spatial distribution of morphologically distinct regions may be discerned (Figure 1). Morphological differentiation of these chromosomes should be regarded as being a consequence, rather than a prerequisite, of the localized process of RNA transcription. Three states of chromatin organization can be distinguished.

1) The continuous deoxyribonucleoprotein (DNP) fiber which maintains the linear integrity of all other chromosomal structures and which proceeds, presumably continuously, from one end of the chromosome to the other. The two fibers, which correspond to sister chromatids (100), can be seen as granulated fused thread of less than 10 nm width (111), which joins regions of condensed

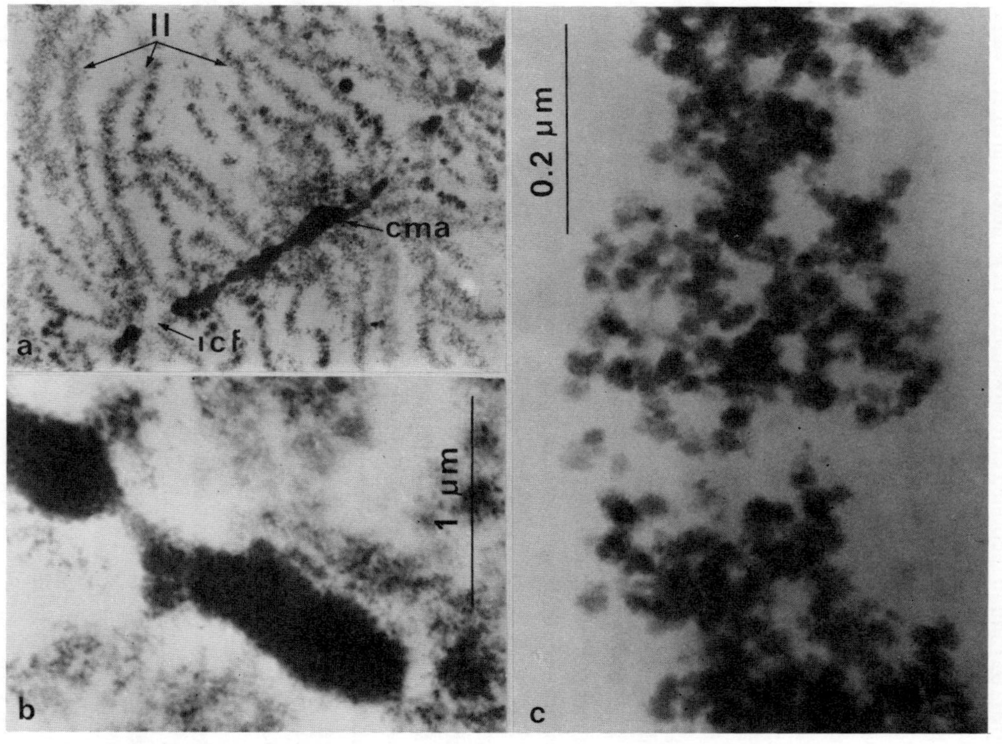

Figure 1. High voltage electron micrographs of the structural features of lampbrush chromosomes. *a*, the relative arrangement of lateral loops (II), the chromomeric axis (cma), and the interchromomeric fibril (icf). *b*, higher magnification of the chromomeric axis showing the honeycomb structure of the condensed DNP. *c*, higher magnification of part of a lateral loop showing the 20-nm RNP beads which constitute the loop matrix. Partly reproduced from Malcolm and Sommerville (105) with permission of Springer-Verlag.

chromatin. This thread is referred to as the interchromomeric fibril and shows no sign of transcriptional activity.
2) Periodic condensates along the axis of the chromosome of the paired DNP fibers into very densely packed granules called chromomeres (Figure 1*a*). These measure, on average, 0.5–2 μm across. When viewed by high voltage electron microscopy (105) and in sectioned chromosomes (111), the chromomeres can be seen to consist of a highly ordered packing arrangement of a 5-nm unit fibril, apparently supercoiled to give 30-nm-thick bundles with a spacing between the bundles of about 50 nm (111). The overall arrangement gives a honeycomb appearance (Figure 1*b*). However, it has been suggested that this particular arrangement, although completely regular within the chromomere and between chromomeres, is an artifact due to swelling during chromosome isolation (111). In sectioned nuclei the chromomeres appear to be more compact. Nevertheless, a highly ordered packing of the 5-nm DNP fibrils in condensed chromatin is inferred from the various observations. Thus, the chromomeric structures appear to be regions of highly condensed chromatin which are not engaged in transcriptional activity. The amount of the genome sequestered in this way is 95% or more, at least in *Triturus*, and, as a consequence, the chromomeres are the major chromosomal site of histones as demonstrated by specific staining (101) and by the use of immunofluorescent probes directed against isolated histone fractions (114, 115).
3) Lateral loops which extend as symmetrical pairs from the regions of condensed chromatin. From enzyme digestion experiments, the continuity of each loop is shown to be due to a continuous DNP fibril (101). Furthermore, an analysis of the kinetics of deoxyribonuclease fragmentation is compatible with the interpretation that the loop axis contains a single DNA double helix (62). Therefore, in the loops, the sister chromatids which are apparently fused in the interchromomeric region and highly condensed in the chromomeres are extended laterally from the chromosome axis as pairs of single chromatids. Associated with the lateral loop axes are large amounts of a normally diffuse ribonucleoprotein (RNP) matrix, which consists of strings of 20-nm particles (105, 111) (Figure 1*c*). Larger RNP structures of up to 0.5-μm diameter, which are associated with certain loop pairs and which become prevalent in the later stage of lampbrush chromosome development, can be seen to consist of aggregates of similar 20-nm particles (105). Also, the structure of morphologically distinct loop pairs (102, 110) can be accounted for in terms of aggregates of 20-nm unit RNP particles (111). The molecular composition of the RNP particles, and hence the visible matrix of the lateral loops, is >95% protein and <5% RNA (104). The protein derived from lampbrush loop RNP constitutes a heterogeneous size range of nonhistone polypeptides (103, 104, 114–117), and the use of immunofluorescent probes directed against nuclear nonhistone proteins shows the lateral loops to be the main, if not the sole, chromosomal location of these proteins (114, 115). On the basis of their morphological state

and their chemical constitution alone, lateral loops obviously represent regions of transcriptionally active chromatin.

In *Triturus* there are at least 5×10^3 lampbrush loops per haploid chromosome complement (102); in *Notophthalmus* the estimated number is about 10^4 loops (100). The loops have axial length which may average 20–30 μm, depending upon the genus and the stage of lampbrush development, although loops of over 100 μm in axial length are not unusual. These lengths do not necessarily represent the total length of DNA double helix present, for loops can be manually stretched up to 2.5 times their original length without any apparent breakage in their axis (102), although there may be various interpretations of this observation. In any event, the lateral loops of *Triturus* lampbrush chromosomes contain less than 5% of the chromosomal DNA.

In *Xenopus* the lateral loops are only a few micrometers in length whereas in higher C value organisms loop lengths tend to be longer (81). In fact, there appears to be a rough proportionality between DNA content and loop length, although the reason for such a relationship would be difficult to explain unless it is an unavoidable consequence of nucleotide sequence duplication. A tacit assumption from early studies on the morphology of lampbrush chromosomes is that a loop pair and an associated chromomere correspond to a unit of genetic activity (110) and in some way parallel the proposed structure of the dipteran polytene chromosomes (100). However, certain observations are not concordant with this interpretation. For instance, electron micrographs reveal that often one region of condensed chromatin (a chromomere) bears more than one pair of lampbrush loops (103, 111, 118). If this is typical of all stages of lampbrush chromosome development, then a chromomere cannot be regarded as a unitary structure. Also, the length of a lampbrush loop, disregarding the associated condensed chromatin from which it extends, is far too great to contain nucleotide sequences coding for only one protein. This consideration stimulated the proposal that loops as genes (and possibly also their associated chromomeres) contain multiple repeats of coding sequences. As already argued (see under "Phylogeny, C Value, and Karyotypes of Amphibia"), this may not represent the general state of organization of genetic sequences. Nevertheless, for various reasons, individual loops can be regarded as heritable structures.

1) A given individual can be homozygous or heterozygous for a particular loop morphology at readily distinguishable marker loci (119).
2) These individual peculiarities of loop morphology are transmitted in a regular Mendelian fashion to the next generation (110).
3) Within a natural population of *Triturus* the relative number of individuals homozygous or heterozygous for particular loop morphologies are consistent with Hardy-Weinberg expectation (110).
4) In subspecies hybrids, loop morphologies peculiar to one subspecies are inherited in the heterozygous state in F_1 hybrids (119).

Further evidence for the restriction of genetic activity to the lateral loops of lampbrush chromosomes and the recognition of certain loop pairs as containing specific genes derives from various procedures in visualizing the process of transcription (see under "Visualization of RNA Transcription"). However, it is worthwhile first to consider what is known of the genetic organization of amphibian genomes with respect to the distribution of particular DNA sequences and to examine how this sequence organization relates to the general structure of lampbrush chromosomes.

DNA Base Sequence Organization

The most striking morphological feature of lampbrush chromosome structure is the even distribution of transcriptional units, in the form of lateral loops, along the entire length of all the chromosomes. Transcriptional activity, which is extensive in this sense, is a feature of all amphibian lampbrush chromosomes irrespective of genome size. The lateral loops are separated, one from the other, by large masses of condensed chromatin which themselves are fairly evenly spaced along the length of the chromosomes. It is difficult to conceive of lampbrush chromosomes being engaged in much less than maximal transcriptional activity. It is of interest then to consider how the different types of genomic sequence (see under "Phylogeny, C Value, and Karyotypes of Amphibia") might be organized with respect to one another and with the lampbrush chromosome as a whole.

Although interspersion of intermediate repetitive sequences with nonrepetitive sequences has been shown to be a general feature of chromosome organization (84–86, 120), it is of interest to know whether this type of sequence arrangement is also a feature of the small percentage of the genome that is transcribed from the lateral loops. The only information on this point is that informational RNA is transcribed during the lampbrush stage from both repetitive (8) and nonrepetitive (64) sequences in *Xenopus* oocytes and that the transcribed repetitive sequences are derived from the major intermediate repetitive DNA component (121). Therefore, the sequences transcribed into RNA on the lateral loops may be representative of the general sequence organization proposed for the majority of the genome. The primary transcript RNA derived from *Triturus* oocytes also has a component which is transcribed from intermediate repetitive sequences (75) (see Figure 10), although here again direct evidence is lacking about the arrangement of sequences in the RNA transcript. The lampbrush loops of *Triturus* are, on average, long enough to contain many coding sequences. As already suggested, each lateral loop may contain repeats of the coding sequence (75) which have diverged during the course of evolution. Although several of these related sequences may be transcribed, only one or a few from each loop are eventually translated into protein.

In terms of the proportions of different frequency classes of amphibian DNA, there is generally a greater number of nonrepetitive sequences than could be accommodated in the lateral loops alone. It is a necessary consequence that considerable numbers of nonrepetitive sequences are contained in the condensed chromatin of the chromomeres and are never transcribed, at least in oocytes. Also, the percentage of intermediate repetitive sequences in the genome is highly variable between amphibian taxa (54, 58), and it is difficult to see what consequence, if any, this variability has on the structure of lampbrush chromosomes.

In the genus *Plethodon,* species with larger genomes tend to have a higher percentage of repetitive DNA (60). In an attempt to rationalize this situation, the increase in the proportion of repetitive sequences has been correlated with an increase in the total number of chromomeres per haploid set of lampbrush chromosomes (68). However, the implication of this finding is that there is a proportionate increase in the number of lampbrush loops with increased genome size, a view that is not compatible with the observed constancy in the number of genes expressed during oogenesis in Amphibia of substantially different C value (65). Either chromomere number is not a reliable index of lampbrush loop number (see under "Spatial Distribution of Chromatin in Lampbrush Chromosomes") or the correspondence between loop number and gene number does not hold. It has further been suggested that in *Plethodon* species there may exist a relationship between number of chromomeres and number of families of repetitive sequences, a family of repeats being confined to a half-chromomere and its associated loop (69). However, the constitution of the intermediate repetitive component of the genome is not well enough characterized to warrant such an interpretation, nor is the chromomere established as anything more than a region of condensed chromatin. Instead of an increase in number of "chromomeres" with a supposed increase in number of families of repetitive sequences, it is perhaps more likely that there is a constancy in the number of "chromomeres" with an increase in the internal repetition frequency of each family of sequences.

Apart from the above speculation, there are several precedents for the existence of gene families as blocks of repetitive sequences. Some of these have been well characterized in terms of genetic organization.

Ribosomal Genes The sequences from which 18 S and 28 S rRNA are transcribed in *Xenopus* are the most extensively analyzed genes in eukaryotes. These sequences have undergone extensive tandem duplication, there being about 450–800 repeats in the nucleolar organizer of *Xenopus.* The main features of the structural organization of the repeating unit of rDNA are shown in Figure 2*A;* for the details, reference should be made to Reeder (28) and Tobler (30). However, there are three recent observations that are important and should be mentioned here.

1) Although considerable variation in the number of ribosomal genes in individuals of the same species has been known for some time in *Xenopus* (54, 122) and in *Ambystoma* (123), individuals with significantly different numbers have been considered to be genetic exceptions to the norm of the species. However, Hutchison and Pardue (124) have recently shown that in four individuals of *Notophthalmus* the number of ribosomal genes, as measured by

filter hybridization, was different in each, ranging from 0.008% to 0.018% of the genome complementary to 18 S and 28 S rRNA. Variability in chromosomal location and activity of nucleolar organizers has also been observed and is discussed under "Cytological Hybridization."

2) The polarity of transcription of the 40 S rRNA precursor has now been unequivocally established as shown in Figure 2*A*—that is, 5′ end - transcribed spacer - 18 S gene - transcribed spacer - 28 S gene - 3′ end. This has been demonstrated by analyzing properties of the fragments of rDNA and ribosomal transcription complexes derived by cleavage with the restriction endonuclease EcoRI. The rDNA fragments cloned in a DNA plasmid of *Escherichia coli* were partially digested with exonucleases specific for 5′ and 3′ ends to give single-strand tails. These were then hybridized separately with 18 S or 28 S rRNA and orientated by monitoring radioactivity and by electron microscopy (127). Fragments of ribosomal transcriptional matrices were orientated by direct electron microscopic examination (128).

3) By examining EcoRI fragments of *Xenopus* rDNA and by heteroduplex mapping of these fragments cloned in *E. coli*, it has been found that the nontranscribed spacer is heterogeneous for length (129). Furthermore, there appears to be less heterogeneity in adjacent rDNA genes derived from oocyte chromosomes, as well as from oocyte nucleoli, than in adjacent rDNA genes derived from somatic cells (130), a finding that is consistent with the proposed rolling circle model for rDNA amplification (131–133). These differences in spacer lengths are found to be genetic properties of individual toads, and each pattern of spacer length is inherited in a stable fashion (134).

5 S RNA Genes In the genome of *Xenopus* there are more than 2×10^4 repeats of sequences containing the genes for 5 S RNA (135, 136), and these are located as separate gene clusters near the telomeres of the long arms of most, if not all, of the chromosomes (107, 137). Purified DNA containing the 5 S RNA genes has been isolated from *Xenopus laevis* (136) and *Xenopus mulleri* (138). The main features of the structural organization of the repeating unit of 5 S DNA are shown in Figure 2*B*. The spacer region alternating between the 5 S RNA genes is about 600 base pairs in *X. laevis* and about twice that length in *X. mulleri*. All of the 5 S RNA genes of this purified DNA are transcribed during oogenesis (139), whereas only a subset of the 5 S genes is active in somatic cells (140, 141). It has recently been demonstrated by analyzing the cleavage products of the restriction endonucleases Hind III and Hae III that the AT-rich spacer, which itself contains tandem repeats of a family of closely related sequences approximately 15 base pairs long (139), is variable in length by regular increments of 14 base pairs (142). Furthermore, fragments of 5 S DNA, produced by partial digestion with Hind III and cloned in *E. coli*, have been shown by denaturation mapping to contain adjacent genes which are heterogeneous in length (143). The fact that adjacent AT-rich spacers can be heterogeneous in length, whereas the 5 S genes themselves are maintained as

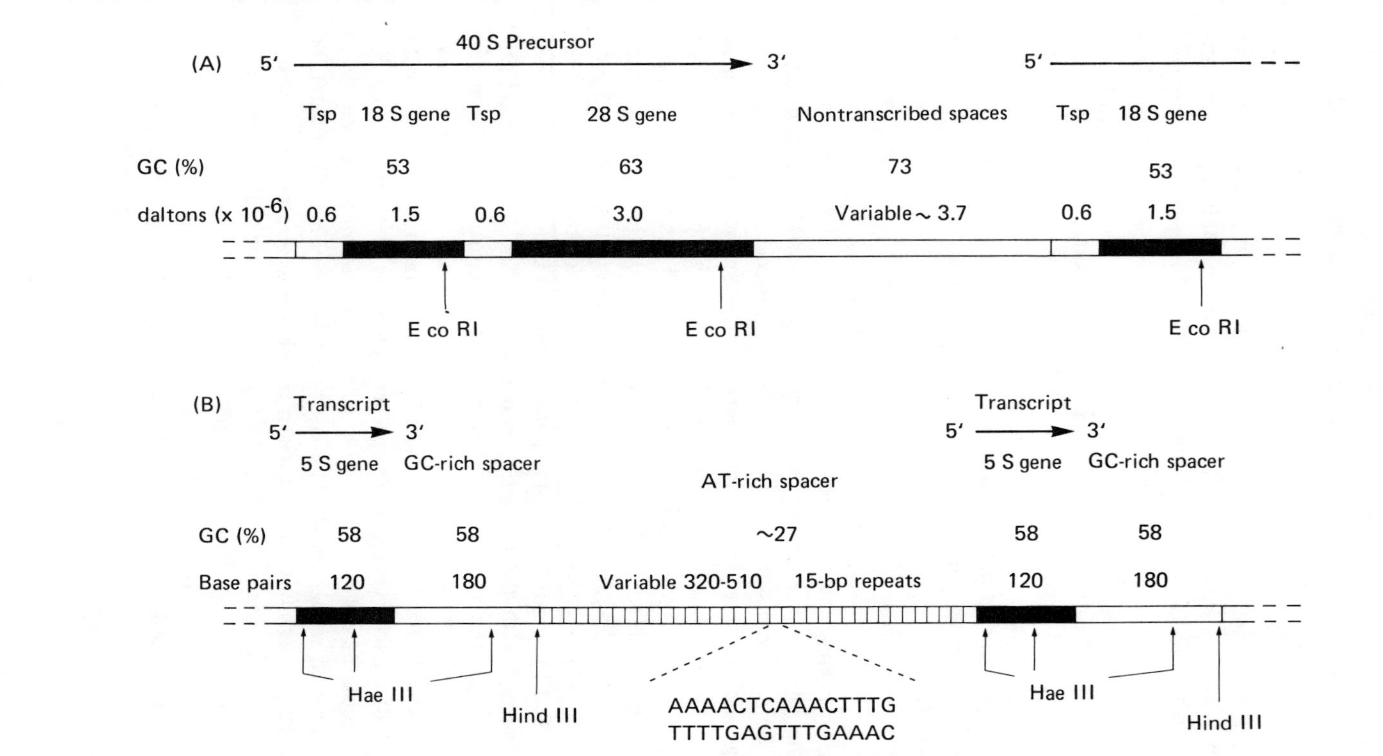

Figure 2. Diagrams of sequence organization in repetitive genes of *X. laevis*. *A*, the 18 S–28 S rRNA gene repeat (125–230). *B*, the 5 S RNA gene repeat (135, 136, 138, 139, 142).

homogeneous sequences, poses problems in suggesting a mechanism to explain the correction of tandemly repeated genetic sequences during the parallel evolution of genes and their spacers. From the results summarized above, Carroll and Brown (142, 143) suggest a process of extensive unequal crossing over between the AT-rich spacers which would lead to the elimination or fixation of variants (144). Such a process would have to occur between the nonhomologous chromosomes, which contain 5 S genes, during a stage of meiosis when the telomeres are clustered (107, 137). However, the location of 5 S genes at other chromosomal positions in *Notophthalmus* and *Triturus* (see under "Cytological Hybridization") presents different problems. Furthermore, there are difficulties in visualizing how the estimated number of 5 S gene repeats could fit into single loop chromomere complexes at their designated chromosomal sites. In addition, it would be expected that transcription would be maximal during oogenesis, i.e., all 5 S sequences situated on lampbrush loops, which is an impossible situation (see under "Cytological Hybridization").

Transfer RNA Genes The haploid genome of *Xenopus* contains approximately 8,000 DNA sequences that are complementary to transfer RNA (tRNA) (145, 146), with an average of about 200 genes for each tRNA species (145). However, the gene number can vary by more than a factor of two for different aminoacyl-tRNA species, and even for different isoaccepting tRNAs (147). There is a general interspersion of tRNA genes with spacer sequences which are estimated to be about 10 times the length of the tRNA gene (147). The reiterated genes for $tRNA_1{}^{Met}$, $tRNA_2{}^{Met}$, and $tRNA^{Val}$ can be separated from one another when high molecular weight DNA ($\sim 9 \times 10^7$) is banded in neutral CsCl gradients, suggesting that there is no intermingling of different tRNA genes (146). Recently, DNA containing the reiterated genes for $tRNA_1{}^{Met}$ has been partially purified by centrifugation in actinomycin-CsCl and Ag^+-Cs_2SO_4 gradients, and digestion with the restriction endonucleases Eco RI and Hpa I generates fragments of little, if any, length heterogeneity, indicating a regular spacing of tandemly repeated tRNA genes (148). However, the exact spatial arrangement of tRNA genes and their spacer sequences has not yet been determined. If the repeat length of a tRNA gene plus its spacer is about 10^3 base pairs, then 200 such repeats could be easily accommodated by a loop-chromomere complex. In fact, this number of sequences would constitute a lampbrush loop of about 60 μm axial length, dimensions not unusual in the chromosomes of urodeles. It is important that tRNA genes, like 5 S RNA genes, be maximally active during oogenesis in order to accumulate the 4 ng of tRNA found in mature oocytes.

Histone Genes There is little information on the numbers and organization of histone genes in the genomes of Amphibia. In sea urchins the organization of histone genes has been studied in some detail, the genes coding for the different histones being intermingled in clusters of less than 7,000 base pairs, with a reiteration of 300–1,000 clusters in various sea urchin taxa (149). Estimates of the degree of reiteration of histone genes in *Xenopus* have been put at 10–20 (150) and 30–50 (151). If, as seems likely, histone genes are clustered in all eukaryotes, then 3.5×10^5 base pairs would be the maximal sequence length for histone genes in *Xenopus*. This number of sequences could be easily arranged within one loop-chromomere complex.

Repetitive Sequences of Unusual Base Composition Gould et al. (152) have studied the effect of various restriction endonucleases on the lampbrush chromosomes of *Notophthalmus*. Endonucleases Hind, Hae, EcoB, EcoK, and Endo I all cause extensive fragmentation of the vast majority of lampbrush loops. However, a notable exception is the resistance of the giant loops on chromosome II to digestion of HaeIII. Because these loops are sensitive to other nucleases but refractile to HaeIII, the sequence 5′GGCC must be systematically absent from the DNA axis of these loops. In a random nucleotide sequence, one Hae-sensitive site is expected to occur once every 256 base pairs, thereby resulting in about 10 scissions per μm of DNA. The fact that the giant loops are about 300 μm in length and completely refractile to the action of HaeIII strongly implies that the axial DNA of these loops consists of a large number of tandem repeats which happen to lack the sequence 5′GGCC. Furthermore, autoradiographic observations indicate that, although adenine, cytidine, and uridine are incorporated into RNA transcripts of giant loops, little incorporation of guanosine is detectable (152). Apparently there is a marked asymmetry in the distribution of bases in the DNA duplex, most of the cytosine residues being confined to the nontranscribed strand. This would account for the Hae resistance, but a possible function for the extensive transcription of RNA sequences lacking guanosine remains unexplained.

Visualization of RNA Transcription

Because of their large size in many amphibian taxa and their extensive involvement in RNA synthesis, lampbrush chromosomes provide a unique opportunity for directly observing the process of transcription. Much of what is known about the genetic activity of lampbrush chromosomes derives simply from the adaptation and exploitation of various cytological techniques.

General Morphology The lateral loops of lampbrush chromosomes, when viewed with phase contrast optics, are generally described as having progressively more RNP matrix associated with the loop axes as they are followed from one end to the other (102, 110). The points at which the loop meets with the condensed chromatin of the chromosome axis are referred to as the "thin" and "thick" insertions. Such an asymmetric and polarized distribution of RNP is less obvious when loops are examined by electron microscopy (105, 111) (see Figure 1). The relative accumulation of RNP toward the thick insertion often falls far short of the amount expected from the transcription of the preceding length of loop axis. In fact, in some loops there is little obvious polarization of RNP matrix. Therefore, the original interpretation of uninterrupted transcription of RNA around the loop, with the generation of increasingly long RNP transcripts, is not altogether borne out by these later observations. As is seen more clearly from studies of spread transcriptional matrices (see under "Chromosome Spreads/Transcriptional Matrices"), the transcription and processing of RNP of the loops are by no means a simple operation.

An additional mechanism involving transcription on lampbrush loops was inspired largely by loop asymmetry, but also by an apparent inverse relationship between chromomere size and loop length and an unusual pattern of labeling seen on some giant loops (see under "Incorporation of RNA Precursors"). The proposal was that the DNA axis itself moved by being continuously spun out from the chromomere at the thin insertion and retracted back into the chromomere at the thick insertion (102, 110, 153), the implication being that virtually all of the genome is eventually transcribed. From what is known about the involvement of RNA polymerase in loop transcription (154) and the low percentage of the genome actually transcribed into RNA during oogenesis (10, 75), such a mechanism seems unlikely, or at best is a feature of only a few unusual loop-chromomere complexes such as the giant loops on chromosome II of *Notophthalmus* (100) and the giant granular loops on chromosome XII of *Triturus* (44, 153). The only method by which any movement of the DNA axis can be demonstrated is by pulse labeling the loop DNA in some way.

Incorporation of RNA Precursors The general autoradiographic pattern of [^{3}H]uridine uptake by lampbrush chromosomes, labeled by injecting the animals (44, 153) or by incubating the excised ovary (100) or isolated oocytes (155), is that the vast majority of lateral loops become labeled along their entire length within 1–2 hr of administering the radioisotope. This labeling pattern is consistent with the interpretation that RNA synthesis occurs at the same rate at all points along the loops. In all of these studies, however, autoradiographic grains are also seen over the regions of condensed chromatin—the chromomeres. This effect has been variously interpreted as being due to collapsed loops lying over the chromosome axis, the presence of short loops not easily resolved by light microscopy, and the accumulation of labeled products at the thick insertions of the loops. However, autoradiographic studies alone do not exclude the possibility of RNA synthesis in the chromomeres or alternatively of transcriptional products being secondarily attached to the chromomeres.

The giant loops of *Notophthalmus* and *Triturus* are unique in several respects.

1) In *Triturus* they are sequentially labeled. Transcription is apparently restricted to the region of the loop adjacent to the thin insertion (45, 100, 153). This implies that polymerase-attached transcripts can travel around the remainder of the loop without concomitant RNA synthesis; they are either carried around by movement of the DNA axis or the polymerase molecules can slip around without being involved in further transcription.
2) An extended period of about 10 days is required for pulse-labeled RNA to travel around the length of the loop. This is slow compared to the transcription rate normally accepted for RNA polymerase ($\sim 2 \times 10^3$ nucleotides/min at

25°C; see under "RNA Polymerase Activity"), which would take at most a few hours to traverse the length of DNA comprising the giant loop axis.
3) As already mentioned, the sequence 5′GGCC is apparently absent from the giant loops of *Notophthalmus* (152).
4) In addition, the pattern of incorporation of RNA precursors is highly unusual, guanosine being present at low levels in the RNA transcript (156). Because the function of these loops remains an anomaly and because the transcription products are highly atypical, the properties of the giant loops should not be used to propose a general model for lampbrush loop activity.

Cytological Hybridization Although it is at times difficult to reconcile the observations from cytological hybridization experiments with theoretical considerations of nucleic acid/chromosome in vitro interactions, the usefulness of this technique, especially in demonstrating the chromosomal location of specific reiterated sequences, should not be underestimated. Moreover, data from cytological studies on lampbrush chromosomes can be used to extend our understanding not only of genetic organization but also of genetic activity within these chromosomes. Nevertheless, autoradiographic identification of particular genetic sequences can be considered to be genuine only when the following criteria are established.

1) The distribution of silver grains should be restricted to only some of the structures in the preparation, for instance, on a limited number of lateral loops or chromomeres or both.
2) The hybridization reaction should be similar on the two lampbrush chromosome homologs, although heterologous reactions have been reported to occur on metaphase chromosomes (124, 157, 158).
3) There should be no corresponding reaction with nondenatured or nuclease-treated control preparation. Spurious hybridization under nonoptimal conditions has been reported (159).
4) There should be no equivalent reaction with a heterologous labeled nucleic acid.
5) The bound nucleic acid should form a stable complex with the complementary chromosomal component.
6) The observations should be compatible with biochemical and genetic studies.

Originally, labeled RNA was used as a probe for the detection of specific chromosomal or extrachromosomal sequences, the RNA being derived either from cells labeled in culture (160) or from polymerase-transcribed DNA (70). In more recent studies the technique has been extended to the use of labeled DNA as a probe for chromosomal RNA transcripts, as well as for their DNA templates. Obviously, this adaptation is of great advantage where particular DNA sequences can be isolated, labeled, and then located on distinguishable chromosomal structures. To date, several specific genetic loci have been identified in this way.

The labeled probe has been prepared in a number of ways: by in vitro iodination of specific DNA fractions (78, 158); by nick-translation of specific DNA fractions (161); and by reverse-transcribed complementary DNA (cDNA) with the use of both total polyadenylated RNA and preparations enriched for specific mRNA as templates (162).

Satellite and Other Repetitive Sequences Hybridization of complementary RNA (cRNA) to lampbrush chromosome preparations of *Notophthalmus* shows that repetitive sequences are widely distributed throughout the genome (70). However, some localization is apparent. The highly repetitive (satellite) sequences are restricted to the centromeric heterochromatin, whereas other repetitive sequences (excluding ribosomal) are located at the telomeric regions of the chromosomes and at the "transcriptionally inactive" sphere loci. It is interesting to note that the sphere loci are transcriptionally inactive during the lampbrush stage, yet during hormone-induced maturation the dense spheres are replaced by prominent lateral loops which are active in RNA transcription (163). The nature of the transcriptional products derived from these particular repetitive sequences is unknown.

Satellite sequences of other genera, such as the GC-rich satellite of *Plethodon,* also hybridize exclusively in the region of the centromeres (72) which in this genus are large, loopless aggregates of DNP showing no signs of transcriptional activity (164). It has also been reported that the intermediately repetitive sequences of *Plethodon* hybridize to many sites on all chromosomes, although a general interspersion of repetitive sequences is not obvious (165). In addition to the centromeric localization of satellite sequences, an AT-rich satellite of *X. mulleri* has been found to hybridize over discrete interstitial sites on the short arms of most chromosomes (107).

Ribosomal Sequences Cytological hybridization of 18 S and 28 S rRNA to lampbrush chromosomes of *Triturus marmoratus* (166) demonstrates that the genes are located at the cytologically identifiable nucleolus organizer, that is, at a subterminal position on the long arm of chromosome X (167). In addition, ^{125}I-labeled *Xenopus* rDNA hybridizes to lampbrush chromosomes of *Triturus c. carnifex* at a position near the middle of the short arm of chromosome IX (158), again confirming the cytologically identifiable nucleolar organizer (168). In these closely related *Triturus* species, the nucleolar organizer on the lampbrush chromosome is represented by what is probably a transcriptionally inactive chromomere (158). This is the situation that would be expected if ribosomal sequences are being effectively transcribed from the many extrachromosomal nucleoli. Yet the nucleolar organizers described in other urodeles, such as *Notophthalmus* (169), *Ambystoma* (170), and *Plethodon* (164), consist of morphologically conspicuous lampbrush structures, often in the form of large lateral loop-chromomere complexes. It is not understood why, in only some instances, there should be apparent transcription of ribosomal sequences from the lampbrush chromosomes. This raises the question of differential control in the expression of chromosomal and extrachromosomal ribosomal genes. Furthermore, there is evidence for considerable heterozygosity in the amount of rDNA available for hybridization in mitotic preparations of *Plethodon* (157), of *Triturus* (158), and of *Notophthalmus* (124). In *Notophthalmus* it has been shown that the variable pattern of hybridization to nucleolar organizers is a characteristic of individual animals. At least two of the three nucleolar organizers may show approximately equal hybridization over both homologs in some animals and quite different levels of hybridization over the two homologs in other animals. That the observed heterozygous effects are due to differences in the number of ribosomal genes, rather than to differences in nucleolar activity or accessibility of the labeled RNA, has been confirmed by measurement of saturation hybridization on filters containing purified DNA isolated from the different animals (124). Similar heterozygous effects are also seen at the five nucleolar organizer loci in *Triturus vulgaris* (106).

5 S RNA Sequences In Amphibia, as in other eukaryotes, multiple 5 S gene copies are clustered at loci which differ from the nucleolar organizer. Cytological hybridization experiments have shown that the actual number of loci can vary considerably (Table 2). For instance, the 5 S genes of *X. laevis* and *X. mulleri* occur at multiple sites which are located at the telomeres on the long arm of most, if not all, of the chromosomes (107, 137). On the other hand, the 5 S genes of *Notophthalmus viridescens*, which also occur at multiple sites, are restricted to loci near the centromeres of chromosomes I, II, VI, and VII (78). Furthermore, an additional site, on the long arm of chromosome X, has been detected in metaphase preparations (124) although this site does not label consistently in lampbrush preparations (78). The simplest situation exists in *T. marmoratus* (166) and in *T. c. carnifex* (106, 166). Here the 5 S genes are located at a single site which is at an intermediate position on the long arm of chromosome X, that is, in the same position as the variable fifth site in *Notophthalmus*. Thus, the 5 S genes can be organized in various ways—either entirely clustered or in blocks at several sites in the genome; and the gene clusters can be located at apparently any position along the chromosome—at or near the telomere, proximal to the centromere, or at an intermediate position.

The interesting feature of the cytological hybridization experiments involving urodeles is that the 5 S sequences can be located on distinct morphological structures of lampbrush chromosomes. The single site of hybridization of [^{3}H]rRNA to the chromosomes of *Triturus* is on, or near, the chromomeres bearing a pair of characteristic dense matrix loops (166). In *Notophthalmus*, labeled 5 S RNA is used to identify the chromomeric location of the 5 S genes, whereas labeled 5 S DNA is used to establish that the lateral loops deriving from these chromomeres are active in transcribing 5 S RNA sequences (78). Arising from these observations are three basic questions. Are the long spacer sequences transcribed? Can all the numerous 5 S DNA sequences plus their spacers be accommodated within a single chromomere-loop complex? What are the evolutionary implications of such variable 5 S genetic locations between different taxa?

Table 2. Location of 5 S RNA genes by slide hybridization

Organism	Probe	Chromosomes: Number	Chromosomes: Location	Reference
X. laevis	^{3}H-labeled 5 S RNA from kidney cells	All	Telomeres of long arms	137
X. mulleri	^{3}H-labeled cRNA transcribed from 5 S DNA	All	Telomeres of long arms	107
N. viridescens	^{125}I-labeled 5 S RNA from *N. viridescens* ovaries	I, II, VI, VII	Chromomeres adjacent to centromeres	78
	^{125}I-labeled 5 S DNA from *Xenopus* erythrocytes	I, II, VI, VII	Loops adjacent to centromeres	78
	^{3}H-labeled cRNA transcribed from *Xenopus* 5 S DNA	I, II, VI, VII	Pericentric	124
		X	Intermediate position on long arm	
T. marmoratus	^{3}H-labeled 5 S RNA from *Xenopus* kidney cells	X	Chromomere of dense matrix loop at intermediate position on long arm	166
T.c. carnifex	^{3}H-labeled 5 S RNA from *Xenopus* kidney cells	X	As in *T. marmoratus*	166
	^{125}I-labeled 5 S RNA from *T.c. carnifex* ovaries	X	As in *T. marmoratus*	106, 115

The 5 S RNA sequences appear to be transcribed from lampbrush loops of fairly normal morphology in both *Triturus* and *Notophthalmus,* yet, if the alternating 5 S DNA spacers were not transcribed, then only short transcripts would be produced and abnormally fine matrix loops would result. Also, if the spacers were not transcribed there would be, at most, twice as many transcripts as genes. This is because a 5 S gene of 120 base pairs could accommodate only one or two polymerase molecules. However, cytological hybridization studies indicate that there is a far greater abundance of available 5 S RNA sequences than 5 S DNA sequences in the lampbrush loops (78), indicating longer transcript lengths. On the other hand, there has been no biochemical demonstration of a high molecular weight precursor to 5 S RNA. Indeed, several lines of evidence argue against this possibility: a fraction of 5 S RNA has been isolated which has a 5′-terminal triphosphate (141, 171); no precursor of 5 S RNA has been detected, even after short pulse labeling, in somatic cells (172, 173); total [^{3}H] RNA reacted with strand-separated 5 S DNA hybridizes with the L strand only and is competed for entirely by 5 S RNA (138). Therefore, this situation is unlike transcription of the high molecular weight (40 S) precursor of 18 S and 28 S rRNA which includes the linking spacer (125, 136). Nevertheless, the 40 S precursor contains only one copy of each rRNA sequence. The only evidence for a 5 S RNA precursor molecule is the presence of a few additional nucleotides at the 3′ end of newly synthesized *Xenopus* oocyte 5 S RNA (141). Thus, the conflict between cytological and biochemical data remains unresolved, although there may exist mechanisms peculiar to the transcription of 5 S sequences in oocytes.

The problem of whether a chromomere-loop complex could accommodate all of the 5 S DNA sequences also presents difficulties. In *Notophthalmus,* 3×10^5 5 S genes (78), plus alternating spacer sequences 5–20 times the length of the gene, would account for about 2.2×10^8–7.8×10^8 nucleotide pairs or 0.54–1.9% of the genome, which is a vastly disproportionate load for 4 out of about 10^4 loop pairs. Perhaps the value of 0.045% of the DNA hybridized to 5 S RNA by filter hybridization is an overestimate, as may be all estimations based on this technique. For instance, substantially lower values have been derived from kinetic data, as compared to filter saturation values, for the number of human ribosomal genes (174).

A consideration of the number and distribution of 5 S genes, as well as of nucleolar organizers, in the karyotypes of different amphibian taxa might prove to be useful in investigating their phylogenic relatedness (166), but more data are required.

A further problem, that of differential control in the transcription of 5 S RNA sequences in oocytes and somatic cells, is also related to the chromosomal location of these sequences. This aspect is discussed under "Noncoordinate Synthesis and Control Mechanisms."

Messenger RNA Sequences Labeled cDNA, transcribed from polyadenylated mRNA, hybridizes to only several of the many thousands of loop pairs of *T. c. carnifex* lampbrush chromosome preparations (162). The autoradiographic pattern shows grains over the matrix around the entire length of the loops, with occasional denser labeled regions at one insertion which probably corresponds to localized accumulation of RNA transcripts. These observations can be accounted for by one or more of the following explanations: 1) there exists an abundant class of cDNA sequences due to either the abundance of a restricted number of mRNA sequences or preferential transcription of several types of mRNA; 2) repetitive coding sequences exist as tandem duplicates around several loops; 3) there is some physical restriction to hybridization with the RNA of the vast majority of loops. This last suggestion raises the general problem of supposed hybridization to RNA transcripts which are known to be associated with large amounts of protein (see under "Role of Proteins Specifically Associated with Transcriptional Products"). Perhaps some of the difficulties in interpretation will be resolved when these reactive genetic loci are identified.

Preliminary observations have in fact revealed the location of one class of coding sequence. Labeled cDNA transcribed from *Xenopus* mRNA enriched for globin sequences apparently hybridizes to only two loop pairs on the long arm of chromosome I of *T. c. carnifex* (Figure 3). Again the distribution of grains is fairly even around the entire length of the chromosome, which implies that the coding sequences are located at the 5′ end of the transcript if these loops, which are many micrometers in length, are considered to be a single unit of transcription (see under "Chromosome Spreads/Transcriptional Matrices"). The apparent transcription of globin sequences has very important implications for the whole nature of transcriptional activity in lampbrush chromosomes and is direct confirmation of other observations on the transcription of RNA containing sequences coding for globin in nonerythroid tissues (175). If the transcription of sequences whose translation products are not necessary for the proper functioning and development of the oocyte is a general phenomenon, then such activity would greatly support the notion that transcription is a "leaky" process. After all, it is questionable whether at least 10^4 different gene products (10, 65, 75) are essential for oogenesis or even early embryogenesis.

Use of Immunofluorescent Probes The detection of particular chromosomal proteins by means of specific antibody binding is subject to technical restrictions and difficulties in interpretation similar to those discussed for the detection of particular nucleotide sequences by means of slide hybridization. It is first necessary to identify and try to exclude nonspecific effects and then to designate genuine specificity, in terms of antibody binding, to only certain localized lampbrush chromosome structures. Following from the observations that by far the main molecular constituent of lampbrush chromosomes is protein (>95% of the chromosomal mass (89, 104)) and that nearly all nucleic acids have protein associated with them to form characteristic nucleoprotein complexes, it is reasonable to investigate the activity of lampbrush chromosomes by locating the sites of particular proteins. To this end, the following questions have been asked:

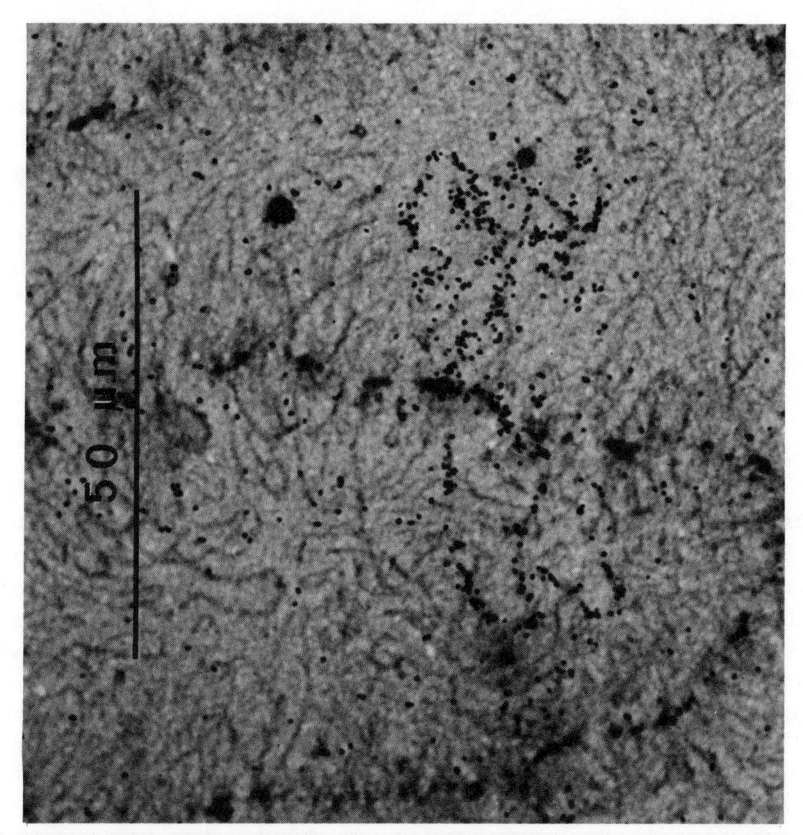

Figure 3. Autoradiograph showing hybridization of [^{3}H]cDNA transcribed from *Xenopus* globin mRNA to two loop pairs on chromosome I of *T. c. carnifex* (162).

1) Are nuclear proteins localized in a discrete manner in lampbrush chromosomes? 2) Are those proteins which are found in particular oocyte RNP fractions also found at the site of RNA transcription, presumably on lampbrush loops? 3) If so, can specific sites of synthesis be detected by virtue of the protein that is uniquely associated with that particular type of RNA transcript? In order to attempt to answer such questions, antisera have been raised in rabbits against several protein fractions derived from the nucleoproteins found in *Triturus* oocytes. These are reacted with lampbrush chromosome preparations which are subsequently treated with sheep anti-rabbit fluorescein-labeled globulin. After washing out unbound serum at each stage, any final specific reaction should be visible on examining the chromosome preparations by ultraviolet microscopy.

Antiserum prepared against total *Triturus* histone reacts strongly with the chromosome axis, in particular with the dense regions of condensed chromatin, and only weakly with the lateral loops, whereas antiserum produced against primary transcript RNP proteins reacts specifically with most, if not all, of the lampbrush loops (114, 115). However, subsequent studies with the use of the immunofluorescent technique have been primarily directed toward the detection of specific genetic loci. To this end, attempts have been made to resolve individual proteins by chromatographic and electrophoretic separation to serve as antigens.

Fractionated histones have so far yielded little additional information on the localization of individual histones. Anti-H2a, anti-H2b, anti-H3, and anti-H4, isolated by salt fractionation and electrophoresis of *Triturus* liver histones, react with the condensed chromatin of the chromomeres, although perhaps not as strongly as low molecular weight proteins, as yet uncharacterized, derived from oocytes themselves (115). These are the general results that one would expect if histones are primarily associated with untranscribed chromatin, although they do not prove that there is no histone association with the transcriptionally active lampbrush loops. Indeed, the only conclusion that can be drawn is that most of the histone is located where most of the DNA is located—in the densely packed chromatin of the chromomeres. The more diffuse loops present a problem with respect to histone localization. The only way of possibly detecting the small proportion of histone that may be associated with the loops is by a more sensitive immunochemical assay, for instance, by the use of ferritin-conjugated antibodies and the detection of their specific chromosome binding by electron microscopy. Histone HI, unlike the other four histones, is not accumulated in amphibian oocytes (12). However, antibodies prepared against histone HI react, rather surprisingly, with a pair of lateral loops on chromosome XI of *Triturus* (115). These loops have a very thin matrix, and anti-HI reacts far more strongly with these than with any other loops or even with chromomeres. The nature of this binding is at present unknown.

Oocyte primary transcript RNP proteins, which constitute a considerable size range of different polypeptides (103, 114, 116) (see Figure 8), have been fractionated by gel filtration (114) and by SDS-acrylamide electrophoresis (115). The immunofluorescent localization of most of these proteins is on nearly every lateral loop. In only one instance is a polypeptide component of total primary transcript RNP restricted to a limited number of loop pairs. A polypeptide in the size range 30–35,000 is located by immunofluorescence on only several loop pairs, which are sited on various chromosomes of *Triturus*, but mainly on chromosome X (115). The function of this protein and the nature of the RNA transcripts to which it is bound remain unknown, although the reaction in itself clearly demonstrates that the RNP is confined to these special loops and is not detected at all in the proximity of the associated chromomeres (75). Furthermore, the whole length of these loops has this particular RNP associated with it, indicating a unity in function of the lateral loops. As far as can be discriminated, all major nonhistone chromosomal proteins are found to be associated with primary transcript RNA. There is no other distinct category of chromatin protein detected in the lampbrush chromosomes of amphibian oocytes apart from RNA polymerase molecules (see under "Chromosome

Spreads/Transcriptional Matrices"). Any other DNA-binding proteins, if present, must be present in very small amounts.

The predominant ribonucleoprotein of early oogenesis is in the form of an RNP particle which sediments at 42 S and contains exclusively 5 S ribosomal RNA, transfer RNA, and their attendant proteins (see under "42 S Particles"). In *Triturus* this RNP fraction contains only two proteins of molecular weight 38,500 and 49,000 (see Figure 8), and the immunofluorescent technique was applied to ascertain whether these proteins associate with low molecular weight RNA species during transcription and, if so, to identify the chromosomal loci. Antiserum prepared against the higher molecular weight protein reacted specifically with one loop pair, a dense matrix locus on chromosome X which, as far as can be told, has the same location as the loop pair that hybridizes 5 S RNA (115). In order to confirm this localization, it would be necessary to demonstrate coincidental immunofluorescent labeling and 5 S RNA (or 5 S DNA) hybridization on the same preparation. As it stands, a particular protein apparently associates with 5 S RNA during the process of its transcription. Antiserum prepared against the lower molecular weight protein from the 40 S RNP reacted with several loop pairs on different chromosomes, but not with the presumptive 5 S locus (115). It can be speculated that these reactive loops are tRNA loci, but this interpretation awaits confirmation by slide hybridization studies. The presumptive tRNA synthesizing loops are often highly extended, especially in chromosome preparations derived from previtellogenic oocytes, and have an even (nonpolarized) RNP matrix (Figure 4).

Chromosome Spreads/Transcriptional Matrices When lampbrush chromosomes are dispersed in dilute saline solution (<0.05 M NaCl) or in distilled water (preferably adjusted to pH 8–9), the lateral loops apparently lose their RNP matrix and become shorter; it is presumed that the DNP axis is eventually retracted into the chromomeric axis of the chromosome (102, 118). However, if such dispersed chromosomes are allowed to spread, attached to carbon-coated grids, and are then stained and viewed by electron microscopy, rather than the loop RNP being released and the axial DNP being condensed, the RNP is seen to have adopted an alternative configuration. Following the early observations of Miller and co-workers (63, 154, 176–179), spread lateral loops have been shown to consist of RNP transcripts emerging as linear fibrils about 5 nm thick from the DNP axis and attached to the DNP axis by virtue of a densely staining particle, 12 nm across, presumed to be the RNA polymerase molecule. Therefore, due to changes in the ionic environment, the beaded RNP of the loop is transformed to linear fibrils. The molecular events involved in this transformation have been described in detail (103). The RNP fibrils within a transcriptional unit, or matrix, are generally closely spaced (25–60 nm between polymerases) and present an uninterrupted gradient of increasing transcript length toward the thick insertion of the loop. Advances in the method of spreading, staining, and shadowing transcriptional matrices have greatly enhanced the visual presentation

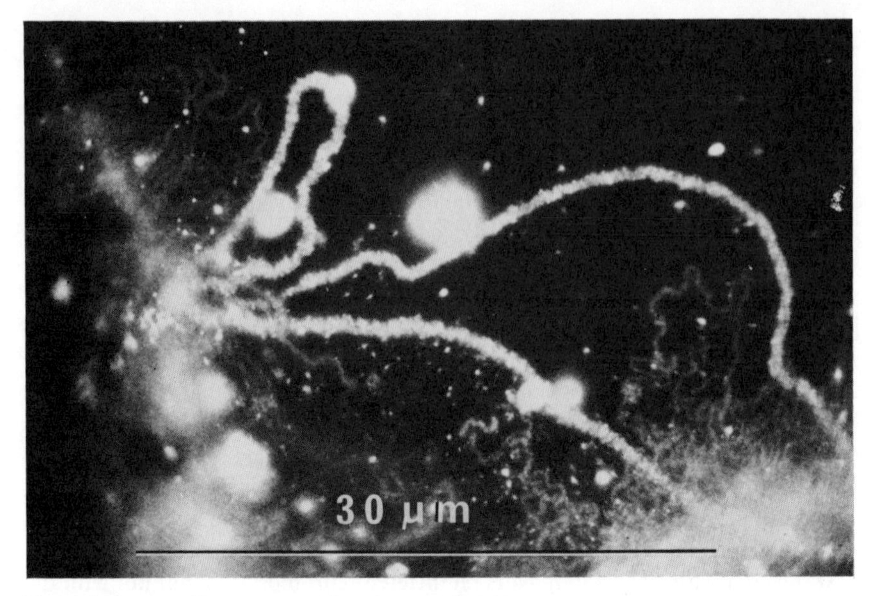

Figure 4. Immunofluorescence of a loop pair which reacts specifically with antiserum prepared against one of the two proteins associated with the 5 S RNA and transfer RNA ribonucleoprotein complex. These loops are tentatively assigned as being one of a set of loop pairs which transcribe *Triturus* oocyte tRNA.

(Figure 5). The application of this technique, probably more than any other, can give direct visual measurements on such important considerations as the extent of genetic transcription; the size of transcripts; the structural organization within a unit of transcription, here a transcriptional matrix of RNP fibrils; and the relationship between different transcriptional matrices. Although initial observations suggested that measurements and interpretations might be straightforward, at present transcription viewed in this way is seen to be a complex process. Some of the observations and difficulties are outlined below.

Bead-fiber Transformation The molecular events involved in the conversion of periodically beaded RNP to linear fibrils have been studied with the use of isolated *Triturus* nuclear RNP (103). In low ionic strength buffer (0.1 mM borate, pH 8.5; 5 mM 2-mercaptoethanol), the compact beaded material gradually unravels and the RNP extends in length. Although some protein is released in the process, there are no specific proteins obviously peculiar to either the beaded or the fibrillar configuration. It is proposed that RNA secondary structure generates the beaded structure and that the integrity of the bead is maintained by means of protein-protein interaction.

Extent of Transcription If the conditions of chromosome spreading are not sufficiently severe, the bulk of the chromatin remains as condensed masses of chromomeric DNP. However, in distilled water, adjusted to pH 8.5–9.0 with a minimal amount of borate buffer, the process of dispersion progresses until the

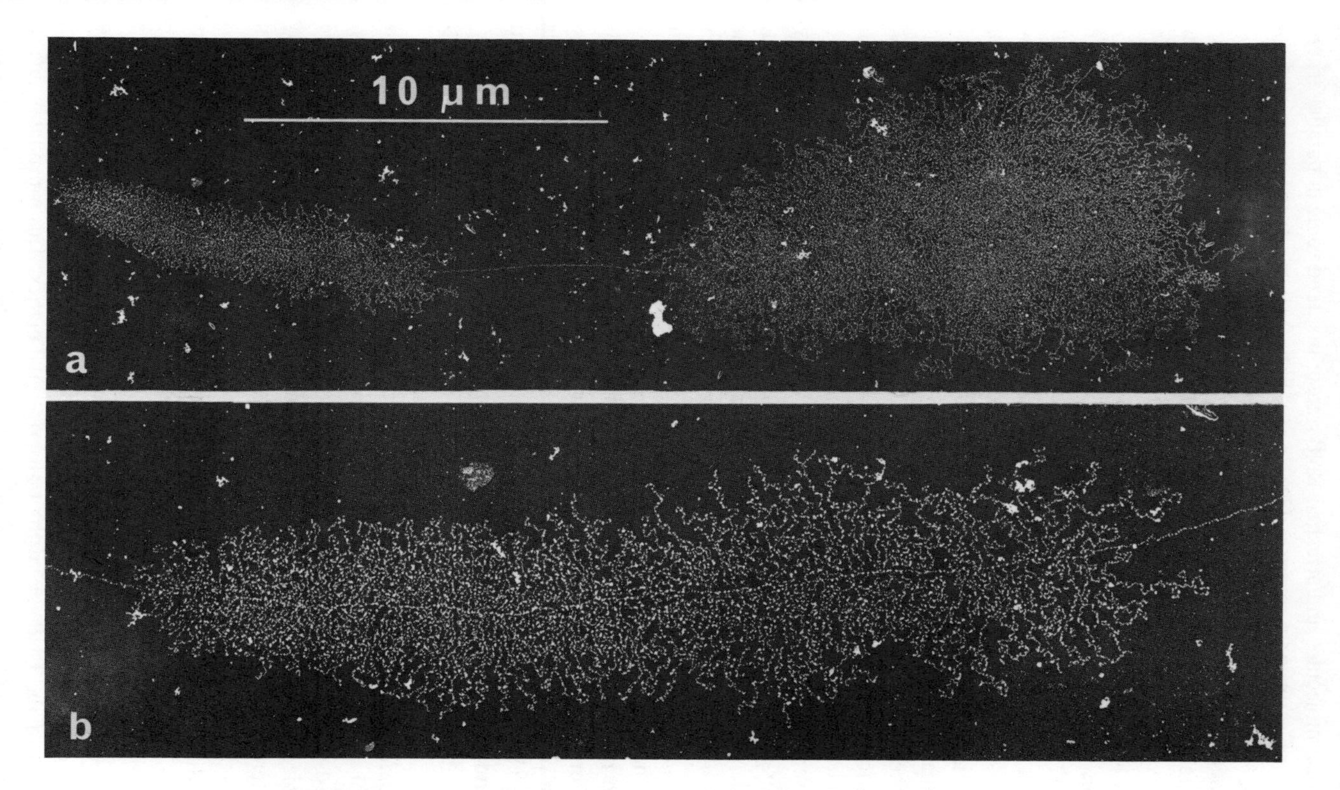

Figure 5. Electron micrographs of spread lampbrush chromosome preparations of *Pleurodeles poireti*. *a*, two adjacent transcriptional matrices of different dimensions. *b*, higher magnification showing the structural features of one transcriptional matrix. The preparations were stained with PTA and Pt shadowed. Reproduced from Angelier and Lacroix (180) with permission of Springer-Verlag.

chromomeric DNP starts to unravel. At this stage the DNP fibers, free of any polymerase molecules and attached RNP transcripts, tend to lie in parallel arrays near the transcriptional matrices (180, 181). With further dispersion, assisted by centrifugation of the chromatin through the weak borate buffer containing 5 mM 2-mercaptoethanol and onto the surface of an electron microscope grid, the DNP is fairly well distributed, so that measurement can be made of that fraction of the extended DNP axis which bears lateral transcript fibrils. In *Triturus* the percentage of active chromatin measured in this way in small (75). Preliminary estimates would put it at less than 10% and perhaps as low as 2% (181). However, more detailed quantitation is required, and difficulties remain in preparing uniformly dispersed chromatin and in obtaining material spread at low enough density to prevent overlaying of fibers. Because there are no internal markers in spread chromosome preparations, the only definite statement that can be made is that at any one time less than 10% of the genome of *Triturus* is in the process of being transcribed. It is possible that in different chromosome preparations alternative regions of the DNP are active in transcription, although the regularity in the shape of the transcriptional matrices would imply that there exist fixed initiation and termination sites. Although granulations are frequently seen along the length of transcriptionally inactive DNP, it has not been conclusively demonstrated that these represent histone-containing particles analogous to nucleosomes. Nevertheless, the bead dimension of spread inactive chromatin of *Drosophila* embryos is estimated to be 8 nm (182), a size compatible with that of the nucleosome structure (183).

Size of Transcripts In spread lampbrush chromosomes of *Notophthalmus*, RNP fibrils measuring well over 10 μm length have been observed at intermediate positions along the loop axes (177–179), and it is presumed that fibrils considerably longer than these, perhaps up to 100 μm or more in length, would be found at the thick insertion end of some loops. The problem is that the longer the transcripts, the less they tend to extend properly and the more they tend to intertwine, making length measurements difficult. In well spread *Pleurodeles* preparations the RNP fibrils extended to lengths of 3–30 μm at the terminal point of transcription (180), and similar measurements have been recorded with the use of *Triturus* lampbrush chromosomes (75, 181). It must be emphasized, however, that continuous, covalently bonded RNA molecules may not extend throughout these very long lengths. The integrity of the RNP fibril could well be maintained by virtue of the large amounts of associated protein should the RNA be nicked at points along the transcript length. In experiments involving the sizing of isolated primary transcript RNA, it was impossible to decide whether the RNA had been reduced in length by processing events or whether the RNA was broken during the isolation procedure (75). As already mentioned, the length of the terminal RNP fibers is, in many instances, less than the length of the DNP axis of the transcriptional matrix. This is probably due to incomplete extension of the RNP which is naturally packaged into periodic 20-nm beads, thereby causing considerable foreshortening of the fibril (105,

111). However, other more complex structures, such as thickenings, rings, and branching configurations, are commonly found, particularly at the termini of the transcript fibers (103, 154, 180). These could also cause some foreshortening of the lateral fibers, but the main interest lies in the observations that they may well be a result of RNA secondary structures of types commonly found in isolated primary transcript RNP (103) (see under "Relative Sizes of Primary Transcript RNA and Messenger RNA"). Nevertheless, the possible function of such structures, if they are not merely artifacts of preparation, remains to be explained.

Shape of Matrices Transcriptional matrices were originally interpreted as consisting of discrete and uninterrupted linear gradients of increasing RNP fibril length, originating from an initiation point of the DNP axis and extending to a termination point at a fixed distance along the axis (63) (Figure 6*A*). Between these points the polymerase molecules are seen to be closely packed and possibly transcribing at maximal rate, and in this respect they resemble fully active rRNA precursor genes (179, 184). This is the situation that occasionally exists in most preparations and is what might be termed the ideal matrix pattern. Such a matrix is equivalent to one dispersed lampbrush lateral loop. However, in most lampbrush chromosome spreads, several variations are present which indicate that the generation of the mature transcript is often a more complex process (Figure 6).

The most common type of variation is a relative foreshortening of RNP fibrils toward the terminal end of the matrix (Figure 6*B*). This could be accounted for by a) packaging of extra length into the terminal thickenings, although this structure does not appear to grow proportionately bigger and is probably a simple loop-back structure with associated proteins (103); b) random scission at the 5′ end on the transcript, which is unlikely because the transcripts normally remain fairly equal in length; c) processing due to discarding of sequences at the 5′ end, although the constancy in length suggests that cleavage occurs at a fixed distance from the DNP axis rather than by recognition of a specific sequence; and d) nonlengthening of the transcript due to slipping of the RNA polymerase and nontranscription of sequences subsequent to a specified point of the DNP axis. In this way mature transcripts could be held on the chromosomal loops. In some regions of transcription the polymerase density is markedly lower and the RNP fibrils are more widely spaced (Figure 6*C*). This is a fairly common feature of *Triturus* chromosome preparations (181) and may be due to a reduction in transcription activity or to loss of transcripts by mechanical forces exerted during preparation. This phenomenon has been discussed with respect to ribosomal genes (184).

Relative Organization of Matrices Normally, adjacent matrices, which would be represented by individual loops in the lampbrush chromosome state, are fairly well spaced and connected by long lengths of DNP which lack transcripts and presumably derive from condensed chromomeric chromatin (75). Occasionally, however, adjacent matrices are found in close proximity (Figure

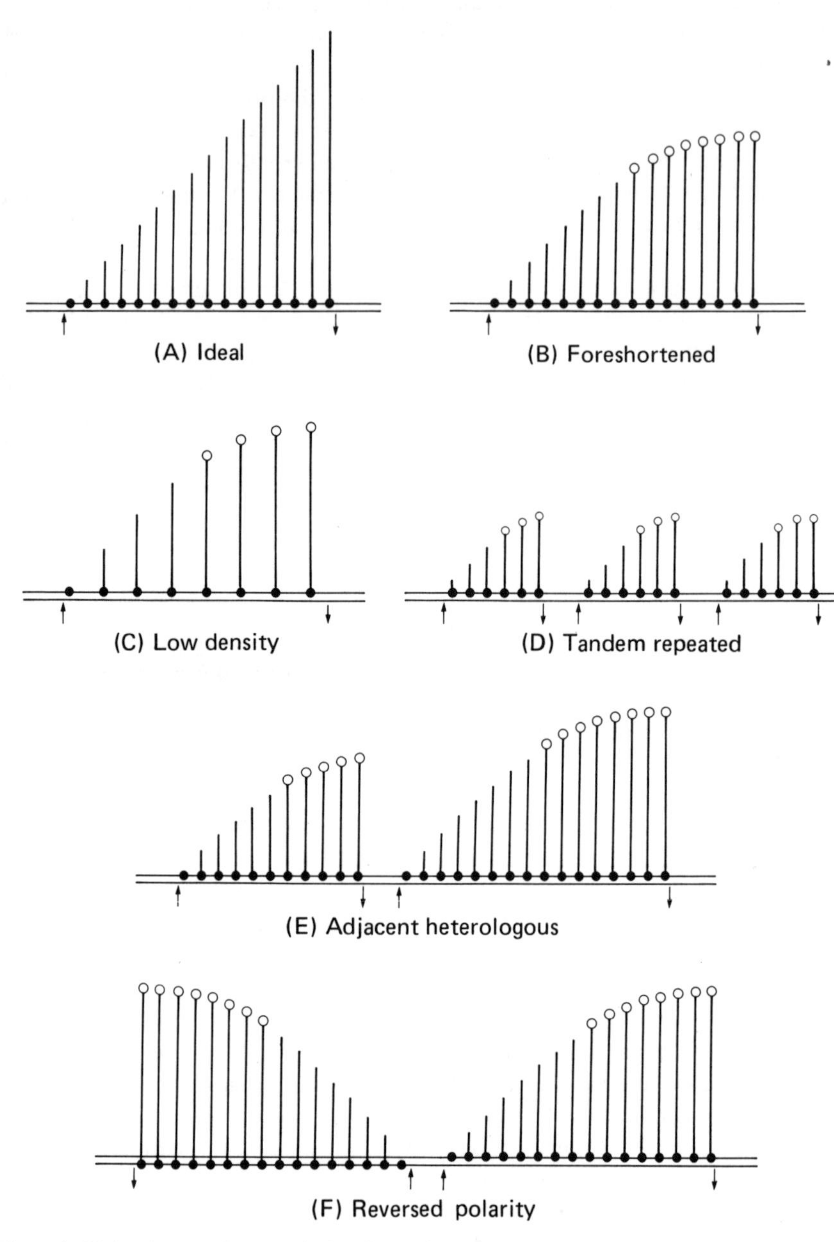

Figure 6. Main shapes of transcriptional matrices observed in spread lampbrush chromosome preparations. Double horizontal lines, DNP; single vertical lines, RNP; •, RNA polymerase; ○, terminal thickening of RNP fiber; ↑, initiation site; ↓, termination site.

6*D*). In both *Pleurodeles* (180) and *Triturus* (181), closely adjacent matrices of different dimensions occur quite frequently (see Figure 5). These matrices of diverse type must be derived from the same lampbrush loop, which strongly suggests that at least some loops are multicistronic. Whether or not the transcriptional products of these matrix complexes have associated functions remains a matter for speculation, although it is probable that their transcription is coordinated. This type of organization is fairly common to the lampbrush loops of various urodeles, where it is seen in long loops with a multiple thin-thick arrangement of the associated RNP (102).

A more familiar situation is that in which adjacent matrices are of similar dimensions, normally with a relatively short nontranscribed spacer region. This is the matrix organization characteristic of the tandem repeats of ribosomal genes of *Notophthalmus* (176, 179), *Triturus cristatus* (75, 181), and *Triturus alpestris* (184). However, it is not known if this type of organization is similar in the tandemly repeated sequences of lampbrush chromosomes, such as 5 S RNA and tRNA genes.

Finally, closely adjacent matrices may have different polarities of transcription. Such an arrangement has been described in *Pleurodeles* (180) and has far-reaching implications: either there is reversed DNA polarity (185) at a point in the loop axis or transcription occurs on both DNA strands in a manner similar to the symmetrical transcription observed in animal viruses (186). However, bidirectional transcriptional units are not seen to overlap nor is oocyte mRNA self-complementary to any substantial extent (187).

In addition to the unusual patterns of transcription described here, similar features have been described in other organisms such as *Acetabularia* (188) and *Drosophila* spermatocytes (189).

TEMPORAL SEQUENCE OF RNA SYNTHESIS DURING OOGENESIS

The mature oocyte of *Xenopus* contains about 4 μg of RNA (3, 7, 11), which is equivalent to the RNA content of some 3×10^5 somatic cells. In order to gain some understanding of the reason for such massive accumulation of RNA, how the synthetic machinery of the oocyte copes with such production, and in what way RNA transcription is controlled during oogenesis, we must first consider the identity of the RNA, the relative proportions of different types, and the relationship between these proportions during oocyte development. Four major categories of RNA can be discriminated according to their function and site of synthesis: 18 S and 28 S ribosomal RNA; 5 S ribosomal RNA; transfer RNA; and RNA transcribed from lampbrush loops, which can be considered in a broad sense as comprising informational RNA. At the completion of oogenesis, the mature oocyte contains 95–96% of its RNA in ribosomes, about 2% as transfer RNA, and the remaining 2–3% as informational RNA (6, 7, 190, 191), much of this in the form of stored messenger RNA (11).

Activity of Nucleolar Organizers in Ribosomal DNA Amplification and Nucleolar Synthesis of Ribosomal RNA

Although outside the scope of this chapter, rDNA amplification is an early event in oogenesis which has a profound effect on the deployment of transcriptional activity and, therefore, the main points should be mentioned. In spite of the observation that amplification occurs during pachytene and involves, in *Xenopus,* the production of about 500 extra copies of each nucleolar organizer (30), this massive amplification does not result in a coincident and corresponding increase in rRNA synthesis. Only later, during diplotene, and specifically about the time of peak lampbrush activity, does the synthesis of 18 S and 28 S rRNA reach its maximal rate (4, 7, 11, 98, 184, 192, 193) (Figure 7). The maximal rate of rRNA transcription is generally considered to be maintained from the onset of vitellogenesis until the oocyte approaches maturity. Originally, biochemical data suggested that there was no detectable rRNA synthesis in small previtellogenic oocytes (6, 7, 190), although later studies with *Xenopus* (191) and *Triturus* (184) have revealed small amounts of labeled 18 S and 28 S rRNA. However, there is always the problem of follicle cell contamination, particularly when manipulating small oocytes and attempting to assay low levels of radioisotope incorporation. Similarly, it was originally found that rRNA synthesis decreased substantially as oocytes completed their growth (2, 3, 192, 193).

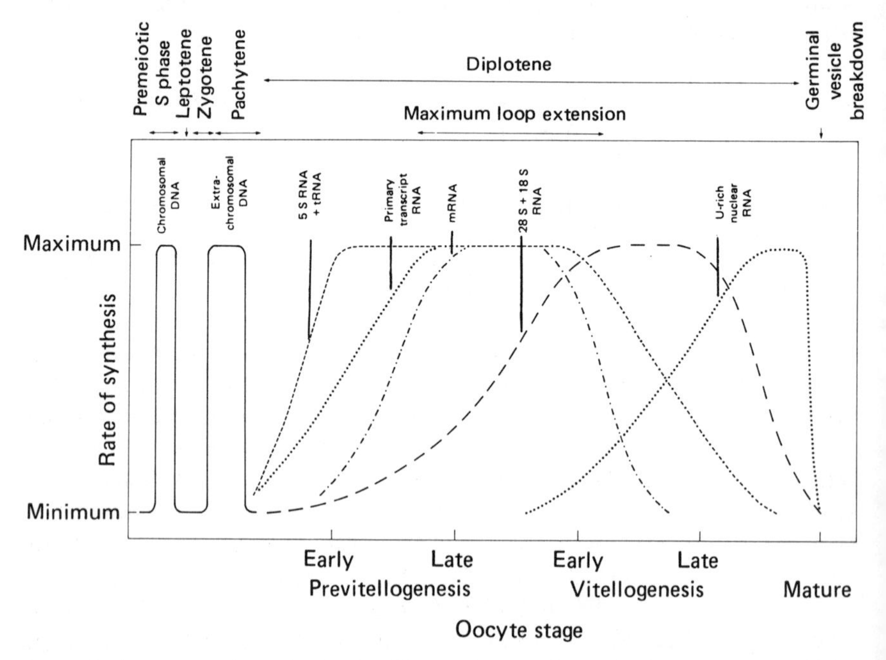

Figure 7. Sequence of synthesis of different forms of nucleic acid during amphibian oogenesis (7, 11, 91, 92, 104).

Indeed, it has been reported that full grown *Xenopus* oocytes contain an inhibitor of rRNA synthesis (194). However, with due consideration to the relationship between endogenous precursor pool levels and the rate of uptake and incorporation of labeled precursors, it has been shown that substantial rRNA synthesis occurs in full grown oocytes of *Rana* (195), *Xenopus* (38), and *Triturus* (184). In fact, it has been suggested that, even up to the time of dissolution of the nuclear membrane at maturation, rRNA synthesis remains maximal (38). The anomaly between apparent continued synthesis with no net increase in rRNA amount in full grown oocytes is explained by the maintenance of a steady state level by the continual replacement of RNA, which is itself basically unstable (39, 196). The half-life of rRNA is quoted as between 9 and 40 days, depending upon the experimental animal examined (39). Autoradiographic studies have to some extent confirmed the biochemical data. It has been shown that nucleoli incorporate labeled RNA precursors in small previtellogenic oocytes (42, 197) and also in full grown oocytes long after the lampbrush chromosomes have contracted and the nucleoli have migrated away from the nuclear membrane prior to maturation (42, 91). However, these experiments do not confirm that it is in fact the 40 S rRNA precursor that is being synthesized, particularly at the late stages of oogenesis.

In view of the difficulty in making a comprehensive statement regarding rRNA synthesis during oogenesis, it may be more instructive to consider one instance in which electron microscopic, autoradiographic, and biochemical data have been correlated in the one experimental system. In *Triturus* the estimated rate of rRNA synthesis, based on these various techniques, is 0.01% in previtellogenic oocyte and 13% in mature oocytes, compared with the 100% rate assigned to midvitellogenic oocytes (184). Furthermore, the regulation in activity of rRNA genes in spread nucleoli is seen to be a function of the frequency of initiation events by RNA polymerase molecules in each transcriptional unit. In other words, activation and inactivation of rRNA genes are not all-or-nothing phenomena; rather, each individual gene can vary independently in its extent of transcription (184).

Kinetic considerations of rDNA activity and rates of ribosome production are considered under "RNA Polymerase Activity." However, it should first be emphasized that amplification of ribosomal genes is of high selective advantage for two fundamental reasons: 1) oocytes could not produce the required number of ribosomes within the normal oogenic period without amplification of rRNA genes; and 2) as an alternative to amplification, increased reiteration of rRNA genes to meet the demand would result in 50% of the genome of *Xenopus* being devoted to the single function of rRNA synthesis.

Synthesis of 5 S RNA and Transfer RNA

Because of similarities in their mode of formation in oocytes, 5 S RNA and tRNA are here considered together. Synthesis of 5 S RNA and tRNA differs from the synthesis of 18 S and 28 S rRNA in several important respects.

1) The genes from which 5 S RNA and tRNA are transcribed are not amplified in any extrachromosomal form (90, 198, 199), but are maintained, in both instances, as clustered repeated sequences of fixed number and at specified sites on the chromosomes of any one animal.
2) Synthesis of 5 S RNA and tRNA starts and reaches a maximal rate in early diplotene (6, 7) (Figure 7). As a result, in early oogenesis, during the previtellogenic stage, 5 S RNA and tRNA constitute the predominant RNA species. In previtellogenic *Xenopus* oocytes, 5 S RNA sequences and tRNA sequences accumulate to such an extent as to constitute 75% of the RNA content (Table 3). Their pattern of synthesis, as would be expected from the location of their genes, largely follows the course of lampbrush chromosome activity.
3) Oocyte 5 S RNA is more stable than oocyte 18 S and 28 S RNA and is also more stable than somatic cell 5 S RNA (31, 200, 201). The same is probably true of oocyte tRNA.

The possible reasons for increased stability of 5 S RNA have been outlined by Denis (31) and are as follows:

a) The concentration of nucleases is probably lower in oocytes than in somatic cells.
b) Protection from nuclease digestion is due to the storage of the majority of 5 S sequences in small oocytes in nucleoprotein particles which sediment at 42 S (6, 201, 202) (see under "42 S Particles"). As it is required, 5 S RNA is progressively released from the 42 S RNP particle and incorporated into ribosomes which start to accumulate in large numbers from about the onset of vitellogenesis (200).
c) Oocyte 5 S RNA differs from somatic cell 5 S RNA in both primary and secondary structure (140, 141, 202, 203). Apart from implying that a different range of 5 S genes are active in oocytes and somatic cells (140, 141), which is an extremely important observation from the point of view of the genetic organization and evolution of different 5 S sequences, and the way in which these sequences may be differentially regulated (see under "Noncoordinate Synthesis and Control Mechanisms"), structural modifications might influence RNA stability. Before being eventually incorporated into ribosomes, 5 S RNA sequences synthesized in previtellogenic oocytes undergo slight modifications at both the 5′ and 3′ ends of the molecules. There is a slow turnover of the 3′-terminal nucleotides which are extra sequences peculiar to newly synthesized oocyte 5 S RNA (204), and there is a loss of one or two phosphate groups from the triphosphorylated 5′-terminal nucleotide (202).

The increased stability of oocyte 5 S RNA and its extensive period of synthesis are probably necessary requirements in view of the fact that the *Xenopus* oocyte contains about 5×10^5 18 S and 28 S rRNA genes and only about 10^5 5 S rRNA genes. At least as many 5 S RNA molecules as 18 S and 28 S rRNA molecules accumulate during oogenesis (6, 7) to form about 1.1×10^{12}

Table 3. RNA content of *Xenopus* oocytes

Stage	RNA per oocyte[a] (μg)	Total noninformation RNA[a] 28 S + 18 S RNA (%)	5 S RNA (%)	tRNA (%)	Total RNA Informational RNA[b] (%)	Poly(A)$^+$ mRNA[a] (%)
Early previtellogenic	0.04	8	45	47	<25	
Late previtellogenic	0.07	49	31	20	12.5	
Early vitellogenic	1.2	80	11	9		~4
Midvitellogenic	3.5	92	5	3		~1
Mature	4.3	94	3	2.5	2–3	0.7–1

[a]Data from ref. 11.
[b]Data from refs. 7, 10, 191.

mature ribosomes (205). Therefore, conditions approaching maximal activity are required in each situation. As will be discussed later (see under "Translational Activity"), only a small percentage of the ribosomes is actually involved in protein synthesis during oogenesis.

Transfer RNA is synthesized and accumulated in oocytes in much the same way as is 5 S RNA (6, 7, 201). Again, most of the tRNA, 90–95% of the molecules present in small oocytes, is incorporated into 42 S RNP particles. If the 42 S particles are of one type, rather than separate but cosedimenting storage compartments for 5 S RNA and tRNA, then the proportion of 5 S RNA to tRNA in the particle is in the molecular ratio of 3:1. Studies on tRNA stability suggest that at least some of the tRNA molecules synthesized in small oocytes persist up to the time of oocyte maturation (200). Although tRNA from *Xenopus* oocytes and from somatic cells have the same average length and contain an equal variety of modified nucleotides, tRNA from small oocytes is structurally different from somatic cell tRNA as measured by reversed phase chromatography (206). As oogenesis proceeds, the oocyte tRNA becomes progressively more like somatic tRNA either by post-transcriptional modification or by changes in the tRNA population. Either oocyte tRNA exists as immature precursor forms which are progressively modified in the course of oogenesis or small oocytes contain a greater number of different isoacceptors than do somatic cells, some of which are lost or diluted during oocyte growth. This latter explanation would require either the repression of certain tRNA genes at a definite stage in oogenesis or the differential degradation of some tRNA species during the later stages of oogenesis (206).

Synthesis of Informational RNA

The main function of lampbrush chromosome activity is generally considered to be the transcription of chromosomal sequences into RNA which contains informational sequences. These sequences are destined to become mRNA molecules, some of which may be incorporated into oocyte polysomes, but the majority of which are stored in a relatively stable form to be used during early embryogenesis. The peak period of synthesis of informational RNA would be expected to be during the phase of maximal extension of the lateral loops which itself coincides with early vitellogenesis (102, 207). Therefore, the oocyte stage, especially with respect to the extent of yolk accumulation, is a useful, albeit approximate, indicator of potential transcriptional activity.

The methodology for the detection of informational sequences transcribed in oocytes has developed with new techniques for selecting out loop-transcript sequences which are generally but a small percentage of the total RNA (4, 7) (Table 3). Early experiments with the use of RNA-DNA hybridization were taken to subsaturation levels but nevertheless demonstrated that the informational RNA transcribed from the repetitive fraction of the genome of *Xenopus* is of high sequence complexity and is synthesized exclusively during the lampbrush stage of oogenesis (8). Furthermore, these same sequences are also present in the embryo after fertilization (8) and may persist until late blastulation (9, 208). It should be remembered that, although the sequences found in oocytes and early embryos are the same, or very similar, it does not follow that they are the same molecules. With longer hybridization times, it was later demonstrated that informational RNA transcribed from the nonrepetitive fraction of the *Xenopus* genome is also synthesized during the lampbrush stage (64) and retains its high sequence complexity till oocyte maturation (10). Recently, techniques have been developed for the isolation of nuclear informational RNA (primary transcript RNA) from amphibian oocytes (75, 104) and polyadenylated messenger RNA (poly(A)$^+$ mRNA) (11, 209).

The formation of potentially coding sequences is a process which can be considered to have several stages. However, it must be remembered that it has not as yet been demonstrated that a simple precursor-product relationship exists between one stage and the next.

1) Primary transcript RNA on the lampbrush loops. The synthesis of these molecules is by definition linked with maximal loop extension. Presumably there is progressive release of transcripts at the termini of the loops, although it is possible that transcripts may be held on the loops after completion of transcription. Indeed, accumulation of RNP is a morphological characteristic of some lampbrush loops (102, 110).
2) Primary transcript RNA released from the chromosomes. If the transcripts are not extensively processed during, or very soon after, transcription, very large molecules of up to 20 μm or more should be detectable in the nucleoplasm free from chromosomes. Such a category of molecules, which has many features in common with chromosome-attached transcripts, has been isolated from *Triturus* oocytes (104). Nuclear RNA of this type may be retained in the nucleoplasm after the condensation of lampbrush chromosomes and even until the dissolution of the nuclear membrane (210).
3) Nuclear RNA as an immediate or early product in the selection of molecules containing the required coding sequences. These would be analogous to the heterogeneous nuclear RNA (hnRNA) molecules of somatic cells; that is, they would tend to be longer and more complex molecules than cytoplasmic mRNA, yet smaller than primary transcript RNA. It would be expected that some of the nuclear RNA molecules are polyadenylated at the 3′ end by this stage.
4) Residual nuclear RNA of unusual base composition. It has been reported that toward late vitellogenesis in *Triturus* oocytes there is an accumulation in the nucleus of RNA with an unusually high (60%) uridylic acid content (104). Similar observations have been made on nuclear RNA of mature *Rana* oocytes which contain 43% uridylic acid (211), on an RNA fraction of vitellogenic *Xenopus* oocytes which contain 41.9% uridylic acid plus an unidentified nucleotide (4), and on nuclear RNA of vitellogenic *Notophthalmus* oocytes which contain variable amounts of between 24.4% and 49.5% uridylic acid (94). The

functional significance of RNA of such peculiar base composition is not understood. Most of the evidence suggests that it is formed during vitellogenesis and accumulates up to the time of oocyte maturation (Figure 7).
5) Cytoplasmic messenger RNA. Four types can be considered: stored poly(A)$^+$ mRNA; polysomal poly(A)$^+$ mRNA; stored poly(A)$^-$; mRNA and polysomal poly(A)$^-$ mRNA. Whether there is any structural difference between the stored and polysome-associated forms is discussed under "Translational Activity." The formation of poly(A)$^+$ mRNA in *Xenopus* oocytes reaches a peak in early vitellogenesis (Figure 7) with little further accumulation throughout the remainder of the lampbrush stage and up to oocyte maturation (11). Although there may be continued synthesis and degradation of this mRNA, the turnover rate is not greater than that for rRNA, at least over a period of 3 days (11). Nevertheless, the possibility of considerable turnover during the extensive lampbrush stage cannot be excluded. Polyadenylated mRNA constitutes 0.7–1.0% of the total RNA of vitellogenic oocytes, but less than 10% of the polyadenylated sequences cosediment with polysomes and are EDTA-sensitive (11). Therefore, polysomal mRNA would appear to be a minor component of the informational RNA contained in oocytes.

Therefore, the synthesis of informational RNA occurs exclusively during lampbrush chromosome activity, although products of various types are stored until fertilization and early embryogenesis. One exceptional set of results is the demonstration that *Xenopus* oocytes synthesize "DNA-like RNA" during hormone-induced ovulation (2, 3, 212). However, the amount of this late-synthesized RNA is small compared with that accumulated during the lampbrush stage (although its rate of synthesis is high), and it is not certain that it is genuinely derived from the oocyte nucleus. In fact, it is probable that in these experiments much of the uptake of labeled precursor occurred after the breakdown of the nuclear membrane (33). Synthesis of RNA on the condensed metaphase chromosomes present at this stage is highly unlikely. Because mitochondrial DNA has a base composition similar to that of the chromosomes, it is most likely that the late synthesis of RNA is attributable to mitochondrial activity (16).

Kinetic Considerations

Noncoordinate Synthesis and Control Mechanisms As stated above, the timing of RNA synthesis in oocytes is dependent upon the transcriptional efficiency of different templates which become available at distinct periods of the oogenic process. Ribosomal 18 S and 28 S sequences are necessarily amplified, and their transcription is directly related to extrachromosomal nucleolar activity. However, for some reason this activity is repressed for the first half of oogenesis. The synthesis of 5 S RNA and tRNA is linked with early lampbrush chromosome activity and continues for the duration of the lampbrush stage. Thus, the site of synthesis and timing of synthetic activity of different ribosomal components are not coordinated (6, 7). Mechanisms which might regulate the eventual production of equimolecular amounts of 18 S and 28 S rRNA and 5 S rRNA have been discussed by Ford (32). Transcription of informational sequences is linked with lampbrush chromosome activity and again is independent of the synthesis of other types of RNA, particularly with respect to the type of RNA polymerase employed (see under "RNA Polymerase Activity"). The major features of nucleic acid synthesis in oocytes are diagrammatically represented in Figure 7.

Whereas extrachromosomal amplification of 18 S and 28 S sequences does not occur in somatic cells, differential control of chromosomal sequences is required in the expression of 5 S genes. Although somatic cells of *Xenopus* transcribe 5 S RNA sequences which are largely conserved, oocytes transcribe 5 S RNA sequences which show more heterogeneity in nucleotide composition but nevertheless constitute only several classes of molecule (140, 141, 171). Although at least 97% of the 24,000 5 S genes are transcribed in oocytes, only a small percentage is transcribed in somatic cells (139). Therefore, the transcription of most 5 S sequences is blocked in somatic cells, and they can be referred to as oocyte-specific. One exception occurs in a transformed kidney cell line in which oocyte-type sequences have apparently been translocated to a site adjacent to the nucleolar organizer (137). As a result, these tissue culture cells synthesize 10–20% of their 5 S RNA as sequences normally exclusive to oocytes (213). It has been suggested that oocyte-specific sequences which are active in lampbrush chromosomes are repressed by being secreted into regions of heterochromatin in somatic cells (140). In the tissue culture line cited above, a likely explanation is that, in being translocated to a site normally transcriptionally active in somatic cells, a gene cluster of oocyte-type 5 S DNA evades repression due to a position effect, rather than to an inherent property of the DNA sequences which might be susceptible to specific diffusible control factors or to a lack of polymerases or other components necessary for transcription. This argument runs into difficulty when one considers the location of 5 S genes in *Triturus*. In this genus, not only is there but one gene cluster, but also its location is at an intermediate position on the long arm of chromosome X (166), a region not normally associated with heterochromatin formation. The problem is not so great in *Notophthalmus* if, as appears from the results of Hutchison and Pardue (124), only one of the five 5 S gene clusters is located at a site away from the centromere. Otherwise, all other 5 S gene clusters detected to date have multiple sites and are located in regions, either centromeric or telomeric (78), which are proximal to regions of heterochromatin formation. However, it would be premature to suggest a general mechanism for the selective repression of 5 S genes.

In contrast to somatic cells, oocytes appear not to repress the transcription of chromosomal genes. There is no evidence to suggest that there are any functional sequences that are permanently situated in regions of condensed chromatin during lampbrush chromosome activity.

RNA Polymerase Activity The identification of nuclear RNA polymerases and their absolute and relative concentrations in oocytes are of fundamental importance in any consideration of transcriptional activity and its control. In eukaryotes there are three basic types of enzyme which can be separated by DEAE-Sephadex chromatography and whose activity can be discriminated on the basis of Mg^{2+} and Mn^{2+} requirements, ionic strength optima, and α-amanitin sensitivity (reviewed ref. 214). The function of the three types can be related to their transcriptive specificities. For instance, polymerase I is located in the nucleolus and presumably transcribes the 40 S rRNA precursor; polymerase II is found in the nucleoplasm and is probably responsible for the transcription of chromosomal sequences to form hnRNA and mRNA; polymerase III is also nucleoplasmic and in certain instances has been shown to be responsible for the synthesis of 5 S RNA and tRNA sequences (215, 216), although a role in the transcription of rDNA has not been excluded.

Xenopus oocytes have been shown to contain extremely high levels of polymerases I, II, and III (217). During oogenesis the relative proportions of the enzymes remain nearly constant although the absolute levels of all forms increase dramatically so that mature oocytes contain 10^4–10^5 times more RNA polymerase activity than do somatic cells (14) (Table 4). These high levels make it unlikely that transcription during oogenesis is limited to any significant extent by a deficiency of polymerase molecules. This interpretation derives support from electron micrographs of nucleolar and lampbrush chromosome spreads prepared from oocytes of *Notophthalmus* (154, 176) and *Triturus* (184) which show tight packing of polymerases, approaching saturation levels, on the DNP axes of regions of transcriptionally active chromatin. Therefore, extremely high levels of RNA polymerase may reflect the high rates of transcription in oocytes. Even when the level of polymerase activity is normalized to account for the large number of genes activated in oocytes (90), there still appears to be an excess of polymerase molecules (14). It is conceivable that high RNA polymerase concentrations are necessary for effective transcription in the large nuclear volumes peculiar to amphibian oocytes. However, it is more likely that polymerases are

Table 4. RNA polymerase activity in *Xenopus* oocytes[a]

	RNA polymerase activity					
	I		II		III	
	(units/10 cells)	%	(units/10^4 cells	%	(units/10^4 cells)	%
Small oocytes	22	31	27	37	24	32
Medium oocytes		37		28		35
Mature oocytes	11,200	33	12,400	36	10,600	31
Cultured kidney cells	0.18	49	0.17	46	0.02	5

[a]Data from ref. 14.

synthesized and stored during oogenesis for later use during early embryogenesis. It has been calculated that the mature oocyte contains levels of RNA polymerases sufficient to provide the egg with enough enzyme to meet its requirements in RNA synthesis up to the gastrula stage (14, 218, 219).

Although there are presumably sufficient quantities of both polymerase I and II to meet synthetic requirements at all stages of oogenesis, there appears to be a disproportionately high level of polymerase III (Table 4). Relative to total polymerase activity, it has been estimated that about 35% (14) or even as much as 70% (220) of the activity is accounted for by polymerase III. It has been suggested that the high level of polymerase III in *Xenopus* oocytes is due to selective accumulation and storage (217); otherwise it may be responsible for the transcription of amplified rDNA rather than of low molecular weight species (221). This latter suggestion is not unreasonable, for in *Xenopus* oocytes rDNA represents more than 50% of the total nuclear DNA content and its transcription accounts for more than 95% of the RNA produced during oogenesis (see under "Temporal Sequence of RNA Synthesis During Oogenesis"). Indeed, there is no a priori reason to assume that any of the RNA polymerases of oocytes have the same template specificities that they appear to have in somatic cells. It can be concluded that gene activity in oocytes is not simply related to or controlled by the cellular level of specific RNA polymerases.

From a knowledge of polymerase spacing, rates of transcription, and the number of genetic units, approximate calculations can be made about the minimal times required to transcribe the numbers of different species of RNA found in mature oocytes. These times should be compatible with the time scale of synthetic activities that occur during oogenesis. The minimal spacing observed between polymerase molecules in transcriptional matrices of *Notophthalmus* is 25 nm ≡ 80 base pairs (154); the maximal rate of transcription of RNA in eukaryote cells at 25°C is less than 2×10^3 nucleotides/min/polymerase (222); and the maximal rate of transcript formation (polymerase release) is $2 \times 10^3/80 = 25$ transcripts/min. The general formula for the minimal time required to form a population of RNA molecules during oogenesis is as follows:

$$\text{Time}_{\text{min}} = \frac{\text{No. transcripts accumulated}}{\text{maximal rate of release} \times \text{No. genes}}$$

For ribosome formation in *Xenopus*, the number of ribosomes in the mature oocyte is 1.1×10^{12} (205) and the number of ribosomal genes is 5×10^5 (77); therefore,

$$\text{Time}_{\text{min}} = 1.1 \times 10^{12}/25 \times (5 \times 10^5) = 8.8 \times 10^4 \text{ min} \sim 60 \text{ days}$$

This value is reasonable because the minimal time for RNA synthesis during hormone-induced oogenic growth in *Xenopus* is 38 days (193).

For 5 S RNA formation in *Xenopus*, the number of 5 S RNA molecules in mature oocytes is approximately 10^{12} (7) and the number of 5 S genes is approximately 10^5 (136); therefore,

$$\text{Time}_{\text{min}} = 10^{12}/25 \times 10^5 = 4 \times 10^5 \text{ min} \sim 300 \text{ days}$$

For tRNA formation in *Xenopus*, the number of tRNA molecules in oocytes is approximately 2.4×10^{12} (7) and the number of tRNA genes is approximately 3.2×10^4 (147); therefore,

$$\text{Time}_{\text{min}} = 2.4 \times 10^{12}/25 \times (3.2 \times 10^4) = 3 \times 10^6 \text{ min} \sim 2 \times 10^3 \text{ days}$$

This value is not compatible with the time scale of oogenesis. There obviously must be some mechanism to accelerate the rate of transcription of tRNA sequences, and perhaps to a lesser extent 5 S sequences, in oocytes.

For mRNA formation in *Xenopus*, the number of less abundant mRNA molecules in mature oocytes is approximately 10^6 (for each of ~1.5×10^4 mRNA species (209). Assuming transcription from single copy sequence,

$$\text{Time}_{\text{min}} = 10^6/25 \times 4 = 10^4 \text{ min} \sim 7 \text{ days}$$

The number of higher abundant class of mRNA molecules in mature oocytes is approximately 3×10^7 (for each of ~500 mRNA species (209)). Assuming transcription from single copy sequence,

$$\text{Time}_{\text{min}} = 3 \times 10^7/25 \times 4 = 3 \times 10^5 \text{ min} \sim 200 \text{ days}$$

This latter value falls within the normal time scale of oogenesis (~1 year). Nevertheless, some of the high abundance mRNA species may be transcribed from repetitive genetic sequences.

Effect of Inhibitors It would seem reasonable to state that lampbrush loop formation and morphology are solely a consequence of the extent of transcription at discrete sites in the genome. As such, an individual loop would be generated by the binding of polymerase molecules, presumably polymerase II in the vast majority of instances, at some appropriate initiation site. The DNP forming the loop axis would be maintained in a decondensed state by the continual activity of closely packed polymerase molecules.

When intact oocytes of urodeles are treated with inhibitors of DNA-dependent RNA synthesis, namely actinomycin D (104, 153, 155, 223) and α-amanitin (224), incorporation of [^{3}H]uridine is blocked, the loop RNP is released, the loops shorten, and eventually the DNP axes are completely retracted into the chromomeric DNP. A similar effect is obtained on treating oocytes with 3-deoxyadenosine (cordycepin) (225). On the other hand, treatment with inhibitors of protein synthesis, puromycin and cyclohexamide, although blocking incorporation of [^{3}H]phenylalanine into the loop RNP, does not affect the morphology of the lateral loops (155, 225). It is questionable whether many of the observed effects on loop structure are due to specific and primary action of the inhibitors rather than to a more general disturbance in oocyte metabolism.

The binding of actinomycin D and α-amanitin directly to the loop axes has not been demonstrated. The only evidence that RNA synthesis per se is required for the maintenance of loop structure is that *N*-β-aminoethyl actinomycin C_3, which has binding properties similar to actinomycin D but does not effectively inhibit RNA synthesis, has no effect itself on loop structure and may even protect the loops from the action of actinomycin D (155). Actinomycin D, at low concentration, binds specifically to deoxyguanosine (226) and subsequently prevents further transcription from the DNA template. Likewise, α-amanitin binds specifically to RNA polymerase II (see 214) and inactivates the enzyme. Lampbrush loop retraction, if brought about by the direct action of these inhibitors of RNA synthesis, becomes explicable only if the DNP active in transcription is first freed of its RNP and polymerases. It can be questioned whether polymerase progression along the template is sufficient to bring about concomitant release of polymerases and transcripts. Although it is conceivable that actinomycin D and α-amanitin have this effect, the action of cordycepin is more difficult to explain. Again, it is possible that any unstabilizing effect, such as inhibition of RNA chain elongation by a precursor analog, is sufficient to disturb the kinetic state of transcription to the extent that there is complete release of the transcription complex. An alternative explanation is that an RNA fraction derived from the normal processing of hnRNA, a post-transcriptional step particularly sensitive to the action of cordycepin, is required to react with the chromosome in order to maintain loop chromatin in a decondensed state (225). An interesting additional observation is that some loops, particularly those with a dense or fibrillar matrix, are less sensitive, and even totally refractile, to α-amanitin treatment (224). It would be interesting to know whether any of these loops are involved in transcription of 5 S RNA and tRNA and, therefore, refractile to α-amanitin due to utilization of RNA polymerase III.

It is not surprising that inhibition of protein synthesis has no effect on loop structure because even treatment periods of 10 hr may be insufficient to cause depletion of available nuclear proteins. It has also been reported that arginine-rich histone and poly(L-lysine) inhibit RNA synthesis and cause rapid loop retraction, whereas lysine-rich histone is only weakly inhibitory and does not affect loop structure (155). Because these experiments were performed on isolated chromosomes, the effects could be due to a general perturbation such as pH change. Also, interpretation is complicated by the findings that lysine-rich histones normally suppress transcription (227), and that lysine-rich histone HI is present at only low levels in *Triturus* (see Figure 8).

RIBONUCLEOPROTEIN PRODUCTS OF TRANSCRIPTION

As in other eukaryotic cells, the different species of RNA become associated with specific proteins during, or soon after, transcription. The binding of protein to oocyte transcripts may be more extensive than in other cells; for instance, oocyte primary chromosomal transcripts contain a higher proportion of protein

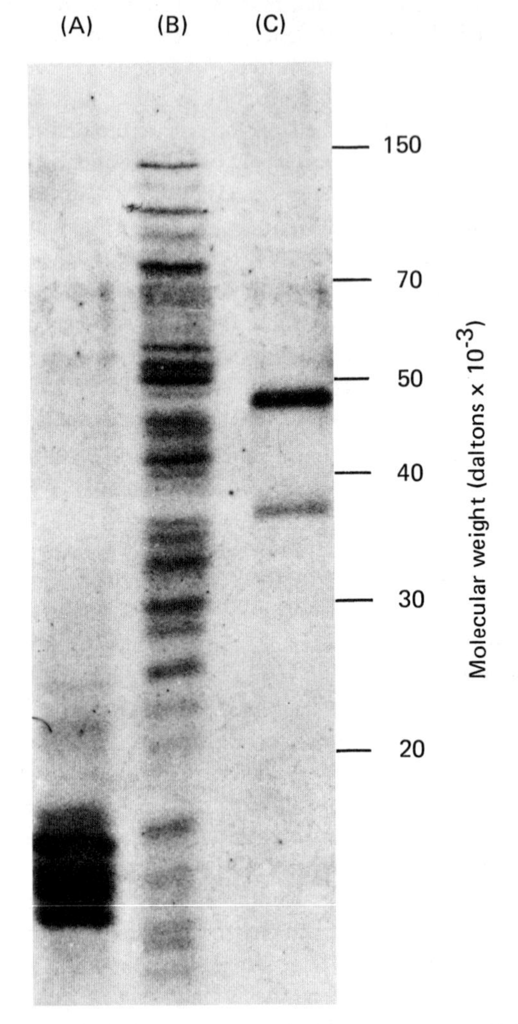

Figure 8. SDS-acrylamide gel electrophoresis of (*A*) *Triturus* histones, (*B*) *Triturus* oocyte primary transcript RNP proteins, and (*C*) the proteins found in association with *Triturus* oocyte 5 S RNA and tRNA as RNP particles which sediment at about 40 S.

to RNA than do the nuclear RNP components of somatic cells (103, 104). In addition, most of the transfer RNA molecules of oocytes are sequestered, along with 5 S rRNA, into a nucleoprotein particle which sediments at 42 S in *Xenopus* (6, 201).

Ribosome Formation

The main features of ribosome formation in amphibian oocytes are mentioned here to serve as a comparison with the formation of other ribonucleoproteins.

Briefly, the synthesis of ribosomal structural proteins is regulated during oogenesis at a rate coordinated with the transcription of 18 S and 28 S RNA (46). During midvitellogenesis, more than 30% of the protein synthesized by *Xenopus* oocytes is estimated to be ribosomal structural protein (46). Ribosomal precursor nucleoprotein is located in the nucleoli as granular elements (176, 228) and can be isolated from the oocyte nuclei of *Notophthalmus* and *Ambystoma* as RNP particles which sediment at 50–55 S and at 30 S (99). Few, if any, 78 S ribosomes are found in the nucleus. It would appear that ribosomes are formed in oocytes in much the same way as in somatic cells, although their assembly may be, at some level, under hormonal control (46).

42 S Particles

In previtellogenic oocytes of *Xenopus* about 90% of the tRNA and 50% of the 5 S rRNA are contained in RNP particles sedimenting at 42 S (6, 201). Because tRNA and 5 S RNA are always present in constant proportions within the 42 S particle (3 molecules of tRNA per molecule of 5 S RNA) (6, 201), it seems likely that there is one type of 42 S particle rather than two cosedimenting forms. Although it has been reported that there are two major density components detected by measurement of buoyant density of 42 S RNP in CsCl after formaldehyde fixation (32), it is possible that this result is due to dissociation prior to fixation. Density and chemical analyses indicate that the protein to RNA ratio is approximately 3:1 (32, 201).

Because 5 S RNA is progressively incorporated into ribosomes during oocyte growth, it can be asked whether any of the proteins of the 42 S particle are also incorporated into ribosomes. When the proteins of 42 S RNP particles of *Xenopus* are analyzed by two-dimensional chromatography on acrylamide gels, one major and two minor basic protein spots are detected (229). None of these correspond with any of the 37 spots produced by the 60 S subunit or the 30 spots produced by the 40 S subunit of *Xenopus* ribosomes. The possibility has not been excluded that the 42 S RNP proteins are precursors of smaller ribosomal proteins. The molecular weights of the 42 S RNP proteins of *Xenopus*, as estimated by sodium dodecyl sulfate (SDS)-acrylamide electrophoresis, are 73,000 and 63,000 for polypeptides derived from the major spot and 73,000 and 46,000 for the less basic minor spots (229).

These molecular weight values are somewhat different from those obtained with the RNP particles sedimenting at about 40 S which are derived from previtellogenic oocytes of *Triturus* (115). Here there are only two protein components, present roughly in the proportion of 3:1 (Figure 8). The major component has a molecular weight of 49,000, whereas the minor component has a molecular weight of 38,500. No further heterogeneity has, as yet, been detected by virtue of charge difference. The most interesting aspect of the study of these proteins is the demonstration that antibodies produced against them react with specific loops of the lampbrush chromosomes (115) (see under "Use of Immunofluorescent Probes").

Informational Primary Transcript Ribonucleoprotein

Large nuclear RNP complexes can be extracted from homogenates of *Triturus* oocytes by differential centrifugation and can be isolated subsequently as a narrow zone of rapidly sedimenting material in 30–55% sucrose gradients (104). Prior treatment of the RNP with 5 mM EDTA and 0.1–1 M NaCl has little or no effect on the gradient profile or on the composition of the material recovered from the peak, yet may serve to minimize the binding of extraneous protein and dissociate any polyribosomes which might cosediment with the nuclear RNP. For various reasons, this isolated material (Figure 9*a*) is believed to represent lampbrush chromosome-derived RNP.

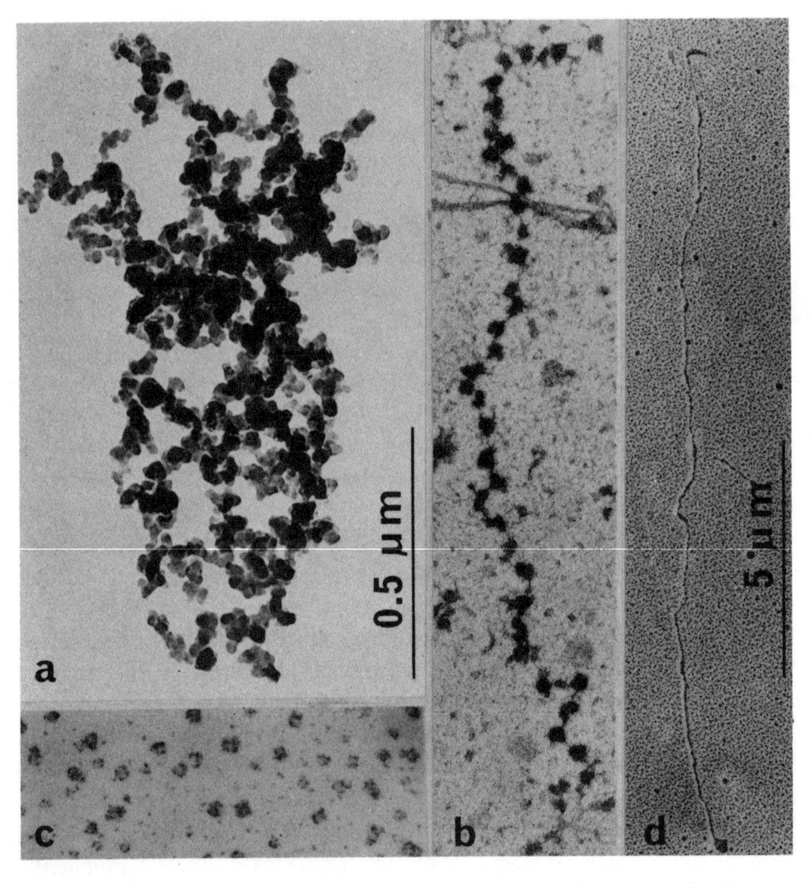

Figure 9. Electron micrographs of *Triturus* primary transcript RNP. *a*, an isolated aggregate; *b*, string of 20-nm beads after treatment with 85% formamide; *c*, individual beads after exposure to 1 μg/ml of ribonuclease; *d*, individual transcript treated with 0.1% SDS-4 M guanidine prior to spreading on 60% formamide. Partly reproduced from Malcolm and Sommerville (103, 105) with permission of The Company of Biologists Limited and Springer-Verlag.

1) It is rapidly labeled when oocytes are incubated in the presence of radioactive precursors of both RNA and protein (104).
2) The protein to RNA ratio is greater than 30:1, a ratio similar to that for chromosome-bound transcripts (104).
3) A substantial increase in the amount of labeled RNA and protein in this material results from treating the oocytes with 25 μg/ml of actinomycin D, a treatment known to release RNP transcripts from the chromosomes (104).
4) It consists of linear arrays of 20-nm particles which are similar in appearance to chromosomal RNP (105, 116) (Figure 9*b*). These linear strings of beads are reduced to monoparticles by low concentration of ribonuclease (Figure 9*c*), but not by deoxyribonuclease (103, 105).
5) The beaded RNP can be converted into long fibrils by treatment with dissociative agents and in this state resembles the RNP fibrils of spread chromosomes in structure and dimension (75, 103) (Figure 9*d*).
6) The RNA extracted from this material has a base composition which is significantly different from rRNA (104); it has a heterogeneous range of sedimentation coefficients (75, 103, 104), and it contains some nucleotide sequences which are homologous to oocyte polyadenylated mRNA sequences (75).
7) The protein derived from this material consists of a large number of different polypeptides with a considerable diversity of molecular weight (75, 114–116). The polypeptide profile on SDS-acrylamide gels covers the range of 10,000–150,000 daltons and has little or no coincidence with the polypeptide profile of any other cell fraction, including nucleolar proteins (116), ribosomal proteins, 42 S RNP proteins, and DNP histones (103) (see Figure 8), but it is similar to the polypeptide profile of proteins extracted from the nucleoplasm and the chromosomes of manually isolated oocyte nuclei (117).
8) Antiserum prepared against the RNP proteins reacts specifically with the lateral loops of lampbrush chromosomes and not with the condensed DNP of the chromomeres or with nucleoli (114, 115).

However, the isolated primary transcript RNP cannot be considered to reflect ideally the proportions of mature transcripts that exist in the oocyte nucleus. For instance, there may be premature release of transcripts from lampbrush chromosomes when the oocytes and their nuclei are ruptured, although in the preparation of chromosome spreads the transcripts appear to be tenaciously bound. Also, because an initial step in isolation is differential centrifugation of the ruptured oocytes, there may be some selection against extremely large transcripts which might sediment with the yolk platelets and against small transcripts which might be left in the supernatant. However, analysis of the supernatant reveals this to contain little material analogous to the major RNP fraction (104). Aggregation between RNP complexes is observed to occur soon after rupturing the oocyte nucleus (230). This aggregation is due to the formation of disulfide linkages and is largely prevented by the addition of chemical reducing agents. However, even in the presence of 5 mM 2-mercaptoethanol, there is a slow increase in the state of aggregation of isolated nuclear RNP (104). Although the isolation of a total nuclear transcript RNP fraction requires certain precautions, this material has proved useful in analysis of the general composition and structure of initial products of the transcription process. Nevertheless, it would be methodologically simpler to isolate for study a specific subset of nuclear transcripts, for example, on the basis of selection of polyadenylated nuclear RNP by poly(U)-Sepharose binding, as has been done with HeLa cell RNP (231).

Messenger Ribonucleoprotein

The mature oocyte of *Xenopus* contains a store of polyadenylated RNA, most of which is localized in the cytoplasm in the form of RNP particles (11). In structure and composition these cytoplasmic RNP particles are quite unlike any form of nuclear RNP which might serve as a putative precursor. The only known common feature lies in the content of similar coding sequences (75).

Oocyte poly(A)$^+$ RNA is detected in RNP particles which sediment mostly slower than ribosomes and over the range 25–80 S (11). The sedimentation of these RNP particles is largely insensitive to EDTA treatment. Less than 10% of the poly(A)$^+$ RNA sediments in a manner consistent with its association with polyribosomes. This is in marked contrast to similar experiments performed with the use of HeLa cells in which 75–80% of the poly(A)-containing material is found to cosediment with polyribosomes and membrane-bound ribosomes. However, after EDTA treatment, the HeLa mRNP sediments over a range similar to that of the mRNP extracted from oocytes. Furthermore, the poly(A)$^+$ RNA purified from oocyte RNP particles sediments over the same range as does HeLa cell mRNA, that is, mostly between 10 S and 40 S. It is interesting to note that the size distribution of oocyte mRNA, as well as the total amount per oocyte, remains approximately unchanged from the time of early vitellogenesis until the oocytes reach maturity. Of this type of RNA, it has been estimated by means of manual enucleation of oocytes that at least 80% is localized in the cytoplasm. The protein to RNA ratio and the types of polypeptide present have not been determined for the small quantities of mRNP that are easily available, so no relationship with nuclear RNA can be established by using these criteria.

The minor component of cytoplasmic mRNP is that which is associated with polyribosomes and is presumably active in synthesizing oocyte proteins which are not derived from an exogenous source. Apparently, much of this protein enters the nucleus, where it is stored, or associates with newly transcribed RNA, or is in some way active in genetic control. Polyribosomes, although containing only a small fraction of the available ribosomes of the oocyte, can be detected and separated from nuclear RNP and free cytoplasmic mRNP by sucrose gradient centrifugation (232). Polyribosome-associated RNP is then released from the isolated ribosomal complex by EDTA treatment. There are several reasons for believing that the isolated material is in fact mRNP: it is rapidly labeled with

[^{3}H]uridine at times before there is any detectable labeling of ribosomal subunits and it is heterogeneous in size, consisting of particles most of which sediment between 30 S and 80 S. In this respect it is similar to free cytoplasmic mRNP (11). In addition, its sedimentation coefficient is related directly to the size of the polyribosomal complex from which it is derived; its protein to RNA ratio lies between 4:1 and 2:1, values which are similar to those of mRNP isolated from other systems, and it is active, while still complexed with ribosomes, in cell-free protein synthesis (232).

Role of Proteins Specifically Associated with Transcriptional Products

The proteins which specifically bind different forms of RNA transcript may constitute a relatively simple form of association, as in the case of two major polypeptides forming a complex with two small RNA species in the 42 S RNP particle (6, 115, 229), and two major proteins complexing with nucleolar RNA (233). Alternatively, there may be a complicated form of association, for example, one in which many polypeptides form a complex with a heterogeneous collection of information-containing RNA transcripts (103, 114–116). A third situation is one in which many proteins form a complex with a few RNA species, as in ribosomes (46, 234). However, it must be borne in mind that the number of polypeptides detected by electrophoresis may be an underestimate in that many additional minor components may be present at relatively low concentration, or it may be an overestimate in that some components may be fortuitously bound during RNP isolation.

Nevertheless, the common feature is that all forms of RNA transcribed in oocytes have specific classes of protein associated with them and that this protein is present in large amounts; the ratio of the mass of protein to RNA is always at least 2:1 in RNP particles, and in primary transcript RNP the ratio reaches the exceedingly high level of at least 30:1. Such high levels of protein are difficult to account for and must serve in a structural capacity. Mature and stable RNP complexes, like ribosomes, contain about equal proportions of protein and RNA. Higher protein levels may be required in precursor forms of RNA to facilitate processing at specific sites, to protect sequences from indiscriminate enzyme attack, or to direct transport from the nucleus to the required location in the cytoplasm. In considering the extensive modification undergone by RNA transcripts, the appropriate nucleotide sites for scission, phosphorylation, polyadenylation, and methylation must be made available by complex binding arrangements with proteins. The modification enzymes themselves are required only at low concentration, and if they are indeed integral to the structure of the RNP it is unlikely that they would be detected by the conventional techniques used to analyze protein constitution.

In experiments measuring the exit to the cytoplasm of labeled nuclear proteins from nuclei transplanted into the oocytes of *Xenopus* (235) and of *Rana* (236), it is shown that there is little transfer of radioactivity to the cytoplasm after the transplanted nucleus has become active. Although it is unlikely that under normal conditions RNA is translocated through the nuclear pore complexes in a naked state, there may, nevertheless, be considerable protein exchange at this level, resulting in a different set of proteins being associated with transcription products when they reach the cytoplasm. However, this is unlikely to be true for the 42 S particles, if, as it appears, the same proteins can be detected at the level of chromosomal transcription (115). Because there is no evidence for protein synthetic activity in eukaryote nuclei in general, the large amounts of highly heterogeneous proteins found in association with nuclear transcription products must be derived from the oocyte cytoplasm. Whether or not they are all synthesized there is another question, for there may be a significant contribution from exogenous sources, as there is for other oocyte proteins (see under "General Features of Oogenesis"). The consideration of the source of nuclear proteins is especially important with respect to the early events in lampbrush chromosome activity and poses the question: Where do the proteins come from to complex with the first transcribed sequences which are probably themselves coding for the translation of nuclear proteins? There exists the possibility that the RNA transcripts derived from early lampbrush loop activity require only one or a few proteins to be successfully transcribed and processed. Loops of this type may be represented by the restricted number that are detected by immunofluorescence with antibody prepared against a 30–35,000 dalton protein (114). In addition, this protein may well be derived, at least initially, from a site of synthesis external to the oocyte.

If, as seems likely, the subsequent heterogeneous collection of proteins is required at a high initial rate of synthesis, the mRNA sequences coding for these proteins may be transcribed from (repetitive?) sequences on specific loops, such as those loops that can be detected by slide hybridization of cDNA to lampbrush chromosomes (162) (see under "Cytological Hybridization"). Due to similarities in localization, it is possible that the specifically immunofluorescent and cDNA-hybridizing loops are one and the same.

It is interesting to note that *Xenopus* oocytes synthesize large quantities of the two proteins which are found to be associated with the 5 S RNA and the tRNA in the 42 S particles during previtellogenesis, whereas little of these proteins is synthesized at later stages of oogenesis (237). Also, the lampbrush loops that are believed to be transcribing 5 S and tRNA sequences, on the basis of specific immunofluorescence directed against their RNP proteins (115), are especially active (extended) in early oogenesis. So here, at least, an early protein synthetic activity coincides with an early transcriptional activity.

The transfer of cytoplasmic protein into nuclei was first demonstrated by using brain nuclei transplanted into *Xenopus* eggs whose cytoplasm had previously been labeled with ^{3}H-amino acids and which had further protein synthesis suppressed by puromycin treatment (238, 239). This active uptake is correlated with nuclear activity, especially in the sense of DNA synthesis. Although the identity of the nuclear proteins cannot be specified, DNA polymerase has been considered a likely possibility (240). Nevertheless, RNA synthesis may also be a consequence of protein uptake. Similar experiments carried out with the use of *Rana* have shown that a significant proportion of the proteins synthesized both before and during oocyte maturation is destined to become nuclear proteins (195, 241). The main drawback with nuclear transplantation experiments is that the identity of the proteins remains unknown. An alternative approach is the microinjection of ^{125}I-labeled specified proteins into oocytes and eggs (235). With the use of *Xenopus* it is observed that labeled oocyte soluble protein, labeled histones, and labeled bovine serum albumin all migrate from the cytoplasm to the nucleus. In addition, there is considerable nuclear accumulation of the oocyte proteins and histones, suggesting selective uptake. In considering the uptake of those proteins which associate with lampbrush chromosomes and become integrated with the RNA transcripts, there is a slight problem concerning size. In both isolated nuclear RNA (103, 114–116) and manually isolated chromosomes (117), there are proteins with molecular weights of 150,000 or more. Because the nuclear membrane presents a strong barrier to injected proteins of molecular weights in excess of 69,000 (242), there must exist some special mechanism for the selective uptake of these large proteins. Selectivity in nuclear uptake has been shown to operate at least at the level of small similarly charged proteins, for there is a pronounced accumulation of histone but not of lysozyme (235). Furthermore, radioactively labeled nuclear proteins of heterogeneous size, and some exceeding 130,000 daltons, re-enter and accumulate in the recipient oocyte nucleus after injection into the cytoplasm (243). Therefore, although the nuclear membrane presents a barrier to the entry of foreign (235, 242) or cytoplasmic (243) proteins, all nuclear proteins, irrespective of size and charge, are selectively accumulated back into the nucleus after being introduced into the oocyte cytoplasm.

STRUCTURE AND SEQUENCE COMPLEXITY OF INFORMATIONAL RNA

The high transcriptional activity of lampbrush chromosomes, as evidenced by cytological studies (see under "Visualization of RNA Transcription"), implies that a large number of different RNA sequences is being transcribed in oocytes. Furthermore, it appears that each type of sequence is transcribed many times in the course of oogenesis and that many of the initial products of transcription are very large, each RNA molecule potentially containing enough information to code for many proteins. An approximate description of the situation as it pertains to the lampbrush chromosomes of *Triturus* is that there are about 10^4 different and evenly distributed sites of transcription (loops) per haploid chromosome complement, that the axial length of a region of transcription, and possibly also the length of its associated RNA transcripts, may extend to 40 μm or more, and that, in all, these values total about 5% of the genome being active (in transcription) (102). Therefore, the genetic complexity of transcriptionally active DNA in *Triturus* oocytes may be as much as 10^9 nucleotide pairs, that is, about 200 times the size of the genome of *E. coli*. It is highly unlikely that the oocyte and the developing embryo require this degree of sequence complexity for translational purposes. The most reasonable explanation is that the vast majority of these sequences are not required in the form of cytoplasmic mRNA and are, therefore, eventually degraded in the nucleus. The role of these apparently superfluous sequences is a matter for speculation. A consequence of reduction in sequence complexity may be the reduction in molecular size of the individual nuclear RNA transcripts rather than the discarding of complete primary transcripts. Therefore, a consideration of the relative molecular size of precursor and product is important. However, the most informative studies relating to these problems are derived from the use of molecular hybridization techniques.

In terms of total RNA contained in *Xenopus* oocytes during the lampbrush stage, it has been shown by RNA-DNA hybridization that both repetitive and nonrepetitive DNA sequences are transcribed (8, 64). It has been reported that the transcribed repetitive DNA amounts to more than 3.5% of the genome and that much of the RNA transcribed from this fraction is conserved throughout oogenesis (8, 9, 208). However, no conclusion can be reached as to the sequence complexity of the repetitive component.

On the other hand, estimates of the sequence complexity and hence the potential genetic informational content of oocytes can be derived by hybridizing RNA from lampbrush stage and mature oocytes with the isolated nonrepetitive fraction of the *Xenopus* genome (10, 64). At least 1.2% of the total genome complexity is present in the RNA of mature oocytes which is transcribed from nonrepetitive sequences. This value is equivalent to about 2×10^7 nucleotide pairs, that is, about 4.5 times the size of the genome of *E. coli.* Although this estimate is considerably lower than the maximal complexity value given for transcription in *Triturus*, it is still high, containing enough information to code for 1.7×10^4 diverse structural genes of an average of 1.2×10^3 nucleotide pairs.

Furthermore, the RNA content of the mature *Xenopus* oocyte is 4 μg (3, 7, 11) and about 1% of this is polyadenylated mRNA (11). Therefore, taking the total informational sequence complexity to be 2×10^7 nucleotides, each mRNA species would be represented on average by 3.7×10^6 copies per oocyte (4×10^{-8} g $= 7.4 \times 10^{13}$ nucleotides; copies $= 7.4 \times 10^{13}/2 \times 10^7$).

These calculations indicate that during oogenesis more than 10^4 genes are active in transcribing about 4×10^6 copies of each. However, no allowance is made for the possibility that some of the transcribed sequence complexity may be noncoding, that there may be vastly different abundance classes of different types of mRNA, and that a considerable proportion of the mRNA may be nonadenylated. These points can be covered by considering independently the contribution made by nuclear RNA and mRNA. First, however, the size relationship between nuclear RNA and cytoplasmic mRNA is discussed.

Relative Sizes of Primary Transcript RNA and Messenger RNA

Most of the rapidly labeled RNA from *Triturus* oocyte nuclei has a sedimentation coefficient in excess of 40 S (104), and the maximal value may be 100 S or more (75, 103). Also, in *Xenopus* oocyte nuclei it has been demonstrated that 75% of [^{3}H]uridine and [^{3}H]guanosine incorporation, over a 24-hr period of previtellogenesis, is into RNA which migrates on electrophoresis more slowly than does a 40 S marker (191). In spite of the high rate of incorporation into high molecular weight RNA, the predominant RNA species during this previtellogenic period are 5 S RNA and tRNA, which constitute 75–80% of the total oocyte RNA (6, 7, 190). Thus, the kinetics of accumulation of high molecular weight RNA differs from that of the low molecular weight species and suggests that at least part of the nuclear RNA is an unstable component. Nevertheless, a level of about 12% of the total RNA is maintained as high molecular weight RNA throughout previtellogenesis (7).

Whereas primary transcript RNA has a sedimentation coefficient of greater than 40 S, rapidly labeled cytoplasmic RNA of *Xenopus* (191) and polyadenylated mRNA of both *Xenopus* (11) and *Triturus* (75) have a size range of between 10 S and 40 S. This alone does not imply that the nuclear RNA is a high molecular weight precursor of mRNA because the mRNA could have an independent and quite different synthesis. However, it has been demonstrated that cDNA transcripts of oocyte mRNA can hybridize to high molecular weight primary transcript RNA of *Triturus* (75). It would appear, therefore, that at least part of the high molecular weight nuclear RNA is a precursor containing mRNA sequences. It may be a general feature of eukaryote cells that there is a high molecular weight precursor to many types of mRNA which is included in the population of heterogeneous nuclear RNA molecules (reviewed ref. 244).

As already implied (see under "Phylogeny, C Value, and Karyotypes of Amphibia"), there may exist a relationship between genome size and the average size of primary transcript molecules. Certainly with respect to lampbrush loop length, higher C value Amphibia tend to have longer chromosomal transcriptional units. For instance, the lateral loops of *Triturus* are on the average 5–10 times longer than those of *Xenopus*, that is, by a proportion commensurate with the difference in genome size (81). However, the relative sizes of primary transcript RNA from different amphibian taxa have not as yet been adequately studied. It is interesting to note that in the Diptera, in which *Aedes* has a 5–6-fold larger genome than does *Drosophila*, an *Aedes* cell line produces hnRNA which is 2–2.5-fold larger than that produced by a *Drosophila* line (245). However, whether this exemplifies a general phenomenon remains to be seen.

Sequence Complexity and Abundance Classes of Messenger RNA

A substantial fraction of the mRNA of amphibian oocytes is polyadenylated (11, 209, 237) and can be isolated on the basis of binding to columns containing oligo(dT)-cellulose (11) or poly(U)-Sepharose (75, 246). Poly(A)$^+$ mRNA is especially useful because it can be easily transcribed with RNA-dependent DNA polymerase to give radioactive single-stranded copies (cDNA) which can be used as a probe to detect mRNA sequences in nucleic acid preparations by molecular hybridization. This cDNA can be used in two main ways—first, to measure the reiteration frequency of the DNA from which the mRNA was originally transcribed and, second, to measure the sequence complexity of the mRNA populations in oocytes.

The kinetics of hybridization of cDNA with excess chromosomal DNA is approximate to the kinetics of reassociation of single copy sequences in *Xenopus* (65). Even with amphibian taxa such as *Triturus*, which have no clearly distinguishable single copy DNA component, the poly(A)$^+$ mRNA sequences are preferentially transcribed from what would seem to be (in kinetic terms) a nonrepetitive fraction of the genome (65). However, it has been reported that the rate of hybridization of cDNA with DNA from *Triturus*, as well as the rate of reassociation of nonrepetitive DNA, is slightly faster than would be expected from a consideration of the genetic complexity of this genus (75) (Table 5). In addition, the thermal stability of the hybrids is significantly lower than that of native sonicated DNA. These observations have led to the suggestion that, although most mRNA sequences are represented only once per haploid genome, there may exist in higher C value Amphibia several sequences which have deviated, by base substitution, from an ancestral coding sequence (75). In addition, more than one of the related sequences may be capable of coding for the functional protein, but only one is expressed in the form of mRNA in an individual oocyte. It might be expected that a group of such related sequences is sited on the same lampbrush loop.

The kinetics of hybrdization of cDNA, or, alternatively, nonrepetitive DNA, with excess unlabeled RNA (an RNA-driven reaction) gives a direct measure of the sequence complexity of the RNA (247, 248). Although the hybridization kinetics of cDNA with its template RNA does not reflect a homogeneous population of mRNA molecules, the average half-reaction rate ($R_0t\frac{1}{2}$) gives a complexity value for *Triturus* oocyte poly(A)$^+$ mRNA of about 2×10^7 nucleotides (65, 75) (see Table 5). Furthermore, this complexity value, which is equivalent to $1–2 \times 10^4$ diverse coding sequences, is similar for both *Triturus* and *Xenopus*, in spite of the fact that the former has a 7-fold greater DNA content (65). Therefore, the absolute number of transcriptionally active genes in amphibian oocytes from different taxa may be about the same, irrespective of genome size. It is interesting to note that the value of $1–2 \times 10^4$ for the number of different genetic sequences approximates the number of lampbrush loops as estimated by cytological examination (102). Therefore, there is on the average one type of coding sequence per lampbrush loop (on the basis of mRNA sequence complexity), as well as one type of coding sequence per haploid genome (on the basis of transcription from nonrepetitive DNA). If the loops are considered to be one transcriptional unit, although there may be complexity

Table 5. Hybridization analysis of gene expression in *Triturus* oocytes

Labeled probe	Driver	$R_0 t_{1/2}$ (mol/s/l)	$C_0 t_{1/2}$ (mol/s/l)	Analytical complexity (nucleotides)	mRNA sequence dilution (%)	Genome expressed (%)	Reannealing (%)	Tm (°C)
cDNA transcribed from oocyte poly(A)$^+$ mRNA	Oocyte poly(A)$^+$ mRNA	9		~2×10^7	100	~0.1	76	80
cDNA transcribed from oocyte poly(A)$^+$ mRNA	Oocyte primary transcript RNA	3.2×10^2		~7×10^8	2.8	~4	69	78
cDNA transcribed from oocyte poly(A)$^+$ mRNA	Liver DNA		6.5×10^3		<0.14		77	77.5
Nonrepetitive liver DNA[b]	Nonrepetitive liver DNA[b]		1.2×10^4				67	79
Nonrepetitive component of sonicated liver DNA			~3×10^4[c]					86

[a]Data from ref. 75.
[b]Component reassociating before $C_0 t$ 5×10^2 removed by hydroxylapatite binding.
[c]Theoretical values based on a genome size of 2.1×10^{10} nucleotide pairs and ~45% of the genome nonrepetitive.

within this unit (see under "Chromosome Spreads/Transcriptional Matrices"), then it seems likely that each loop codes for one, or very few, proteins. On the other hand, these similarities in numbers may be purely coincidental, and the clustering of several different types of coding sequence within only a proportion of the lampbrush loops remains a possibility. For more direct information on the genetic sequence organization of loops, it is necessary to analyze primary transcript RNA.

A more detailed analysis of the sequence complexity of *Xenopus* oocyte poly(A)$^+$ mRNA reveals two abundance classes, about 50% of the RNA being represented by 500 mRNA species, the other 50% being represented by 1.5×10^4 mRNA species (209). The 30-fold difference in abundance is taken into consideration in the calculations of transcription rates (see under "RNA Polymerase Activity"). In considering the identity and function of these mRNA molecules, it is interesting to note that the sequences and their distribution between abundance classes are similar throughout oogenesis and up to the stage of the fertilized egg. Furthermore, the mRNA sequences present at each oogenic stage are the same because the RNA from any one oocyte size will drive to completion with identical kinetics the hybridization reaction with cDNA transcribed from the mRNA of any other oocyte size. The only observed change in mRNA formation during oogenesis is that previtellogenic oocytes have a greater proportion of some high abundance species than do postvitellogenic oocytes. The identity of the high abundance mRNA species is not known. Nevertheless, most of the mRNA species are not peculiar to oocytes because at least 90% of the oocyte sequences are found in *Xenopus* tadpoles, adult somatic cells, and tissue culture cells (209).

Sequence Complexity and Sequence Organization in Primary Transcript RNA

The sequence complexity of polyadenylated nuclear RNA in mouse Friend cells has been estimated from the kinetics of hybridization of cDNA to excess amounts of its nuclear RNA template (249), whereas the sequence complexity of hnRNA in sea urchin gastrulae has been estimated from the kinetics of hybridization of labeled nonrepetitive DNA to excess amounts of the RNA (250). In both of these systems, the sequence complexity of the nuclear RNA is shown to be greater than that of the polyribosomal mRNA population—by a factor of 5 in Friend cells and by a factor of 10 in sea urchin gastrulae. Moreover, Friend cell nuclear RNA consists of at least two abundance classes, and population number difference may also exist in the nuclear RNA of sea urchin gastrulae.

A third approach has been used in the estimation of primary transcript RNA complexity in *Triturus* oocytes (75). This involves the measurement of mRNA sequence dilution by comparing the kinetics of hybridization of cDNA to its poly(A)$^+$ mRNA template with the kinetics of hybridization of this same cDNA to primary transcript RNA. This technique gives a measure of the fraction of the transcript RNA which contains sequences homologous to poly(A)$^+$ mRNA, but it gives no information on the nature and complexity of the remaining sequences. This later aspect has been studied by hybridizing primary transcript RNA with excess chromosomal DNA (75) and is described later.

The extent of hybridization of cDNA with oocyte nuclear RNA is almost as great as the extent of hybridization of the cDNA with its own template mRNA, indicating that similar nucleotide sequences are present in both types of RNA. However, the average reaction rate ($R_0t\frac{1}{2}$) with nuclear RNA is 35 times slower than with mRNA (Table 5). Since high molecular weight RNA is the putative precursor of mRNA, this result suggests that the mRNA sequences comprise only a small percentage (3% or less) of the nuclear transcript. That is, only 3% of the RNA concentration is effective in binding cDNA. The low kinetic rate of the reaction is not due to any restriction conferred by the structure or method of isolation of the nuclear RNA because labeled nuclear RNA itself hybridizes extensively with DNA in vast excess (Figure 10). If a maximum of 3% of the

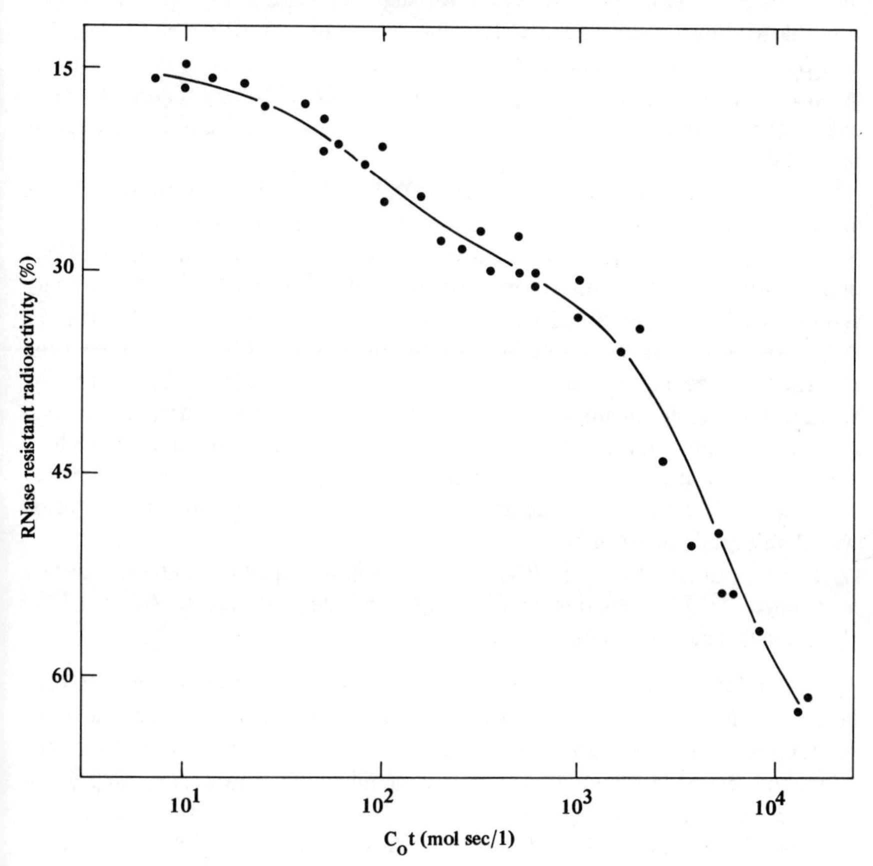

Figure 10. Kinetics of hybridization of ^{125}I-labeled primary transcript RNA with excess *Triturus* DNA. Reproduced from Sommerville and Malcolm (75) with permission of Springer-Verlag.

nuclear RNA contains coding sequences of 1.2×10^3 nucleotides, then the average total transcript will have a potential sequence complexity in excess of 4×10^4 nucleotides or a length equivalent of 16 μm or more. This value approaches the cytologically observed average length of the chromosomal loops and substantiates the view that at least some of the loops are transcribed in toto. Many of the longer loops undoubtedly contain multiple transcriptional matrices (see under "Chromosome Spreads/Transcriptional Matrices"). Nevertheless, the size measurements of primary transcript RNA by means of sedimentation analysis (75, 103, 104) and electron microscopy (75) are in fair agreement with molecular size derived from kinetic complexity analysis, even allowing for the probability that some primary transcripts contain several potentially coding sequences. Uncertainty remains about how average sequence complexity values relate to the organization of sequences within the transcript length. For instance, it has been implied that informational sequences are sited at the 3′ end of the loop transcript (75). The only reason for suggesting this is the general observation that coding sequences are situated at the 3′ end of hnRNA molecules, that is, adjacent to the polyadenylated tail. However, there may exist differences between oocyte primary transcript RNA and hnRNA of eukaryote somatic cells; one of these differences lies in the initial size of transcripts found in the oocytes of urodeles.

The rate of hybridization of oocyte cDNA with total chromosomal DNA is about 20 times slower than the reaction with primary transcript RNA. Taking into account the difference in nucleation rates of DNA-RNA and DNA-DNA duplex formation, the simplest interpretation of these data is that the primary transcript RNA is transcribed from about 2–4% of the genome of *Triturus* (75). A slightly higher value is obtained by hybridizing ^{125}I-labeled nonrepetitive DNA with increasing concentrations of primary transcript RNA. The saturation value in these experiments is equivalent to about 4% of the nonrepetitive DNA of *Triturus* being transcribed in oocytes (75). However, there appears to be a component of primary transcript RNA that is transcribed from moderately repetitive DNA; therefore, the saturation value of 4% genome transcription would be a minimal estimate.

In fact, primary transcript RNA is complex in its sequence content, as can be seen when ^{125}I-labeled oocyte RNA is hybridized with excess *Triturus* DNA (Figure 10). There are three main features.

1) About 15% of the radioactivity is found in RNA that is ribonuclease-resistant because it forms double-stranded hairpin loops. The regions of internal RNA duplex are about 100 base pairs long, and when converted into single strand fragments they are found to reassociate rapidly with values suggesting a low sequence complexity (103).
2) Most of the radioactivity is found in RNA that hybridizes with DNA at about the rate expected of slowly reassociating DNA. This fraction represents se-

quences which have a low repeat frequency and presumably contains most of the coding sequences.

3) About 15% of the radioactivity is found in RNA that hybridizes to DNA at a rate similar to that found for the reassociation of intermediately repetitive DNA. Whether these repetitive sequences are interspersed with low repeat frequency sequences in the primary transcript or whether they represent a subset of RNA molecules that is transcribed solely from repetitive DNA remains to be seen. Covalent linkage between the different RNA frequency classes has not, as yet, been demonstrated.

In conclusion, interpretations of molecular hybridization data are compatible with cytological observations that imply that a low percentage of the genome (less than 5% in *Triturus*) but many different genes ($1-2 \times 10^4$ in all taxa) are transcribed during amphibian oogenesis. However, the structure of the primary transcript molecule appears to be complicated and may contain duplex regions, interspersed intermediately repetitive sequences, and deviated informational sequences, in addition to one, or a few, protein coding sequences.

TRANSLATIONAL ACTIVITY

Informational RNA can be correctly designated as such only by its translation into protein. In amphibian oocytes the large and complex mRNA population, which results from extensive lampbrush chromosome activity (11, 65, 75, 209), may be considered to comprise two types—one active in oocytes in synthesizing proteins required for successful oocyte development, the other stored in an inactive condition ready to be activated after fertilization and to provide the proteins required for successful early embryogenesis (9, 209). Several questions relating to translational control are of the utmost importance in understanding the significance of oogenesis and the potential of the unfertilized egg. What factors restrict translation during oogenesis? Do transcribed and stored sequences code for different proteins? Are different types of mRNA transcribed and translated at different stages of oogenesis? Do poly(A)$^+$ and poly(A)$^-$ sequences code for different proteins in oocytes? What effect does polyadenylation have on translational activity in oocytes? In order to investigate some of these problems, oocyte mRNA has been translated in cell-free systems, and the products of synthesis have been compared with oocyte endogenous synthesis (237, 246). Furthermore, the translational potential of oocytes and eggs has been extensively tested by microinjection of heterologous (251–254) and homologous (255) mRNA.

RNA isolated from oocytes has been shown to be functional mRNA by various procedures. Initiation sites for translational activity can be titrated by the ability of mRNA to bind a 40 S Met-tRNA$_F$ complex to the 60 S ribosomal subunit, thereby forming a complete 80 S initiation complex. Poly(A)$^+$ RNA from *Xenopus* oocytes has been shown to be effective in initiating translational complexes in a rabbit reticulocyte cell-free system, whereas poly(A)$^-$ RNA has been shown to be less effective (237). Both poly(A)$^+$ RNA and nonribosomal poly(A)$^-$ RNA from the oocytes of *Xenopus* (237), *Triturus* (246), and *Notophthalmus* (256) have been shown to be efficient templates for protein synthesis in the wheat germ cell-free system. When analyzed by SDS-acrylamide gel electrophoresis, the products of translation can be seen to comprise a large number of different polypeptides in the molecular weight range of 10,000–150,000. The presence of the methylated 5′-terminal sequence $m^7G^{5'}ppp^{5'}N$ has been thought to be essential for translation of many viral and mammalian mRNA sequences (257, 258). However, it is now apparent that this 5′ modification (capping) is not essential for the translation of all mRNA molecules in all systems (259), although it may still be essential in the wheat germ system. Therefore, the possibility exists that capping could provide a control mechanism in the translation of oocyte mRNA. However, the general finding is that the extent of stimulation of translation by poly(A)$^+$ mRNA and poly(A)$^-$ mRNA from both *Xenopus* (237) and *Triturus* (246) is unaffected by the presence of *S*-adenosylmethionine, which normally stimulates methylation, or *S*-adenosylhomocysteine, which normally inhibits methylation. Furthermore, the SDS-acrylamide gel profile of polypeptides synthesized in the cell-free system is identical in the presence or absence of both of these compounds. Thus, methylation of any form of oocyte mRNA (stored mature stage poly(A)$^+$ RNA; previtellogenic stage mRNP; polysomal mRNA) is not required as an in vitro modification for effective translation in the wheat germ system, and it is concluded that all types of oocyte mRNA are methylated in vivo (237, 246). This conclusion is supported by the observation that translation of *Triturus* oocyte mRNA is inhibited by 7-methylguanosine-5′-phosphate, an inhibitor of capped mRNA (260), to an extent similar to the inhibition of capped HeLa poly(A)$^+$ mRNA (261). Therefore, the restriction in translation of oocyte mRNA in vivo appears not to reside at the level of methylation of the 5′ terminus.

The effective translation of oocyte mRNA in the wheat germ cell-free system gives further information on the mRNA sequence content of oocytes. Darnbrough and Ford (237) have drawn the following conclusions with respect to *Xenopus* oogenesis.

1) There is no difference between the coding properties of poly(A)$^+$ RNA extracted from oocytes at all stages of development, from previtellogenesis to maturity.
2) In previtellogenic oocytes, the in vitro products of poly(A)$^+$ RNA extracted from both polysomes and stored mRNP are identical.
3) Although there are problems in isolating genuine nonpolyadenylated mRNA, there is no evidence for any product of poly(A)$^-$ RNA not being synthesized by poly(A)$^+$ RNA.

4) There is no evidence for any changes in the coding properties of poly(A)$^-$ mRNA in the course of oogenesis.

However, when the endogenous translation of oocyte mRNA is examined, a completely different situation is seen to pertain; the pattern of protein synthesis changes markedly during oogenesis. For example, previtellogenic oocytes of *Xenopus* synthesize large quantities of the two proteins associated with 5 S RNA and tRNA in the 42 S RNP particles (see under "42 S Particles"), whereas very little of these proteins is made at later stages of oogenesis (237). Although the change in pattern of synthesis is less from vitellogenesis to maturity, this pattern of endogenous translation does not correspond with the in vitro synthetic pattern.

The main unresolved question, then, concerns the mechanism whereby only certain mRNA sequences are translated in vivo. Endogenous protein synthesis in oocytes in relatively low (11, 232, 262, 263). This is exemplified by the observations that less than 1% of the ribosomes in full grown *Xenopus* oocytes is in the form of polysomes (263) and that less than 5% of the poly(A)$^+$ RNA is associated with ribosomes at any one stage of oogenesis (11). Therefore, there is no restriction in the quantitative availability of ribosomes or mRNA. In fact, the complete protein synthetic machinery of amphibian oocytes can be readily activated by the introduction of heterologous mRNA and apparently has a vastly excessive translational capacity (251, 252). As already discussed, the control of translation probably does not involve adenylation at the 3′ terminus or selective capping at the 5′ terminus of oocyte mRNA. The selection of only certain mRNA sequences to be translated apparently operates at two levels–first, in the uptake of mRNA sequences to form polysomes, a process that is probably regulated by the type of associated protein and the general structure of the mRNP, and, second, in the translation of mRNA already contained in polysomes, which is differentially expressed during oogenesis. Darnbrough and Ford (237) have suggested that the successful translation of polysomal mRNA could be achieved by a specific tRNA deficiency, binding of some "repressor" molecule, or a structural deficiency in the mRNA. Therefore, the changing pattern of protein synthesis during oogenesis would be determined by relieving the deficiency or removing the inhibitor. This suggestion is not unfounded, for, as already discussed (see under "Synthesis of 5 S RNA and Transfer RNA"), the population of tRNA molecules differs between early and full grown oocytes (206).

One specific category of proteins studied by cell-free translation of oocyte mRNA is histone (256). By translational activity, three classes of mRNA are detected in *Xenopus* and *Notophthalmus* ovary RNA: poly(A)$^+$ nonhistone mRNA, poly(A)$^-$ histone RNA, and, surprisingly, poly(A)$^+$ histone mRNA. In contrast to some of the experiments cited above, Ruderman and Pardue (256) found that ovary RNA apparently lacks an abundant poly(A)$^-$ nonhistone mRNA. Histone mRNA is detected in very small, previtellogenic oocytes of

Xenopus and accumulates to a level higher than that found in sea urchin eggs. Apparently a large fraction of the *Xenopus* oocyte histone mRNA is translated during oogenesis and stored in the nucleus, later to become associated with embryonic nuclear DNA (12).

Quantitation of mRNA levels is subject to limitations in isolation procedures. For instance, the separation of poly(A)$^+$ and poly(A)$^-$ mRNA is not entirely satisfactory since it is difficult to discriminate between poly(A)$^-$ RNA and mRNA with short poly(A) tails (209, 237). In addition, it must be pointed out that there are several objections to basing specific mRNA levels on in vitro translational activity: a) although high abundance mRNA sequences are detected by this technique, less abundant sequences may constitute the bulk of the mRNA; 2) there may be intrinsically differential translational activity relating to variations in mRNA structure; 3) different sequences may be preferentially synthesized under different incubation conditions; and 4) detection of in vitro synthesized proteins is biased in favor of low molecular weight proteins.

CONCLUDING REMARKS

Although amphibian oocytes are cells which are highly specialized in preparation for fertilization and embryogenesis, they may nevertheless demonstrate some general principles of eukaryotic gene activity. It is, therefore, pertinent to list the special features of lampbrush chromosomes and their activity and, where possible, to indicate what may be more general phenomena.

1) There is a general increase in gene repetition frequency with increased DNA content. This is particularly evident in the repeat frequency of rRNA (65) and 5 S RNA (78) genes in various Amphibia.

2) The extent to which a coding sequence is represented is dependent upon the selection pressure that has been exerted on that particular block of sequences. Some gene families are highly conserved, for example, histone-coding sequences, whereas in other instances the related sequences have evidently diverged to such an extent that they are no longer detected as being homologous, with possibly only one or a few sequences retaining their coding function (75). Reassociation kinetic data are difficult to resolve in terms of the Britten-Kohne interpretation (59), especially in examining the DNA from high C value Amphibia.

3) The number of different genes expressed during oogenesis tends to be constant in spite of massive variations in DNA content (65). This number of 1–2 $\times 10^4$ genes is roughly correlated with the number of cytologically observable lampbrush loops (102), although multicistronic loops are not uncommon (180, 181). The number of genes expressed during oogenesis may approach the maximal number of coding sequences present in eukaryote cells (79, 80).

4) The transcription of informational sequences in oocytes is "leaky." For instance, globin mRNA is apparently transcribed in lampbrush chromosomes (162) when there is no obvious requirement for globin molecules, even during early embryogenesis. However, it has not as yet been shown that stable globin

mRNA exists in isolated oocytes. Also, although the yolk protein, vitellogenin, is synthesized by liver cells and transported to oocytes in vast amounts, a low level of vitellogin transcription-translation apparently occurs endogenously (21). That the phenomenon of "leaky" transcription pertains to other eukaryote cells has been demonstrated by the transcription of globin mRNA in nonerythroid tissues (175).

5) Amphibian oocytes participate in extensive genome transcription. For instance, the sequence complexity of nuclear RNA in *Triturus* oocytes is about 35 times the complexity of cytoplasmic mRNA (75). The extent of transcription may be positively correlated with genome size in various eukaryote taxa (81, 245). Consequently, it is likely that there is considerable post-transcriptional control in the selection of potential mRNA sequences.

6) Post-transcriptional processing is probably controlled by the large amounts of nonhistone protein that associate with oocyte primary transcript RNA (104). The lower protein to RNA ratio found in the nuclear RNP of other systems may reflect a lesser degree of stable information storage.

7) Transcription and translation, especially in early oogenesis, are largely geared to production of nuclear proteins (241). Transcription at the vast majority of chromosomal loci is dependent upon the initial production of RNA-binding proteins.

8) It is possible that transcriptional control is to some extent affected by the feedback of end product protein to bind its coding nascent RNA. The only indication of such an effect is the specific binding of histone HI to a single loop pair of the lampbrush chromosomes of *Triturus*, a genus which does not synthesize significant amounts of this protein during oogenesis (115).

9) The oocyte nucleus serves as a repository for various proteins which are responsible for the control of gene expression during embryogenesis. Examples are histones (12), RNA polymerases (14), and o^+ protein (50). The deployment of these proteins after germinal vesicle breakdown and during early cleavage is probably important in determination and differentiation.

10) Translational control is particularly evident in oocytes. Of the vast amounts of stored mRNA which are capable of in vitro translation, only a small percentage is actually translated during oogenesis (237, 246).

ACKNOWLEDGMENTS

I thank Professor H. G. Callan, Dr. Peter Ford, Dr. David Malcolm, Dr. Robert Old, and Dr. Michael Rosbash for communicating and allowing me to use unpublished data. I am indebted to David Malcolm and Robert Old for many useful discussions during the course of writing this chapter.

REFERENCES

1. Daneholt, B. (1974). Int. Rev. Cytol. 39:417.
2. Brown, D. A., and Littna, E. (1964). J. Mol. Biol. 8:669.

3. Brown, D. A., and Littna, E. (1964). J. Mol. Biol. 8:688.
4. Davidson, E. H., Allfrey, V. G., and Mirsky, A. E. (1964). Proc. Natl. Acad. Sci. U.S.A. 52:501.
5. Brown, D. A., and Littna, E. (1966). J. Mol. Biol. 20:95.
6. Ford, P. J. (1971). Nature 233:561.
7. Mairy, M., and Denis, H. (1971). Dev. Biol. 24:143.
8. Davidson, E. H., Crippa, M., Kramer, F. R., and Mirsky, A. E. (1966). Proc. Natl. Acad. Sci. U.S.A. 56:856.
9. Crippa, M., Davidson, E. H., and Mirsky, A. E. (1967). Proc. Natl. Acad. Sci. U.S.A. 57:885.
10. Davidson, E. H., and Hough, B. R. (1971). J. Mol. Biol. 56:491.
11. Rosbash, M., and Ford, P. J. (1974). J. Mol. Biol. 85:87.
12. Adamson, E. C., and Woodland, H. R. (1974). J. Mol. Biol. 88:263.
13. Grippo, P., and Lo Scavo, A. (1972). Biochem. Biophys. Res. Commun. 48:280.
14. Roeder, R. G. (1974). J. Biol. Chem. 249:249.
15. Dawid, I. B. (1965). J. Mol. Biol. 12:581.
16. Dawid, I. B. (1966). Proc. Natl. Acad. Sci. U.S.A. 56:269.
17. Chase, J. W., and Dawid, I. B. (1972). Dev. Biol. 27:504.
18. Balinsky, B. I., and Devis, R. J. (1963). Acta Embryol. Morphol. Exp. 6:55.
19. Wallace, R. A., and Jared, D. W. (1969). Dev. Biol. 19:498.
20. Wallace, R. A., Jared, D. W., and Nelson, B. L. (1970). J. Exp. Zool. 175:259.
21. Wallace, R. A., Nickol, J. M., Ho, T., and Jared, D. W. (1972). Dev. Biol. 29:255.
22. Edstrom, J.-E., and Lambert, B. (1975). Prog. Biophys. Mol. Biol. 30:57.
23. Davidson, E. H. (1968). Gene Activity in Early Development. Academic Press, New York.
24. Meyer, G. F. (1963). Chromosoma 14:207.
25. Hess, O., and Meyer, G. F. (1963). J. Cell Biol. 16:527.
26. Hess, O. (1965). Chromosoma 16:222.
27. Morescalchi, A. (1973). *In* A. B. Chiarelli and E. Capanna (eds.), Cytotaxonomy and Vertebrate Evolution, p. 233. Academic Press, New York.
28. Reeder, R. H. (1974). *In* M. Nomura, A. Tissieres, and P. Lengyel (eds.), Ribosomes, p. 489. Cold Spring Harbor Press, New York.
29. Birnstiel, M. L., Chipchase, M., and Speirs, J. (1971). Prog. Nucleic Acid Res. Mol. Biol. 11:351.
30. Tobler, H. (1975). *In* R. Weber (ed.), The Biochemistry of Animal Development, Vol. III, p. 91. Academic Press, New York.
31. Denis, H. (1974). *In* J. Paul (ed.), MTP International Review of Science, Biochemistry Series One, Biochemistry of Cell Differentiation, pp. 9, 95. Butterworths, London.
32. Ford, P. J. (1973). *In* J. D. Biggars and A. W. Schuetz (eds.), Oogenesis, p. 167. University Park Press, Baltimore.
33. Smith, L. D. (1975). *In* R. Weber (ed.), The Biochemistry of Animal Development, Vol. III, p. 1. Academic Press, New York.
34. Thomas, C. (1969). J. Embryol. Exp. Morphol. 21:165.
35. Brachet, J., Denis, H., and de Vitry, F. (1964). Dev. Biol. 9:398.
36. Briggs, R., Green, E. U., and King, T. J. (1951). J. Exp. Zool. 116:455.
37. Ecker, R. E., and Smith, L. D. (1968). Dev. Biol. 18:232.
38. LaMarca, M. J., Smith, L. D., and Strobel, M. (1973). Dev. Biol. 34:106.

39. Leonard, D. A., and LaMarca, M. J. (1975). Dev. Biol. 45:199.
40. Smith, L. D., Ecker, R. E., and Subtelny, S. (1968). Dev. Biol. 17:627.
41. Callan, H. G., personal communication.
42. Ficq, A. (1961). Symposium on Germ Cells and the Earliest Stages of Development, p. 121. Fondatione A. Baselli, Milan.
43. Gall, J. G. (1963). *In* M. Locke (ed.), Cytodifferentiation and Macromolecular Synthesis, p. 119. Academic Press, New York.
44. Gall, J. B., and Callan, H. G. (1962). Proc. Natl. Acad. Sci. U.S.A. 48:562.
45. Ford, P. J. (1966). Biochem. J. 101:369.
46. Hallberg, R. L., and Smith, D. C. (1975). Dev. Biol. 42:40.
47. Humphrey, R. R. (1966). Dev. Biol. 13:57.
48. Briggs, R., and Cassens, G. (1966). Proc. Natl. Acad. Sci. U.S.A. 55:1103.
49. Briggs, R. (1972). J. Exp. Zool. 181:271.
50. Brothers, A. J. (1976). Nature 260:112.
51. Swift, H. (1950). Proc. Natl. Acad. Sci. U.S.A. 36:643.
52. Goin, O. B., Goin, C. J., and Bachmann, K. (1967). Copeia 233.
53. Sexsmith, E. (1968). Ph.D. thesis, University of Toronto.
54. Straus, N. A. (1971). Proc. Natl. Acad. Sci. U.S.A. 68:799.
55. Wallace, H., and Birnstiel, M. L. (1966). Biochim. Biophys. Acta 144:296.
56. Bachmann, K. (1970). Histochemie 22:289.
57. Olmo, E. (1973). Caryologia 26:43.
58. Morescalchi, A., and Serra, V. (1974). Experientia 30:487.
59. Britten, R. J., and Kohne, D. E. (1968). Science 161:529.
60. Mizuno, S., and Macgregor, H. C. (1974). Chromosoma 46:239.
61. Morescalchi, A. (1975). *In* T. Dobzhansky, M. K. Hecht, and W. C. Steere (eds.), Evolutionary Biology, Vol. 8, p. 339. Plenum Publishing Corporation, New York.
62. Gall, J. G. (1963). Nature 198:36.
63. Miller, O. L., Jr. (1965). Natl. Cancer Inst. Monogr. 18:79.
64. Davidson, E. H., and Hough, B. R. (1969). Proc. Natl. Acad. Sci. U.S.A. 63:342.
65. Rosbash, M., Ford, P. J., and Bishop, J. O. (1974). Proc. Natl. Acad. Sci. U.S.A. 71:3746.
66. Morescalchi, A., Olmo, E., and Serra, V. (1974). Experientia 30:619.
67. Ullrich, F. H. (1966). Chromosoma 18:316.
68. Vlad, M., and Macgregor, H. C. (1975). Chromosoma 50:327.
69. Macgregor, H. C., Mizuno, S., and Vlad, M. (1976). *In* P. L. Pearson and K. R. Lewis (eds.), Chromosomes Today, Vol. V, p. 331. John Wiley and Sons, New York.
70. Barsacchi, G., and Gall, J. G. (1972). J. Cell Biol. 54:580.
71. Macgregor, H. C., and Kezer, J. (1971). Chromosoma 33:167.
72. Macgregor, H. C., Horner, H., Owen, C. A., and Parker, I. (1973). Chromosoma 43:329.
73. Walker, P. M. B., Flamm, W. G., and McLaren, A. (1969). *In* A. Lima-de-Faria (ed.), Handbook of Molecular Cytology, p. 52. North-Holland, Amsterdam.
74. Callan, H. G. (1967). J. Cell Sci. 2:1.
75. Sommerville, J., and Malcolm, D. B. (1976). Chromosoma 55:183.
76. Paul, J. (1972). Nature 238:444.
77. Buongiurno-Nardelli, M., Amaldi, F., and Lava-Sanchez, P. A. (1972). Nature (New Biol.) 238:134.
78. Pukkila, P. J. (1975). Chromosoma 53:71.
79. Bishop, J. O. (1974). Cell 2:81.

80. Lewin, B. (1975). Cell 4:77.
81. Malcolm, D. B., and Sommerville, J., unpublished results.
82. Britten, R. J., and Davidson, E. H. (1969). Science 165:349.
83. Georgiev, G. P. (1969). J. Theor. Biol. 25:473.
84. Davidson, E. H., Hough, B. R., Amenson, C. S., and Britten, R. J. (1973). J. Mol. Biol. 77:1.
85. Chamberlin, M. E., Britten, R. J., and Davidson, E. H. (1975). J. Mol. Biol. 96:317.
86. Davidson, E. H., Galau, G. A., Angerer, R. C., and Britten, R. J. (1975). Chromosoma 51:253.
87. Britten, R. J., and Davidson, E. H. (1971). Q. Rev. Biol. 46:111.
88. Ohno, S. (1972). Brookhaven Symp. Biol. 23:366.
89. Izawa, M., Allfrey, V. G., and Mirsky, A. E. (1963). Proc. Natl. Acad. Sci. U.S.A. 50:811.
90. Brown, D. D., and Dawid, I. B. (1968). Science 160:272.
91. Macgregor, H. C. (1968). J. Cell Sci. 3:437.
92. Coggins, L. W., and Gall, J. G. (1972). J. Cell Biol. 52:569.
93. Hanocq-Quertier, J., Baltus, E., Ficq, A., and Brachet, J. (1968). J. Embryol. Exp. Morphol. 19:273.
94. Edstrom, J.-E., and Gall, J. G. (1963). J. Cell Biol. 19:279.
95. Gall, J. G. (1966). Natl. Cancer Inst. Monogr. 23:475.
96. MacGillivray, A. J., and Rickwood, D. (1974). *In* J. Paul (ed.), MTP International Review of Science, Biochemistry Series One, Biochemistry of Cell Differentiation, Vol. 9, p. 301. Butterworths, London.
97. Ecker, R. E. (1972). *In* S. Bonotto, R. Goultier, R. Kirchmann, and J. R. Maisin (eds.), Biology and Radiobiology of Anucleate Systems, Vol. 1, p. 165. Academic Press, New York.
98. Davidson, E. H., and Mirsky, A. E. (1965). Brookhaven Symp. Biol. 18:77.
99. Rogers, M. E. (1968). J. Cell Biol. 36:421.
100. Gall, J. G. (1963). *In* M. Locke (ed.), Cytodifferentiation and Macromolecular Synthesis, p. 119. Academic Press, New York.
101. Macgregor, H. C., and Callan, H. G. (1962). Q. J. Microsc. Sci. 103:173.
102. Callan, H. G. (1963). Int. Rev. Cytol. 15:1.
103. Malcolm, D. B., and Sommerville, J. (1977). J. Cell Sci. 38:475.
104. Sommerville, J. (1973). J. Mol. Biol. 78:487.
105. Malcolm, D. B., and Sommerville, J. (1974). Chromosoma 48:137.
106. Rudak, E. (1976). Ph.D. thesis, University of St. Andrews.
107. Pardue, M. L. (1974). Cold Spring Harbor Symp. Quant. Biol. 38.
108. Callan, H. G., and Lloyd, L. (1975). *In* R. C. King (ed.), Handbook of Genetics, Vol. IV, p. 57. Plenum Press, New York.
109. Giorgi, F., and Galleni, L. (1972). Caryologia 25:107.
110. Callan, H. G., and Lloyd, L. (1960). Philos. Trans. R. Soc. Lond. (Biol. Sci.) 243:135.
111. Mott, M. R., and Callan, H. G. (1975). J. Cell Sci. 17:241.
112. Riemann, W., Muir, C., and Macgregor, H. C. (1969). J. Cell Sci. 4:299.
113. Muir, C., and Whitley, J. E. (1972). J. Cell Sci. 10:335.
114. Scott, S. E. M., and Sommerville, J. (1974). Nature 250:680.
115. Sommerville, J., Crichton, C., and Malcolm, D. B., manuscript in preparation.
116. Sommerville, J., and Hill, R. J. (1973). Nature (New Biol.) 245:104.
117. Maundrell, K. (1975). J. Cell Sci. 17:579.
118. Gall, J. G. (1956). Brookhaven Symp. Biol. 8:17.

119. Callan, H. G., and Lloyd, L. (1956). Nature 178:355.
120. Davidson, E. H., and Britten, R. J. (1973). Q. Rev. Biol. 48:565.
121. Hough, B. R., and Davidson, E. H. (1972). J. Mol. Biol. 70:491.
122. Miller, L., and Brown, D. D. (1969). Chromosoma 28:430.
123. Sinclair, J. H., Carrol, C. R., and Humphrey, R. R. (1974). J. Cell Sci. 15:239.
124. Hutchison, N., and Pardue, M. L. (1975). Chromosoma 53:51.
125. Dawid, I., Brown, D. D., and Reeder, R. H. (1970). J. Mol. Biol. 51:341.
126. Birnstiel, M. L., and Grunstein, M. (1972). Fed. Eur. Biochem. Soc. Proc. 23:349.
127. Dawid, I. B., and Wellauer, P. K. (1976). Cell 8:443.
128. Reeder, R. H., Higashinakagawa, T., and Miller, O., Jr. (1976). Cell 8:449.
129. Wellauer, P. K., Dawid, I. B., Brown, D. D., and Reeder, R. H. (1976). J. Mol. Biol. 105:461.
130. Wellauer, P. K., Reeder, R. H., Dawid, I. B., and Brown, D. D. (1976). J. Mol. Biol. 105:487.
131. Bird, A., Rogers, E., and Birnstiel, M. (1973). Nature (New Biol.) 242: 226.
132. Hourcade, D., Dressler, D., and Wolfson, J. (1973). Proc. Natl. Acad. Sci. U.S.A. 70:2926.
133. Rochaix, J.-D., Bird, A., and Bakken, A. H. (1974). J. Mol. Biol. 87:473.
134. Reeder, R. H., Brown, D. D., Wellauer, P. K., and Dawid, I. B. (1976). J. Mol. Biol. 105:507.
135. Brown, D. D., and Weber, C. S. (1968). J. Mol. Biol. 34:661.
136. Brown, D. D., Wensink, P. C., and Jordan, E. (1971). Proc. Natl. Acad. Sci. U.S.A. 68:3175.
137. Pardue, M. L., Brown, D. D., and Birnstiel, M. L. (1973). Chromosoma 42:191.
138. Brown, D. D., and Sugimoto, K. (1973). J. Mol. Biol. 78:397.
139. Brownlee, G. G., Cartwright, E. M., and Brown, D. D. (1974). J. Mol. Biol. 89:703.
140. Ford, P. J., and Southern, E. M. (1972). Nature (New Biol.) 241:7.
141. Wegnez, M., Monier, R., and Denis, H. (1972). FEBS Lett. 25:13.
142. Carroll, D., and Brown, D. D. (1976). Cell 7:467.
143. Carroll, D., and Brown, D. D. (1976). Cell 7:477.
144. Smith, G. P. (1974). Cold Spring Harbor Symp. Quant. Biol. 38:507.
145. Birnstiel, M. L., Sells, B. H., and Purdom, I. F. (1972). J. Mol. Biol. 63:21.
146. Clarkson, S. G., Birnstiel, M. L., and Purdom, I. F. (1973). J. Mol. Biol. 79:411.
147. Clarkson, S. G., Birnstiel, M. L., and Serra, V. (1973). J. Mol. Biol. 79:391.
148. Clarkson, S. G., and Kurer, V. (1976). Cell 8:183.
149. Kedes, L. H. (1976). Cell 8:321.
150. Birnstiel, M. L., Gross, K., Schaffner, W., and Telford, J. (1975). Fed. Eur. Biochem. Soc. Proc. 38:3.
151. Destree, O. H. J., personal communication.
152. Gould, D. C., Callan, H. G., and Thomas, C. A., Jr. (1976). J. Cell Sci. 21:303.
153. Snow, M. H. L., and Callan, H. G. (1969). J. Cell Sci. 5:1.
154. Miller, O. L., Jr., Beatty, B. R., Hamkalo, B. A., and Thomas, C. A., Jr. (1970). Cold Spring Harbor Symp. Quant. Biol. 35:505.
155. Izawa, M., Allfrey, V. G., and Mirsky, A. E. (1963). Proc. Natl. Acad. Sci. U.S.A. 63:378.

156. Hartley, S. (1976). Ph.D. thesis, University of St. Andrews.
157. Macgregor, H. C., and Kezer, J. (1973). Chromosoma 42:415.
158. Hennen, S., Mizuno, S., and Macgregor, H. C. (1975). Chromosoma 50:349.
159. Moar, M. H., Purdom, I. F., and Jones, K. W. (1975). Chromosoma 53:345.
160. Gall, J. G., and Pardue, M. L. (1969). Proc. Natl. Acad. Sci. U.S.A. 63:378.
161. Macgregor, H. C., and Mizuno, S. (1976). Chromosoma 54:15.
162. Old, R. W., and Callan, H. G., unpublished results.
163. Mancino, G., Barsacchi, G., and Nardi, I. (1968). Atti. Accad. Naz. Lincei. 45:180.
164. Kezer, J., and Macgregor, H. C. (1973). Chromosoma 42:427.
165. Mizuno, S., Andrews, C., and Macgregor, H. C. (1976). Chromosoma 58:1.
166. Barsacci-Pilone, G., Nardi, I., Batistoni, R., Andronico, F., and Beccari, E. (1974). Chromosoma 46:135.
167. Nardi, I., Ragghianti, M., and Mancino, G. (1972). Chromosoma 37:1.
168. Mancino, G., Nardi, I., and Ragghianti, M. (1972). Experientia 28:856.
169. Gall, J. G. (1954). J. Morphol. 94:283.
170. Callan, H. G. (1966). J. Cell Sci. 1:85.
171. Brownlee, G. G., Cartwright, E., McShane, T., and Williamson, R. (1972). FEBS Lett. 25:8.
172. Perry, R. P., and Kelley, D. E. (1968). J. Cell Physiol. 72:235.
173. Weinberg, R. A. (1973). Annu. Rev. Biochem. 42:329.
174. Young, B. D., Hell, A., and Birnie, G. D. Cell, in press.
175. Humphries, S., Windass, J., and Williamson, R. (1976). Cell 7:267.
176. Miller, O. L., Jr., and Beatty, B. R. (1969). Genetics (Suppl.) 61:133.
177. Miller, O. L., Jr., and Bakken, A. H. (1972). Acta Endocrinol. 168:155.
178. Miller, O. L., Jr., and Hamkalo, B. A. (1972). Int. Rev. Cytol. 33:1.
179. Miller, O. L., Jr., Beatty, B. R., and Hamkalo, B. A. (1972). *In* J. D. Biggars and A. W. Schuetz (eds.), Oogenesis, p. 119. University Park Press, Baltimore.
180. Angelier, N., and Lacroix, J. C. (1975). Chromosoma 51:323.
181. Malcolm, D. B., and Sommerville, J., unpublished results.
182. McKnight, S. L., and Miller, O. L., Jr. (1976). Cell 8:305.
183. Olins, A. D., and Olins, D. E. (1974). Science 183:330.
184. Scheer, U., Trendelenburg, M. F., and Franke, W. W. (1976). J. Cell Biol. 69:465.
185. Wolff, S., Lindsley, D. L., and Peacock, W. J. (1976). Proc. Natl. Acad. Sci. U.S.A. 73:877.
186. Kamen, R., Lindstrom, D. M., Schure, H., and Old, R. W. (1975). Cold Spring Harbor Symp. Quant. Biol. 39:187.
187. Malcolm, D. B., and Old, R. W., unpublished results.
188. Spring, H., Scheer, U., Franke, W. W., and Trendelenburg, M. F. (1975). Chromosoma 50:25.
189. Glätzer, K. H. (1975). Chromosoma 53:371.
190. Thomas, C. (1970). Biochim. Biophys. Acta 224:99.
191. Thomas, C. (1974). Dev. Biol. 39:191.
192. Brown, D. D. (1976). Curr. Top. Dev. Biol. 2:47.
193. Scheer, U. (1973). Dev. Biol. 30:13.
194. Crippa, M., Tocchini-Valentini, G. P., and Andronico, F. (1972). *In* J. D. Biggars and A. W. Schuetz (eds.), Oogenesis, p. 193. University Park Press, Baltimore.

195. Smith, L. D., and Ecker, R. E. (1970). Curr. Top. Dev. Biol. 5:1.
196. LaMarca, M. J., Stroble-Fidler, M. C., Smith, L. D., and Keem, K. (1975). Dev. Biol. 47:384.
197. Van Gansen, P., and Schram, A. (1974). J. Cell Sci. 14:85.
198. Evans, D., and Birnstiel, M. L. (1968). Biochim. Biophys. Acta 166:274.
199. Wegnez, M., and Denis, H. (1972). Biochimie 54:1069.
200. Mairy, M., and Denis, H. (1972). Eur. J. Biochem. 25:535.
201. Denis, H., and Mairy, M. (1972). Eur. J. Biochem. 25:524.
202. Wegnez, M., and Denis, H. (1973). Biochimie 55:1129.
203. Denis, H., Wegnez, M., and Willem, R. (1972). Biochimie 54:1189.
204. Denis, H., and Wegnez, M. (1973). Biochimie 55:1137.
205. Perkowska, E., Macgregor, H. C., and Birnstiel, M. A. (1968). Nature 217:649.
206. Denis, H., Mazabraud, A., and Wegnez, M. (1975). Eur. J. Biochem. 58:43.
207. Duryee, W. (1950). Ann. N. Y. Acad. Sci. 50:920.
208. Crippa, M., and Gross, P. R. (1969). Proc. Natl. Acad. Sci. U.S.A. 62:121.
209. Rosbash, M., personal communication.
210. Sommerville, J., unpublished results.
211. Finamore, F. J., and Volkin, E. (1958). J. Biol. Chem. 236:443.
212. Brown, D. D., and Littna, E. (1966). J. Mol. Biol. 20:81.
213. Ford, P. J., and Mathieson, T. (1976). Nature 261:433.
214. Rutter, W. J., Goldberg, M. I., and Perriard, J. C. (1974). *In* J. Paul (ed.), MTP International Review of Science, Biochemistry Series One, Biochemistry of Cell Differentiation, Vol. 9, p. 267. Butterworths, London.
215. Price, R., and Penman, S. (1972). J. Mol. Biol. 70:435.
216. Weinmann, R., and Roeder, R. G. (1974). Proc. Natl. Acad. Sci. U.S.A. 71:1790.
217. Roeder, R. G., Reeder, R. H., and Brown, D. D. (1970). Cold Spring Harbor Symp. Quant. Biol. 35:727.
218. Roeder, R. G. (1972). *In* M. Sussman (ed.), Molecular Genetics and Developmental Biology, p. 163. Prentice-Hall, Inc., New York.
219. Wasserman, P. M., Hollinger, T. C., and Smith, L. D. (1972). Nature (New Biol.) 240:208.
220. Long, E., Dina, D., and Crippa, M. (1976). Eur. J. Biochem. 66:269.
221. Wilhelm, J., Dina, D., and Crippa, M. (1974). Biochemistry 13:1200.
222. Kafatos, F. C., and Gelinas, R. (1974). *In* J. Paul (ed.), MTP International Review of Science, Biochemistry Series One, Biochemistry of Cell Differentiation, Vol. 9, p. 223. Butterworths, London.
223. Mancino, G., Barsacchi, G., and Nardi, I. (1968). Atti. Accad. Naz. Lincei. 45:591.
224. Mancino, G., Mardi, I., Corvaja, N., Fiume, L., and Marinozzi, V. (1971). Exp. Cell Res. 64:237.
225. Fiume, L., Nardi, I., Bucci, S., and Mancino, G. (1972). Exp. Cell Res. 75:11.
226. Sobel, H. M. (1973). Prog. Nucleic Acid Res. Mol. Biol. 13:153.
227. Georgiev, G. P. (1969). J. Theor. Biol. 25:473.
228. Miller, O. L., Jr. (1966). Natl. Cancer Inst. Monogr. 23:53.
229. Delaunay, J., Wegnez, M., and Denis, H. (1975). Dev. Biol. 42:379.
230. Hill, R. J., Maundrell, K., and Callan, H. G. (1973). Nature (New Biol.) 242:20.
231. Kumar, A., and Pederson, T. (1975). J. Mol. Biol. 96:353.
232. Sommerville, J. (1974). Biochim. Biophys. Acta 349:96.

233. Hill, R. J., Maundrell, K., and Callan, H. G. (1974). J. Cell Sci. 15:145.
234. Ford, P. J. (1971). Biochem. J. 125:1091.
235. Gurdon, J. B. (1970). Proc. R. Soc. Lond. (Biol.) 176:303.
236. Ecker, R. E., and Smith, L. D. (1971). Dev. Biol. 24:559
237. Darnbrough, C., and Ford, P. J. Dev. Biol. 50:285.
238. Arms, K. (1968). J. Embryol. Exp. Morphol. 20:367.
239. Merriam, R. W. (1969). J. Cell Sci. 5:333.
240. Gurdon, J. B. (1969). Dev. Biol. (Suppl.) 3:59.
241. Ecker, R. E., and Smith, L. D. (1971). Dev. Biol. 24:559.
242. Bonner, W. M. (1975). J. Cell Biol. 64:53.
243. Bonner, W. M. (1975). J. Cell Biol. 64:431.
244. Lewin, B. (1975). Cell 4:11.
245. Lengyel, J., and Penman, S. (1975). Cell 5:281.
246. Old, R. W., manuscript in preparation.
247. Bishop, J. O. (1969). Biochem. J. 113:805.
248. Birnstiel, M. L., Sells, B. H., and Purdom, I. F. (1972). J. Mol. Biol. 63:21.
249. Getz, M. J., Birnie, G. D., Young, B. D., MacPhail, E., and Paul, J. (1975). Cell 4:121.
250. Hough, B. R., Smith, M. J., Britten, R. J., and Davidson, E. H. (1975). Cell 5:291.
251. Gurdon, J. B., Lane, C. D., and Woodland, H. R. (1971). Nature 233:177.
252. Moar, V. A., Gurdon, J. B., Lane, C. D., and Marbaix, G. (1971). J. Mol. Biol. 61:93.
253. Woodland, H. R., Ford, C. C., Gurdon, J. B., and Lane, C. D. (1972). *In* M. Sussman (ed.), Molecular Genetics and Developmental Biology, p. 393. Prentice-Hall, Inc., New Jersey.
254. Lane, C. D., and Knowland, J. (1975). *In* R. Weber (ed.), The Biochemistry of Animal Development, Vol. III, p. 145. Academic Press, New York.
255. Berridge, M. V., and Lane, C. D. (1976). Cell 8:283.
256. Ruderman, J. B., and Pardue, M. L., in press.
257. Both, G. W., Furuichi, Y., Muthukrishnan, S., and Shatkin, A. J. (1975). Cell 6:185.
258. Muthukrishnan, S., Both, G. W., Furuichi, Y., and Shatkin, A. J. (1975). Nature 255:33.
259. Rose, J. K., and Lodish, H. F. (1976). Nature 262:32.
260. Hickey, E. D., Weber, L. A., and Baglioni, C. (1976). Nature 261:71.
261. Salditt-Georgieff, M., Jelinek, W., Darnell, J. E., Furuichi, Y., Morgan, M., and Shatkin, A. (1976). Cell 7:227.
262. Cox, R. A., Ford, P. J., and Pratt, H. (1970). Biochem. J. 119:161.
263. Woodland, H. R. (1974). Dev. Biol. 40:90.

International Review of Biochemistry
Biochemistry of Cell Differentiation II, Volume 15
Edited by J. Paul

4
Programmed Information Flow in the Sea Urchin Embryo

E. S. WEINBERG
Department of Biology,
Johns Hopkins University,
Baltimore, Maryland, U.S.A.

From the beginnings of modern experimental embryology right up to the most current fashions of gene organization and transcriptional regulation, the sea urchin embryo has been the subject of intense interest. There is a voluminous literature on both classical and biochemical investigations, and fortunately this material is now summarized in two comprehensive monographs (1, 2). Therefore, this review will not present a historical view of the biochemistry of sea urchin development, but rather will focus on some of the more interesting recent findings on the storage and utilization of informational molecules in the embryo. The major proportion of work to be discussed has been done with five sea urchin species, *Paracentrotus lividus* from the Mediterranean, *Psammechinus miliaris* from the North Sea, *Arbacia punctulata* from the Atlantic coast of the United States, and *Strongylocentrotus purpuratus* and *Lytechinus pictus* from the United States' Pacific coast. Since there appear to be only minor differences in the informational basis of development in the five species, the ensuing discussion will omit mention of the species in which each experiment was done.

The unfertilized sea urchin egg is relatively quiescent, but upon fertilization or artifical activation there is an increase of enzyme activity, change in cell surface components, and a large augmentation in protein synthesis (1, 2). The zygote goes through a sequence of 6 cleavages resulting in a ball of about 60 cells (morula). Upon further divisions, the cells become arranged as a hollow sphere of several hundred to a thousand cells (blastula). The embryo surface develops cilia, there is rotation within the fertilization membrane, and then the embryo hatches at about the tenth cleavage. Several hours later a set of cells migrates into the balstocoel and aggregates into clumps which develop into centers of spicule formation (mesenchyme blastula stage). This is followed by invagination and cell differentiation (gastrula). Later stages of development which involve spicule growth, coelom formation, digestive tract development, and formation of other tissues are termed prism and pluteus stages. This whole process takes from 24–72 hr depending on the species (3, 4). The great increase in cell number takes place with little change in embryo size, protein content (5–7), or RNA content (8–11). Strictly speaking, there is no growth, only rapid cell division, cellular movement, and tissue differentiation.

The temporal control of these events is quite precise, being predictable for each species. Needless to say, there is an obligatory sequence of these processes which in only a few cases can be circumvented. Perhaps even more impressive is the exact nature of the spatial determination in the embryo. Thus, two major problems of development that can be studied with the system are temporal control (the determination of when and in what order processes occur) and spatial control (the determination of where processes occur). The evidence to be discussed deals mainly with the former problem, which has been avidly studied with biochemical methods. The latter problem appears to be a much more difficult concept to approach, but there is some evidence which is presented at the end of this review.

ACTIVATION OF THE EGG

When the sea urchin egg is fertilized by sperm, or activated artificially, an impressive sequence of events takes place in the cell. Changes in structure and permeability, activation of enzymes, stimulation of metabolic pathways, and increase in the rate of protein synthesis have been the best studied changes (12, 13). Epel (14) has made a distinction between "early" and "late" responses to fertilization. Within 1 min of fertilization a set of responses such as Na^+ influx (15), activation of enzymes such as NAD kinase (16), and cortical granule changes (17) takes place. About 4–10 min later a second group of changes can be monitored, such as the appearance of phosphate transport (18), development of K^+ conductance (15), activation of protein synthesis (14), and DNA synthesis (19). Since these changes occur prior to or accompanying the activation of protein synthesis, and since some have been shown to occur in the presence of inhibitors of protein synthesis (20–23), it is unlikely that any of these processes depends on the synthesis of new proteins by the fertilized egg. The egg thus contains a store of proteins and other factors which are quickly activated on fertilization. They are present as a result of the differentiated programmed process of oogenesis.

Although the unfertilized egg is relatively inactive in the synthesis of protein (20, 24, 25), it is clear now that, if changes in permeability and endogenous pool sizes are considered, the egg does synthesize some protein. Polysomes can clearly be identified and shown to be active in the unfertilized egg (26, 27). Upon fertilization there is a dramatic 5–30-fold increase in protein synthesis (14, 26–28). Although there is no accumulation of protein during development through the gastrula (5–7), considerable synthesis does take place at all times during development (7, 29–32).

Protein synthesis also increases when the egg is only partially activated. When the egg is treated with NH_4^+, the early responses of cortical reactions and Na^+ influx changes do not occur, but late changes in K^+ conductance, DNA synthesis, and protein synthesis do occur (13, 33–35). Thus, the events occurring at fertilization are not all obligatorily linked. Activation of protein synthesis also can be shown to be independent of K^+ conductance. The acidification of sea water, a treatment which abolishes K^+ conductance, does not affect the synthesis of protein (13).

STORAGE OF MESSENGER RNA IN THE UNFERTILIZED EGG

This topic has been extensively reviewed in this series (36) and elsewhere (1, 2, 37–39). It is worth summarizing the evidence for stored maternal messenger

RNA (mRNA), however, with respect to recent experiments. The evidence for the presence of the mRNA in the unfertilized egg comes mainly from a) the rapid activation of protein synthesis and mobilization of ribosomes into polysomes shortly after fertilization; b) the observation that, even in the absence of RNA synthesis, protein synthesis will be activated and will continue through embryonic development to blastula; c) direct demonstration of particular stored mRNAs by translation of egg RNA in cell free systems and by competition-hybridization experiments; d) the demonstration of qualitative similarity of proteins made in the unfertilized egg and zygote, and e) the presence in the oocyte of most mRNA sequences found in the gastrula polysomes.

Mobilization of Ribosomes into Polysomes

The rapid activation of protein synthesis within a few minutes of fertilization occurs without a concomitant dramatic activation of RNA synthesis (40), although the rate of incorporation of [^{3}H]adenosine into RNA does increase during the early cleavage events. Some, but not all, of the early incorporation of nucleosides into RNA seems to be due to end labeling of transfer RNA (tRNA) (41–45), although there are indications of mRNA synthesis even before the first cell division (27, 46). Polysomes rapidly accumulate after the first division (47–50). The amount of mRNA necessary for the observed 30-fold increase in polysomes during the first 30 min of development was estimated to be 3×10^{-12} g per embryo by Humphreys (40). He measured the rate of newly synthesized RNA that entered the polysomes during this period and found it to be only 0.16×10^{-12} g per embryo, or only 5.3% of the mRNA utilized. Most of the mRNA involved in protein synthesis, even within 30 min of fertilization, was shown to have its origin in the unfertilized egg.

Events in Absence of Nuclear Transcription

The use of inhibitors of RNA synthesis and the removal of the nucleus have both indicated that protein synthesis in the preblastula embryo does not depend on newly synthesized RNA. Actinomycin D eliminates most transcription (74–95%) in newly fertilized eggs, but does not interfere with protein synthesis until the blastula stage (42, 51, 52), at which time the embryo ceases to develop. Most of the actinomycin-resistant uridine incorporation is due to the turnover of the pCpCpA end sequence in tRNA. Actinomycin D might have effects on mRNA half-life and initiation of protein synthesis in addition to the inhibition of transcription (53–55). It is fortunate that there is another way to rule out the role of newly synthesized RNA in early protein synthesis and embryonic development. Enucleate half eggs (merogones) can be prepared by centrifugation in sucrose gradients (56). These egg fragments can be chemically activated and observed to go through a series of cleavages. Activated anucleate and nucleate fragments both show incorporation of precursor into protein (57–59) or into polysomes (60). Whole eggs and non-nucleate fragments were shown to have a similar time course of incorporation of amino acids into protein (7, 61). These two approaches, actinomycin D treatment and enucleation, have recently been brought together to test whether the actinomycin D used in earlier experiments had any direct effect on protein synthesis itself (62). Not only did actinomycin D treatment (20 μg/ml) of enucleate half eggs result in no change in the level of protein synthesis, but also there was no change in the spectrum of proteins that could be resolved by disc gel electrophoresis. The inhibitor, therefore, does not appear to affect the utilization or stability of the mRNA in the early embryo when used under the conditions of Sargent and Raff (62). Both actinomycin D and enucleation experiments argue strongly for a maternal mRNA population which supports most of the protein synthesis in the preblastula embryo.

Demonstration of Stored Messenger RNAs

A direct way of proving the existence of the stored mRNA would be to extract a translatable messenger of known specificity. This has been done for the histone mRNAs (63, 64). There have been several studies showing the ability of egg RNA to stimulate amino acid incorporation into protein in cell-free systems (65–67), but the products of the reactions were not characterized. Ruderman and Pardue (68) have recently shown that the egg RNA codes for a variety of proteins, as characterized by acrylamide gel electrophoresis of the cell-free translation products. The only specific mRNAs thus far identified by their translational activity are those coding for histones.

The histone gene messenger system is discussed in greater detail in the last section of this review. It is sufficient for now to know that nuclear proteins (presumably including histones) account for a large fraction of the newly synthesized protein in early cleavage (69, 70). The histone proteins can be translated in vitro on templates from isolated embryonic RNA (64, 71–74) or unfertilized egg RNA (64, 75). Histone mRNA is synthesized in the cleaving embryo and can be resolved into several bands on polyacrylamide gels (71, 76–82) which have been shown to coincide with histone translational activity when tested in wheat germ cell-free systems (72, 82). The bulk of this RNA, visualized by the staining of polyacrylamide gels, is shown to be of similar size in both eggs and embryos (83). Thus, although histone mRNA appears to be synthesized in the embryo, a set of very similar RNAs is shown also to reside in the unfertilized egg by both translational and electrophoretic criteria. Additional evidence for the presence of histone mRNA in the egg is the ability of egg RNA to compete with labeled embryonic histone mRNA in RNA-DNA hybridization (83–85). A similar sized ribonucleoprotein fraction of the egg cytoplasm was shown to contain RNA both coding for histones in the in vitro translation assay and competing with [^{3}H]histone mRNA in the hybridization assay (75, 84). These results bear out the predictions that there are maternal histone mRNAs based on the production of histones in actinomycin D-treated embryos (86, 87).

Although other specific mRNAs have yet to be identified by these techniques, a number of distinct proteins have been shown to be synthesized in actinomycin D-treated embryos and, thus, are inferred to be translated, at least in part, on stored maternal mRNA. Examples include microtubular proteins (88, 89), ribosomal proteins (90), and nucleotide reductase (91). The presence of tubulin mRNA in the egg is further indicated by the rapid rate of tubulin synthesis in activated merogones (89). Experiments by Barrett and Angelo (92) with sea urchin hybrids suggest the maternal nature of the mRNA coding for the hatching enzyme (responsible for the digestion of the fertilization membrane). The hatching enzymes of two species of sea urchin can be differentiated by their sensitivity to Mn^{2+}. Reciprocal hybrids of the two species each have the maternal type of hatching enzyme. Evidence in addition to the actinomycin experiments is needed before the maternal nature of the mRNAs for some of these proteins is concluded.

Types of Protein Synthesized

Yet another indication of the presence of stored egg mRNA is the finding that most of the identifiable proteins synthesized in the embryo, even at gastrula, are also synthesized in the egg. Brandhorst (93) showed this quite convincingly with the use of the powerful two-dimensional gel method of O'Farrell (94). Over 400 [^{35}S]methionine-containing proteins were identified, with all but a few of the proteins synthesized in the egg also synthesized in the zygote. This general pattern was maintained until blastula. Although this result could be explained by turnover of egg-type mRNAs during development, as well as by utilization of egg mRNAs for translation in the embryo, this seems highly unlikely in view of Humphreys' results (40) on the rate of mRNA synthesis in the early embryo. In fact, the work of Brandhorst might be seen as the qualitative confirmation of Humphreys' quantitative results on total mRNA populations. Earlier results based on the poorer resolution of proteins by electrophoresis on acrylamide disc gels showed that there were no detectable differences in the pattern of proteins synthesized in unfertilized eggs and early cleavage embryos (95, 96), and the changes that did occur at hatched blastula were prevented by actinomycin treatment (97). In another study which used an antibody assay for determination of newly synthesized sea urchin antigens, little change was found in the pattern of synthesis of antigen until hatching, both in the presence and absence of actinomycin (98, 99).

Types of Messenger RNA in Egg

The last evidence for stored maternal mRNA to be discussed is the demonstration that most of the different kinds of sequences found in gastrula polysomal mRNA are also present in the oocyte (100). These results were obtained by fractionating the single copy DNA into two populations: those with complementarity to gastrula mRNA (mDNA) and those lacking such sequences. Oocyte total RNA hybridized to the gastrula mDNA as well as the gastrula mRNA did. These results are discussed again in the section dealing with changing populations of RNA in the embryo. Ovary polysomal mRNA hybridized to a somewhat lesser extent, indicating that perhaps not all of the oocyte mRNA sequences are present on polysomes. These results do not, however, demand that exactly the same molecules are present in the egg and late embryo. Therefore, the evidence is somewhat circumstantial that these egg sequences are stored for later use since they might just be turning over during embryogenesis.

The basic tenet of the original hypothesis of stored mRNA (101–104) is borne out at least in a quantitative sense. However, the questions of how translation is activated and where the templates are located within the egg still remain to be answered. The process of activation of protein synthesis has been approached by two nonexclusive hypotheses: the increasing availability of stored mRNA for initiation of translation or the increasing competence of the translational apparatus itself, or both. The former idea implies that the mRNA is compartmentalized or stored in some "masked" form, whereas the latter idea implies a defect that should be observable in the ability of egg ribosomes to translate available mRNAs. Much more has been done to test the second hypothesis than the first.

Competence of Egg Ribosomes

It is known that, whereas egg ribosomes are active in translating artificial mRNAs such as poly(U), they are relatively inactive in endogenous protein synthesis (105–115). The response to the poly(U) may not be indicative of the actual translation of natural mRNAs in the egg. Soluble factors from embryo homogenates do not markedly improve the in vitro endogenous synthesis of egg ribosomes, implying that the defect is not in factors such as tRNAs, amino acyl-tRNA synthetases, and amino acids (111, 112, 116). It was, therefore, important to test whether the egg ribosomes were active in translating natural eukaryotic mRNAs. Differences in endogenous rates of synthesis when using ribosomes of eggs or embryos in cell-free systems (117) were even more pronounced when microsomes of eggs and embryos were used (112). A more recent investigation has used globin mRNA to compare the egg and embryo ribosomal competence (118). Clegg and Denny found that both egg and zygote ribosomes synthesized α- and β-globin in response to rabbit globin mRNA in a Krebs II ascites cell-free system. In most cases, the egg ribosomes had an even higher activity than the zygote ribosomes. These experiments demonstrated similar competence of the two ribosome classes not only by assaying for the amount of incorporation of amino acids into protein, but also by the separation and analysis of the α- and β-globin synthetic products. Both kinds of ribosomes had similar Mg^{2+} and K^+ optima, and both were equally stimulated in their translation of globin mRNA by a globin ribosomal wash fraction. When the ribosomes were isolated as whole microsomes a small difference was noted in the endogenous protein synthesis in the Krebs ascites system, the embryo being somewhat more active than the egg preparation. But on the addition of globin mRNA both egg and embryo preparations were stimulated to the same extent. These results do not support earlier suggestions of the presence of an inhibitor on egg ribosomes, based on trypsin activation of egg ribosomes (109) or high salt wash activation (119). This inhibitory activity was found to be present only on unfertilized eggs, and it decreased poly(U)-directed polyphenylalanine synthesis in both sea urchin (119, 120) and rabbit reticulocyte cell-free systems (120). Hille (121), using another sea urchin species, found only a slight inhibitory activity of poly (U)-primed translation with 1 M NH_4Cl salt washes of both egg and embryo ribosomes. Kedes and Stavy (113) found no difference in the poly(U)-primed translation with egg and blastula ribosomes and hybrid mixtures of the two. There is no evidence, therefore, for the presence of an egg ribosomal factor which inhibits the translation of natural mRNAs. Thus, the ribosomes themselves do not appear to be the limiting factor in translational control in the egg.

There is some preliminary evidence which hints at mechanisms of activation. When NH_4^+ activates the metabolism of unfertilized eggs there is a release to the medium of nondialyzable factors which, on addition to partially activated eggs, decreases the rate of protein synthesis (122, 123). Candidates for the factors involved include a glycoprotein with a molecular weight of 150,000 and smaller nondialyzable factors. Because the earliest events seen in egg activation are at the egg surface (12), this result becomes particularly interesting.

The "masking" hypothesis of mRNA sequestration is very attractive since egg ribosomes and soluble factors seem to be competent in translation, the protein spectrum synthesized is very much the same in the egg and zygote, and the rate of translation of mRNAs on polysomes in the egg and zygote is quite alike (27, 124, 125). Since the average number of ribosomes per polysome is about the same in eggs and embryos (27, 124), the mRNAs that are translated are probably being used as efficiently in the two cases. In fact, it has been estimated that the time an amino acid remains in a nascent chain is the same or even less in the egg than in the embryo (27, 125). As development proceeds, more mRNA and ribosomes are found in the polysomes, and this mRNA must be almost entirely of maternal origin (40). It appears, therefore, that the translational efficiency of egg ribosomes is not limiting. Rather, the defect seems to be the unavailability of the egg mRNA for initiation of translation. Little, however, is known of the mRNA-protein interactions in the egg or embryo. In addition to polysomal mRNA, there are ribonucleoproteins (RNP) of heterogeneous mass which seem to contain mRNA (126). Nonpolysomal RNP has been studied in the developing sea urchin embryo (126, 127) and has been postulated here and in other organisms to be a repository of stored information or "informosomes" (128). Little is known, however, about egg RNP, and in the developing sea urchin embryo there is now good evidence against the RNP being a storage form for newly synthesized mRNA to be utilized at a later embryonic stage (129) (see under "Ribonucleoprotein"). As mentioned earlier, histone mRNA has been found to reside in an egg postribosomal fraction, presumably as RNP (75, 84). The proteins adhering to the RNA have not yet been characterized and the chance of nonspecific protein-RNA interactions during cell fractionation (130) must be considered. Even in the case of histone mRNA, much remains to be done to indicate whether the messenger is really "masked." Thus, in the 15 years since Hultin's (20) first translation experiments with sea urchin ribosomes, although much has been learned, no final answer is available to the question of what causes activation of protein synthesis upon fertilization.

PROGRAMMED CHANGES IN PROTEIN SYNTHESIS

The determination of the embryo's behavior at any one time results from many levels of control. One of the more immediate factors is the stage-specific presence of various proteins. Examples are known of the appearance of enzyme activities at particular embryonic stages, but very little information exists about the control of synthesis and turnover of any specific proteins. New proteins are obviously required for development since embryos quite quickly stop all change when puromycin or cycloheximide is administered (22, 23, 131–133). These experiments do not indicate whether there must be synthesis of new kinds of protein at various stages or whether there is simply the replenishment of protein species which are turning over. Treatment of the embryos with actinomycin D from fertilization onward results in arrest as late as in blastulae (51, 52). This implies that no new mRNAs are needed to maintain early development (see above), but gives no indication of selective translation of stored mRNAs in these early stages.

Most of the information available on this point comes from analysis of the spectrum of total or newly synthesized proteins by gel electrophoresis or immunological assay. Serological assays of limited resolution detected new antigens appearing at the time of mesenchyme blastula, and these changes were blocked by actinomycin D (98, 99, 134). Early attempts to use polyacrylamide disc gel electrophoresis showed no change (96, 135) or only minor changes (95, 136) in the pattern of synthesis in different embryonic stages of the relatively few resolvable proteins. A more exacting analysis by Terman (97) indicated significant changes in the distribution of proteins synthesized in the zygote and mesenchyme blastula.

The work of Brandhorst (93), referred to above in the context of proteins synthesized upon activation of the egg, provides a detailed view of the constancy or change of the protein synthetic pattern. When soluble methionine-labeled proteins are subjected to isoelectric focusing in one direction and electrophoresis through a sodium dodecyl sulfate (SDS)-polyacrylamide gel in a second direction (94), over 400 spots can be resolved by autoradiographical methods. They are a subset of the total proteins, being methionine-containing soluble polypeptide chains. These proteins are probably synthesized on the more abundant mRNAs, and the translation products of the less frequent messenger would not be visualized. The gel patterns are very similar in preparations from pulse-labeled unfertilized eggs, zygote, blastula, gastrula, and pluteus. Although some of the spots have different intensities in the unfertilized egg and zygote, all but a few are in identical positions in these two stages. Despite the 5–30-fold increase in the rate of protein synthesis after fertilization, there appear to be few qualitative changes in the proteins being made. The protein populations continue to be very similar until hatching, with changes in some spots seen. Even at gastrula, when many new spots appear and changes in intensity occur, the majority of spots are in the same positions, and the total increase in spots from egg to pluteus is only 20% of the total. During this period, although most differences are due to the appearance of new spots, some species disappear and, in a few cases, appear and disappear. Quite an interesting finding is that the stained protein pattern is said to be identical at all stages, and these proteins fail to show label. Most of the proteins made during oogenesis, therefore, are conserved throughout development and are not made during embryogenesis. One intensely stained band is said to be unlabeled in the egg, but is intensely labeled in later stages. According to these results, most of the embryo proteins are synthesized at all stages of embryogenesis and in the unfertilized egg, but probably not during oogenesis. Many proteins do first appear, however, at about the time of gastrulation, and a number of proteins have stage-specific programs during earlier embryonic stages.

Protein synthesis has been said to be under translational control during early development (1, 2). This conclusion is based on the finding that the stage-specific translation patterns of preblastulae persist even in the presence of actinomycin D (95, 97, 137). However, these results are dependent on a very poor resolution of newly synthesized proteins. It will be most interesting when the O'Farrell gel system is applied to actinomycin D-treated embryos or activated merogones. Judgment as to whether the major changes in protein synthesis up to hatching are under translational (i.e., post-transcriptional) control must await these experiments. The actinomycin D or merogone experiments will not shed light on translational controls in postblastula stages since the embryos will begin to suffer from lack of needed new mRNAs. The clearest test of translational control will probably come from studies on specific proteins such as the histones. Shifts in the proportion of various histones and presence of new histone species might be correlated with the presence of the histone mRNAs as assayed in cell-free systems (see below). Other changes in the synthesis of specific proteins which might eventually be used in this respect include the synthesis of a collagen-like protein in the mesenchyme blastula (138, 139) and the stage-specific, actinomycin-resistant appearance of enzymes such as nucleotide reductase (91) and hatching enzyme (92).

PROGRAMMED CHANGES IN RNA POPULATIONS

Although fertilized eggs treated with actinomycin D or activated merogones continue to synthesize protein and go through cleavage, they do arrest at blastula (51, 52, 59, 140). Normal cleaving embryos do in fact synthesize mRNA, which appears on polysomes even during the first cleavage (40, 46). Although this transcription is evidently not essential for development to blastula, the RNA is utilized for protein synthesis in the cleaving embryo.

A role for the synthesis of pregastrula mRNA is suggested by actinomycin experiments on different stages of sea urchin and sand dollar embryos (140, 141). If actinomycin is continuously applied from any time after fertilization up to hatching, the embryos fail to develop past the swimming blastula. If, however, actinomycin is applied to mesenchyme blastulae, gastrulation takes place even though most of the RNA synthesis is inhibited. Similarly, when the drug is continuously applied to gastrulae, differentiation into prisms occurs. While these results might be explained by changes in permeability to the drug or to a loss of actinomycin sensitivity of particular transcription processes, the most straightforward interpretation is that mRNAs produced at one stage are necessary and sufficient to maintain development to a later stage. These observations indicate that the developmental changes in the embryo may be programmed by changes in the presence and location of mRNAs. A series of investigations have attempted to measure the degree of change in RNA populations in different embryonic stages.

In the 1960s a number of measurements based on RNA-DNA competition–hybridization were made to determine when the population of RNA sequences changed during development. Labeled RNA, obtained from a particular stage of development, was hybridized to DNA which had been affixed to cellulose nitrate filters. Unlabeled RNA from different stages was tested for its ability to compete against this hybridization. The greater the competition, the more alike the labeled and unlabeled RNA preparations were. Unfortunately, these hybridization methods were sufficient to detect only RNAs transcribed from reiterated DNA sequences (142). Since most of the mRNAs transcribed in the gastrula, for example, are from single copy DNA (143), it is probable that these filter hybridizations were mostly measuring heterogeneous nuclear RNA (hnRNA) which is complementary in part to reiterated DNA (144). Glisin et al. (145) found that the hybridization of labeled blastula RNA could be competed for by unlabeled RNA from blastulae or earlier stages, but not competed for as well by unlabeled gastrula RNA. These results implied that the blastula RNA sequences were present at earlier stages, but only partially present in the gastrula. Whitely et al. (146), using somewhat different but equally limited hybridization methods, found that labeled prism RNA was competed for less and less well as the unlabeled competitor RNA was taken from earlier stages. The labeled prism RNA was shown to contain new sequences not present in the earlier embryos. These experiments and additional similar studies with sand dollar and sea urchin embryos (147–150) have been recently reviewed (151). The main conclusion of these experiments are that each developmental stage has a characteristic population of reiterated transcripts, that some of these sequences present in the egg continue to be transcribed at all stages, and that most of the change in sequence that does take place occurs at or just prior to gastrulation, but there also are detectable shifts at earlier stages.

Britten and Davidson's group has made great strides in understanding the organization of sequences in DNA, especially in the sea urchin genome (143, 144, 152, 153). Since the gastrula mRNA sequences were found to be almost entirely complementary to single copy DNA (143), it was important to use hybridization methods which could assay the representation and overlap of single copy DNA in the mRNA populations and total RNA of different embryonic stages. Galau et al. (154) measured the extent of the single copy DNA sequences which were represented in gastrula mRNA by hybridizing puromycin-released polysomal RNA (to eliminate any non-mRNA) to purified single copy DNA under conditions of increasing R_0t (concentration of RNA $\times$ time of hybridization). Since the DNA was labeled, an accurate saturation value could be obtained. The result indicated that 1.35% of this DNA had hybridized, so that, if asymmetric transcription was assumed, this represented 2.7% of the single copy DNA, amounting to 1.7×10^7 nucleotide pairs. If the average length of mRNA is assumed to be 1,200 nucleotides (155), the gastrula mRNA complexity corresponds to about 14,000 different genes.

In addition to serving as a measure of the amount of DNA sequence represented in the RNA population, these hybridizations give interesting information about the abundance of particular RNA sequences. It is well known that the rate of RNA hybridization to DNA, when the RNA is in excess, can give an indication of the concentration of particular sequences (156, 157). When the rates of the hybridizations of Galau et al. (154), discussed above, were analyzed in this manner, the RNA concentrations were found to be considerably lower than would be predicted by dividing the amount of mRNA per embryo by the number of sequences predicted from the complexity measurements. The conclusion was drawn that a major portion of the mRNA (92%) is composed of a very limited number of sequences and that the hybridization reaction is really driven by only 8% of the mRNA, representing 1.4×10^4 diverse sequences. The most abundant sequences would not hybridize with enough DNA to be seen as a distinct component of the reaction. Galau et al. (154) calculate that, if the amount of mRNA per embryo is 7×10^{10} nucleotides (40, 158), the number of mRNA molecules per embryo would be approximately 5.8×10^7, of which only 4.7×10^6 would be in the complex class. On the basis of 1.4×10^4 different species, there would be about 340 copies of each per embryo, or less than one copy per gastrula cell. Similar methods were used to show that the saturation reaction with gastrula hnRNA is driven by 2.5% of the nuclear RNA (159). The total amount of hnRNA per nucleus is estimated to be 2×10^8 nucleotides (160, 161). The complexity of the hnRNA determined by Hough et al. (159) from the saturation level of hybridization is 1.74×10^8 nucleotides. Thus, there is less than one molecule of each type of hnRNA of the complex class per nucleus at gastrula.

Returning to the assay of the degree of change of RNA populations during development, the problem was now to test whether the sequences present at one stage are present in the mRNA preparations from other stages. Galau et al. (100)

have recently presented a set of elegant experiments which answers this question for most of these sequences (the very abundant mRNAs, perhaps a few hundred kinds, would not be resolvable by this method). The single copy sea urchin DNA of *S. purpuratus* was fractionated into two populations, one complementary to gastrula mRNA (mDNA) and the other with virtually no sequence complementarity to gastrula mRNA (null DNA). Each of these DNA fractions was hybridized to mRNAs of ovary, blastula, gastrula, pluteus, exogastrula, and three adult tissues, as well as to total RNA from oocytes. The maximal percentage of the two DNA preparations that can be hybridized with each of these RNAs was determined. When gastrula mRNA was used, 57% of the mDNA (a complexity of 17×10^6), but none of the null DNA, was hybridized. For comparison, only 0.67% of the unfractionated single copy DNA was hybridized with gastrula mRNA. When adult tissue mRNAs were used, 7–12% of the mDNA and 0–0.3% of the null DNA hybridized. These values indicated that a sequence complexity of $2.1–3.5 \times 10^6$ nucleotides, or $1.5–3.0 \times 10^3$ different mRNAs, was shared between the gastrula and each adult tissue. The various adult tissues shared most of this subset of sequences since there was no additivity of hybridization. Two of the adult tissues, coelomocyte and tubefoot, did not seem to have sequences represented in the null DNA although the intestinal mRNA hybridized to about 0.35% of this fraction. Since the null DNA made up most of the complexity of the single copy DNA, this saturation value represented a complexity of 3.7×10^6 nucleotides. Intestinal mRNA thus had a total complexity of 5.8×10^6 nucleotides, 36% of which was also found in the gastrula mRNA. By these criteria there appears to be considerable overlap in mRNA populations in adult and embryonic tissues and a more restricted set of sequences in the adult tissues.

Galau et al. (100) next looked at the hybridization of ovary mRNA and total oocyte RNA to the two DNA preparations and found a large overlap in sequence with gastrula mRNA. The oocyte RNA hybridized to the mDNA to the same extent as gastrula mRNA did. Virtually all the gastrula mRNA sequences must have been present in the oocyte (but they need not have been exactly the same molecules), an observation referred to in an earlier section as an indication (but not a proof) of stored maternal mRNA. In addition, the oocyte RNA hybridized to 1.4% of the null DNA, indicating the presence of RNA with a complexity of 20×10^6 nucleotides, in addition to the set shared with gastrula. The total oocyte RNA complexity, 3.7×10^7 nucleotides, was very similar to the value of 3.0×10^7 found for oocyte RNA of another sea urchin, *A. punctulata* (162), and was close to measurements of complexity made for oocytes of other species (163, 164). An interesting point is that the total RNA of the oocyte, although having a greater complexity than embryonic mRNA, had a much lower complexity than gastrula hnRNA (159). Ovary polysomal RNA hybridized to a smaller percentage of both mDNA and null DNA than to total oocyte RNA, with the result that, of a total complexity of 16.5×10^6 nucleotides in the ovary mRNA, 10.4×10^6 were shared and 6×10^6 not shared with the gastrula mRNA. The ovary mRNA had a more restricted set of sequences than the oocyte total RNA.

The sequences shared with gastrula mRNA were, therefore, not all present on polysomes in the oocyte. The rate constant of the hybridization reactions shows that there is about a 5-fold greater number of each mRNA per embryo in the oocyte than in the polysomal RNA per gastrula.

Hybridization with the other embryonic mRNAs gave interesting results as well. Galau et al. (100) showed that about 70% of the gastrula polysomal sequences was present in blastula polysomal mRNA (9.1×10^6 nucleotides), whereas a fairly large population of blastula polysomal mRNA (15×10^6 nucleotides) was not complementary to the gastrula sequences. The total complexity of blastula polysomal mRNA (27×10^6 nucleotides) and was only slightly lower than the complexity of total oocyte RNA (33×10^6 nucleotides) and was greater than the complexity of gastrula mRNA (13×10^6 nucleotides) or ovary mRNA (16×10^6 nucleotides). The experiments do not indicate to what extent overlap may exist in the ovary, oocyte, and blastula null DNA sequences. It is tempting to think that a set of oocyte sequences is transferred to polysomes from the cytoplasm during cleavage and another overlapping set of these sequences disappears from the polysomes in postblastula stages. Such hypotheses must await future experiments.

In the pluteus mRNA there appeared to be no null DNA sequences, and 82% of the gastrula mRNA sequences was still present. These sequences, being shared with the gastrula set, must have been in the oocyte as well. Exogastrula mRNA gave identical results, allowing Galau et al. (100) to conclude that the failure to gastrulate does not involve the loss of more than 2,000 different mRNAs.

These difficult but extremely informative experiments show the relative constancy of a set of sequences in polysomal RNA from the oocyte to pluteus. The role and fate of the polysomal and nonpolysomal sequences that hybridize to the null DNA cannot yet be determined. The total sequence complexity of the mRNA present in the various stages amounts to 6% of the single copy DNA or $2–3 \times 10^4$ different genes. Although all the detectable pluteus polysomal mRNA sequences are found in the oocyte, the shifts from blastula to gastrula and gastrula to pluteus do involve the change in thousands of different mRNAs present on the polysomes. These changes may reflect changes in subcellular localization of the molecules and/or selective differences in stability, synthesis, and degradation of RNA species. The results do not yet indicate which RNAs are transcribed at which stage, but they do show how and when the transcripts are used in the embryo. The data speak mainly of the presence of the class of complex, infrequent mRNAs, although over 90% of the mRNA is composed of only a small number, perhaps a few hundred, of sequences (154). These abundant mRNAs might be regulated quite differently from the complex class assayed by the saturation hybridization experiments. Yet the results of Brandhorst (93) give indications that this class (in so far as it is all translated with similar efficiency) may also be a reasonably constant population throughout development since the proteins synthesized at any one time are a reflection of the most abundant polysomal mRNAs. There are significant changes at both oogenesis and gastrulation that might indicate that more switching is involved in the abundant than in the complex RNAs. The experiments of Galau et al. (100) would be insensitive to changes in concentration of individual mRNAs of the complex class. Changes in the intensities of the spots in Brandhorst's work (93) indicate that the abundant mRNAs may be present in different amounts at relative stages. If regulatory signals in the sea urchin embryo are of a quantitative rather than a qualitative nature, the hybridization experiments on total mRNA populations might not be conclusive. Another possibility is that a great amount of qualitative variation in the RNA of different stages may reside in the hnRNA, which represents a much higher degree of single copy sequence, 28.5% in the gastrula (159). In fact, this may be one interpretation of the experiments of Whitely's group (146–151), although his approach is limited to reiterated transcripts present in abundant quantities. What is surprising, however, is that even if hnRNA populations shift during development, the polysomal mRNA populations, which would be thought to have a greater effect on cellular action by virtue of the protein populations they code for, are fairly constant. It is important not to lose sight of the fact that we are talking about thousands of different RNA species, most of which may be required for mundane "housekeeping" activities. The RNAs required for a developmental change might be a very small proportion of the total sequences.

One approach to unraveling these problems is to look at the regulation of synthesis and translation of particular well defined mRNAs. The best understood case, the histone mRNAs, is discussed in a later section. Histones serve as a good example since both qualitative and quantitative changes in histone synthesis do occur.

In summary, a very large part of the single copy genome complexity is present as RNA in the sea urchin gastrula. About 28% of the single copy genome is represented in the hnRNA, and a far more restricted class of sequences is present on polysomes, 2.7% of the single copy DNA. The vast majority of both types of sequence is present in concentrations of less than one molecule per cell, but the major amount of RNA is found in a limited number of abundant sequences. Assay of variations in kinds of RNA sequences in different stages shows that most of the gastrula and prism polysomal mRNA sequences are found to be present in the oocyte, but transient changes in polysomal mRNA sequences are observed during development.

All stages of the embryo are active in transcription with considerable energy resources involved in this process. Instantaneous synthetic rates of $7.2–13.0 \times 10^{-15}$ g of nucleotide per embryo per min have been measured for the combined populations of hnRNA and mRNA (160, 165, 166). The instantaneous rate of synthesis is higher at blastula than at pluteus (160), and there is little or no ribosomal RNA synthesis at the earlier stage (45, 167–170). Despite the evidence presented above that showed a relative constancy in polysomal mRNA sequences, the synthesis of hnRNA and mRNA is a substantial process at every developmental stage.

TRANSCRIPTIONAL EVENTS

Structural RNAs

The level of synthesis of 18 and 28 S ribosomal RNA (rRNA) before hatching blastula is either very low or nonexistent (45, 167–170). Since there is extensive synthesis during cleavage and blastula stages of hnRNA of size spanning the 18 S and 28 S region of gradients (160, 171–174) it is very difficult to judge whether or not small amounts of rRNA are being produced in the early embryo. Emerson and Humphreys (175, 176) have claimed that there is a constant rate of rRNA synthesis on a per nucleus basis, until the embryo begins feeding after pluteus (177). However, Sconzo et al. (167, 168) present convincing evidence, with another sea urchin species, that there is an activation of rRNA synthesis after hatching, which cannot be explained by the increase in number of nuclei. The final resolution of this problem will have to come from hybridization experiments with the use of purified sea urchin ribosomal DNA probes (178, 179). In any case, the bulk of rRNA in the embryo at all stages up to prism is stable RNA synthesized in the oocyte. When rRNA is made, it appears to be transcribed as a larger precursor sedimenting at 33 S (180) and is accompanied by the presence of RNA polymerase I (181, 182). The ribosomal RNA synthesis and processing appear to be analogous to those of other eukaryotic cells (183).

Although early reports indicated that 5 S ribosomal RNA and tRNA were labeled only after mesenchyme blastula, except for tRNA pCpCpA end labeling (41–45, 169, 170), a more recent study shows that cleaving embryos incorporate [^{3}H] guanosine into both tRNA and 5 S RNA (184, 185). The radioactive label was recovered as [^{3}H] GMP from hydrolyzed tRNA so that the incorporation was shown to be not merely due to pCpCpA end labeling. Methylation of tRNA during preblastula stages has also been observed (185). Virtually nothing more is known about the transcription of these types of RNA in the embryo. It seems probable, therefore, that the 5, 18, and 28 S RNAs are not synthesized coordinately. A set of five bands, somewhat larger than the 5 S RNA, can be resolved on polyacrylamide gels (185). Some of these small RNAs are synthesized during sea urchin embryo cleavage, but at least one is synthesized only at hatching. One of these probably corresponds to the "5.8 S" RNA, which is associated with the 28 S rRNA in the ribosome (186) and which is coded for by part of the ribosomal 18 + 28 S gene complex, at least in *Xenopus laevis* (187). These structural RNAs all must be investigated with more modern techniques.

Heterogeneous Nuclear RNA and Messenger RNA

Characteristics of Nuclear and Polysomal RNA Perry (183) has recently reviewed the characteristics of hnRNA and mRNA and the extent to which a precursor-product relationship exists between them. From a collection of data on a wide variety of eukaryotes, it has been said that a portion of the highly complex, rapidly turning over, large hnRNA may serve as a precursor to the 5–10-fold less complex, more stable, smaller mRNA. The evidence for this statement with specific sequences, however, is quite limited. The best studied example is the globin mRNA, which has been found to be represented as a larger sequence in the hnRNA (188–193). The most recent of these experiments (191–193), which are quite elegant, show the precursor to be 14–15 S. there is no evidence for sea urchin hnRNA serving as a precursor to any specific mRNA, but the characteristics of the two RNA classes are very much like those of other eukaryotic organisms.

Although most of the newly synthesized RNA never leaves the nucleus (129, 161, 171, 194), polysomes are found to contain internally labeled mRNA within several minutes of addition of $[^3H]$adenosine to the embryos (129). About 5–10% of the newly synthesized RNA has a cytoplasmic location after 5–10 min of incubation (161). The two RNA populations have a difference in stability, with the nuclear RNA showing a half-life of decay of 7–12 min, whereas the polysomal mRNA decays with a half-life of 65–70 min (129, 161). The rates of synthesis and turnover of the RNA populations are considered further below.

As previously discussed, the hnRNA appears to be over 10-fold as complex as the polysomal mRNA of gastrulae (153, 159). Both populations are similar to the extent that most of the different sequences are present at less than one copy per cell. Although fluctuations in the complex class of mRNAs at different stages of development have been discussed (100), nothing is known about hnRNA sequence changes. A further difference between the two populations is the greater extent of repetitive DNA sequences present in hnRNA (144).

The size of newly synthesized hnRNA and mRNA has been measured by many workers. Kung (174) used denaturing conditions of separation in looking at the size distribution of these RNAs. He used denaturing dimethyl sulfoxide-sucrose gradients or acrylamide-formamide gels to measure the size of labeled nuclear and cytoplasmic RNA from hatching blastulae. It was found that the size of the newly synthesized nuclear RNA was quite heterogeneous, with a broad distribution from $0.5–3.0 \times 10^6$ daltons, whereas the cytoplasmic RNA was somewhat smaller. Similar observations on the cytoplasmic and hnRNA from a variety of stages were made by Nemer's group (195–198). Many examples exist of a somewhat larger size distribution on nondenaturing gradients of the two RNA classes (50, 161, 172, 173, 175, 199). These gradients may lead to overestimates of size since aggregation can occur in nondenaturing medium, even after a denaturation step prior to the gradient centrifugation (195). A large cytoplasmic RNA class up to 15×10^6 daltons has been seen even with the use of denaturing gels in blastulae and gastrulae of *P. lividis* (200).

Rate of Synthesis and Turnover When the kinetics of incorporation of radioactive nucleosides into RNA was first studied, it was noted that the rate of incorporation, linear at first, gradually reached a plateau (170, 201). Kijima and Wilt (165) were the first to take advantage of the knowledge that the overall amount of RNA is constant during development (158), and they made the inference that the decreasing rate of incorporation meant that the RNA was turning over. They also understood the importance of measuring the specific activity of the precursor pool. Rates of synthesis at different developmental stages could be calculated from the percentage of total GMP in RNA that turns over during a set time period. From these calculations, they concluded that the rate of transcription per nucleus was actually decreasing during development since the percentage of embryonic RNA turned over does not increase as fast as cell number. Since autoradiographic experiments showed that the major part of the RNA remained in the nucleus, the rapid turnover rates measured at each stage are mostly a reflection of hnRNA metabolism, but cytoplasmic decay of the RNA could not be ruled out. Aronson and Wilt (194) isolated the pulse-labeled RNA from nuclear and cytoplasmic fractions and showed the rapid labeling and fast decaying component to be restricted to the nucleus.

A similar approach was taken by Brandhorst and Humphreys (160, 161). They used a novel measurement of ATP pool size by the luciferase assay (176) and a decay analysis of the accumulation of the radioactive RNA to find both the instantaneous rate of RNA synthesis and the half-life of the RNA, assuming steady state level. The first set of experiments (160) analyzed total RNA synthesis at blastula and pluteus stages. Two stability classes were found in each case, a rapidly turning over component with a half-life of 5–7 min (35% of the steady state level of newly synthesized RNA) and a more stable class of the remaining RNA with a half-life of 1 hr. In the second set of experiments (161) the RNA was extracted from nuclear and polysomal fractions. Using mesenchyme blastulae, they found that the nuclear RNA had a half-life of 7 min, the polysomal 80 min, and the total RNA had a biphasic decay with 7- and 80-min components. There is a slow accumulation of counts in the cytoplasm until, at steady state, most of the radioactivity is in polysomes. Wu and Wilt (202) examined the kinetics of incorporation into total RNA and poly(A)-containing RNA in cleavage, blastulae, and gastrulae. The accumulation curves were very much the same for the RNA populations in each stage, but there was some evidence for greater representation of the poly$(A)^+$ RNA in the cytoplasm. An unexplained feature was the linear incorporation over several hours into total and poly$(A)^+$ RNA during cleavage.

Dworkin and Infante (129), using similar methods, showed the half-lives of nuclear and polysomal RNA to be 12 and 65 min, respectively. Instantaneous rates of synthesis were also comparable to previous data. Grainger and Wilt (166) used a completely different approach, measuring density shifts of RNA labeled with ^{13}C and ^{15}N precursors to measure the rate of RNA synthesis. Results were presented only for total RNA and only one decay component could be seen, with a half-life of 23 min, a rate of synthesis similar to that found by other workers. However, longer half-lives of mRNAs were determined when a pulse-chase approach was used. Nemer et al. (196) showed that [^3H]uridine-labeled RNA decayed over a period of several hours during a cold uridine chase. More recently Galau et al. (203) measured the mRNA half-life to be about 5 hr by using [^3H]guanosine as the precursor and by following the incorporation over a 40-hr period.

In conclusion, there appears to be a rapid rate of RNA synthesis amounting to a turnover of 0.7–1.34% of the total RNA every 10 min (165) at stages from midcleavage to pluteus. At any one time up to 5-fold more precursor is incorporated into the nuclear, rapidly turning over component than into the more stable cytoplasmic RNA (129, 161). These results are impressive, considering the relative constancy of the population of mRNA present in the cell at different developmental times. Such synthetic activity may be indicative of a subtle quantitative control in which changes in development result from different relative concentrations of the various mRNAs. Perhaps even more dramatic is the intense synthetic activity associated with the hnRNA during cleavage and blastula formation, as well as in later stages.

Polyadenylation of RNA Following egg activation, some of the stored RNA is polyadenylated in the cytoplasm (204–206), even in activated enucleated merogones (205). The level of poly(A) in the zygote rises 2-fold within 30–120 min of fertilization, and the increase occurs predominantly in the polysome fractions (205, 207). The labeling is unaffected by doses of actinomycin that inhibit most uridine incorporation or by ethidium bromide, which inhibits mitochondrial transcription (205). Polyadenylated RNA molecules do exist in the egg, but their size distribution is somewhat different from those in the zygote (208). It is not clear to what extent the zygotic polyadenylation occurs on the preadenylated RNAs. Since the polyadenylation can occur in NH_4^+-activated eggs, it is not causally linked with events such as cortical granule breakdown, cell division, or cell cycle events (35).

The only functional role thus far shown for polyadenylated mRNA is that the stability of globin mRNA injected into *X. laevis* oocytes and eggs is greater if the mRNA has poly(A) at its 3′ end (209, 210). It is interesting to examine earlier data on the synthesis and turnover of polyadenylated RNA in the sea urchin embryo. There do not seem to be great differences in the turnover or rate of synthesis of polyadenylated RNAs in polysomes or in total RNA, nor between polyadenylated and nonpolyadenylated total RNAs (202). Polyadenylation does not seem to play a role in the activation of maternal mRNA since, when the poly(A) formation was blocked with cordycepin, no effect was observed on embryonic protein synthesis (211).

There is now a substantial amount of evidence for the presence of three classes of mRNAs in the embryo, characterized with respect to their poly(A) content: poly(A)-containing, poly$(A)^-$nonhistone, and poly$(A)^-$histone mRNAs (195–197, 212). These classes are potentially interesting since the polyadenylation may provide an easy identification mechanism that can be exploited for purposes of control. The poly $(A)^-$ mRNAs were shown clearly not to be degradation products of the poly$(A)^+$ class by mixing experiments and by demonstration of different sequences by hybridization (195). The poly$(A)^+$ and non-histone poly$(A)^-$cytoplasmic RNAs showed identical kinetics of emergence into the cytoplasm, and both were shown to be transcribed from mostly single copy DNA (195). The poly$(A)^-$ mRNA contained a large amount of histone mRNA sedimenting at 9–10 S, but also had a broad distribution of sequence sizes, as analyzed in formamide gradients, with a mean of 21–22 S in the gastrula (195). The poly(A)$^+$ mRNAs were a bit smaller on the average.

The ratios of the three classes have been carefully monitored through development (197). Polysomal RNA was purified by EDTA release from polysomes to be sure that mRNA was being assayed. The incorporation of precursor into 9 S poly(A)$^-$ mRNA is very high during the early cleavage, amounting to as much as 60% of the non-4 S newly synthesized mRNA. During early blastula this fraction decreases to about 30% and tapers off slowly during later development, reaching a level of about 12%. Much of this material is obviously histone mRNA, but no characterization is made to prove this. The poly(A)$^-$RNA greater than 9 S remains at about 30% of the pulse-labeled RNA throughout development. On the other hand, the poly(A)$^+$ mRNA is a very small percentage of the labeled material during cleavage, but during the blastula period the amount rises from 5% to over 50% of the total counts in mRNA. This percentage stays fairly constant from then on. Since the poly(A)$^+$ and poly(A)$^-$ populations are different in sequence, at least in late blastula (195), the shift in amounts of the two classes appearing on polysomes at a key time during development may be of significance. The hybridization experiments (195) may, however, only monitor the more abundant RNA populations since the labeled complementary DNA (cDNA) made from poly(A)$^+$ mRNA was tested for hybridization to the unlabeled poly(A)$^+$ and poly(A)$^-$ mRNA preparations. This approach, combined with that of Galau et al. (100), should give a fairly complete idea of whether the poly(A)$^+$ and (A)$^-$ mRNAs differ qualitatively or quantitatively.

There is also information about the degree of polyadenylation of hnRNA in the embryo. Dubroff and Nemer (198) have identified three classes of hnRNA based on their ability to bind to poly(U)-Sepharose and oligo(dT)-cellulose. The three classes, α, β, and γ, have a short internal poly(A) element, a long poly(A) element at the 3$'$ terminus, and no poly(A), respectively. The total hnRNA had a mean sedimentation of 36 S, only 2.5 times the length of the mRNAs of the embryos, but the β class taken alone was smaller, sedimenting at 31 S. The three classes show different levels of incorporation of [^{3}H]uridine in a 1-hr pulse (much longer than the turnover time, so probably indicative of the steady state amount) at different developmental times (213). The γ class is always the most prominent, usually comprising about 64% of the radioactivity, but showing a rise to 76% during the late cleavage–early blastula. The β class shows a corresponding decrease during early blastula and increases from about 10% to 32% at mesenchyme blastula. The α class varies from 5% to 15% with the maximum reached in the midblastula. Comparisons of these shifts with the changes in mRNA populations suggest a possible relationship between a portion of the γ-hnRNA with poly(A)$^-$ histone mRNA and a part of the β-hnRNA with the poly(A)$^+$ mRNA. The α-hnRNA is most prominent during the time that transcription is expected to be required for the later developmental events.

Poly(A)$^-$ and poly(A)$^+$ mRNAs from sea urchin eggs and embryos have been shown to be active in a wheat germ cell-free protein synthesis system (68, 217). Ruderman and Pardue (68) have shown that some of the poly(A)$^-$ and poly(A)$^+$ mRNAs code for proteins of the same size. An egg poly(A)$^-$ mRNA, coding for a 60–80,000-M.W. protein is not detectable, however, in poly(A)$^-$ RNA from the embryo, and other differences between the size of proteins synthesized on the poly(A)$^-$ and poly(A)$^+$ templates have been observed. A less specific criterion, the ratio of incorporation of individual amino acids into proteins synthesized on the two RNA populations, also indicates some differences in the populations (212). The finding that amphibian oocytes contain poly(A)$^-$ and poly(A)$^+$ histone mRNA (68, 214) completes the negation of the generalization that all histone mRNA is poly(A)$^-$ and all nonhistone mRNA, poly(A)$^+$. In fact, the demonstration of differences in sequence, size, and time of synthesis of these various classes of RNA may eventually provide insight into the translational controls presumed to exist during sea urchin development. The poly(A) content might be a recognition signal for the mobilization of particular sequences for translation.

Ribonucleoprotein Cytoplasmic RNA is present in polysomes and as a free ribonucleoprotein complex. The RNP particles, first found in fish embryos (215), were also described in sea urchin embryos (127). Extensive characterization of the RNP in sea urchin embryos has shown that they contain RNA of heterogeneous size and are not merely RNAs bound to the small ribosomal subunit (126). These particles have been termed "informosomes" and have been postulated to represent mRNAs in transit between the nucleus and the polysomes (104). In some cases specific mRNAs have been identified in free RNPs. Such examples include the already mentioned case of sea urchin egg histone mRNA (75, 84) and α-globin mRNA (216), myosin mRNA (217), and actin mRNA (218) from other species.

The sea urchin embryo, with its activation of protein synthesis on fertilization and its program of utilization of mRNAs during early development, is a system in which RNPs have been thought to be a repository of information for delayed utilization (128). A recent analysis of the kinetics of synthesis and turnover of RNA in embryonic RNP indicates that this view is incorrect (129). The bulk of the free RNP containing newly synthesized RNA in the developing embryo is shown to be neither a storage form of stable mRNA nor a precursor vehicle for polysomal mRNA. There is no constant relationship between the amount of newly synthesized RNA in polysomes and free mRNP in various developmental stages, and the ratios of poly(A)$^+$ RNA are not always the same in the two classes of particles. The most convincing evidence is that the instantaneous rate of accumulation of label into RNP is much lower than into polysomes (5.0×10^{-17} g versus 2.2×10^{-16} g of RNA per min per nucleus). Since the rate of newly synthesized RNA appearing in RNP is less than that in polysomes, the RNP cannot serve as a simple precursor. The turnover of the

RNA in RNP is faster than in polysomes, however. The half-life for RNA in RNP is 40–50 min, in polysomes 80–90 min, and in nuclei 20 min. These data are inconsistent with the view of RNP as the location of "masked" stable RNA, at least for RNA synthesized by the embryo. Dworkin and Infante (129) propose that the RNP contains an excess of mRNA which cannot be accommodated by the polysomes. There may be some translational selection at this point and the extra amount in the RNP preferentially decays. The RNP containing the maternal mRNA may, however, be a storage form of RNA to be used at particular times during development. Results on the nature of RNPs containing newly synthesized RNA give no information on this point.

There has been some discussion that regulation of protein synthesis may occur via different translational rates on particular polysome classes. Early reports that "light" polysomes in the cleaving embryo are inactive came from the observation that, although these polysomes contained most of the newly synthesized polysomal RNA during cleavage, most of the incorporation of amino acids occurred in heavier polysomes (49, 127). It is now known that the larger polysomes attain a higher specific activity of incorporated amino acids since they contain longer nascent protein chains than the lighter polysomes (50), and the newly synthesized RNA accumulating on the light polysomes is mostly small histone mRNA (50, 76). There is no convincing evidence, therefore, for a class of inactive polysomes.

A set of observations by Nemer's group, however, has indicated that different polysomes may translate mRNAs at different rates. The poly(A)$^+$ mRNAs, although similar in size during different developmental stages, appear to be found in smaller polysomal classes in early development and shift to larger polysomes at blastula (196, 197). There seems to be a switch from incomplete to complete ribosomal loading, perhaps indicating a more frequent initiation of new polypeptide chains. The distribution of poly(A)$^-$ mRNAs among different sized polysomes does not seem to change during development, and, in fact, at gastrula they seem underloaded in comparison with the poly(A)$^+$ class. The poly(A)$^-$ histone mRNAs are preferentially found in the small polysomes, as expected from their size. These observations may indicate that different mRNA populations have a level of translational control based on selective, stage-dependent initiation rates.

Control of Transcription Little specific information is available about the control of transcription in sea urchin embryos, but it is worth mentioning that, for two reasons, the sea urchin is providing one of the most powerful systems in which to study this problem. First, the sequence organization of the sea urchin genome is the best studied of any eukaryote. Repetitive and nonrepetitive sequences are interspersed, with about 50% of the genome consisting of 300–400 nucleotide repetitive sequences interspersed with unique sequences approximately 1,000 nucleotides in length (152). The remaining part of the genome consists of longer period interspersed organization, tandemly repeated DNA, and long stretches of single copy DNA. Hybridization experiments with gastrula mRNA show that messengers are almost entirely transcribed from single copy sequences (143), whereas the majority of hnRNA molecules may be transcribed from both reiterated and single copy DNA (144). There are indications that this pattern is also true for blastula RNA, although the analysis is not as extensive and the histone mRNA was surprisingly not measured as repetitive transcript (219). This pattern of sequence organization and transcription fits in well with ideas on redundant recognition elements residing alongside structural gene sequences (155, 220, 221). Davidson et al. (153) have found that single copy DNA sequences adjacent to the short repetitive elements are greatly enriched in mRNA coding sequences. These single copy "repeat-contiguous" sequences are the genes for 80–100% of the mRNAs found in gastrula, indicating the importance of the interspersed repeats. Further work along these lines obviously will use recombinant DNA techniques enabling one to follow the sequence organization and transcriptional patterns of individual specific genes. Because of the previous work on genome organization and developmental regulation, the sea urchin will be most useful for these experiments. A second reason for interest in transcriptional controls in the sea urchin embryo is the large amount of information available on sea urchin histone genes (see below). No other eukaryotic genes coding for mRNA are as well studied. Transcriptional and translational regulation of histone synthesis in the embryo can be approached experimentally with the use of pure histone mRNA and histone gene probes.

Mitochondrial RNA

Much of the RNA synthesis in the very early cleaving embryo takes place within the mitochondria in the cytoplasm (46, 222–225). Evidence comes from autoradiographic analysis (46), the use of activated enucleated egg fragments (222, 223), isopycnic centrifugation of sea urchin embryo homogenates (224), ethidium bromide sensitivity (46, 224, 225), and rifampicin sensitivity (225). As early as the first cell cycle, mitochondrial and nuclear RNA synthesis can be detected (46, 225). There appears to be an increase in incorporation of precursors into mitochondrial RNA within the first 10 min after fertilization, but, since sufficient uptake and pool studies have not been done, one cannot yet conclude that there is a true increase in synthetic rate at fertilization (223, 225).

Mitochondrial RNA was shown to be of a variety of sizes with possible species corresponding to rRNA, mRNA, and tRNA (222–224, 226). Devlin (227) has recently presented a comprehensive report on the kinds of RNA synthesized by the mitochondria at different stages of development. He separated mitochondrial RNA into poly(A)$^+$ and poly(A)$^-$ fractions and analyzed them by electrophoresis on polyacrylamide gels. The poly(A)$^+$ RNAs were resolved into eight species with molecular weights varying from $3.1–5.6 \times 10^5$. These RNAs were synthesized by anucleate fragments as well as whole embryos, with the same ratio of incorporation into the eight fractions. Although the percentage of newly synthesized poly(A)$^+$ RNA found in the cytoplasm shifted from 13% in cleavage to 39% in mesenchyme blastula, the percentage of incorporation into poly(A)$^+$ RNA within the mitochondria remained fairly constant at 49–53%. The eight poly(A)$^+$ species were found to be synthesized in the same proportion at all developmental stages, and the rate of incorporation per embryo remained fairly constant, although the cytoplasmic poly(A)$^+$ RNA synthesized per embryo increased by a factor of over 100. These experiments provide excellent evidence for the mitochondrial origin of these RNAs. In addition, Devlin (227) was able to show that these RNAs were found on mitochondrial polysomes and that the poly(A)$^-$ RNAs made in the mitochondria were almost certainly mitochondrial rRNA. Devlin estimates that if the molecular weights of these RNAs are summed, account is made of tRNAs probably synthesized within mitochondria, asymmetric transcription is assumed, and a value of 9.5×10^6 daltons is used for the mitochondrial genome (228), over 96% of the information in the mitochondrial DNA would be utilized.

SPATIAL INFORMATION

This review has been concerned almost entirely with the temporal aspects of developmental change. Yet an equally important aspect of development is the determination of the position of differentiation in the embryo. There is an extensive literature on specific cell movements and the fate of the cleavage products in the sea urchin embryo (38). Perhaps the best known example is Hörstadius' work on the fate of the micromeres, mesomeres, and macromeres of the 16-cell sea urchin embryo (229, 230). The micromeres, 4 cells at the vegetal pole, will divide into cells which migrate into the blastocoel at the mesenchyme blastula stage and form skeletal elements. Mesomeres, 8 cells in the animal half of the 16-cell embryo, and the macromeres, 4 large cells between the other two cell types, also have different developmental potentials. The segregation of determinants at the 16-cell stage and the ability to isolate the three cell types in preparative amounts (231, 232) have made this a favorable system for investigating the biochemical basis of localization of potential for determination.

A basic assumption has been that the expression of the determinants involves some form of differential gene action. Much of the work with the system has involved the comparison of RNA populations and protein synthesis in the various cell types. The chemical basis of the determinants, however, may reside in substances so rare as to make detection virtually impossible. The various cells of the 16-cell embryo all have reasonably similar rates of [^{3}H]uridine and [^{3}H]leucine incorporation (199, 231, 233), although at the 32-cell stage the incorporation of [^{3}H]uridine is lower in the micromeres (233). The three cell types all appear to synthesize microtubular proteins (234). An interesting observation has been made by Mizuno et al. (232) using competition-hybridization assays to study the degree of similarity of the RNA in the various cell types of sand dollar embryos. The reactions, done with DNA loaded on filters, only measure the more abundant transcripts from reiterated DNA, and competition kinetics is sometimes difficult to interpret since the more abundant sequences in the competitor RNA could be the most effective in the competition reaction. The experiments showed that labeled RNA from the 16-cell embryo was competed for equally well by unlabeled RNA from micromeres, mesomeres, and macromeres, by 16-cell embryo RNA, and only slightly less by unfertilized egg RNA. The unlabeled egg-type RNAs thus seem to be equally distributed, and perhaps there is a small addition in the number of sequences accumulated after fertilization. However, when labeled micromere RNA was competed for with these unlabeled RNAs, unfertilized egg RNA was a significantly poorer competitor than 16-cell embryo RNA. This difference was not observed when labeled mesomere and macromere RNA was used. There appear to be new sequences synthesized in the micromeres that are not present in the egg or in the macromeres and mesomeres. It will be of great interest to use more detailed hybridization analysis to determine whether these are true qualitative differences and whether the cells also show differences in single copy transcripts. Also useful will be analysis of the newly synthesized proteins with highly resolving two-dimensional gels.

Although little has been learned about the chemical basis of prelocalization of determinants in embryos thus far, the sea urchin embryo should provide an excellent system for further study.

HISTONE GENE SYSTEM

Histone Messenger RNAs

The advances in identification of histone mRNAs and characterization of histone genes have been summzried in a recent review (235). Since this information provides perhaps the best system for understanding specific regulatory events in the embryo, it is worth relating some of the advances in the histone system and pointing out several informational transitions which can be approached. The initial observations that led to the identification of histone mRNAs were made with HeLa cells (236–238) and not the sea urchin embryo. Soon thereafter, similar experiments by Nemer and Lindsay (39) and Kedes et al. (69) and Kedes and Gross (76, 218) showed that the light polysomes of the cleaving embryo were making a histone-like product and contained a newly synthesized RNA sedimenting at 9 S. The nascent products made on these polysomes were histone-like in their electrophoretic mobility and cation exchange chromatographic elution profile (70). Newly synthesized RNA on these polysomes could be resolved into specific fractions by gel electrophoresis (76–82) and was thought to correspond to the mRNA coding for the different histone proteins. A more convincing indication that this was indeed true came from sequencing studies and translations in cell-free systems. The various gel-purified RNA fractions were shown to contain different sequences by fingerprinting techniques and the fastest migrating bands were characterized as H4 mRNA by the sequencing of T1 ribonuclease digestion products (71, 79, 80, 240). Initial attempts to characterize the RNA in the bands by analysis of products of translations in cell-free systems allowed an identification of H1 and H4 mRNAs, but the remaining histone mRNAs were not resolved sufficiently on the gels (71, 72). Gross et al. (82) elegantly separated the individual mRNAs on urea-acrylamide gels and were able to assign a coding specificity to each band of [^{3}H] RNA. The identification of these mRNAs as the template for each of the major histone proteins is now complete, based on work with the isolated (see below) histone genes. The H2A, H3B, and H3 mRNAs have been shown to hybridize to DNA that, when sequenced, contains the expected coding specificity for the histones (241, 242).

Histone Genes

Parallel to the work on the characterization of the mRNAs, much was being learned about the organization of the histone genes. Kedes and Birnstiel (243) made many of the key observations by hybridization of the ^{3}H-9 S gradient-purified RNA to sea urchin DNA. The genes were estimated to be 400-fold reiterated from the kinetics of hybridization in conditions of DNA excess, although the results were not indicative of a single class of reiterated DNA. Kedes and Birnstiel also showed that the DNA complementary to the mRNA had a higher buoyant density than the bulk of the DNA. When sonicated DNA was used, the buoyant density of the complementary sequences markedly increased as if some noncomplementary AT-rich sequences were being released from association with the GC-rich histone sequences. Because the 9 S RNA hybridized to the DNA over several log units of the DNA C_0t curve, it was important to extend these observations with the use of gel-purified RNA fractions. Weinberg et al. (77) showed that each fraction hybridized to DNA of the same buoyant density, several fractions removed from the main DNA band on a CsCl equilibrium density gradient. In addition, the hybrids were shown to have a very high melting temperature indicative of a GC content of about 54%, with little low melting material. These RNA fractions, containing few if any contaminating sequences, show one component, second order kinetics when hybridized to DNA in conditions of DNA excess, and up to 90% of the RNA anneals to the DNA. Four of the RNA fractions hybridized identically, with indications that the genes are reiterated 1,000–1,200 fold. Reiteration values of 300–1,200 have been found for histone genes in a variety of sea urchin species (77, 79, 240, 243, 244). These experiments with individual histone mRNAs were highly suggestive of a tandemly repeating structure in which the genes for the various histones are intermittently clustered.

This model became more probable when each mRNA fraction was hybridized to greatly enriched histone DNA (245). The histone DNA could be purified more than 100-fold by repeated equilibrium centrifugations in actinomycin D-CsCl gradients (245, 246). This enriched DNA, even though selected by many fractionation steps, showed no loss of any of the sequences hybridizable to the histone mRNAs. In fact, the enriched histone DNA hybridized with equal efficiency to each of four gel-purified histone mRNAs (245). More recently the model of clustering of the five histone genes has received complete verification from experiments which used restriction endonucleases. Partially purified histone DNA was digested with restriction enzymes, and the products were analyzed by electrophoresis on Agarose (78, 81, 247, 248). Although there are a wide variety of nonhistone DNA sequences present, cleaved into a distribution of fragments visible as a smear on the gel, definite bands can be identified. Using *S. purpuratus* DNA, Weinberg et al. (247) and Kedes et al. (78) showed that Eco RI digested the histone DNA into two visible fragments of 1.8–2.2 and 4.2–4.8 kb, and Hind III digestion yielded a single fragment of 6.0–7.0 kb. Both groups showed that these fragments hybridize with 9 S mRNA and that the smaller RI fragment contains the Hind III digestion site. The 6–7-kb repeating structure of the histone DNA was thus demonstrated and was further indicated by the sizes of the histone DNA obtained by partial digestion with Hind III (247). These experiments were paralleled by work with *P. miliaris* histone DNA done by Birnstiel's group (81, 248). These investigators also demonstrated a 6-kb repeating unit by restriction enzyme digestions including conditions of partial digestion. Clustering of the different histone genes in the 6–7-kb repeat unit was demonstrated by these three groups by hybridizing the individual histone mRNAs to the restriction enzyme digestion products of histone DNA (78, 81, 247, 248) or of the DNA of cloned plasmids containing histone gene fragments (78, 249). All the histone mRNA fractions hybridized to the 6–7-kb unit fragment in both species and preliminary maps of the distribution of the various histone genes were presented (78, 81, 247).

Further mapping was carried out with the use of mRNA hybridizations to fragments of restriction enzyme-digested *P. miliaris* histone DNA (248) and to restriction enzyme-treated plasmid DNA which contained the *S. purpuratus* histone gene fragments (250). Overlapping fragments from digestions with several restriction enzymes could be oriented with respect to the mRNAs with which they hybridized. The order of the genes was the same in the two species: H1, H4, H2B, H3, and H2A. The polarity of the coding strand was determined by exonuclease and strand separation experiments (250, 251), and transcription must occur in the H1→H4→H2B→H3→H2A direction. In addition, the noncoding sequences inferred from Kedes and Birnstiel's original hybridizations were demonstrated by electron microscopic visualization of RNA-DNA hybrids (252) and R loops (253) and suggested by denaturation maps (254). These sequences alternate with each histone gene in the repeating structure but have yet to be determined to be "spacer" sequences in the sense of being non-transcribed or non-functional.

The hundreds of histone gene repeats appear to be quite similar by the criteria of sharp melting and high RNase resistance of the hybrids formed with the mRNAs (77, 79, 244). The finding of a discrete size of histone repeat by restriction enzyme analysis also speaks for the homogeneity of the genes (78, 81,

247, 248). Yet our laboratory has recently demonstrated that both enriched histone DNA and unfractionated DNA contain a considerable variety of histone gene repeat lengths as seen by hybridizations of histone RNA probes to restriction enzyme-produced fragments resolved on Agarose gels (255). Within a single individual urchin a major fraction of the histone genes may even be several kilobases larger than the modal 6.3 kb repeat, and the predominant size classes can be quite different from individual to individual. These length differences, however, can be shown to result from variations of length of small regions of the repeat, the major portion of the repeat being identical in all units. The histone genes, therefore, appear to be quite similar in sequence within a species. Hybrids made with the mRNA of one species and the DNA of another species, however, melt at a lower temperature than the homogenous hybrids (77, 244, 256). The lower melting temperature has been attributed to base substitutions occurring at the third base position of codons which would not change the specificity for a particular amino acid sequence (77, 244, 256). This hypothesis has turned out to be true since Grunstein et al. (80) have shown that *S. purpuratus* and *L. pictus* H4 mRNAs contain numerous base substitutions allowable from the degeneracy of the genetic code. Rates of base substitution of $3–6 \times 10^{-9}$ substitutions per codon per year were estimated from the melts of hybrids formed with DNAs and RNAs of closely related sea urchins (77, 244) and from direct sequence comparisons (80). Histone genes of widely diverse species, however, do have enough sequence complementarity to form hybrids (243, 244, 256).

Program of Histone Synthesis

The histone genes and their products are useful markers in the study of information flow in the embryo. As has been mentioned earlier, the histone mRNAs are the only specific stored messenger to be isolated from the egg. The stored mRNAs are judged to include most of the histone mRNAs synthesized by the embryo on the basis of three criteria. First, the in vitro translation products of egg and preblastula histone mRNA are very similar when analyzed by SDS or acetic acid-urea polyacrylamide gels (63, 64, 73, 75). Second, unlabeled egg RNA can compete with labeled embryonic histone mRNA in hybridizations with DNA (83–85). Competitions for the hybridization of total labeled 9 S mRNA (84–85) or for gel-separated mRNAs corresponding to H1, H4, and a mixture of H2A, H2B, and H3 (83) were all effective. Third, comparison of the patterns of ^{32}P-labeled morula RNA, stained unlabeled morula RNA, and stained unlabeled egg RNA on polyacrylamide gels revealed a great similarity of bands in the 9 S region (83). Correspondence was seen for bands representing H4, H1, and perhaps other histone mRNAs. Some bands in this region were stained but not labeled. The identity of these unlabeled moieties as histone mRNAs is yet to be established.

During cleavage a large fraction of the new polysomal RNA is histone mRNA (76, 172), and much of the protein synthesized appears to be histones (69). All five classes of histones can be shown to be synthesized from the two-cell stage to pluteus (87), although there have been reports of the absence of synthesis of some histones in the early embryo (257). The newly synthesized histones extracted from chromatin, however, do not seem to be present in equal amounts (74, 87, 257, 258). In addition, dramatic shifts occur in the ratio of histones synthesized during development, especially for the ratio of histones identified as H2A and H2B by SDS and acetic acid urea polyacrylamide gels (74, 87, 259). The identification of H2A and H2B histones of early stages, however, is questionable using these gel systems. The shifts may be in the types of H2A and H2B histones, as resolved by Triton X-100 gels (258). The best studied shift is the transition from a cleavage-type H1 to a gastrula H1, of different mobility on SDS and acetic acid-urea gels (64, 74, 87, 257, 259). The new H1 appears at hatching in *A. punctulata* and at gastrula in *L. pictus* and *S. purpuratus* (87, 257). These shifts might reflect changes in transcription of the mRNAs, stability and processing of the transcript, or selective translation of mRNAs in the cell. The last possibility seems unlikely since the shifts seen in vivo are also observed if mRNA is extracted from the various stages and translated in cell-free systems (64, 74). A comparison of the ratio of histones or the type of H1 histone synthesized in vivo, or made in vitro on extracted mRNAs, shows that the patterns are very much the same (67, 74). The gastrula-type H1 mRNA does not appear to be present as more than 1% of the blastula-type H1 mRNA in the egg by in vitro translation criteria. These results might also be interpreted as a sequestration of the mRNA in a form which is not translatable by the wheat germ cell-free system, but there is no evidence for this at all. Little information is available on the relationship of the amount or synthesis of each histone mRNA and the in vivo or in vitro translation of each histone. According to Lifton and Kedes (83), egg H1 and H4 histone mRNAs should be most prevalent in the early embryo, being the main stored species. Histones labeled in vivo during the 1–16-cell stage with [^{3}H]leucine or during morulae with [^{3}H]lysine or [^{14}C]arginine do not include very much H4, however (74, 87, 258). The patterns of synthesis of individual histone mRNAs at different stages are difficult to determine from the literature since precise labeling times are usually not given and the mRNA identification is often absent. However, from patterns of radioactivity on polyacrylamide gels, it is obvious that the newly synthesized histone mRNAs from any one stage are on polysomes to different extents (76–83). This may be due to differential transcription or to selective loading of newly synthesized mRNAs on polysomes. It will be interesting to quantitatively determine the amount of each histone mRNA present at each developmental stage and correlate this with the level of synthesis of the mRNAs and of the individual histones. This analysis should provide detailed information on the level of controls resulting in differential synthesis of the various histones. Especially interesting will be a study of transcription from the repeating unit since the gene organization invites speculation of a polycistronic transcript.

Throughout this discussion the histones have been spoken of as five distinct polypeptide chains. Recent evidence, however, shows that each histone class may be a family of very similar proteins with very minor variations in amino acid sequence (258). For example, an H2A protein synthesized during cleavage contains one methionyl residue, whereas H2A variants synthesized after blastula do not contain methionine. The use of Triton-X gels by Cohen, Zweidler, and co-workers (258, 260) has allowed the resolution of potential variants for H2A and H2B as well as H1 (258). The synthesis of the variants follows a stage-specific program. Recently our laboratory, in collaboration with Cohen and Newrock (261), has shown that mRNAs isolated from different developmental stages code in a wheat germ system for these variants. The patterns on Triton-X gels of the in vitro products are identical with the patterns of histones synthesized in vivo at the various embryonic stages. It, therefore, appears that these variants are coded for by distinct mRNAs. It may be possible to resolve the different mRNAs as bands in polyacrylamide gels or as minor components of a T1-ribonuclease digest of mRNA fractions (71, 240). Since there are more than five types of histone mRNAs, there must be more than five kinds of genes. The histone gene repeat unit is large enough to accomodate additional structural genes, and some of this extra sequence could code for the variants. The repeats might, on the other hand, be heterogeneous in sequence, coding for one variant in one repeat, another variant in another 6-kb unit. A third possibility is that different variants are coded for by as yet unidentified histone gene units which might not cross-hybridize with the major mRNAs used as probes. Such alternatives exist also for the coding of the quite different sperm histones (259). Resolution of these alternatives must await identification of the sequences encoded in the various multiple repeats and identification and isolation of the mRNAs coding for each variant.

The histone gene system is highly regulated with quantitative and qualitative shifts in the mRNAs and proteins being synthesized. Histone mRNA is found as both stored mRNA and newly synthesized embryonic mRNA. The egg mRNA is activated at fertilization and shifts to a polysome location. The histone regulation, in these respects, contains analogies to the regulation of total information usage in the embryo, and study of its control should be useful in understanding developmental change.

REFERENCES

1. Giudice, G. (1973). Developmental Biology of the Sea Urchin Embryo. Academic Press, New York.
2. Stearns, L. (1974). Sea Urchin Development, Cellular and Molecular Aspects. Dowden, Hutchison, and Ross, Inc., Stroudsburg, Pennsylvania.
3. Hinegardner, R. (1967). *In* F. H. Wilt and N. K. Wessells (eds.), Methods in Developmental Biology. T. Y. Crowell Company, New York.
4. Okazaki, K. (1975). *In* G. Czihak (ed.), The Sea Urchin Embryo–Biochemistry and Morphogenesis. Springer-Verlag, Berlin.
5. Ephrussi, B., and Rapkin, L. (1928). Ann. Physiol. Physiochim. Biol. 3:386.
6. Gustafson, T., and Hjelte, M. B. (1951). Exp. Cell Res. 2:474.
7. Fry, B. J., and Gross, P. R. (1970). Dev. Biol. 21:125.

8. Schmidt, G., Hecht, L., and Thanhauser, S. J. (1948). J. Gen. Physiol. 31:203.
9. Elson, D., Guatafson, T., and Chargaff, E. (1954). J. Biol. Chem. 209:285.
10. Tocco, G., Orengo, A., and Scarano, E. (1963). Exp. Cell Res. 31:52.
11. Comb, D. G., Katz, S., Branda, R., and Penzino, C. J. (1965). J. Mol. Biol. 14:195.
12. Epel, D., Pressman, B. C., Elsaesser, S., and Weaver, A. M. (1969). *In* G. M. Padilla, G. L. Whitson, and I. L. Cameron (eds.), The Cell Cycle: Gene Enzyme Interactions. Academic Press, New York.
13. Epel, D., Steinhardt, R., Humphreys, T., and Mazia, D. (1974). Dev. Biol. 40:245.
14. Epel, D. (1967). Proc. Natl. Acad. Sci. U. S. A. 57:899.
15. Steinhardt, R. A., Lundin, L., and Mazia, D. (1971). Proc. Natl. Acad. Sci. U. S. A. 68:2426.
16. Epel, D. (1964). Biochem. Biophys. Res. Commun. 17:69.
17. Runnstrom, J. (1966). Adv. Morphog. 5:221.
18. Whitely, A. H., and Chambers, E. L. (1966). J. Cell Physiol. 68:309.
19. Hinegardner, R. T., Rao, B., and Feldman, D. F. (1964). Exp. Cell Res. 36:53.
20. Hultin, T. (1961). Exp. Cell Res. 25:405.
21. Wilt, F. H., Sakai, H., and Mazia, D. (1967). J. Mol. Biol. 27:1.
22. Black, R. E., Baptist, E., and Piland, J. (1967). Exp. Cell Res. 48:431.
23. Brachet, J., Denis, H., and de Vitry, F. (1964). Dev. Biol. 9:398.
24. Nakomo, E., and Monroy, A. (1958). Exp. Cell Res. 14:236.
25. Monroy, A., and Vitorelli, M. (1962). J. Cell Comp. Physiol. 60:285.
26. Tyler, A., Piatigorsky, J., and Okazaki, H. (1966). Biol. Bull. 131:204.
27. Humphreys, T. (1969). Dev. Biol. 20:435.
28. MacKintosh, F. R., and Bell, E. (1967). Biochem. Biophys. Res. Commun. 27: 425.
29. Giudice, G., Vitorelli, M. L., and Monroy, A. (1962). Acta Embryol. Morphol. Exp. 5:113.
30. Berg, W. E. (1965). Exp. Cell Res. 40:460.
31. Neifakh, A. A., and Krigsgaber, M. R. (1968). Dokl. Biol. Sci. 183:639.
32. Berg, W. E., and Mertes, D. H. (1970). Exp. Cell Res. 60:218.
33. Steinhardt, R., and Mazia, D. (1973). Nature (Lond.) 241:400.
34. Mazia, D., and Ruby, A. (1974). Exp. Cell Res. 85:167.
35. Wilt, F., and Mazia, D. (1974). Dev. Biol. 37:422.
36. Paul, J. *In* J. Paul (ed.), MTP International Review of Science, Biochemistry Series One, Vol. 9, Biochemistry of Cell Differentiation. Butterworths, London.
37. Gross, P. R. (1967). Curr. Top. Dev. Biol. 2:1.
38. Davidson, E. H. (1968). Gene Activity in Early Development. Academic Press, New York.
39. Nemer, M. (1967). Progr. Nucleic Acid Res. Mol. Biol. 7:243.
40. Humphreys, T. (1971). Dev. Biol. 26:201.
41. Glisin, V. R., and Glisin, M. V. (1964). Proc. Natl. Acad. Sci. U. S. A. 52:1548.
42. Gross, P. R., Malkin, L. I., and Moyer, W. A. (1964). Proc. Natl. Acad. Sci. U. S. A. 51:407.
43. Wilt, F. H. (1963). Biochem. Biophys. Res. Commun. 11:447.
44. Wilt, F. H. (1964). Dev. Biol. 9:299.
45. Nemer, M. (1963). Proc. Natl. Acad. Sci. U. S. A. 50:230.
46. Selvig, S. E., Greenhouse, G. A., and Gross, P. R. (1972). Cell Diff. 1:5.

47. Monroy, A., and Tyler, A. (1963). Arch. Biochem. Biophys. 103:431.
48. Rinaldi, A., and Monroy, A. (1969). Dev. Biol. 19:73.
49. Infante, A. A., and Nemer, M. (1967). Proc. Natl. Acad. Sci. U. S. A. 58:681.
50. Kedes, L. H., and Gross, P. R. (1969). J. Mol. Biol. 42:559.
51. Gross, P. R., and Cousineau, G. H. (1963). Biochem. Biophys. Res. Commun. 10:321.
52. Gross, P. R., and Cousineau, G. H. (1964). Exp. Cell Res. 33:368.
53. Singer, R. H., and Penman, S. (1972). J. Mol. Biol. 78:321.
54. Goldstein, E. S., and Penman, S. (1973). J. Mol. Biol. 80:243.
55. Yund, M. A., Kafatos, F. C., and Regier, J. C. (1973). Dev. Biol. 33:362.
56. Harvey, E. B. (1936). Biol. Bull. 71:101.
57. Brachet, J., Ficq, A., and Tencer, R. (1963). Exp. Cell Res. 32:168.
58. Tyler, A. (1963). Am. Zool. 3:109.
59. Denny, P. C., and Tyler, A. (1964). Biochem. Biophys. Res. Commun. 14:245.
60. Burny, A., Marbaix, G., Quertier, J., and Brachet, J. (1965). Biochim. Biophys. Acta 103:526.
61. Tyler, A. (1966). Biol. Bull. 130:450.
62. Sargent, T. D., and Raff, R. A. (1976). Dev. Biol. 48:327.
63. Gross, K. W., Jacobs-Lorena, M., Baglioni, C., and Gross, P. R. (1973). Proc. Natl. Acad. Sci. U. S. A. 70:2614.
64. Arceci, R. J., Senger, D. R., and Gross, P. R. (1976). Cell 9:171.
65. Slater, D. N., and Spiegelman, S. (1966). Proc. Natl. Acad. Sci. U. S. A. 56:164.
66. Piatigorsky, J., and Tyler, A. (1970). Dev. Biol. 21:13.
67. Jenkins, N., Taylor, M. W., and Raff, R. A. (1973). Proc. Natl. Acad. Sci. U. S. A. 70:3287.
68. Ruderman, J., and Pardue, M. (1977). Dev. Biol., in press.
69. Kedes, L. H., Gross, P. R., Cognetti, G., and Hunter, A. L. (1969). J. Mol. Biol. 45:337.
70. Moav, B., and Nemer, M. (1971). Biochemistry 10:881.
71. Grunstein, M., Levy, S., Schedi, P., and Kedes, L. H. (1973). Cold Spring Harbor Symp. Quant. Biol. 38:717.
72. Levy, S., Wood, P., Grunstein, M., and Kedes, L. (1975). Cell 4:239.
73. Gross, K. W., Ruderman, J., Jacobs-Lorena, M., Baglioni, C., and Gross, P. R. (1973). Nature (New Biol.) 241:272.
74. Ruderman, J., and Gross, P. R. (1974). Nature (Lond.) 247:36.
75. Gross, K. W., Jacobs-Lorena, M., Baglioni, C., and Gross, P. R. (1973). Proc. Natl. Acad. Sci. U. S. A. 70:2614.
76. Kedes, L. H., and Gross, P. R. (1969). Nature (Lond.) 223:1335.
77. Weinberg, E. S., Birnstiel, M., Purdom, I. F., and Williamson, R. W. (1972). Nature (Lond.) 240:225.
78. Kedes, L. H., Cohn, R., Lowry, J., Chang, A. C. Y., and Cohen, S. N. (1975). Cell 6:359.
79. Grunstein, M., and Schedl, P. (1976). J. Mol. Biol. 104:323.
80. Grunstein, M., Schedl, P., and Kedes, L. H. (1976). J. Mol. Biol. 104:351.
81. Birnstiel, M. L., Gross, K., Schaffner, W., and Telford, J. (1975). FEBS Proc. Tenth Paris Meeting 38:3.
82. Gross, K. W., Schaffner, W., Telford, J., and Birnstiel, M. (1976). Cell 8:479.
83. Lifton, R. P., and Kedes, L. H. (1976). Dev. Biol. 48:47.

84. Skoultchi, A., and Gross, P. R. (1973). Proc. Natl. Acad. Sci. U. S. A. 70:2840.
85. Farquhar, M. N., and McCarthy, B. J. (1973). Biochem. Biophys. Res. Commun. 53:515.
86. Crane, C. M., and Villee, C. A. (1971). J. Biol. Chem. 246:719.
87. Ruderman, J., and Gross, P. R. (1974). Dev. Biol. 36:286.
88. Raff, R. A., Greenhouse, G., Gross, K. W., and Gross, P. R. (1971). J. Cell Biol. 50:516.
89. Raff, R. A., Colot, H. V., Seleg, S. E., and Gross, P. R. (1972). Nature (Lond.) 235:211.
90. Kaulenas, M. S., and Unsworth, B. R. (1974). Mol. Gen. Genet. 135:231.
91. Noronha, J. M., Sheys, G. H., and Buchanan, J. M. (1972). Proc. Natl. Acad. Sci. U. S. A. 69:2006.
92. Barrett, D., and Angelo, G. M. (1969). Exp. Cell Res. 57:159.
93. Brandhorst, B. P. (1976). Dev. Biol. 52:310.
94. O'Farrell, P. H. (1975). J. Biol. Chem. 250:4007.
95. Terman, S. A., and Gross, P. R. (1965). Biochem. Biophys. Res. Commun. 21:595.
96. MacKintosh, F. R., and Bell, E. (1969). Science 164:961.
97. Terman, S. A. (1970). Proc. Natl. Acad. Sci. U. S. A. 65:985.
98. Westin, M., Perlmann, H., and Perlmann, P. (1967). J. Exp. Zool. 166:331.
99. Westin, M. (1969). J. Exp. Zool. 171:297.
100. Galau, G. A., Klein, W. H., Davis, M. M., Wold, B. J., Britten, R. J., and Davidson, E. H. (1976). Cell 7:487.
101. Gross, P. R. (1964). J. Exp. Zool. 157:21.
102. Brachet, J., Denes, H., and DeVitry, F. (1964). Dev. Biol. 9:398.
103. Hultin, T. (1964). Dev. Biol. 10:305
104. Spirin, A. S. (1966). Curr. Top. Dev. Biol. 1:1.
105. Wilt, F. H., and Hultin, T. (1962). Biochem. Biophys. Res. Commun. 9:313.
106. Nemer, M. (1962). Biochem. Biophys. Res. Commun. 8:511.
107. Tyler, A. (1963). Am. Zool. 3:109.
108. Nemer, M. (1962). Biochem. Biophys. Res. Commun. 8:511.
109. Monroy, A., Maggio, R., and Rinaldi, A. M. (1965). Proc. Natl. Acad. Sci. U. S. A. 54:107.
110. Tyler, A. (1967). Dev. Biol. (Suppl.) 1:170.
111. Stavy, L., and Gross, P. R. (1967). Proc. Natl. Acad. Sci. U. S. A. 57:735.
112. Stavy, L., and Gross, P. R. (1969). Biochim. Biophys. Acta 182:193.
113. Kedes, L. H., and Stavy, L. (1969). J. Mol. Biol. 43:337.
114. Castañeda, M. (1969). Biochim. Biophys. Acta 179:381.
115. Vitorelli, M. L., Caffarelli-Mormino, I., and Monroy, A. (1969). Biochim. Biophys. Acta 186:408.
116. Hultin, T. (1964). Exp. Cell Res. 34:608.
117. Maggio, R., Vitorelli, M., Rinaldi, A. M., and Monroy, A. (1964). Biochem. Biophys. Res. Commun. 15:436.
118. Clegg, K. B., and Denny, P. C. (1974). Dev. Biol. 37:263.
119. Metafora, S., Felicetti, L., and Gambino, R. (1971). Proc. Natl. Acad. Sci. U. S. A. 68:600.
120. Gambino, R., Metafora, S., Felicetti, L., and Reisman, J. (1973). Biochim. Biophys. Acta 312:377.
121. Hille, M. B. (1974). Nature (Lond.) 249:556.

122. Johnson, J. D., and Epel, D. (1975). J. Cell Biol. 67:193a.
123. Epel, D. (1975). J. Gen. Physiol. 66:5a.
124. Denny, P. C., and Reback, P. (1970). J. Exp. Zool. 175:133.
125. MacKintosh, F. R., and Bell, E. (1969). J. Mol. Biol. 41:365.
126. Infante, A. A., and Nemer, M. (1968). J. Mol. Biol. 32:543.
127. Spirin, A. S., and Nemer, M. (1965). Science 150:214.
128. Spirin, A. S. (1969). Eur. J. Biochem. 10:20.
129. Dworkin, M., and Infante, A. A. (1976). Dev. Biol. 53:73.
130. Baltimore, D., and Huang, A. S. (1970). J. Mol. Biol. 47:88.
131. Hultin, T. (1961). Experientia 17:410.
132. Wilt, F. H., Sakai, H., and Mazia, D. (1967). J. Mol. Biol. 32:543.
133. Rebhun, L., White, D., Sander, G., and Ivy, N. (1973). Exp. Cell Res. 77:312.
134. Sevaljevic, L., and Ruzdijic, S. (1971). W. Roux Archiv. 168:187.
135. Spiegel, M., Ozaki, H., and Tyler, A. (1965). Biochem. Biophys. Res. Commun. 21:135.
136. Spiegel, M., Spiegel, E. S., and Meltzer, P. S. (1970). Dev. Biol. 21:73.
137. Ellis, C. H., Jr. (1966). J. Exp. Zool. 163:1.
138. Pucci-Minafra, I., Casano, C., and La Rosa, C. (1972). Cell Diff. 1:157.
139. Golob, R., Chetsanya, C. J., and Doty, P. (1974). Biochim. Biophys. Acta 349:135.
140. Giudice, G., Mutalo, V., and Donatuti, G. (1968). W. Roux Archiv. 161:118.
141. Barros, C., Hand, G. S., Jr., and Monroy, A. (1966). Exp. Cell Res. 43:167.
142. Melli, M., and Bishop, J. O. (1969). J. Mol. Biol. 40:117.
143. Goldberg, R. B., Galau, G. A., Britten, R. J., and Davidson, E. H. (1973). Proc. Natl. Acad. Sci. U. S. A. 70:3516.
144. Smith, M. J., Hough, B. R., Chamberlain, M. E., and Davidson, E. H. (1974). J. Mol. Biol. 85:103.
145. Glisin, V. R., Glisin, M. V., and Doty, P. (1966). Proc. Natl. Acad. Sci. U. S. A. 56:285.
146. Whitely, A. H., McCarthy, B. J., and Whitely, H. R. (1966). Proc. Natl. Acad. Sci. U. S. A. 55:519.
147. Whitely, H. R., McCarthy, B. J., and Whitely, A. H. (1970). Dev. Biol. 21:216.
148. Whitely, A. H., and Whitely, H. R. (1972). Dev. Biol. 29:183.
149. Mizuno, S., Whitely, H. R., and Whitely, A. H. (1973). Differentiation 1:339.
150. Whitely, H. R., Mizuno, S., Lee, Y. R., and Whitely, H. R. (1975). Am. Zool. 15:141.
151. Whitely, H. R., and Whitely, A. H. (1975). Curr. Top. Dev. Biol. 9:39.
152. Graham, D. E., Neufeld, B. R., Davidson, E. H., and Britten, R. J. (1974). Cell 1:127.
153. Davidson, E. H., Hough, B. R., Klein, W. H., and Britten, R. J. (1975). Cell 4:217.
154. Galau, G. A., Britten, R. J., and Davidson, E. H. (1974). Cell 2:9.
155. Davidson, E. H., and Britten, R. (1973). Q. Rev. Biol. 48:565.
156. Birnstiel, M. L., Sells, B. H., and Purdom, I. F. (1972). J. Mol. Biol. 63:21.
157. Lewin, B. (1975). Cell 4:77.
158. Whitely, A. H. (1950). Am. Naturalist 83:249.
159. Hough, B. R., Smith, M. J., Britten, R. J., and Davidson, E. H. (1975). Cell 5:291.

160. Brandhorst, B. P., and Humphreys, T. (1971). Biochemistry 10:877.
161. Brandhorst, B. P., and Humphreys, T. (1972). J. Cell Biol. 53:474.
162. Anderson, D. M., Galau, G. A., Britten, R. J., and Davidson, E. H. (1976). Dev. Biol. 51:138.
163. Davis, F. C. (1975). Biochim. Biophys. Acta 390:33.
164. Davidson, E. H., and Hough, B. R. (1971). J. Mol. Biol. 56:491.
165. Kijima, S., and Wilt, F. H. (1969). J. Mol. Biol. 40:235.
166. Grainger, R. M., and Wilt, F. H. (1976). J. Mol. Biol. 104:589.
167. Sconzo, G., Pirrone, A. M., Mutolo, V., and Giudice, G. (1970). Biochim. Biophys. Acta 199:435.
168. Sconzo, G., Pirrone, A. M., Mutolo, V., and Guidice, G. (1970). Biochim. Biophys. Acta 199:441.
169. Comb, D. G. (1965). J. Mol. Biol. 11:851.
170. Comb, D. G., Katz, S., Branda, R., and Pinzino, C. J. (1965). J. Mol. Biol. 14:195.
171. Aronson, A. I., Wilt, F. H., and Wartrovaara, J. (1972). Exp. Cell Res. 72:309.
172. Hogan, B., and Gross, P. R. (1972). Exp. Cell Res. 72:101.
173. Peltz, R. (1973). Biochim. Biophys. Acta 308:148.
174. Kung, C. S. (1974). Dev. Biol. 36:343.
175. Emerson, C. P., and Humphreys, T. (1970). Dev. Biol. 23:86.
176. Emerson, C. P., and Humphreys, T. (1971). Anal. Biochem. 40:254.
177. Humphreys, T. (1973). *In* S. J. Coward (ed.), Developmental Regulation: Aspects of Cell Differentiation. Academic Press, New York.
178. Patterson, J. B., and Stafford, D. (1970). Biochemistry 9:1278.
179. Patterson, J. B., and Stafford, D. (1971). Biochemistry 10:2775.
180. Sconzo, G., Vitrano, E., Bono, A., DiGiovanni, L., Mutolo, V., and Giudice, G. (1971). Biochim. Biophys. Acta 232:132.
181. Roeder, R. G., and Rutter, W. G. (1969). Nature (Lond.) 224:234.
182. Roeder, R. G., and Rutter, V. G. (1970). Biochemistry 9:2543.
183. Perry, R. P. (1976). Annu. Rev. Biochem. 45:605.
184. O'Melia, A. F., and Villee, C. A. (1972). Nature New Biol. 239:51.
185. Fredericksen, S., and Hellung-Larsen, D. (1974). Exp. Cell Res. 89:217.
186. Sy, J., and McCarty, K. S. (1970). Biochim. Biophys. Acta 199:86.
187. Spiers, J., and Birnstiel, M. (1974). J. Mol. Biol. 87:237.
188. Imaizumi, T., Diggelmann, H., and Scherrer, K. (1973). Proc. Natl. Acad. Sci. U. S. A. 70:1122.
189. Spohr, G., Imaizumi, T., and Scherrer, K. (1974). Proc. Natl. Acad. Sci. U. S. A. 70:1122.
190. Macnaughton, M., Freeman, K. B., and Bishop, J. O. (1974). Cell 1:117.
191. Ross, J. (1976). J. Mol. Biol. 106:403.
192. Curtis, P. J., and Weissmann, C. (1976). J. Mol. Biol. 106:1061.
193. Kwan, S.-P., Wood, T. G., and Lingrel, J. B. (1977). Proc. Natl. Acad. Sci. U. S. A. 74:178.
194. Aronson, A. I., and Wilt, F. H. (1969). Proc. Natl. Acad. Sci. U. S. A. 62:186.
195. Nemer, M., Graham, M., and Dubroff, L. M. (1974). J. Mol. Biol. 89:435.
196. Nemer, M., Dubroff, L. M., and Graham, M. (1975). Cell 6:171.
197. Nemer, M. (1975). Cell 6:559.
198. Dubroff, L. M., and Nemer, M. (1975). J. Mol. Biol. 95:455.
199. Hynes, R. O., Greenhouse, G. A., Minkoff, R., and Gross, P. R. (1972).
200. Sconzo, G., Albanese, I., Rinaldi, A. M., LoPresti, G., and Giudice, G. (1974). Cell Diff. 3:297.

201. Siekevitz, R., Maggio, R., and Catalano, C. (1966). Biochim. Biophys. Acta 129:145.
202. Wu, R. S., and Wilt, F. H. (1974). Dev. Biol. 4:352.
203. Galau, G. A., Lipson, E. D., Britten, R. J., and Davidson, E. H. (1977). Cell 10:415.
204. Slater, D. W., Slater, I., and Gillespie, D. (1972). Nature (Lond.) 240:333.
205. Wilt, F. H. (1973). Proc. Natl. Acad. Sci. U. S. A. 8:2345.
206. Slater, I., and Slater, D. W. (1974). Proc. Natl. Acad. Sci. U. S. A. 71:1103.
207. Slater, I., Gillespie, D., and Slater, D. W. (1973). Proc. Natl. Acad. Sci. U. S. A. 70:406.
208. Slater, D. W., Slater, I., Gillespie, D., and Gillespie, S. (1974). Biochem. Biophys. Res. Commun. 60:1222.
209. Huez, G., Marbaix, G., Hubert, E., Leclercq, M., Nudel, V., Soreq, H., Salomon, R., LeBleu, B., Revel, M., and Littauer, U. Z. (1974). Proc. Natl. Acad. Sci. U. S. A. 71:3143.
210. Marbaux, G., Huez, G., Burny, A., Cleuter, Y., Hubert, E., Leclercq, M., Chantrenne, H., Soreq, H., Nudel, U., and Littauer, U. Z. (1975). Proc. Natl. Acad. Sci. U. S. A. 72:3065.
211. Mescher, A., and Humphreys, T. (1974). Nature (Lond.) 249:138.
212. Fromson, D., and Verma, D. P. S. (1976). Proc. Natl. Acad. Sci. U. S. A. 73:148.
213. Dubroff, L. M., and Nemer, M. (1976). Nature (Lond.) 260:120.
214. Levinson, R. G., and Marcu, K. B. (1976). Cell 9:311.
215. Spirin, A. S., Belitsima, N. V., and Ajtkhozhin, M. A. (1964). Zh. Obshch. Biol. 25:321.
216. Jacobs-Lorena, M., and Baglioni, C. (1972). Proc. Natl. Acad. Sci. U. S. A. 69:1425.
217. Buckingham, M. E., Caput, D., Cohen, A., Whalen, R. G., and Gross, F. (1974). Proc. Natl. Acad. Sci. U. S. A. 71:1466.
218. Bag, J., and Sarkar, S. (1975). Biochemistry 14:3800.
219. McColl, R. S., and Aronson, A. I. (1974). Biochem. Biophys. Res. Commun. 56:47.
220. Britten, R. J., and Davidson, E. H. (1969). Science 165:349.
221. Britten, R. J., and Davidson, E. H. (1971). Q. Rev. Biol. 46:111.
222. Gray, S. P. (1970). J. Mol. Biol. 47:615.
223. Chamberlain, J. P. (1970). Biochim. Biophys. Acta 213:183.
224. Chamberlain, J. P., and Metz, C. B. (1972). J. Mol. Biol. 64:593.
225. Cantatore, P. (1974). Cell Diff. 3:45.
226. Selvig, S. E., Gross, P. R., and Hunter, A. L. (1970). Dev. Biol. 22:343.
227. Devlin, R. (1976). Dev. Biol. 50:443.
228. Pikó, L., Blair, O. G., Tyler, A., and Vinograd, J. (1968). Proc. Natl. Acad. Sci. U. S. A. 59:838.
229. Hörstadius, S. (1939). Biol. Rev. 14:132.
230. Hörstadius, S. (1935). Pubbl. Staz. Zool. Natoli 14:251.
231. Hynes, R. O., and Gross, P. R. (1970). Dev. Biol. 21:383.
232. Mizuno, S., Lee, Y. R., Whitely, A. H., and Whitely, H. R. (1974). Dev. Biol. 37:18.
233. Spiegel, M., and Rubinstein, N. A. (1972). Exp. Cell Res. 70:423.
234. Hynes, R. O., Raff, R. A., and Gross, P. R. (1972). Dev. Biol. 27:150.
235. Kedes, L. H. (1976). Cell 8:321.
236. Borun, T. W., Scharff, M. D., and Robbins, E. (1967). Proc. Natl. Acad. Sci. U. S. A. 58:1977.

237. Robbins, E., and Borun, T. W. (1967). Proc. Natl. Acad. Sci. U. S. A. 57:409.
238. Gallwitz, D., and Mueller, G. C. (1969). J. Biol. Chem. 244:5947.
239. Nemer, M., and Lindsay, D. T. (1969). Biochem. Biophys. Res. Commun. 35:156.
240. Grunstein, M., Schedl, P., and Kedes, L. (1973). *In* B. A. Hamkalo and J. Papaconstantinou (eds.), Molecular Cytogenetics. Plenum Press, New York.
241. Sures, I., Maxam, A., Cohn, R. H., and Kedes, L. H. (1976). Cell.
242. Birnstiel, M. L., Schaffner, W., and Smith, H. O. (1977). Nature 266:603.
243. Kedes, L. H., and Birnstiel, M. (1971). Nature (New Biol.) 230:165.
244. Farquhar, M. N., and McCarthy, B. J. (1973). Biochemistry 12:4113.
245. Birnstiel, M., Telford, J., Weinberg, E., and Stafford, D. (1974). Proc. Natl. Acad. Sci. U. S. A. 71:2900.
246. Weinberg, E. S., and Overton, G. C. *In* G. Stein and L. Kleinsmith (eds.), Methods in Cell Biology. Academic Press, New York. In press.
247. Weinberg, E. S., Overton, G. C., Shutt, R. H., and Reeder, R. H. (1975). Proc. Natl. Acad. Sci. U. S. A. 72:4815.
248. Schaffner, W., Gross, K., Telford, J., and Birnstiel, M. (1976). Cell 8:471.
249. Kedes, L. H., Chang, A. C. Y., Houseman, D., and Cohn, S. N. (1975). Nature (Lond.) 255:533.
250. Cohn, R. H., Lowry, J. C., and Kedes, L. H. (1976). Cell 9:147.
251. Gross, K., Schaffner, W., Telford, J., and Birnstiel, M. (1976). Cell 8:479.
252. Wu, M., Holmes, D. S., Davidson, N., Cohn, R., and Kedes, L. H. (1976). Cell 9:163.
253. Holmes, D. S., Cohn, R. H., Kedes, L. H., and Davidson, N. (1977). Biochemistry 16:1504.
254. Portmann, R., Schaffner, W., and Birnstiel, M. (1976). Nature (Lond.) 264:31.
255. Overton, G. C., and Weinberg, E. W., manuscript in preparation.
256. Birnstiel, M. L., Weinberg, E. S., and Pardue, M. L. (1973). *In* B. Hamkalo and J. Papaconstantinou (eds.), Molecular Cytogenetics. Plenum Press, New York.
257. Seale, R. L., and Aronson, A. I. (1973). J. Mol. Biol. 75:647.
258. Cohen, L., Newrock, K. M., and Zweidler, A. (1975). Science 190:994.
259. Easton, D., and Chalkley, R. (1972). Exp. Cell. Res. 72:502.
260. Alfageme, C. R., Zweidler, A., Mahowald, A., and Cohen, L. H. (1974). J. Biol. Chem. 249:3729.
261. Newrock, K., Cohen, L., Hendricks, M., Donnelly, R., and Weinberg, E., manuscript in preparation.

International Review of Biochemistry
Biochemistry of Cell Differentiation II, Volume 15
Edited by J. Paul

5
Biosynthesis of Eye Lens Protein

H. BLOEMENDAL
Department of Biochemistry,
University of Nijmegen,
Nijmegen,
The Netherlands

It is generally accepted that macromolecules give the living cells their specific properties. During development, changes in structure and function of those molecules take place which, in a number of cases, can be followed by the available techniques in molecular biology. Due to their great potential for structural variations, the proteins are among the best candidates to serve as "markers" in such fundamental processes as differentiation, growth, and aging. In this connection it has been stressed many times that the biogenesis of lens proteins, named crystallins, may play an important role as a model system.

The vertebrate lens is a well defined structure. Throughout the whole lifespan of the organ it forms long fiber cells which arise from a monolayer of epithelial cells situated at the anterior part of the lens (Figure 1). Its final structure resembles an onion with concentric layers of long fibers around a central part, called the nucleus or core. This nucleus is nothing other than fiber cells formed during the embryonic and fetal stages. Fibers formed in the later stages are arranged at the periphery or cortex of the organ. It is this structure which records the events of lens cell differentiation.

A very specific aspect of the lens is that it never sheds its cells. Therefore, strictly speaking, there is no phenomenon comparable to "normal" cell death, and all changes observed in the inner fibers can be attributed to aging.

When the epithelial cells differentiate into fibers they lose most of the cell organelles, with the exception of the plasma membranes and the polyribosomes in the outer fiber layers. Since the fibers lack nuclei, it can be concluded that while crystallin synthesis continues the messengers involved are stabilized (2, 3).

Although there is a great similarity in properties among the lenses of the various mammals, some differences do exist. For instance, the human eye lens grows throughout life, albeit at a slowly decreasing rate. That of the cow, rabbit, and rat grows rapidly during the first half of the life of the animal, although

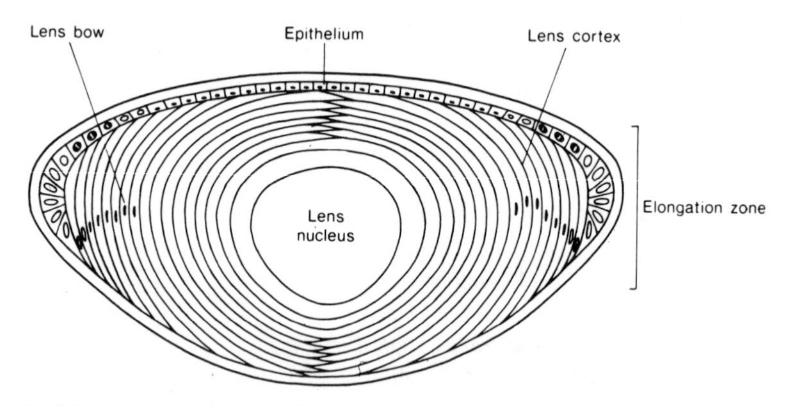

Figure 1. Schematic cross-section of a mammalian eye lens. The avascular lens which is surrounded by a capsule is composed of 1) an outer single layer of cuboidal cells, the epithelium; 2) a zone of differentiation which contains cells that are developing into fibers; and 3) a number of discrete layers of which the outer ones are composed of hexagonal cells, the nucleated fibers. These fibers are displaced toward the center of the lens, where they change into oblong fibers lacking nuclei and most other cell organelles. The lens nucleus comprises the oldest fibers, closely packed, which originated at the embryonic stage. It was thought that mitotic activity in the fibers is irreversibly inhibited. However, recent observations revealed that the nuclei containing fiber cells are able to dedifferentiate into replicating epithelium-like cells when cultured in vitro (1).

almost no increase in weight can be observed thereafter. Another difference is the hardening of the mammalian lens nucleus, caused by gradual dehydration with growing age. The human lens, however, maintains a water content of about 65% throughout life (4).

The lens of the chick eye has been used many times for the study of various aspects of differentiation. For instance, the distribution of nucleic acid synthesis at different stages of differentiation has been investigated by various groups (5–8). The identification of chick crystallins and their changes while the animal ages has been reviewed by Clayton (9, 10). Piatigorsky and collaborators studied crystallin synthesis during the formation of embryonic chick lens fibers in vivo and in vitro (11). In later studies, properties, synthesis, and post-translational modification of δ-crystallin were investigated (12, 13). However, the latter investigations have, at least for the time being, some limitations since more detailed chemical data, such as the amino acid sequence of the individual crystallin subunits, are still lacking. In this respect the calf lens is a better tool.

It is believed that the crystallins have only structural significance. These proteins, of which about 90% are soluble in water or buffer, can easily be separated by gel filtration into four main classes, named α-, β_H-, β_L-, and γ-crystallin (14) (Figure 2). The water-insoluble part of the eye lens is called

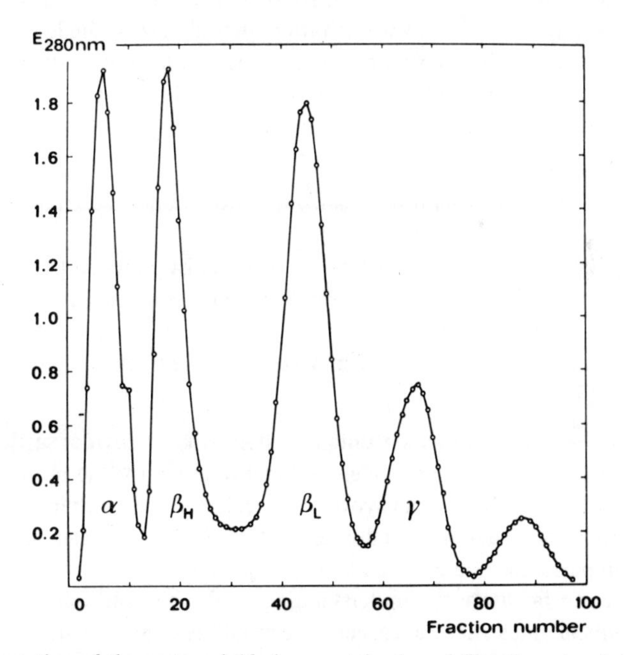

Figure 2. Separation of the water-soluble lens proteins by gel filtration. A column (100 × 2.5 cm) was packed with Sephadex G-200 and equilibrated in a buffer containing 0.05 M Tris-HCl, pH 7.6, 0.05 M NaCl, and 0.001 M EDTA. Elution was performed with the same buffer. Protein (200 mg) was loaded onto the column. Elution was for 40 hr. (The fraction behind γ-crystallin contains material of low molecular weight, in particular nucleotides.)

albuminoid (15–18). This fraction also contains some α-crystallin which is absorbed or attached to membranes (19). The structure-function relationship of biological membranes constitutes one of the most fundamental problems of molecular biology. This notion led several groups working in the field of lens research to initiate studies on various aspects of lenticular plasma membranes (20–29, cf. 27). The morphological feature of lens cell elongation is paralleled by the formation of plasma membranes. The fibers, as compared to the epithelial cells from which they originate, have a very high surface to volume ratio and are almost completely devoid of other membranous entities, such as endoplasmic reticulum. The lens, therefore, is one of the best sources for the isolation of plasma membranes. The plasma membranes of the lens fibers also represent a very suitable model for studying the molecular organization of intercellular junctions since large areas of the fiber surface are connected by these structural elements (23, 26, 27). Until recently, attempts to characterize specific constituents of the fiber membranes were restricted to the fraction which remains after extraction of the albuminoid by urea or guanidine hydrochloride (20, 22, 24, 25, 29). Better defined membrane preparations have been obtained after gradient centrifugation (21, 23, 27).

One may ask whether membrane constituents are preformed, for instance, in the elongation zone (cf. Figure 1) or, alternatively, whether de novo synthesis of membranes may occur in the whole outer cortical part of the lens. The question then arises whether differentiation is accompanied by the synthesis of new lens fiber membrane proteins different from those of epithelial membranes.

For two reasons this presentation is mainly devoted to the description of the biosynthesis of α-crystallin:

1) Hitherto, calf α-crystallin has been the most intensively studied lens protein (cf. 10, 30–32).
2) From a quantitative point of view, it is the most important crystallin species since approximately 75% of the total protein synthesis in vitro is due to α-crystallin production (33), whereas organ culture experiments revealed that about 50% of the total synthesis in vivo is directed toward the production of α-crystallin (34).

Obviously, the lens represents a unique system which provides sufficient quantities of a specific protein and enough material to allow for the purification of the messenger RNA involved in its synthesis. Incidentally, this chapter also refers to the formation of β- and γ-crystallin as well as to the synthesis of noncrystallin proteins, in particular the plasma membrane proteins.

In order to facilitate the understanding of the biosynthesis, a brief account of the chemical and structural aspects of crystallins is given first.

α-CRYSTALLIN

Most studies have been done on bovine α-crystallin. Since it has the highest molecular weight and also the highest electrophoretic mobility of the water-soluble lens proteins, it can easily be isolated from a lens supernatant (15,000 × *g*) either by gel filtration (14) (cf. Figure 2), by ultracentrifugation (35, 36), or by preparative electrophoresis (37, 38). In contrast with what has been thought previously, α-crystallin is not a homogeneous polymeric protein, but a mixture of different sized aggregates (36, 39).

In 1962, Bloemendal et al. were able to show that α-crystallin is made up of different subunits, held together by noncovalent bonds (40). Upon exposure to 7 M urea, the high molecular aggregate with an average molecular weight of 800,000 dissociates into four major polypeptide chains, named αA_1, αA_2, αB_1, and αB_2 (41, 42) (Figure 3). It is discussed later that only αA_2 and αB_2 are primary gene products (43, 44). In aged α-crystallin additional polypeptides can be found. However, it has been demonstrated that these chains are products of post-translational modification (45, 46). Meanwhile, the complete amino acid

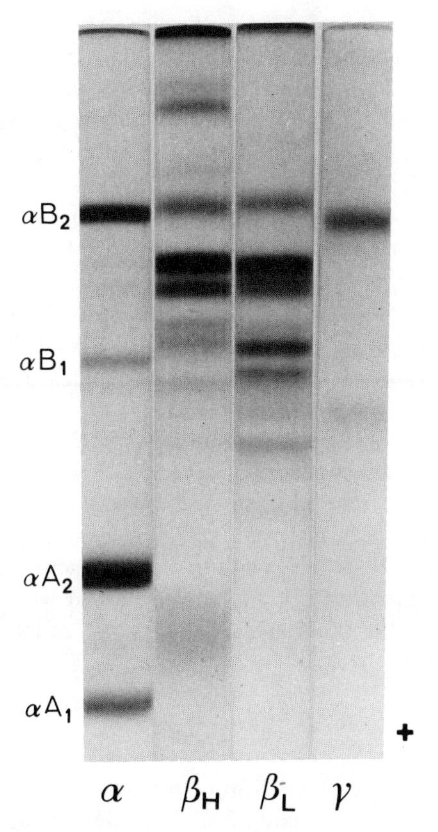

Figure 3. 7 M Urea-polyacrylamide gel electrophoresis pattern of the polypeptide chains of the separated water-soluble crystallins. Electrophoresis was performed in 10% acrylamide gels containing 7 M urea and 100 mg per liter of dithiothreitol at pH 8.6. Samples containing 40–50 μl of protein were layered on top of the gels after 1 hr of pre-electrophoresis. The electrophoretic separation was for 5 hr at 5 mA per gel tube. The gels were stained with Amido black.

sequence of both αA_2 and αB_2 has been established (47, 48). In Figure 4 the high degree of homology of the two chains (approximately 60%) can be seen.

β-CRYSTALLIN

The β-crystallins are characterized by a molecular weight which is intermediate between that of α- and γ-crystallin. On gel filtration columns, β-crystallin is separated into two fractions of higher (β_H) and lower (β_L) molecular weight (210,000 and 52,000, respectively) (Figure 2). Immunoelectrophoresis of these fractions yields at least eight precipitation arcs (14, 49). Upon addition of dissociating reagents such as urea, guanidine-HCl, or sodium dodecyl sulfate, a decrease in the sedimentation coefficient of β-crystallin is observed. The average molecular weight of the individual polypeptides drops to about 25,000 (40, 50).

Alkaline polyacrylamide gel electrophoresis in the presence of 6 M urea reveals nine bands for β_H and eight for β_L. Several chains migrate with identical mobility. The major polypeptide designated βB_p (principal β chain) is shared by both β fractions. A close relationship between a number of β-crystallin polypeptides, characterized by different electrophoretic mobilities, is suggested by the great similarity in amino acid composition, molecular weight, and tryptic peptides. Presumably, as in α-crystallin chains, some β-polypeptides arise by post-translational modification such as deamidation. The characteristic urea-polyacrylamide gel pattern of β_H and β_L is shown in Figure 3. It can be seen that both fractions have the most predominant band in common. The main difference is the presence of two basic chains and an acidic chain in β_H and two chains in the neutral region in β_L. The complexity of β-crystallins has not only been observed in calf lens, but also in various other mammalian species (51, 52) and in chick lens (53).

Sodium dodecyl sulfate gel electrophoresis reveals the occurrence of five and four size classes of polypeptides in β_H and β_L, respectively (14). Zigler and Sidbury (54) reported somewhat different results. Recent progress in calf β-crystallin research has been reviewed by Bloemendal and Herbrink (14) and by Harding and Dilley (32). Studies on the primary structure of the βB_p chain have only been started recently (55). A chemical feature which α- and β-crystallin have in common is the acetylated NH_2-terminal amino acid residue of the individual polypeptide chains.

γ-CRYSTALLIN

γ-crystallin comprises about one-fifth of the water-soluble protein of calf lens. The γ-crystallins, although appearing homogeneously in ultracentrifugation studies, are a mixture of related monomeric proteins, which occur in different quantities. On polyacrylamide gels the predominant polypeptide is easily visualized (cf. Figure 3). The primary structure of the major γ component has been reported by Croft (56). In contrast to the α- and β-crystallins, the γ-crystallin α-CRYSTALLIN SEQUENCE : HOMOLOGY BETWEEN αA_2 AND αB_2 POLYPEPTIDE CHAIN

```
αA2  ac-Met-Asp-Ile-Ala-Ile-Gln-His-Pro-Trp-Phe-Lys-Arg-Thr-Leu-Gly-Pro-Phe-    -Tyr-Pro-Ser-Arg-Leu-Phe-Asp-Gln-Phe-Phe-Gly-Glu-
αB2  ac-Met-Asp-Ile-Ala-Ile-His-His-Pro-Trp-Ile-Arg-Arg-Pro-Phe-Phe-Pro-Phe-His-Ser-Pro-Ser-Arg-Leu-Phe-Asp-Gln-Phe-Phe-Gly-Glu-

-Gly-Leu-Phe-Glu-Tyr-Asp-Leu-Leu-Pro-Phe-Leu-Ser-Ser-Thr-Ile-Ser-Pro-Tyr-Tyr-    -Arg-Gln-    -Ser-Leu-Phe-Arg-         -Thr-Val-
-His-Leu-Leu-Glu-Ser-Asp-Leu-Phe-Pro-        -Ala-Ser-Thr-Ser-Leu-Ser-Pro-Phe-Tyr-Leu-Arg-Pro-Pro-Ser-Phe-Leu-Arg-Ala-Pro-Ser-Trp-

-Leu-Asp-Ser-Gly-Ile-Ser-Glu-Val-Arg-Ser-Asp-Arg-Asp-Lys-Phe-Val-Ile-Phe-Leu-Asp-Val-Lys-His-Phe-Ser-Pro-Glu-Asp-Leu-Thr-Val-
-Ile-Asp-Thr-Gly-Leu-Ser-Glu-Met-Arg-Leu-Glu-Lys-Asp-Arg-Phe-Ser-Val-Asn-Leu-Asn-Val-Lys-His-Phe-Ser-Pro-Glu-Glu-Leu-Lys-Val-

-Lys-Val-Gln-Glu-Asp-Phe-Val-Glu-Ile-His-Gly-Lys-His-Asn-Glu-Arg-Gln-Asp-Asp-His-Gly-Tyr-Ile-Ser-Arg-Glu-Phe-His-Arg-Arg-Tyr-
-Lys-Val-Leu-Gly-Asp-Val-Ile-Glu-Val-His-Gly-Lys-His-Glu-Glu-Arg-Gln-Asp-Glu-His-Gly-Phe-Ile-Ser-Arg-Glu-Phe-His-Arg-Lys-Tyr-

-Arg-Leu-Pro-Ser-Asn-Val-Asp-Gln-Ser-Ala-Leu-Ser-Cys-Ser-Leu-Ser-Ala-Asp-Gly-Met-Leu-Thr-Phe-Ser-Gly-Pro-Lys-Ile-Pro-Ser-Gly-
-Arg-Ile-Pro-Ala-Asp-Val-Asp-Pro-Leu-Ala-Ile-Thr-Ser-Ser-Leu-Ser-Ser-Asp-Gly-Val-Leu-Thr-Val-Asp-Gly-Pro-Arg-Lys-Gln-

-Val-Asp-Ala-Gly-His-Ser-Glu-Arg-Ala-Ile-Pro-Val-Ser-Arg-Glu-Glu-Lys-Pro-            -Ser-Ser-Ala-Pro-Ser-Ser-COOH  173  RESIDUES
-        Ala-Ser-Gly-Pro-Glu-Arg-Thr-Ile-Pro-Ile-Thr-Arg-Glu-Glu-Lys-Pro-Ala-Val-Thr-Ala-Ala-Pro-Lys-Lys-COOH  175  RESIDUES
```

HOMOLOGY : 55 %

Figure 4. The primary structure of the A_2 and the B_2 chains of calf α-crystallin. Notice the high degree of homology between the two polypeptides.

polypeptides have a free NH_2-terminal amino acid. Björk, who initiated systematic studies upon the γ-crystallins from calf lens, observed that much larger amounts are found in the nucleus than in the cortex (57, 58). The best separation of the γ-crystallins from the other crystallins is achieved by gel filtration on Sephadex G-75, as has already been demonstrated by Björk (57). However, a β-crystallin component of low molecular weight, β_S (59), is eluted together with the γ-crystallin fraction. The separation can be improved by high speed centrifugation of the total lens extract prior to gel filtration (60). In that case β_S emerges as a separated peak, whereas a minor β fraction, β_M, is separated from β_L (Figure 5). Analysis on alkaline urea-polyacrylamide gel shows that β_M and β_S are also different from γ-crystallin in their polypeptide composition (Figure 6). Further fractionation of the individual γ-crystallins is achieved on SE-Sephadex (Figure 7).

It has been claimed that γ-crystallin has a relatively high level of bound carbohydrate (61). The carbohydrate moieties are characterized by a large amount of glucose, a feature which is rather uncommon with glycoproteins (62).

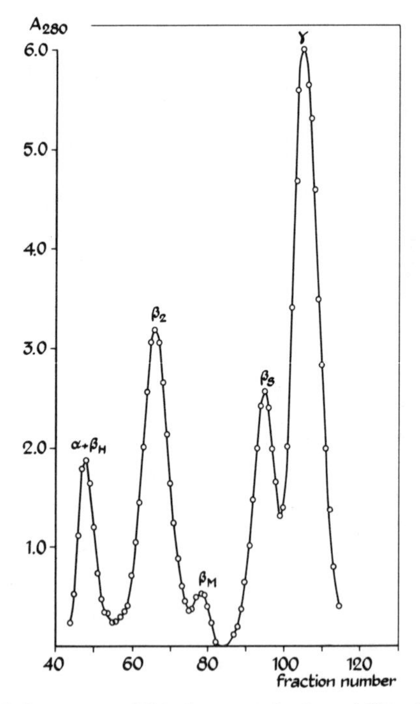

Figure 5. Separation of the water-soluble lens proteins by gel filtration. A column (100 × 3 cm) was packed with Sephadex G-75. Further conditions were as described under Figure 2. The sample solution was centrifuged at 220,000 × *g* for 16 hr prior to gel filtration, which resulted in removal of a major part of α- and β_H-crystallin. Thereafter, elution was performed at a flow rate of 30 ml/hr.

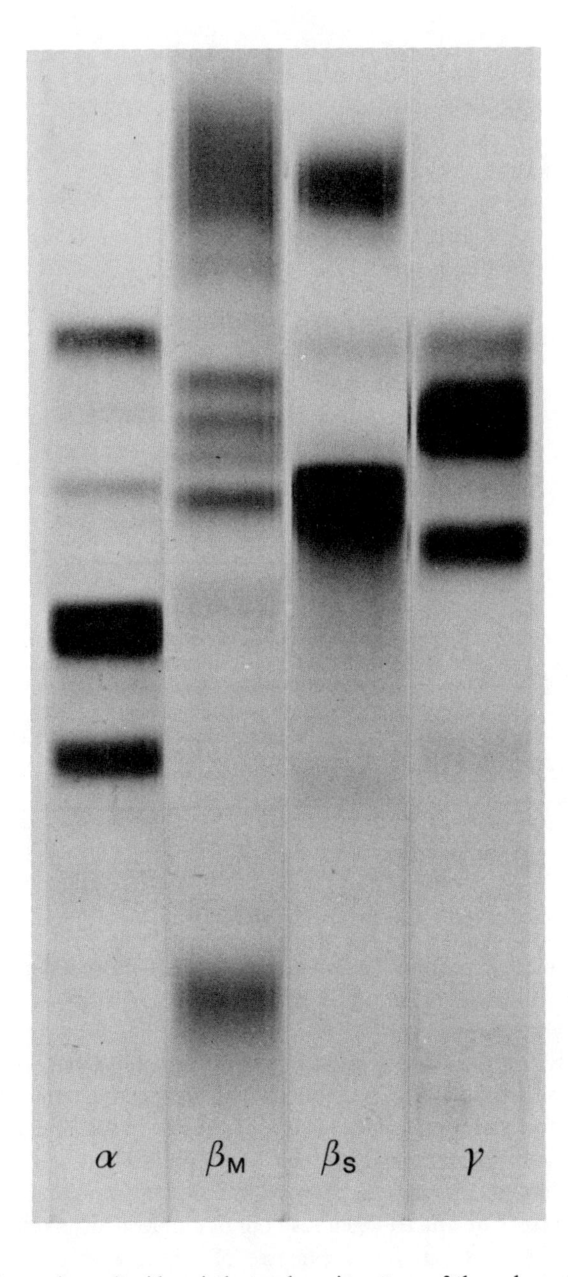

Figure 6. 7 M Urea-polyacrylamide gel electrophoresis pattern of the polypeptide chains of the separated low molecular weight crystallins. Conditions were as described in the legend to Figure 3. For comparison, the pattern of α-crystallin is also shown.

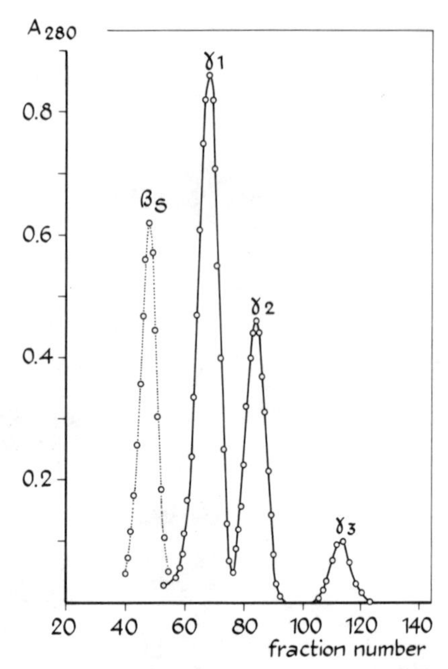

Figure 7. Separation of the γ-crystallin components by chromatography on Sulfoethyl (SE)-Sephadex. The experiment was performed as described by Björk (57) with minor modifications. The column dimensions were 40 × 1.5 cm. Elution at a flow rate of 30 ml/hr was carried out with a linear sodium acetate gradient ranging from 0.2–0.5 M at pH 5. The peak fractions emerged at the following salt concentrations: β_S at 0.24 M, γ_1 at 0.28 M, γ_2 at 0.30 M, and γ_3 at 0.34 M.

SYNTHESIS OF LENS PROTEIN

Biosynthesis of Crystallins in Intact Organ

In the introductory section it was emphasized that mammalian eye lenses grow to a greater or lesser extent throughout life. This growth is accompanied by protein biosynthesis.

The biosynthesis of lens proteins can be studied in a variety of systems. For instance, whole animals can be injected with radioactive amino acids, followed by analysis after various intervals of time of the newly synthesized products. The disadvantage of this method is obvious. For technical reasons one can only use rather small laboratory animals, but even in this case considerable dilution of the labeled precursors by the endogenous pool of amino acids takes place. Nevertheless, protein synthesis in the eye lenses of whole animals has been studied incidentally. Rats were injected intraperitoneally with [^{35}S]methionine. One day after injection the major part of the incorporated radioactivity was detected in the cortex. Only 3% of the total label could be found in the insoluble lens protein. After 7 weeks almost half of the radioactivity was found in the latter fraction, suggesting that this protein arises by post-translational aggregation of pre-existing components. An indication that the protein is virtually not broken down was the observation that no loss of radioactivity took place between the 4th week and the 8th week (63).

Lens protein biosynthesis can also be studied in organ culture. A number of data showed that de novo synthesis was virtually restricted to the epithelium and the outer cortex of the lens (34, cf. 31). Moreover, it was concluded that the synthesis of α-, β-, and γ-crystallin occurred at almost similar rates. However, refinement of the analytical methods revealed that only α-, β_L-, and γ-crystallin have been synthesized in relatively equal amounts, whereas β_H synthesis is extremely low (64). Electrophoretic analysis on polyacrylamide gels containing 6 M urea showed that the "typical" β_H polypeptides, namely, the two highly basic β chains and β_A, have not been synthesized.

Incubation of whole lenses in the absence and presence of actinomycin D showed that crystallin messenger is stable in the fibers, but not in the epithelium (65). Clayton et al. (66) have recently shown that in normal chick lens both the soluble and the membrane-specific proteins are stable in the presence of actinomycin D. In this type of experiment the assumption has been made that messenger RNA (mRNA) concentration is the limiting factor in protein biosynthesis (67). Stewart and Papaconstantinou (3, 68) showed that in the presence of a dose of actinomycin D which inhibits RNA synthesis the incorporation of amino acids into protein was strongly inhibited in the rapidly dividing cells from embryonic bovine lens. After 4 hr 65% inhibition was observed. With fiber cells the inhibition was only 22%. The corresponding values for cultured calf lens were 50% and 13%, respectively. In contrast, actinomycin D had almost no effect on cultured adult bovine lenses as far as amino acid incorporation into proteins of epithelial cells was concerned. On the other hand, the striking observation was made that the specific activity of fiber cell proteins increased to about 50%. This stimulatory effect of the drug has also been reported in other eukaryotic systems (69, 70).

The effects of actinomycin on the incorporation of tritium-labeled leucine into total lens protein, α-crystallin aggregates, and the individual α-polypeptide chains are summarized in Table 1. Obviously, the stoichiometry of polypeptide assembly is significantly affected by the addition of the drug. A shift toward preferential incorporation of radioactive leucine via the αA_2 chains was also reported for fibrogenesis in vivo (71). It cannot be excluded that the stability of the messenger in the lens fibers is due to the occurrence of the potent ribonuclease inhibitor which has been demonstrated in calf lenses (72, 73).

An interesting observation related to lens fiber formation has been made by Ortwerth et al. (74). When the lens epithelium differentiates into fibers a 2-fold increase in phenyl-transfer RNA (tRNA) takes place. This doubling is not a consequence of increased synthesis of a tRNA species which is already present, but by the onset of the formation of another major phenyl-tRNA species. This phenomenon might be related to the preferential synthesis of α-crystallin in the

Table 1. Effect of actinomycin D on incorporation of radioactive leucine into total protein, α-crystallin (aggregate), and individual polypeptides[a]

			α-Crystallin polypeptides				
			Relative concentration		Relative radioactivity		
	Total protein	α-Crystallin (aggregate) (dpm/mg of protein)	αB_2 (%)	αA_2 (%)	αB_2 (%)	αA_2 (%)	$\alpha A_2/\alpha B_2$ (%)
Control	116,583	126,203	24.5	51.7	37.2	53.7	1.44
Actinomycin D	105,058	117,619	27.8	50.6	31.9	60.6	1.90

[a]Control adult lenses were incubated for 4 hr at 37°C in Hank's salt solution containing 26 μCi/ml of L-(4,5-^{3}H)leucine (Schwarz BioResearch, 53.9 Ci/mM). A similar incubation was performed in the presence of 25 μg/ml of actinomycin D (Merck, Sharp and Dohme) in order to bring down RNA synthesis to less than 1% of the control. At the end of the incubation period, epithelial cells were removed and the protein was isolated. The radioactivity was measured by scintillation counting. (From Delcour et al. (70a).)

lens cortex since phenylalanine is one of the most abundant amino acid residues in α-crystallin (cf. Figure 4).

Crystallin Synthesis in Cell Culture

Although rather fruitful studies have been undertaken with cultured epithelium from chick lens (cf. 75), the results with mammalian cells in culture have not as yet been very exciting. Morphologically the epithelial cells undergo elongation, but unfortunately this feature does not seem to be accompanied by the synthesis of any specific crystallin polypeptide (76). In addition, dexamethasone-provoked elongation does not result in de novo crystallin synthesis (77).

The tissue culture experiments seem to be in contrast with the results obtained in whole lenses in that γ-crystallin is associated with cell elongation in the latter system (78). Adult bovine lens cells produce in vitro collagen and glycoproteins (79, 80). Some of these glycoproteins resemble the glycoproteins found in the lens capsule.

Biosynthesis of Crystallins in Homologous Cell-free System

A viable cell-free system which allows the study of the biosynthesis of crystallins can easily be prepared from fresh lenses (64) simply by removing the capsule, the inner cortex, and the nucleus. The remaining epithelial monolayer and the outer cortex are homogenized, and 1 volume of the tissue is suspended in an equal volume of distilled water. The suspension is centrifuged at 15,000 × *g* for 10 min. The supernatant fraction is then divided into small portions which are stored at −70°C. For incubation the lysate is supplemented with appropriate ions, cofactors, and amino acids. That this system indeed carries out all the steps involved in α-crystallin biosynthesis can be demonstrated in the following way.

1) If $[^{35}S]$Met-$tRNA^{f}_{Met}$ is added as the only labeled precursor to the incubation system, all α-crystallin polypeptide chains bear *N*-acetyl-Met–Asp–Ile–Ala at their NH_2 termini (compare the primary structure of α-crystallin chains in Figure 4) (81). This means not only that correct initiation took place, but also that the system still contains the complete machinery which is responsible for NH_2-terminal acetylation. Recently, Granger et al. partially purified the acetylating enzyme from lens and showed that it is specific in that it catalyzes the transfer of the acetyl group from acetyl coenzyme A to the NH_2-terminal amino acid residue, but not the acetylation of the ϵNH_2 group of lysine (82).
2. After 4 min of incubation of the lens lysate with $[^{35}S]$Met-$tRNA^{f}_{Met}$, radioactive α-crystallin can be isolated from the supernatant fraction (200,000 × *g*) in the void volume of a Sephadex G-200 gel filtration column. This means that completed de novo synthesized polypeptide chains are released from the ribosomes, and a high molecular weight aggregate is formed. However, as mentioned earlier, only αA_2 and αB_2 chains can be found in the newly synthesized macromolecule. A similar observation has been reported by Stauffer et al. (83). Furthermore, as in organ culture, with the exception of the β_H

fraction, all other crystallin subclasses are synthesized in the cell-free system (Figure 8).

Biosynthesis of Noncrystallin Protein in Cell-free System

The lens cell-free system synthesizes, in addition to the crystallins, polypeptides which coelectrophorese with lens plasma membrane protein. These lens components escape detection when the electrophoretic analysis after the cell-free

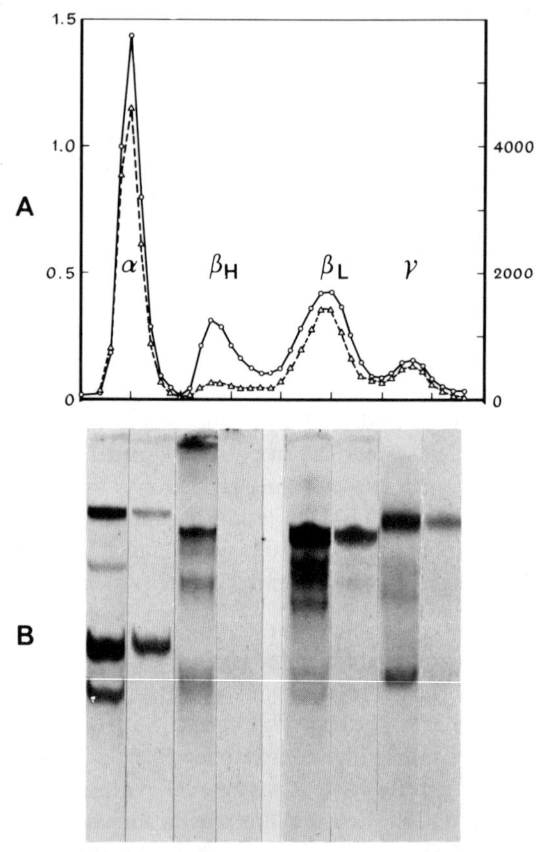

Figure 8. *A,* gel filtration pattern on Sephadex G-200 of the newly synthesized crystallins in the homologous cell-free system. Lens lysate (0.6 ml) (cf. 64) was supplemented with 1 μmol of ATP, 0.5 μmol of GTP, 5 μmol of 2-mercaptoethanol, 10 μmol of creatine phosphate, 50 μg of creatine phosphokinase, 10 μmol of Tris-HCl, pH 7.4, 50 μmol of KCl, and 3 μmol of magnesium acetate. Ten μCi of ^{14}C-labeled amino acid mixture (20 μmol of each) were added. Final volume was 1 ml. Incubations were performed at 30°C for 45 min. Radioactivity was counted in a liquid scintillation counter. 0–0–0–0, protein concentration (mg/ml); △–△–△–△, radioactivity (cpm/ml). The crystallins emerge at the following fraction numbers: α, 30–50; β_H, 50–68; β_L, 68–90; γ, 90–105. *B,* gel electrophoresis pattern in 7 M urea-polyacrylamide of the newly synthesized crystallins in the lens cell-free system. Stained gels and autoradiograph are shown under the corresponding peak fraction. Autoradiography of the dried gels was performed for 6 weeks. (Notice that αB_1, αA_1, and none of the β_H chains are synthesized.)

incubation is carried out on 7 M urea polyacrylamide gels. By this method differences between the individual polypeptides are visualized on the basis of their net charges. It appeared that membrane polypeptides could not be separated in this way from the crystallins. The latter proteins are composed of subunits whose molecular weights vary from 20,000 to about 32,000. Investigations on lens plasma membranes showed that, with the exception of one polypeptide (26,000 daltons), all membrane-specific proteins are located in a region of molecular weight higher than 32,000.

After sodium dodecyl sulfate polyacrylamide gel electrophoresis followed by scintillation autoradiography (84), it can be demonstrated that proteins of higher molecular weight than 32,000 have also been synthesized in the homologous system. This observation is in favor of the assumption that messengers coding for membrane proteins occur in the cortex of the eye lens (85).

Biosynthesis of Lens Protein in Heterologous Systems

α-Crystallin A_2 Messenger The abundance of α-crystallin in the mammalian lens and its rather simple substructure in terms of componant polypeptides, of which only two are primary gene products, render this macromolecule particularly suitable for biosynthetic studies.

Lens polyribosomes (86) are the source of the different messenger RNAs. After dissociation of the polyribosomes by sodium dodecyl sulfate (SDS) treatment, a single run in the zonal centrifuge results in the separation of two size classes of messengers with sedimentation values of 10 S and 14 S, respectively (87). Additional recentrifugation, affinity chromatography on oligo(dt)-cellulose, and semipreparative electrophoresis in polyacrylamide gels containing formamide (88) eventually yield a highly purified 14 S lens mRNA preparation. Analysis of the total population of crystallin messengers on a sucrose gradient (Figure 9), followed by translation of the isolated fractions in different cell-free systems (88–91) or in living oocytes (92), reveals that the mRNA species which encodes the αA_2 chain has a sedimentation value of 14 S, whereas the αB_2 messenger is found in the 10 S region, where the γ- and β-crystallin messengers are also located (cf. Figure 10). It has also been demonstrated that the 14 S and 10 S lens mRNA fractions are found in discrete ribonucleoproteins with sedimentation coefficients of 21 S and 16 S, respectively (95).

From the total amino acid sequence of αA_2 and αB_2 chains, an almost identical molecular weight (20,000) could be derived for both polypeptides. Nevertheless, the αA_2 chain is encoded by a messenger which apparently has a size large enough to code for a protein of about 40,000 daltons. The reason for this "abnormal" length is far from clear. As in other eukaryotic messengers, the 14 S lens mRNA carries at its 3′ end a poly(A) track with an approximate length of 150 adenylic acid residues (96). The nucleotides required to encode the amino acid residues of the αA_2 chain number 519, leaving about 700 nucleotides with an unknown destination. At this moment, one can only speculate about the function of the excess nucleotides. A priori, it cannot be excluded that the 14 S

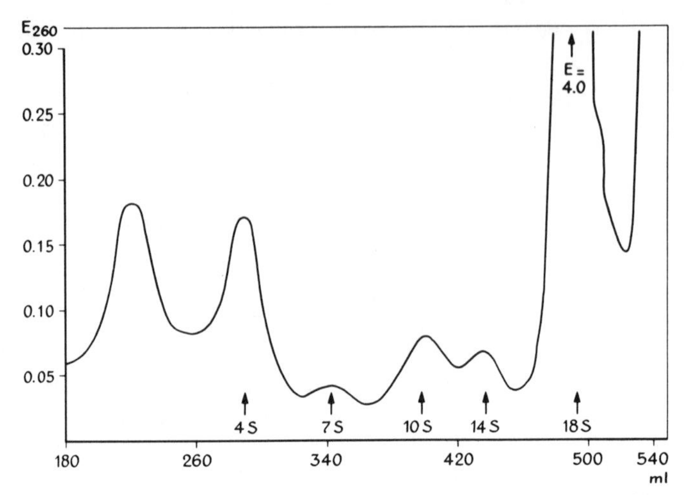

Figure 9. Zonal centrifugation of polysomal RNA from calf lens. Lens polyribosomes were suspended in a medium containing 6% sucrose; 0.05 M Tris-HCl, pH 7.4; and 1.0% SDS, incubated for 5 min at 37°C and diluted twice with the same medium from which SDS was omitted. An exponential gradient from 8% to 28% (w/w) sucrose was applied. All sucrose solutions were pretreated by boiling with 0.02% diethylpyrocarbonate for 30 min. The samples were applied in a volume varying from 10 ml to 25 ml, containing about 3 mg/ml of polyribosomes. Routinely, an overlayer of 200 ml was used. Centrifugation was performed at 2°C in an IEC B XXX zonal rotor for 15 hr at 50,000 rpm.

message codes for a larger precursor of αA_2. However, it has not been possible to demonstrate the occurrence of such a protein; this does not necessarily exclude its existence. If processing of a precursor polypeptide takes place extremely fast, it would escape detection. On the other hand, the amount of extra nucleotides is large enough to account for a bicistronic messenger, coding for two identical or different polypeptide chains. This type of messenger RNA which has never been found preciously in any eukaryotic system should have two initiation sites. Trials by this author's laboratory to demonstrate this by ribosome binding experiments in the presence of radioactive methionyl-$tRNA^{f}_{Met}$ had only negative results. Again, this failure is not absolute proof, but only a rather weak indication that the 14 S mRNA is not bicistronic. A more likely explanation is that the calculated excess nucleotides may be located in regions of secondary structure in lens 14 S mRNA (97). In hemoglobin, mRNA regions of secondary structure have been reported (98). It is interesting that rat αA_2 (99), which differs only in 6 amino acid residues from calf αA_2, is also encoded by a messenger with the "abnormal" sedimentation value of 14 S (100). In the light of the latter interpretation of the occurrence of excess nucleotides, this means that not only the coding region but also the large size of noncoding parts of the 14 S mRNA are strongly conservative in evolution.

Biosynthesis of Noncrystallin Protein The question has been asked whether or not membrane-specific proteins are synthesized in the lens cortex since, a

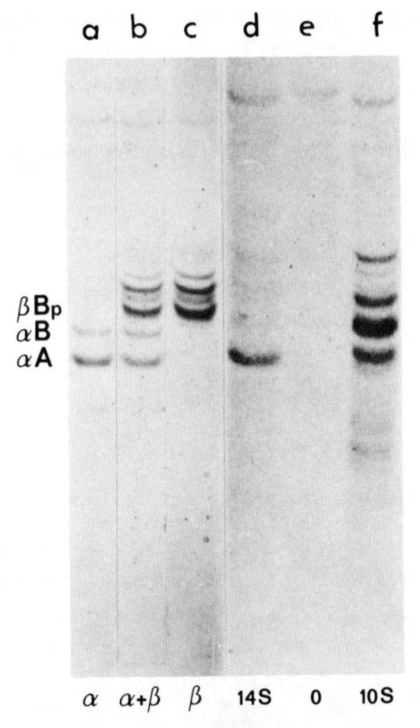

Figure 10. SDS-polyacrylamide gel electrophoresis pattern of the newly synthesized crystallins in the Krebs-ascites lysate. The incubation mixture contained 10 μl of preincubated ascites S-30 (93), 2.5 μmol of magnesium acetate, 100 μmol of KCl, 25 μmol of Tris-HCl, pH 7.5, 6 μmol of 2-mercaptoethanol, 1 μmol of ATP, 0.1 μmol of GTP, 5 μmol of creatine phosphate, 0.2 mg of creatine phosphokinase, and 5 μCi/ml of ^{14}C-labeled amino acids or 10 μmol (10 Ci/mmol) of [^{35}S]methionine plus the remaining unlabeled amino acids. Lens 10 S or 14 S was added at 60 μg/ml. Incubation was at 37°C for 1 hr. Gel electrophoresis in 0.10% SDS-15% acrylamide was performed according to Laemmli (94). For further experimental details, see ref. 102. *a, b,* and *c,* stained gels; *d, e,* and *f,* autoradiographs of the products obtained after addition of 14 S mRNA, without messenger, and with 10 S mRNA, respectively.

priori, it could not be excluded that those proteins originate from the epithelium and remain in the cell during the process of differentiation (27). Observations in the lens cell-free homogenate (see under "Biosynthesis of Crystallins in Homologous Cell-free System") had already indicated that de novo synthesis takes place in the homologous cell-free system. This observation is further sustained by results obtained in a heterologous system. Since the messengers for lens plasma membrane proteins should be found on lens polysomes, this author's group translated the whole population of lens fiber polysomes in a reticulocyte lysate as described by Evans and Lingrel (101). It can be seen that in the heterologous system also not only the crystallin subunits but also a variety of polypeptides with a molecular weight higher than 32,000 have been synthesized

(Figure 11). The protein pattern of lens fiber plasma membranes differs from that of the epithelial membranes. The most prominent component of lens fiber membranes, MP26, is virtually absent in the membrane protein pattern of the epithelium. This means that the process of differentiation not only involves quantitative variations in protein synthesis but also a pronounced qualitative difference in the protein spectrum.

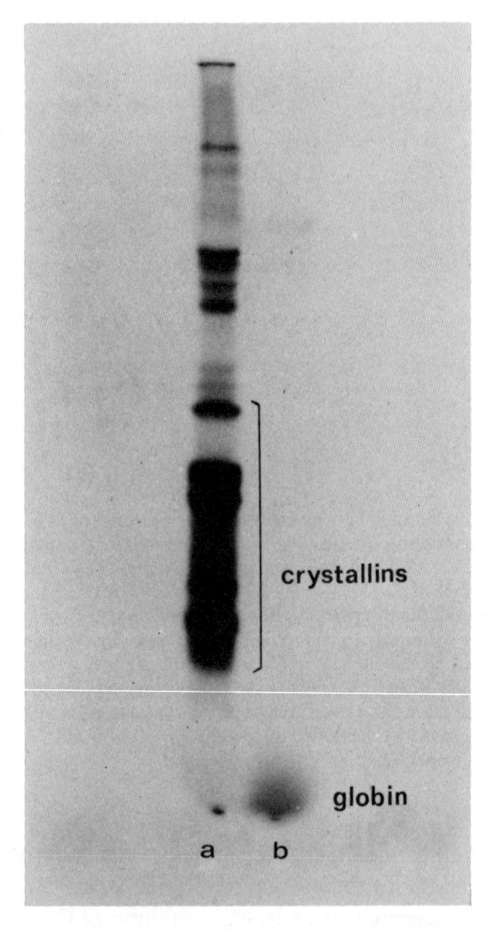

Figure 11. Translation of total lens fiber polyribosomes in a reticulocyte cell-free system. Lens polyribosomes at a concentration of 400 μg/ml were added to the following incubation mixture: 1 μmol of ATP, 0.5 μmol of GTP, 5.0 μmol of 2-mercaptoethanol, 10 μmol of creatine phosphate, 50 μg of creatine phosphokinase, 50 μmol of Tris-HCl, pH 7.4, 50 μmol of KCl, 3 μmol of magnesium acetate, and 0.1 μmol of 20 amino acids. The only labeled amino acid was methionine (200 Ci/mmol). The incubation was carried out at 30°C for 60 min. Analysis of the newly formed products was carried out on SDS gels according to Laemmli (94). Gels were processed for autoradiography as described by Bonner and Laskey (84). The drying procedure has been described by Berns and Bloemendal (102). Notice that in addition to crystallins various other polypeptides in the membrane-protein region are synthesized. The gels are "overexposed" in order to visualize the membrane protein. *a*, 400 μg of lens polysomes added; *b*, no lens polysomes added.

Lens fiber plasma membranes have junctions which appear as a typical pentalayer structure in the electron microscope (23, 26–28). Their physiological function appears to be adhesion and communication between adjacent cells. These junctions, which can be purified by treatment of the isolated plasma membranes with detergents, are characterized by the two major membrane polypeptides MP26 and MP34. The latter constituent seems to be a major intrinsic polypeptide both in the epithelial and in the fiber plasma membranes since it resists both urea and deoxycholate treatment. For the understanding of the biogenesis of the fiber junctions, it would be important to ascertain whether MP34 is synthesized de novo. In the experiments referred to in the preceding section (with the exception of the 14 S mRNA translation product), the only parameter to demonstrate identity of newly synthesized protein was coelectrophoresis with the isolated native protein. Additional evidence was obtained after incubation of the reticulocyte lysate with lens polysomes, followed by immunoprecipitation with a specific antiserum directed against purified lens plasma membranes. Analysis of the precipitate on sodium dodecyl sulfate polyacrylamide gels revealed that MP34 was synthesized preferentially (Figure 12). Control experiments with an antiserum directed against bovine serum albumin showed no radioactivity in the 34,000 region.

It has been suggested that α-crystallin polypeptides might be linked to the plasma membranes (19). Indeed, it is rather difficult to remove crystallin chains completely from purified membrane preparations. However, whether this is a functional interaction rather than an artifactual adsorption has to be established.

AGE DEPENDENCE OF CRYSTALLINS

α-Crystallin

Senescence belongs to that group of problems which have not yet been resolved. One can observe the changes which take place upon aging, but the primary impulses which induce this process have still to be found. Changes in cell structure and functions are frequently described in aging studies. Furthermore, age-related alterations have been proposed during the translation of messengers. For instance, Orgel (105) postulated an error-catastrophy hypothesis according to which a low frequency of errors occur during protein biosynthesis. He predicted that if such "errors" occurred in proteins which participate in the protein synthesis process (e.g., tRNA synthetases or DNA-dependent RNA polymerase), the resultant loss of specificity would induce a rapidly increasing number of errors in proteins synthesized thereafter. It cannot be excluded that in lens cell culture Orgel's theory is applicable. The aging phenomena described in the present review, however, are exclusively post-translational.

It is now generally recognized that not only the biosynthesis of a protein but also its intracellular breakdown is a fundamental process (106). Proteins in animal cells differ considerably in their rate of degradation. Some enzymes are degraded at a very high rate; others have a rather slow turnover. The intracellular

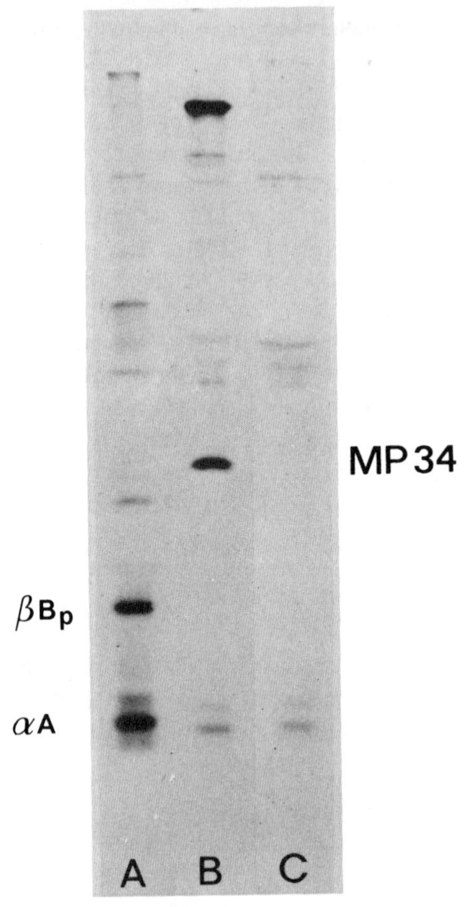

Figure 12. De novo synthesis of a specific major lens plasma membrane protein in the reticulocyte cell-free system. The incubation conditions were as described in the legend to Figure 11. The protein was identified after immunoprecipitation with a specific antiserum directed against lens plasma membranes and subsequent SDS-gel electrophoresis. Immunoprecipitation was performed as described for viral polypeptides (103). The antiserum was prepared with purified lens fiber plasma membranes and tested as described by Van Zaane et al. (104). *A*, translation products of lens polysomes added to the lysate (electrophoretic analysis without immunoprecipitation); *B*, pattern obtained after immunoprecipitation with the antiplasma membrane serum (in addition to MP34, a hitherto unidentified polypeptide of about 120,000 daltons is also detected); *C*, control experiment with a sample obtained after immunoprecipitation with an antiserum directed against bovine serum albumin.

degradative rates seem to be related to certain properties of the protein such as hydrophobicity, charge, size, and especially the compactness of folding. The lens fiber is a good example to show that protein degradation in a cell is highly specific. Although mitochondrial and nuclear proteins are apparently removed during fiber formation, at least α-crystallin is virtually unaffected. This phenomenon is to some extent similar to the maturation changes which occur in reticulocytes. It has been suggested several times that both the degradation and the deamidation of protein are related to cellular aging. This suggestion has been put on a chemical basis by observations made on α-crystallin. Three types of alterations have been found upon aging of α-crystallin: 1) formation of high molecular weight aggregates; 2) deamidation of the polypeptide chains; and 3) a selective degradation which starts from the COOH terminus.

Formation of High Molecular Weight α-Crystallin Several investigators showed that the size of α-crystallin, isolated from lenticular fibers at different stages of maturation, increases as a function of fiber age. Calf lens α-crystallin with a molecular weight higher than 10^7 was first described by Spector et al. (39, 107). The crucial fractionation step was on Bio-Gel A-15m. Later experiments on Agarose of still larger pore size (Bio-Gel A-50m) revealed the presence of aggregates with molecular weight greater than 5×10^7 (108). When the total mixture of the water-soluble lens proteins is subjected to gel filtration on a Bio-Gel A-5m column, the HM α-crystallin emerges in the void volume, whereas the "normal" α-crystallin fraction with an average molecular weight of 800,000 is eluted as a second peak followed by β_H-, β_L-, and γ-crystallin (Figure 13) (109). In the epithelium and outer cortex virtually no HM α-crystallin is detectable, whereas the amount of the large aggregates increases from inner cortex to the nucleus.

On the Agarose A-50m column, HM α-crystallin is separated into three fractions the heaviest component of which has a sedimentation coefficient of approximately 200 S (109). The high molecular weight aggregates are characterized by an increased proportion of partially degraded A and B chains. Aggregation of α-crystallin in the aging lens has been related to the presence of glucose (107) or calcium ions (108, 110). However, the phenomenon has not yet fully been elucidated. Li (111) reported that the increase in size of α-crystallin is associated with the appearance of low molecular weight chromophores. It has been claimed that the "aged" fraction of α-crystallin binds to the plasma membranes (19).

Deamidation of Polypeptide Chains Postsynthetic loss of amide groups has been observed previously in cytochrome *c* and aldolase (112, 113). In this connection Robinson et al. (114) suggested that deamidation initiates the degradation of protein in vivo. This loss of amide groups from asparagine and glutamine would result in conformational changes and hence increased susceptibility to proteolysis (115). The age dependence of the α-crystallin subunit structure was first clearly demonstrated with embryonic α-crystallin. Subunit αA_1, which is originally absent (116, 117), gradually appears with increasing age of the embryo. Although the four major polypeptides are found in adult fiber cells, the epithelial cells contain virtually αA_2 and αB_2 (33, 118, 119). Newly synthesized α-crystallins both in whole lenses (64, 83) and in the lens cell-free system (64) contain only A_2 and B_2 chains. It has been claimed that extensive conversion of A_2 to A_1 should occur during a 1-hr incubation period (119). In our experiments, however, such a rapid conversion has never been observed. In

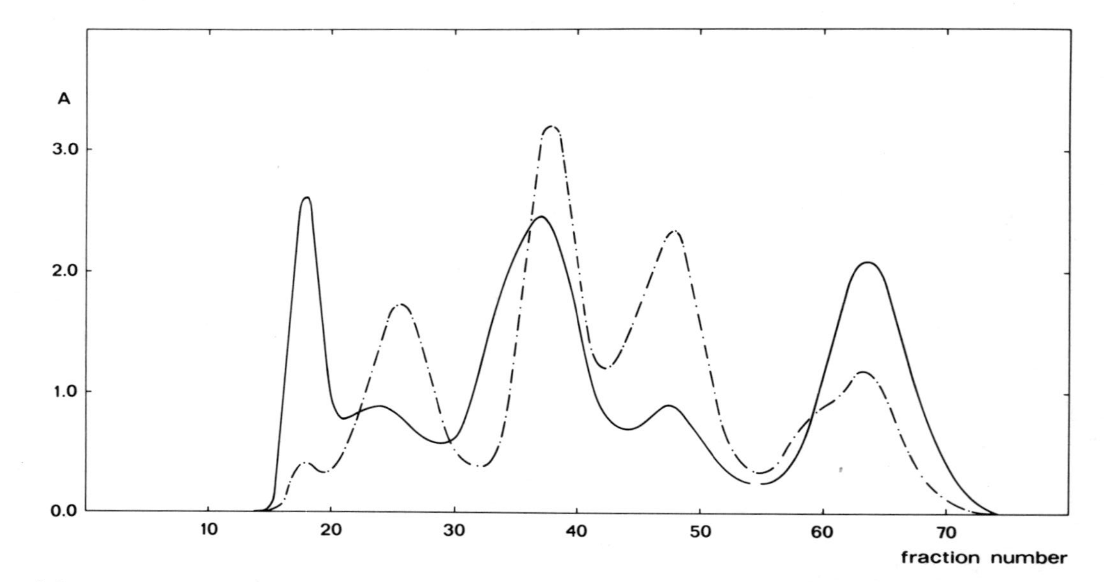

Figure 13. Separation of the water-soluble lens proteins by gel filtration. A column (90 × 2.5 cm) was packed with Bio-Gel A-5m (exclusion limit 5 × 10^6 daltons) and equilibrated in a buffer containing 0.1 M Tris-HCl, pH 7.3. Elution was performed with the same buffer. Protein (200 mg) was loaded onto the column. Fractions of 6.5 ml were collected at a flow rate of 0.5 ml/min. • — •, absorbance at 280 nm of calf cortex; —, absorbance at 280 nm of calf nucleus. (From van Kleef and Hoenders (109).)

contrast, purified αA_2 chains had to be incubated for at least 8 weeks before αA_1 appeared (120). After 2 weeks an α-crystallin polypeptide, named $\alpha A_{1.5}$, was formed of intermediate electrophoretic mobility between αA_2 and αA_1. This finding suggests that the conversion of A_2 to A_1 is presumably a two step deamidation process.

Selective Degradation from COOH Terminus The established amino acid sequence of the COOH-terminal tryptic peptide of αA_2 chains is Glu–Glu–Lys–Pro–Ser–Ser–Ala–Pro–Ser–Ser (cf. Figure 4). In some cases a tryptic peptide was found which differed from this sequence in that it lacked Ser–Ala–Pro–Ser–Ser (45). This finding suggested the occurrence of a shortened αA_2 chain. Further investigations revealed that several shortened polypeptides exist in aged α-crystallin, all lacking discrete numbers of amino acid at the COOH-terminal part of the chain.

A tentative aging scheme based on COOH-terminal degradation is given in Figure 14. It cannot be excluded that proteases are involved in this selective degradation. For instance, a neutral protease has been found in lens and partially characterized (121, 122). The precise function of this enzyme, however, is still unknown.

Effect of Aging on β- and γ-Crystallin

Very little is known about the effect of aging on β- and γ-crystallins. Experiments in this laboratory with whole lenses and the lens cell-free system (64) suggested that β_H is not synthesized directly, but arises by association of β_L components with the specific β_H polypeptides.

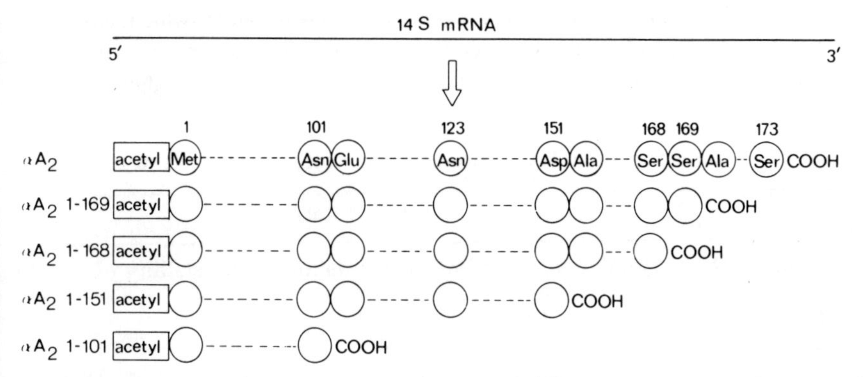

Figure 14. Tentative aging scheme of the major α-crystallin polypeptide A_2. The shortened chains A_2^{1-169} and A_2^{1-151} can already be detected in the embryonic stage. A_2^{1-151}, however, is more abundant in adult nuclear than in embryonic α-crystallin. A_1^{1-101} is not detectable in embryonic α-crystallin. Similar degradation takes place in αA_1, which arises by deamidation from αA_2 (Asn_{123} is converted into Asp_{123}). The migration distance on polyacrylamide-urea gels suggests that a second transition from an asparagine (or glutamine) residue to aspartic acid (or glutamic acid) might have occurred. This is sustained by the observation that in vitro deamidation yields a product $\alpha A_{1.5}$, which migrates at intermediate mobility between αA_2 and αA_2 (120).

Van Kamp (123) undertook a systematic study to investigate the distribution of crystallins as a function of age and location in the lens. For this purpose the organ was divided into three parts, containing the outer, relatively youngest part of the cortex, an inner cortex fraction, and the core. From the epithelium to the lens nucleus, not only increasing quantities of HM α-crystallin but also higher amounts of β_H were found. In subsequent studies (124, 125) calf lenses were divided into nine concentric layers. The soluble lens proteins from these fractions were subjected to gel filtration on Agarose A-5m. Again the amounts of HM α-crystallin and β_H increased from epithelium to nucleus, whereas the β_L and α-crystallin content decreased markedly (Table 2). This phenomenon was investigated on the level of β_H subunit composition (126). The results are shown in Figure 15. The most impressive changes can be seen with the two characteristic β_H polypeptides of 32,000 and 31,000 molecular weight. An increase in the amount of the 31,000 molecular weight polypeptide (βB_{1b}) is accompanied by a gradual disappearance of the 32,000 molecular weight component (βB_{1a}).

Comparison of the protein profiles of cow and calf lens nucleus reveals that during the process of aging the 32,000 molecular weight polypeptide diminishes markedly. This author believes that the 31,000 band arises by some post-translational modification of βB_{1a}. This assumption is sustained by the fact that a messenger for the 32,000 molecular weight component can be found (cf. Figure 16), whereas there is no mRNA for the 31,000 polypeptide detectable. The reason why no β_H formation is observed in vitro may be that the transformation of βB_{1a} into βB_{1b} is a prerequisite for β_H assembly.

As far as γ-crystallin is concerned, Kabasawa and Kinoshita (127) reported that with aging of the bovine lens a new form of γ-crystallin emerges. They found two types of γ-crystallin in the older bovine lens, one of which, existing in the nucleus, was identical with calf γ-crystallin. The second γ species was detected in the cortex and was of greater molecular weight and higher isoelectric

Table 2. Content of soluble lens proteins of nine concentric layers of calf lenses

	Protein content (% of total)				
Layer	HM α-Crystallin	α-Crystallin	β_H-Crystallin	β_L-Crystallin	γ-Crystallin
Epithelium		49.0	15.7	32.9	
II		40.6	14.8	25.6	18.0
III		39.6	13.9	25.2	18.7
IV	2.4	36.3	16.6	23.6	20.8
V	2.8	38.8	16.6	21.8	19.6
VI	2.2	36.4	23.1	21.1	17.5
VII	2.7	36.4	21.5	20.4	19.0
VIII	5.1	35.0	21.0	18.5	21.5
Nucleus	12.6	29.5	19.5	14.9	23.5

From Hoenders et al. (125).

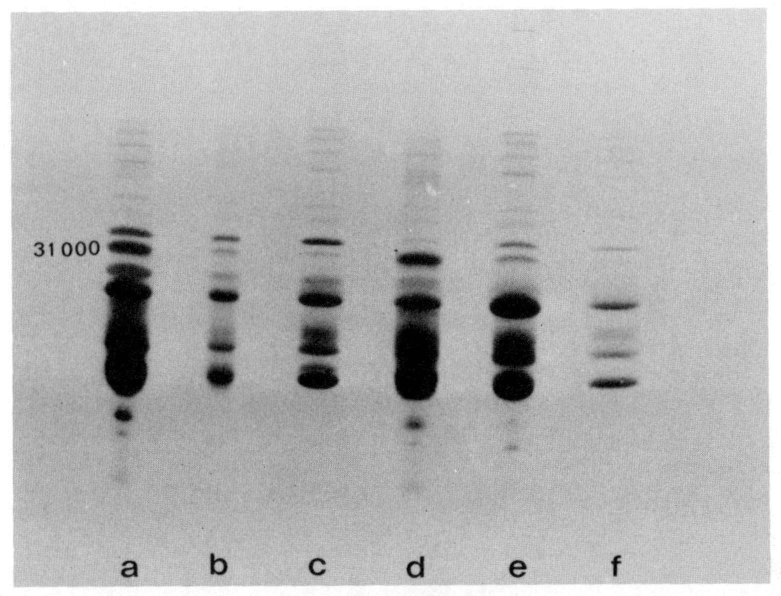

Figure 15. Sodium dodecyl sulfate-polyacrylamide slab gel electrophoresis pattern of the total water-soluble lens proteins from different layers of the eye lens. *a,* calf lens nucleus; *b,* calf lens inner cortex; *c,* calf lens outer cortex; *d,* cow lens nucleus; *e,* cow lens inner cortex; *f,* cow lens outer cortex. Electrophoresis was carried out according to Laemmli (94).

point than calf lens γ-crystallin. However, no data were provided to give these differences a chemical basis.

CONCLUDING REMARKS

The vertebrate eye lens has often been presented as a model system for the study of development, growth, and aging in higher organisms. Accumulating data on the molecular level, especially that acquired from calf lens tissue, show that this statement is actually more than just a suggestion. Of course, part of the available information is not yet rigorous enough to allow definite conclusions about such important problems as the regulatory mechanisms in differentiation. For instance, it is not understood at this moment why bovine epithelial cells bearing all the information for the production of the crystallins apparently lose this ability when cultured on plastic, whereas crystallin synthesis continues upon cultivation of the epithelium on its natural substrate, the lens capsule (128). Another unresolved problem is the size of the 14 S crystallin messenger which directs the synthesis of the αA_2 chain. The αB_2 chain with the same molecular weight and a 60% homology as compared to αA_2 is encoded by a 10 S messenger (a reasonable size relation between messenger RNA and encoded polypeptide). The assumption that the excess nucleotides in the 14 S mRNA species are due to regions of secondary structure might be an oversimplification. The problem

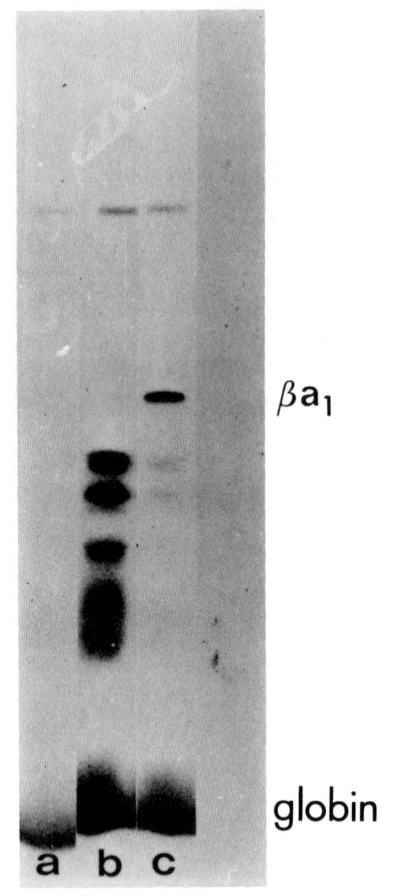

Figure 16. Translation of mRNA fractions in a reticulocyte system. The incubation conditions for gel electrophoresis and autoradiography were as described in the legend to Figure 11. The amount of mRNA added was 40 μg/ml. *a*, control experiment, no messenger added; *b*, translation products formed after addition of the total lens crystallin 10 S mRNA fraction; *c*, translation products formed after addition of the βB_{1a} mRNA (12.5 S). The mRNA was obtained after purification of total polysomal poly(A)-containing RNA on a 15–35% sucrose gradient.

became even more interesting and complicated after it was observed that in rat lens the 14 S mRNA directs, in addition to αA_2, the formation of an αA type of chain which differs from αA_2 in that it contains a sequence of 22 extra amino acid residues in the internal position (100, 129).

Some progress has been made in identifying newly synthesized lens membrane-specific protein in the homologous and heterologous systems after addition of lens polysomes. The fiber plasma membranes are characterized by only a few major polypeptides, two of which also occur in the isolated junctions. Detailed structural studies of these highly interesting components are under way.

Since α-crystallin chains are detected in all purified fiber membrane preparations, one may speculate that these polypeptides exert a certain function which requires their interaction with membrane constituents. Moreover, direct visualization in the electron microscope of isolated plasma membrane preparations strongly suggests that crystallins form stable complexes with other integral plasma membrane constituents and cytoplasmic microfilaments (27). Since the mammalian eye lens in the process of vision has to accommodate, which means that the lens body changes in shape while the total surface of the composing cells remains constant, one has to assume that there exists a contractile apparatus for at least part of the fiber membranes. Which lens proteins are actively involved in such a device is also a matter of speculation. Actin-like protein would be a good candidate for this function. Again, there is not yet enough evidence available for this hypothesis. An indication might be the finding of a major protein component in both the water-soluble fraction and the urea-soluble part of the eye lens which has the same molecular weight as the actin monomers in other tissues. Moreover, this protein is specifically retained on an affinity chromatography column loaded with NDAseI-Sephaxose 4B (130). This purification is based upon the work of Lazarides and Lindberg (130), who were able to show that actin from calf thymus has the properties of the naturally occurring inhibitor deoxyribonuclease I.Characterization of this protein had just been started when this review was completed.

REFERENCES

1. Iwig, M., and Glässer, D. (1976). *In* O. Hockwin (ed.), Documenta Ophthalmologica, pp. 47–55. Dr. W. Junk bv Publishers, The Hague.
2. Papaconstantinou, J., Stewart, J. A., and Koehn, P. V. (1966). Biochim. Biophys. Acta 114:428.
3. Stewart, J. A., and Papaconstantinou, J. (1967). J. Mol. Biol. 29:357.
4. van Heyningen, R. (1976). Invest. Ophthalmol. 15:685.
5. Reeder, R., and Bell, E. (1965). Science 150:71.
6. Hanna, C., and Keatts, H. C. (1966). Exp. Eye Res. 5:111.
7. Modak, S. P., Morris, G., and Yamada, T. (1968). Dev. Biol. 17:544.
8. Modak, S. P., and Persons, B. J. (1971). Exp. Cell Res. 64:476.
9. Clayton, R. M. (1970). Curr. Top. Dev. Biol. 5:115.
10. Clayton, R. M. (1974). *In* H. Davson and L. T. Graham (eds.), The eye, pp. 400–494. Academic Press, London.
11. Piatigorsky, J., de Webster, H. F., and Craig, S. P. (1972). Dev. Biol. 27:176.
12. Piatigorsky, J., Zelenka, P., and Simpson, R. T. (1974). Exp. Eye Res. 18:435.
13. Piatigorsky, J. (1975). Exp. Eye Res. 21: 245.
14. Bloemendal, H., and Herbrink, P. (1974). Ophthalmic Res. 6:81.
15. Mörner, C. T. (1894). Z. Physiol. Chem. 18:61.
16. Ruttenberg, G. (1965). Exp. Eye Res. 4:18.
17. Waley, S. G. (1965). Exp. Eye Res. 4:293.
18. Dische, Z. (1965). Invest. Ophthalmol. 4:759.

19. Bracchi, P. G., Carta, F., Fasella, P., and Maraini, G. (1971). Exp. Eye Res. 12:151.
20. Dische, Z., Hairstone, M. A., and Zelmenis, G. (1967). Protides Biol. Fluids Pro. Colloq. 15:123.
21. Bloemendal, H., Zweers, A., Vermorken, F., Dunia, I., and Benedetti, E. L. (1972). Cell Diff. 1:91.
22. Lasser, A., and Balazs, E. A. (1972). Exp. Eye Res. 13:292.
23. Dunia, I., Sen Ghosh, C., Benedetti, E. L., Zweers, A., and Bloemendal, H. (1974). FEBS Lett. 45:139.
24. Broekhuyse, R. M., and Kuhlmann, E. D. (1974). Exp. Eye Res. 19:297.
25. Alcalá, J., Lieska, N., and Maisel, H. (1975). Exp. Eye Res. 21:581.
26. Benedetti, E. L., Dunia, I., and Bloemendal, H. (1974). Proc. Natl. Acad. Sci. U.S.A. 71:5073.
27. Benedetti, E. L., Dunia, I., Bentzel, C. J., Vermorken, A. J. M., Kibbelaar, M., and Bloemendal, H. (1976). Biochim. Biophys. Acta 457:353.
28. Philipson, T., Hanninen, L., and Balazs, E. A. (1975). Exp. Eye Res. 21:205.
29. Broekhuyse, R. M., Kuhlmann, E. D., and Stols, A. L. H. (1976). Exp. Eye Res. 23:365.
30. Bloemendal, H. (1972). Acta Morphol. Neerl. Scand. 10:197.
31. Waley, S. G. (1969). *In* H. Davson (ed.), The Eye, Vol. I, pp. 299–379. Academic Press, London.
32. Harding, J. J., and Dilley, K. J. (1976). Exp. Eye Res. 22:1.
33. Delcour, I., and Papaconstantinou, J. (1972). J. Biol. Chem. 247:3289.
34. Spector, A., and Kinoshita, J. H. (1964). Invest. Ophthalmol. 3:517.
35. Bloemendal, H., Bont, W. S., Jongkind, J. F., and Wisse, J. H. (1964). Biochim. Biophys. Acta 82:191.
36. Bloemendal, H., Berns, T., Zweers, A., Hoenders, H., and Benedetti, E. L. (1972). Eur. J. Biochem. 24:401.
37. Bloemendal, H., and ten Cate, G. (1959). Arch. Biochem. Biophys. 84: 512.
38. Björk, I. (1960). Biochim. Biophys. Acta 45:372.
39. Spector, A., Li, L.-K., Augusteyn, R. C., Schneider, A., and Freund, T. (1971). Biochem. J. 124:337.
40. Bloemendal, H., Bont, W. A., Jongkind, J. F., and Wisse, J. H. (1962). Exp. Eye Res. 1:300.
41. Bloemendal, H. (1969). Exp. Eye Res. 8:227.
42. Waley, S. G. (1969). Exp. Eye Res. 8:477.
43. Bloemendal, H., Berns, A. J. M., van der Ouderaa, F., de Jong, W. W., (1972). Exp. Eye Res. 14:80.
44. Palmer, W. G., and Papaconstantinou, J. (1969). Proc. Natl. Acad. Sci. U.S.A. 64:404.
45. de Jong, W. W., van Kleef, F. S. M., and Bloemendal, H. (1974). Eur. J. Biochem. 48:271.
46. van Kleef, F. S. M., Nijzink-Maas, M. J. C. M., and Hoenders, H. J. (1974). Eur. J. Biochem. 48:563.
47. van der Ouderaa, F. J., de Jong, W. W., and Bloemendal, H. (1973). Eur. J. Biochem. 39:207.
48. van der Ouderaa, F. J., de Jong, W. W., Hilderink, A., and Bloemendal, H. (1974). Eur. J. Biochem. 49:157.
49. Testa, M., Armand, G., and Balazs, E. A. (1965). Exp. Eye Res. 4:327.

50. Bont, W. S., Jongkind, J. F., Wisse, J. H., and Bloemendal, H. (1962). Biochim. Biophys. Acta 59:512.
51. Maisel, H., and Goodman, M. (1965). Am. J. Ophthalmol. 59:697.
52. Holt, W. S., and Kinoshita, J. H. (1968). Invest. Ophthalmol. 7:169.
53. Truman, D. E., and Clayton, R. M. (1974). Exp. Eye Res. 18:485.
54. Zigler, J. S., and Sidbury, J. B. (1973). Exp. Eye Res. 16:207.
55. Herbrink, P. (1976). Ph.D. thesis, University of Nijmegen.
56. Croft, L. R. (1972). Biochem. J. 128:961.
57. Björk, I. (1964). Exp. Eye Res. 3:254.
58. Björk, I. (1970). Exp. Eye Res. 9:152.
59. van Dam, A. F. (1966). Exp. Eye Res. 5:255.
60. Bloemendal, H., and Zweers, A. (1976). Doc. Ophthalmol. 8:91.
61. Kabasawa, I., and Kinoshita, J. H. (1973).Exp. Eye Res. 16:143.
62. Spiro, R. G. (1970). Annu. Rev. Biochem. 39:599.
63. Fulhorst, H. W., and Joung, R. W. (1966). Invest. Ophthalmol. 5:298.
64. Strous, G. J. A. M., van Westreenen, H., van der Logt, J., and Bloemendal, H. (1974). Biochim. Biophys. Acta 353:89.
65. Papaconstantinou, J. (1967). Science 156:338.
66. Clayton, R. M., Truman, D. E. S., Hunter, J., Odeigah, P. G., and De Pomerai, D. I. (1976). *In* O. Hockwin (ed.), Documenta Ophthalmologica, pp. 27–29. Dr. W. Junk bv Publishers, The Hague.
67. Kafatos, F. C., and Gelinas, R. (1974). *In* J. Paul (ed.), Biochemistry of Cell Differentiation, pp. 223–264. Butterworths, London.
68. Stewart, J. A., and Papaconstantinou, J. (1967). Proc. Natl. Acad. Sci. U.S.A. 58:95.
69. Wattiaux, J. M., Libion-Mannaert, M., and Delcour, J. (1971). Gerontologia 17:289.
70. Palmiter, R. D., and Schimke, R. T. (1973). J. Biol. Chem. 248:1502.
70a. Delcour, J., Odaert, S., and Bouchet, H. (1976). *In* Y. Courtois and F. Reynault (eds.), Colloque INSERM, p. 43.
71. Delcour, J., and Papaconstantinou, J. (1974). Biochem. Biophys. Res. Commun. 57:134.
72. Ortwerth, B. J., and Byrnes, R. J. (1971). Exp. Eye Res. 12:120.
73. van de Broek, W. J. M., Koopmans, M. A. G., and Bloemendal, H. (1974). Mol. Biol. Rep. 1:295.
74. Ortwerth, B. J., Yonuschot, G. R., Heidlege, J. F., and Chu-Der, O. M. Y. (1975). Exp. Eye Res. 20:417.
75. Piatigorsky, J., Rothschild, S. S., and Milstone, L. M. (1973). Dev. Biol. 34:334.
76. van Venrooij, W. J., Groeneveld, A. A., Bloemendal, H., and Benedetti, E. L. (1974). Exp. Eye Res. 18:517.
77. van Venrooij, W. J., Groeneveld, A. A., Bloemendal, H., and Benedetti, E. L. (1974). Exp. Eye Res. 18:527.
78. Papaconstantinou, J. (1965). Biochim. Biophys. Acta 107:81.
79. Hughes, C., Laurent, M., Lonchampt, M. O., and Courtois, Y. (1975). Eur. J. Biochem. 52:143.
80. Lonchampt, M. L., Laurent, M., and Courtois, Y. (1976). INSERM 60:163.
81. Strous, G. J. A. M., van Westreenen, H., and Bloemendal, H. (1973). Eur. J. Biochem. 38:79.
82. Granger, M., de Jong, W. W., Tesser, I., and Bloemendal, H. (1976). Proc. Natl. Acad. Sci. U.S.A. 73:3010.

83. Stauffer, J., Rothschild, C., Wandel, T., and Spector, A. (1974). Invest. Ophthalmol. 13:135.
84. Bonner, W. M., and Laskey, R. A. (1974). Eur. J. Biochem. 46:83.
85. Vermorken, A. J. M., Hilderink, J. M. H. C., van de Ven, W. J. M., and Bloemendal, H. (1975). Biochim. Biophys. Acta 414:167.
86. Bloemendal, H., Schoenmakers, J., Zweers, A., Matze, R., and Benedetti, E. L. (1966). Biochim. Biophys. Acta 123:217.
87. Berns, A. J. M., de Abreu, R. A., van Kraaikamp, M., Benedetti, E. L., and Bloemendal, H. (1971). FEBS Lett. 18:159.
88. Berns, A., Janssen, P., and Bloemendal, H. (1974). Biochem. Biophys. Res. Commun. 59:1157.
89. Berns, A. J. M., Strous, G. J. A. M., and Bloemendal, H. (1972). Nature (New Biol.) 236:7.
90. Mathews, M. B., Osborn, M., Berns, A. J. M., and Bloemendal, H. (1972). Nature (New Biol.) 236:5.
91. Berns, A. J. M., Schreurs, V. V. A. M., van Kraaikamp, M. W. G., and Bloemendal, H. (1973). Eur. J. Biochem. 33:551.
92. Berns, A. J. M., van Kraaikamp, M., Bloemendal, H., and Lane, C. D. (1972). Proc. Natl. Acad. Sci. U.S.A. 69:1606.
93. Mathews, M. B., and Korner, A. (1970). Eur. J. Biochem. 17:328.
94. Laemmli, U. K. (1970). Nature (Lond.) 227:680.
95. Bloemendal, H., Berns, A. J. M., Strous, G., Mathews, M., and Lane, C. D. (1972). Proceedings of the Eighth Meeting of the Federation of European Biochemical Societies, Vol. 27, pp. 237–250. North Holland/American Elsevier, Amsterdam.
96. Piperno, G., Bertazzoni, U., Berns, A. J. M., and Bloemendal, H. (1974). Nucleic Acid Res. 1:245.
97. Favre, A., Bertazzoni, U., Berns, A. J. M., and Bloemendal, H. (1974). Biochem. Biophys. Res. Commun. 56:273.
98. Favre, A., Morel, C., and Scherrer, K. (1975). Eur. J. Biochem. 57:147.
99. de Jong, W. W., van der Ouderaa, F. J., Versteeg, M., Groenewoud, G., van Amelsvoort, J. M., and Bloemendal, H. (1975). Eur. J. Biochem. 53:237.
100. Cohen, L. H., Smits, D. P. E. M., and Bloemendal, H. (1976). Eur. J. Biochem. 67:563.
101. Evans, M. J., and Lingrel, J. B. (1969). Biochemistry 8:829.
102. Berns, A. J. M., and Bloemendal, H. (1974). Methods Enzymol. 30:675.
103. Van Zaane, D., Gielkens, A. L. J., Dekker-Michielsen, M. J. A., and Bloemers, H. P. J. (1975). Virology 67:544.
104. Van Zaane, D., Dekker-Michielsen, M. J. A., and Bloemers, H. P. J. (1976). Virology 75:113.
105. Orgel, L. E. (1963). Proc. Natl. Acad. Sci. U.S.A. 49:517.
106. Goldberg, A. L., and St. John, A. C. (1976). Ann. Rev. Biochem. 45:747.
107. Spector, A., Freund, T., Li, L.-K., and Augusteyn, R. C. (1971). Invest. Ophthalmol. 10:677.
108. Spector, A., and Rothschild, C. (1973). Invest. Ophthalmol. 12:225.
109. van Kleef, F. S. M., and Hoenders, H. J. (1973). Eur. J. Biochem. 40:549.
110. Jedziniak, J. A., Kinoshita, J. H., Yates, E. M., Hocker, L. O., and Benedek, G. B. (1972). Invest. Ophthalmol. 11:905.
111. Li, L.-K. (1974). Exp. Eye Res. 18:383.
112. Flatmark, T., and Sletten, K. (1968). J. Biol. Chem. 243:1623.

113. Midelfort, C. F., and Mehler, A. H. (1972). Proc. Natl. Acad. Sci. U.S.A. 69:1816.
114. Robinson, A. B., Mckerrow, J. H., and Cary, P. (1970). Proc. Natl. Acad. Sci. U.S.A. 66:753.
115. Robinson, A. B., and Rudd, C. J. (1974). Curr. Top. Cellular Reg. 8:247.
116. Schoenmakers, J. G. G., and Bloemendal, H. (1968). Nature (Lond.) 220:790.
117. Delcour, J., and Papaconstantinou, J. (1970). Biochem. Biophys. Res. Commun. 41:401.
118. van Kamp, G. J., Schats, L. H. M., and Hoenders, H. J. (1973). Biochim. Biophys. Acta 295:166.
119. Chen. J. H., Layers, G. C., Spector, A., Schutz, G., and Feigelson, P. (1974). Exp. Eye Res. 18:189.
120. van Venrooij, W. J., de Jong, W. W., Janssen, A., and Bloemendal, H. (1974). Exp. Eye Res. 19:157.
121. van Heyningen, R., and Waley, S. G. (1963). Biochem. J. 86:92.
122. Blow, A. M. J., van Heyningen, R., and Barrett, A. J. (1975). Biochem. J. 145:591.
123. van Kamp, G. J. (1973). Ph.D. thesis, University of Nijmegen.
124. van Kamp, G. J., and Hoenders, H. J. (1973). Exp. Eye Res. 17:417.
125. Hoenders, H. J., van Kamp, G. J., Liem-The, K., and van Kleef, F. S. M. (1973). Exp. Eye Res. 15:193.
126. Vermorken, A. J. M., Herbrink, P., and Bloemendal, H. (1977), in press.
127. Kabasawa, I., and Kinoshita, J. H. (1973). Exp. Eye Res. 16:143.
128. Vermorken, A. J. M., Groeneveld, A. A., Hildreink, J. M. H. C., de Waal, R., and Bloemendal, H. (1977), in press.
129. Cohen, L. H., Westerhuis, L. W. J. J. M., de Jong, W. W., and Bloemendal, H. (1977), in press.
130. Kibbelaur, M., Dunia, I., Benedetti, E. L., and Bloemendal, H. (1977), in press.
131. Lazarides, E., and Lindberg, U. (1974). Proc. Natl. Acad. Sci. U. S. A. 71:4742.

International Review of Biochemistry
Biochemistry of Cell Differentiation II, Volume 15
Edited by J. Paul

6
The Biology of the Friend Cell

P. R. HARRISON

The Beatson Institute for Cancer Research,
Wolfson Laboratory for Molecular Pathology,
Glasgow, Scotland

The original research was supported by grants from the Medical Research Council and the Cancer Research Campaign.

NORMAL ERYTHROPOIESIS

For three basic reasons, erythropoiesis has been recognized for many years as a favorable system for studies of cell differentiation (see ref. 1 for recent review). First, the red blood cell is characterized by a series of well defined functions or products, of which hemoglobin is the classic example. However, more recently other enzyme changes and membrane changes (for example, spectrin (3) and glycophorin (4) formation, loss of H-2 antigen (5), changes in membrane structure (6), and other surface antigen changes (7)) have been quantitatively determined (2). In addition, the development of highly sensitive assays for globin-specifying sequences (8–12) has permitted a closer examination of the manner in which globin gene expression is regulated during erythroid cell maturation (13–16).

Secondly, the ontogeny of red blood cells is now understood in some detail. The manner in which red blood cells are derived from a stem cell capable of producing all the hematopoietic lines has been deduced from cell kinetic and other studies (17, 18). The existence of intermediate stem cells (ECP) which become committed exclusively to the erythroid line under the influence of the organ stroma (19, 20) has been inferred; these committed cells later become sensitive to the hormone erythropoietin (ERC) and then differentiate to produce the recognizable erythroid series. Recently, the existence of both types of committed erythroid stem cell has been demonstrated directly by in vitro cloning techniques (21–23). These stages in the formation of erythroid cells are summarized in Figure 1.

A third reason for the study of erythropoiesis is the existence of normal developmental systems (e.g., fetal liver) which are almost exclusively devoted to red blood cell formation. Despite the fact that these developmental systems contain mainly committed erythroid cells, they have proven very useful for molecular and cytological studies, studies concerning the regulation of hemoglobin synthesis, and, in particular, the role of the hormone erythropoietin (reviewed recently by Harrison (1)). Erythropoietin seems to have a dual role: to maintain the committed erythroid stem cell pool and, at least at high doses, to

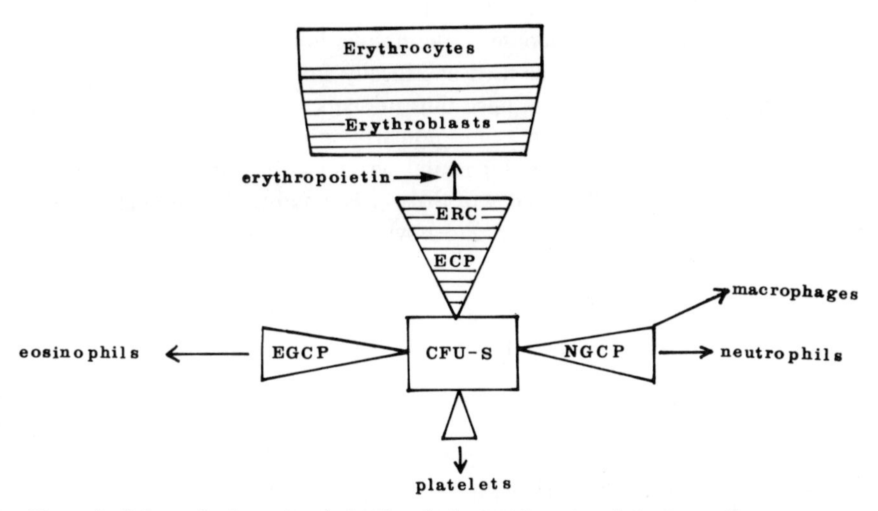

Figure 1. Scheme for hematopoiesis. The pluripotent hematopoietic stem cell compartment (CFU-S) is a self-maintaining population. The first stage in differentiation involves the transformation of CFU-S into cells committed to one or the other of the hematopoietic lines. ECP, erythroid committed precursor cells; NGCP, neutrophilic granuloid committed precursor cells; EGCP, eosinophilic granuloid committed cells. These transit populations undergo a number of cell divisions during their maturation but without the capability for prolonged self-maintenance. The second stage in differentiation triggers the late committed precursors to mature into the recognizable hematopoietic line. The mechanisms or factors which control the different hematopoietic cell line productions are not understood. But the second stage in erythroid differentiation involves the action of the hormone erythropoietin in stimulating proliferation and maturation of late committed erythroid cell precursors (ERC). Modified from Tambourin and Wendling (24) with permission from *Nature.*

enhance the formation of globin messenger RNA (mRNA) in immature erythroblasts.

Nevertheless, such studies have been seriously limited by several considerations. First, the normal developmental systems, with the exception of yolk sac erythropoiesis, differentiate asynchronously, which results in a complex mixture of erythroblasts at many stages of development. Although this mixture of cells can be fractionated to some extent (25–27), this represents a serious limitation on the purity and numbers of cells that can be obtained. Secondly, until recently (28) it has proven impossible to maintain erythroid precursor cells for long periods. This again has limited the types of experiments which can be performed and in particular has prevented any genetic studies. Therefore, for many years laboratories have attempted to obtain long-term cultures of erythroid precursor cells, which could then be induced to undergo erythroid maturation as a synchronous cohort. In general, such attempts have been unsuccessful. However, the demonstration by C. Friend et al. (29) in 1971 that a line of virus-infected leukemic cells could be induced to produce hemoglobin when treated in a specific way seemed to fulfill many of these requirements. Nevertheless, these cells are abnormal in their response to the normal hormonal regulatory mechanisms and, therefore, cannot be used uncritically for studies of erythropoiesis. Friend virus-induced leukemia in mice thus presents a model system in which a specific tumor virus causes a leukemia primarily affecting arguably one of the most thoroughly understood developmental systems. The purpose of this article is, therefore, to evaluate the biology of these cells in terms of their relevance for studies of erythroid differentiation and how the process becomes abnormal in leukemia.

ORIGIN OF FRIEND ERYTHROLEUKEMIC CELLS

Friend Virus-induced Leukemia

Charlotte Friend first reported (30) in 1957 the isolation of a cell filtrate from Ehrlich ascites tumor cells which produced in certain strains of mice a disease characterized by massive enlargement of the spleen and liver (and to a lesser extent the lymph nodes), anemia, the appearance of nucleated red blood cells in the peripheral circulation, and finally death. Microscopic examination revealed that the bone marrow, spleen, liver, and peripheral blood were all markedly infiltrated with abnormal mononucleated cells characterized by fine chromatin structure and a high nucleus to cytoplasm ratio typical of immature erythroblasts. The viral nature of the causative agent was soon established from its physical and other properties, and it became known as "Friend virus."

Subsequently, Mirand (31) isolated a variant of the anemia-producing Friend virus strain (FV-A) which produced leukemia in mice associated with polycythemia (FV-P). The precise relationship between these two strains is not clear: both induce an overproduction of red blood cells (FV-P more so than FV-A), but the red blood cell lifespan is shorter than normal (32) due to extensive cell death in the immature erythroblast compartment within the spleen (33, 34). The anemia produced by FV-A infection is apparently due to hypervolemia (32) rather than to any fundamental difference in the way the two strains affect the regulation of erythropoiesis per se. In both cases, once the infection is established (35–38), the enhanced red blood cell production is unaffected by manipulating the level of erythropoietin, in contrast to the situation in normal mice.

Host Control of Friend Erythroleukemia

A great deal has been learned in recent years about the nature of Friend virus and the manner in which specific host genes affect the course of the disease. This subject has been reviewed in detail elsewhere (39–41), and, therefore, only the most salient points are summarized here (Tables 1 and 2).

Friend virus, like Rauscher leukemia virus, is distinguished from most other murine leukemia viruses (MuLVs) in comprising a complex of two virus components, in a way somewhat analogous to murine sarcoma virus (40). In susceptible mice the number of spleen foci produced after Friend virus inoculation is

Table 1. Virus components of Friend virus complex

Spleen focus-forming virus (SFFV)	Defective erythroid-specific component
Helper virus	One of several murine leukemia viruses (MuLV)
Endogenous virus	Released from DMSO-treated Friend cells during differentiation; may replace usual exogenous helper component

proportional to the virus dose. This implies that under these conditions the spleen focus-forming agent (SFFV) is the only limiting infectious entity. Moreover, the dilution endpoint for spleen focus formation by Friend virus coincides with the endpoints for induction of splenomegaly and polycythemia (42, 43). However, in mice genetically resistant (at the FV-1 locus) to infection by a range of murine leukemia viruses, the dose-response pattern for spleen focus formation is more complex, which implies that focus formation requires infection by at least one MuLV in addition to SFFV. Thus, the defectiveness of SFFV is established by the restriction of the expression of the helper virus by the FV-1 gene (40). The presence of helper MuLVs in Friend virus has also been demonstrated by removal of SFFV by passage in hosts resistant to SFFV (for example, in rats or in mice genetically different at a second locus, FV-2, which is not appreciably linked to FV-1 (39)) and also by dilution of Friend virus beyond the endpoint for spleen focus formation (44, 45). These helper virus preparations regularly lack the capacity to induce erythroleukemia (40). The exclusive role of the defective SFFV in inducing erythroleukemia is also confirmed by the fact that apparently normal SFFV can be rescued from non-virus-producing tumors obtained by inoculation of mice with Friend virus, by superinfection with MuLVs alone (46–48). However, the two strains of Friend virus seem to have different amounts of the two components, which may explain the different pathologies of the disease; polycythemia is associated with FV-P infection and this virus has a high titer of SFFV, whereas the converse is true for the anemia-producing strain of virus (FV-A) (40).

Friend erythroleukemia is also governed by other host genes (39, 41). One gene (Rgv-1) maps in the H-2 region and seems to have a fundamental effect on

Table 2. Host genes affecting Friend leukemia

Gene	Effects
FV-1	Helper virus component of Friend virus complex
FV-2	Spleen focus-forming (SFFV) component of Friend virus
W	Multipotential stem cell
Sl	Microenvironment for stem cell proliferation
Rgv-1	Immune response to virus-associated antigens

the immune response to antigens associated with virus infection. This gene appears to influence the host response to all RNA tumor viruses and is, therefore, not specific to Friend virus. Finally, Friend erythroleukemia is governed by host genes affecting the production of normal hematopoietic cells, either directly (W locus) or indirectly via the hematopoietic stromal microenvironment (Sl locus) (49, 50). Since these genes act by affecting the target cell for Friend virus action, they are considered more fully in the next section.

Target Cell for Friend Virus

Much evidence has accumulated to show that Friend virus exerts its effects in erythroid precursor cells. Stem cells of some kind are required since x irradiation of normal mice suppresses Friend virus induction of leukemia, whereas a syngeneic graft of spleen or bone marrow cells reverses this suppression (51, 52). In contrast, the ability of stromal cells in the spleen to support hematopoietic cell proliferation and differentiation does not seem to be impaired by Friend virus infection (53, 54). Furthermore, susceptibility of mice to FV-P is controlled by the W locus, which affects the formation of multipotential hematopoietic stem cells (50), and by the steel locus (Sl) (49), which affects the hematopoietic microenvironment, particularly in relation to erythroid development. Moreover, the fact that spleen focus formation and replication of FV-P are enhanced in mice subjected to prior stimulation of the committed erythroid cell compartment (55, 56) suggests that it is this cell compartment rather than the multipotential stem cell itself which is the target for Friend virus, the main effect being to allow it to differentiate independent of erythropoietin. This conclusion has recently been supported by two compelling pieces of evidence. First, the elimination of multipotential, but not erythroid-committed, stem cells (by treatment with myleran followed by administration of erythropoietin) does not eliminate Friend virus-induced proliferation of erythroid cells in the spleens of polycythemic mice; no such response is observed in mice in which the committed erythroid cell compartment has not been restored after myleran treatment (57). Secondly, inoculation of lethally irradiated mice with normal spleen cells gives rise to erythroid, granulocytic, and megakaryocytic colonies in the spleen. However, infection of such mice with Friend virus 4 days after grafting yields, in addition, hyperbasophilic colonies (Friend leukemic cells) solely within the erythroid colonies (24), although all types of hematopoietic colonies in the spleen are known to contain multipotential stem cells (19). In addition, experiments in which the timing of injection of virus is varied relative to grafting suggest that it is a relatively late committed erythroid cell which is the target for Friend virus (see Figure 2). Thus, it seems to be clearly established that in vivo the infection and/or transformation of committed erythroid cells is a sufficient event for the establishment of Friend leukemia. Cells from spleens from such leukemic animals differentiate in vitro in the absence of erythropoietin (unlike normal erythroid precursor cells) and produce colonies which are indistinguishable in size or stage of maturation from those from normal cells (58). Friend

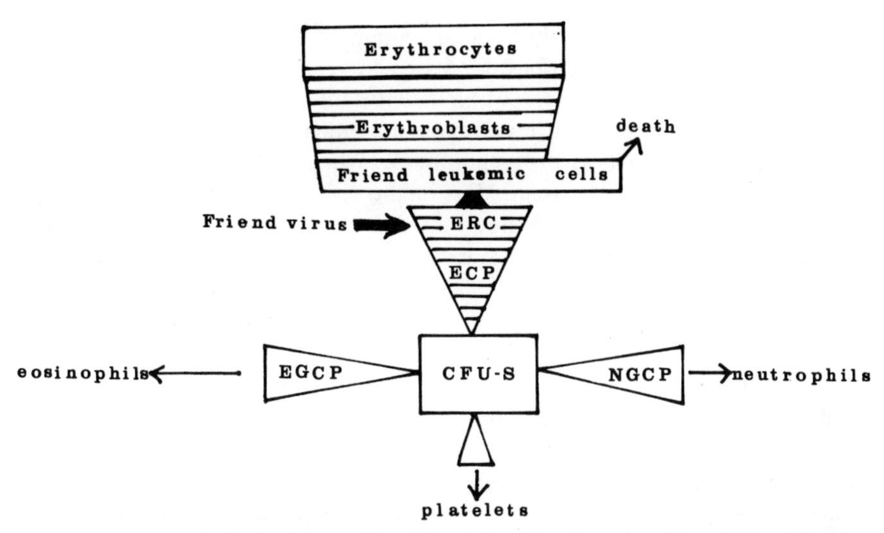

Figure 2. Scheme for leukemogenesis by Friend virus. See legend to Figure 1 for abbreviations. Friend virus may infect most primitive hematopoietic stem cells and replicate therein, but the oncogenic potency of Friend virus is only expressed at the late erythroid committed precursor cell stage. After viral transformation, Friend leukemic cells multiply and differentiate along the erythrocytic pathway, independent of erythropoietin. Friend leukemic cells are not self-maintaining (they die or differentiate). Consequently, the progression of the disease requires a constant recruitment of target cells from the ECP and CFU-S compartments. Modified from Tambourin and Wendling (24) with permission from *Nature.*

virus infection, therefore, confers on the committed erythroid cell the capacity for differentiation independent of erythropoietin, but it does not seem to allow maintenance of this compartment (33, 57–59). This constant depletion of the committed erythroid cell compartment may then affect other stem cell pools (60–62), for example, by feedback effects. However, Friend virus infection may affect other hematopoietic lines (e.g., megakaryocytes (63)) independent of its effect on the erythroid line.

Recently it has proved possible to study the interaction of Friend virus with different hematopoietic cells in vitro. Clarke et al. (59) have shown that infection of bone marrow cells in plasma clots in vitro allows erythropoiesis to proceed without erythropoietin. This in vitro effect also mimics the in vivo situation in being governed by the FV-2 gene, since this erythropoietin-independent erythropoiesis is only observed with bone marrow cells from mice permissive to SFFV at the FV-2 locus. However, other workers have observed that Friend virus infection alters the pattern of differentiation of bone marrow cultures by maintaining blast cells, early granulocytes, and both multipotential and granulocyte-committed stem cells (N. Teich and T. Dexter, personal communication). This change in differentiation pattern is again found only with bone marrow cells from mice genetically sensitive to SFFV at the FV-2 locus. This culture system does not show erythropoietin-independent erythropoiesis after

Friend virus infection, possibly because committed erythroid cells may not be maintained under these conditions.

Thus, one of the main effects of Friend virus infection seems to be to render erythropoiesis independent of erythropoietin, which accounts for the massive production of erythroblasts in Friend disease. Whether these proliferating cells are truly malignant is not certain. Most of the cells in spleens or bone marrow from Friend virus-infected mice show very few chromosome abnormalities (64, 65). Nevertheless, tumor cells arise early after infection with Friend virus, as judged by acquisition of autonomous growth potential and transplantability in subcutaneous as well as hematopoietic sites (66). Experiments with the use of hybrid or heavily irradiated mice which are resistant to Friend virus infection prove that this acquisition of autonomous growth potential is not explained by localized infection of host cells by virus released from the inoculated cells themselves (67, 68). Nevertheless, the capacity for self-renewal of these proliferating spleen cells gradually declines (67, 69), in contrast to the truly autonomous nature of established tumor cell lines initiated from mice surviving 1–2 months after infection (70–72). Since these established lines invariably show chromosome abnormalities and may be monoclonal (73), the formation of truly malignant cells may be relatively rare and may require a second event after the initial "transformation" of committed erythroid cells to erythropoietin independence.

Isolation of Friend Cells in Culture

For these reasons, it is perhaps not surprising that it has proven very difficult to establish cell cultures directly from spleens of mice infected with Friend virus, although with certain strains of Friend virus it has proven possible (74). However, subcutaneous implantation of spleen or liver fragments from mice in advanced stages of the virus-induced leukemia invariably results in tumor formation at the site of inoculation (70, 71, 73, 75). Some of these tumors may show erythroid characteristics (for example, globins or occasionally erythrocyte surface antigens (76) as detected by immunofluorescence), but most are relatively undifferentiated and resemble reticulum cell sarcomas (77). Fortunately, they adapt readily to tissue culture to give suspension cells which retain their tumor-producing properties and release Friend virus (78–80). These cell lines do not normally show extensive maturation (81, 82) (for example, hemoglobin), although some early erythroid markers may be partially expressed, for example, spectrin and glycophorin (H. Eisen and W. Ostertag, personal communication) or carbonic anhydrase (83). However, in some cell lines a minor population of the cells shows evidence of erythroid maturation as judged by iron uptake into heme, absorption spectrum, and benzidine staining (38, 78). In some cases the extent of erythroid maturation can be enhanced by growing the cells intraperitoneally in perfusion chambers (76). But compelling evidence of erythroid maturation was first demonstrated in 1971 by Friend and co-workers (29) by treating the cells with dimethylsulfoxide (DMSO). As is described in detail in a subsequent section, since that time a variety of other chemicals has been discovered which acts similarly.

Two quite different hypotheses can be advanced to explain why both the subcutaneous tumors and Friend cell lines fail to differentiate, in contrast to the "transformed" spleen cells which undergo erythroid maturation both in vivo and in vitro in the basence of erythropoietin (58, 59, 84). It is conceivable that the second event which renders the erythropoietin-independent, committed erythroid cells malignant may also interfere with their ability to differentiate, possibly by altering the integration or expression of the SFFV genome within the cell as postulated by Ostertag and coworkers (85). In fact, about five copies of the combined SFFV/helper virus sequences seem to be integrated in Friend cell DNA, but only some of the sequences are integrated endogenously in normal DBA/2 mouse DNA (86, 87). These studies have been performed with the use of complementary DNA (cDNA) obtained by transcribing the viral RNA genomes present in the Friend virus complex by using its endogenous reverse transcriptase. Characterization of this viral cDNA by titration with labeled viral RNA and other methods indicates that it represents an almost complete copy of the Friend virus genome (87). Part of these integrated Friend virus sequences (presumably those of the endogenous helper component) seems to be transcribed during embryogenesis both in erythroid cells (14-day fetal liver) and other tissues (I. Pragnell, W. Ostertag, K. Gorski, and P. R. Harrison, unpublished results) (Figures 3 and 4). Whether transcription of these viral sequences is restricted to certain stages of development is currently under investigation.

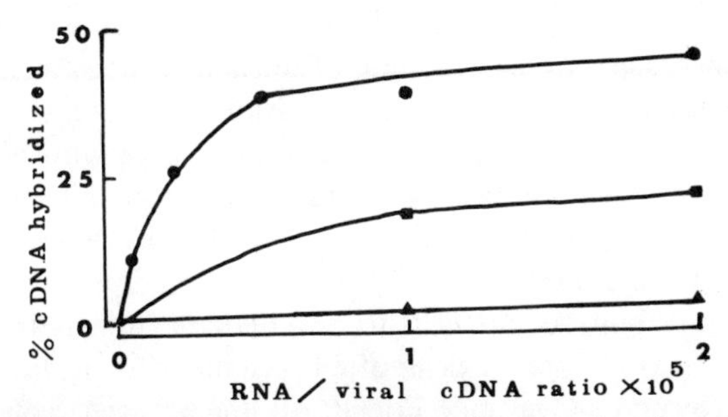

Figure 3. Transcription of Friend virus sequences in normal embryonic mouse tissues. Nuclear (•) or cytoplasmic (▪) RNA isolated from 14-day mouse fetal livers (88) was annealed with Friend virus cDNA as described elsewhere (86, 87). Hybrid formation was assayed by resistance to digestion by single-strand–specific nuclease Sl (11). Increasing ratios of RNA to viral cDNA give increasing titration of viral cDNA until a plateau of about 40% of the viral cDNA is hybridized. But no hybridization to nuclear RNA from human (HeLa) cells is evident (▲). Hybridization of all available complementary sequences is known to be complete (11, 87). Therefore, this experiment shows that about half the viral cDNA (presumably corresponding to the endogenous murine leukemia virus component of Friend virus) is expressed at low levels at the 14-day stage of mouse embryogenesis. The precise level of virus-specific sequences in the nuclear RNA sample can be estimated to be about 20 ppm (87). A lower proportion of virus sequences is present in cytoplasmic RNA. It is possible that some of the virus-specific sequences transcribed in the nucleus may not reach the cytoplasm, if the lower plateau of saturation for cytoplasmic RNA is genuine.

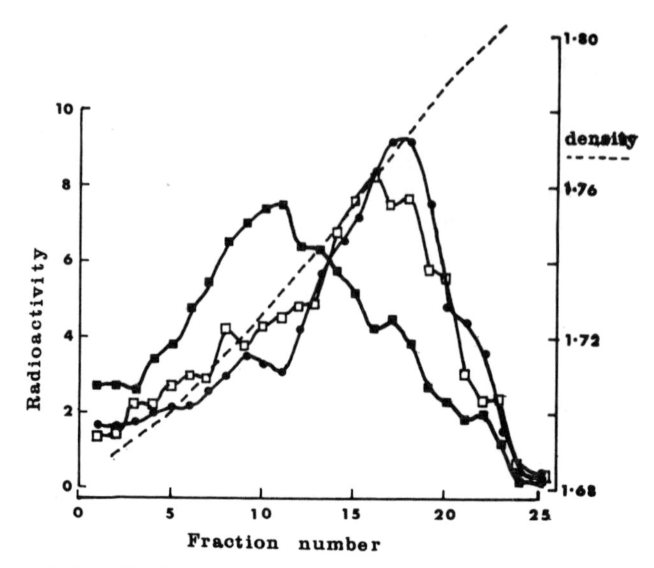

Figure 4. Transcription of Friend virus sequences in normal embryonic mouse tissues. Nuclear RNA from 14-day fetal livers or from the remainder of the 14-day embryos was annealed with Friend virus cDNA as described in Figure 3. The mixture was then banded in CsCl after treatment with ribonuclease (11). About 50–60% of the viral cDNA annealed to total embryo (•) or fetal liver RNA (□) bands at a density characteristic of a RNA-DNA hybrid, whereas no hybridization to ribosomal RNA occurs (■).

During differentiation of Friend cells an endogenous virus is released which is of different host range (tropism) to that of the helper virus present in the virus released from Friend cells during normal logarithmic growth (85). It has been proposed that transformation of normal erythroid cells with SFFV prevents them from differentiating normally, whereas induction of the endogenous virus by DMSO releases the maturational block, possibly by interfering with the action of SFFV (85). However, virus release can be prevented without necessarily affecting the ability of the cells to differentiate (summarized in ref. 86), although intracisternal A-type particles often accumulate even when virus release does not occur (89). Moreover, one Friend cell line has been isolated by Ikawa and co-workers (clone K2) (90) which is not inducible for hemoglobin although virus release is inducible. Yet this clone does produce spectrin (W. Ostertag and H. Eisen, personal communication) and, therefore, differentiation is not completely suppressed in this clone. These studies illustrate how the relationship between expression of virus and erythroid functions may be very complex and requires more sophisticated analysis in terms of the molecular biology of specific viral gene products.

An alternative hypothesis is simply to postulate that the failure of subcutaneous tumors and Friend cell lines to differentiate may reflect the selection by the conditions for actively proliferating cells. Since after Friend virus infection the committed erythroid stem cell compartment is not maintained (see above) and since erythroid maturation involves the formation of a nondividing cell, this selection process would automatically mean that differentiating cells would be lost from the proliferating cell population.

Summary

Inoculation of Friend virus into susceptible strains of mice induces a leukemia which is characterized by massive production of erythroid cells. Friend virus is a complex of two RNA tumor virus components, one of which (SFFV) is responsible for producing erythroleukemia but is defective in the absence of the second helper component, a murine leukemia virus. The host response to the two virus components is controlled by two separate genes. Although Friend virus may infect other hematopoietic cells, it "transforms" committed erythroid stem cells; they then differentiate in the absence of the hormone erythropoietin. However, the erythroid stem cell compartment is not self-maintaining after transformation; normally it is maintained by erythropoietin. This transformation of erythroid stem cells to erythropoietin independence does not necessarily render them malignant. However, cells from spleens from infected mice do form tumors when implanted subcutaneously, and from these tumor cell lines have been obtained in tissue culture (Friend cells). These show little evidence of erythroid maturation beyond the immature erythroblast stage even in the presence of erythropoietin, but they differentiate when treated with DMSO or certain other chemicals. Since Friend cells are most probably derived from committed erythroid stem cells, DMSO-induced differentiation in Friend cells may be relevant to the later stages of normal erythroid maturation, but not to the process of commitment itself.

ERYTHROID MATURATION IN FRIEND CELLS

In 1971, Friend and co-workers reported that one clone of erythroleukemic cells responded to DMSO by producing heme and hemoglobin, accompanied also by morphological changes resembling normal orthochromatic cells (29). Since that time these findings have been confirmed with several independently derived erythroleukemic cell lines and extended to cover a wide range of erythroid functions. This work has demonstrated clearly that DMSO-treated Friend cells undergo changes which resemble closely the process of normal erythroid maturation. This has allowed these cells to be used as a model for elucidating how a whole program of functions are coordinated during the development of a single cell type.

Hemoglobin and Globin

The production of hemoglobin 3-5d after DMSO treatment was demonstrated initially by benzidine staining, by its specific absorption spectrum, and by polyacrylamide gel electrophoresis (29, 91). More thorough studies by ion exchange chromatography, immunoprecipitation, and fingerprint analysis then showed that the globins produced in DMSO-treated cells are the adult α and β chains characteristic of the strain of mice from which Friend cells are derived (73, 92). In most Friend cell lines, α and β chain synthesis is relatively balanced (73, 92). Nevertheless, the ratio of β^{major} to β^{minor} chains synthesized in DMSO-treated Friend cells can vary considerably in independently derived lines by as much as 1.3–9, compared to a value of 4 for normal adult DBA/2 mouse erythrocytes (83). This is an interesting finding since the relative enhancement of a minor β chain is also a characteristic of erythropoiesis in fetal liver (93).

Globin Messenger RNAs

Globin mRNA is present in small amounts in most Friend cell lines before treatment with DMSO. This is shown by the fact that most of the sequences present in cDNA transcribed from adult globin mRNAs are hybridized by excess cytoplasmic RNA from untreated Friend cells (94, 95). However, globin mRNAs begin to accumulate after about 24–30 hr of DMSO treatment and reach a maximal concentration after about 3 days. This has been demonstrated both by molecular hybridization techniques with the use of globin cDNA (94–98) and cell-free translation studies (98). As in normal erythroid cells, accumulation of hemoglobin and globin mRNA is prevented by treatment of cells with bromodeoxyuridine under conditions in which cell proliferation is only marginally affected (99, 100, 101). In our cell lines the increase in rate of heme and globin synthesis during DMSO treatment parallels quite closely the rise in globin mRNA level (Figure 5). The lag in detection of hemoglobin by benzidine staining is, therefore, probably due to insensitivity of the method. Thus, there is no evidence that globin mRNA is stored prior to translation. Other workers (74) have claimed that a burst of synthesis of 9 S RNA precedes globin synthesis in one of their Friend cell lines, but this is not a characteristic of other cell lines of similar origin (W. Ostertag, personal communication).

More detailed studies have shown that α- and β-globin mRNAs do not accumulate coordinately. Using cDNAs transcribed from partially purified α- or β-globin mRNAs, Orkin and co-workers obtained an α- to β-globin mRNA ratio of 3.7 30–35 hr after DMSO treatment; this then decreased progressively to a value of unity in the fully induced Friend cells, i.e., to the value in normal reticulocytes (102). Whether this is characteristic of erythroid maturation in general remains to be seen. In any event, these studies do not exclude the possibility of a switch in β-globin mRNA production from nonadult to adult-type during Friend cell maturation; but this seems unlikely since embryonic or fetal globin mRNAs from yolk sac nucleated erythrocytes do cross-hybridize with the adult α,β-cDNA (P. R. Harrison, unpublished results).

Unfortunately, none of the results discussed above clarifies precisely the molecular level(s) at which globin genes are regulated during Friend cell maturation. Accumulation of globin mRNA could be affected by rates of transcription

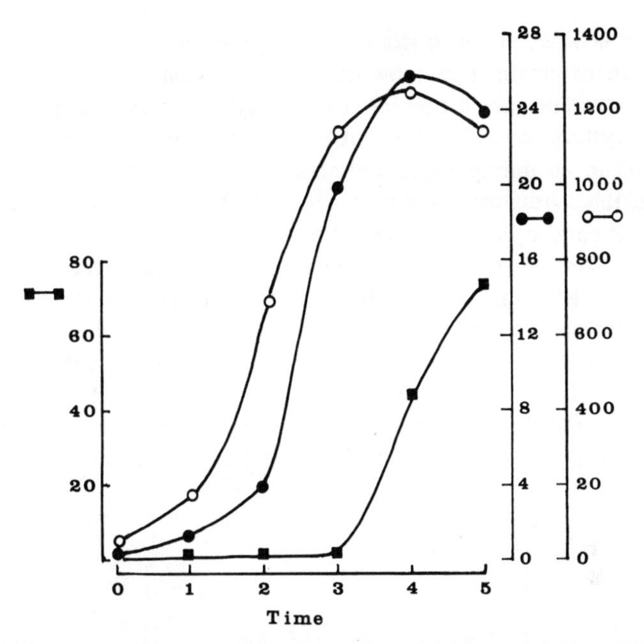

Figure 5. Time course of induction of hemoglobin, globin chains, and globin mRNAs after treatment of Friend cells with DMSO. ▪, percentage of cells staining positive for hemoglobin with benzidine; ○, percentage of the total protein synthesis which coelectrophoreses with adult globin chains; •, globin mRNA content expressed as parts per million of the total cytoplasmic RNA, as measured by titration with globin cDNA (12). *Abscissa*, time in 1.5% DMSO (days).

of the globin genes; the stability, processing, and transport of the primary transcripts into the cytoplasm; and the stability of the mRNAs themselves. Much of this information is not available at the present time. However, recent studies have emphasized that differential mRNA stability is an important factor in how hemoglobin accumulates to the virtual exclusion of other proteins during erythroid maturation. Lodish and Small (103) have demonstrated by cell-free translation studies that differential destruction of nonglobin mRNAs (for example, coding for one of the major nonglobin reticulocyte proteins of 64,000 MW) is a factor in determining why globin accumulates in the final stages of erythropoiesis. Similar evidence has been obtained with the use of DMSO-treated Friend cells. Aviv and co-workers (104) have shown by affinity chromatography with globin cDNA that the half-life of globin mRNA is about 16 hr (i.e., the same as in reticulocytes), whereas the majority of the poly(A)-containing RNA decays more rapidly (half-life ~ 3 hr), although a minor component is more stable (half-life ~ 37 hr). Calculations based on these half-lives show that hemoglobin and globin mRNA should not accumulate to the extent actually found in reticulocytes; therefore, these authors postulate that the most effective way to achieve the required high level of globin mRNA would be by destabilization of the most stable mRNA population.

Another stage at which globin gene expression could be regulated is during processing of globin gene transcripts into functional messenger molecules. In view of the claims for higher molecular weight precursors to globin mRNAs in normal erythroid cells (e.g., refs. 105–108), attempts have been made to detect such precursors during induction of erythroid maturation in Friend cells. In work in this laboratory (N. Affara and P. R. Harrison, unpublished results) we have assayed for globin sequences in different molecular weight classes of nuclear RNA from untreated and DMSO-treated Friend cells, by fractionating the nuclear RNA on 99% DMSO-sucrose gradients or 100% formamide-polyacrylamide gels. The results indicate that in DMSO-treated cells virtually all the globin RNA sequences are of indistinguishable size as globin mRNA in the cytoplasm. However, the majority of globin RNA molecules in nuclear RNA from untreated Friend cells is slightly larger (about 14 S), with a minor component perhaps very much larger (Figure 6). In contrast, Friend virus

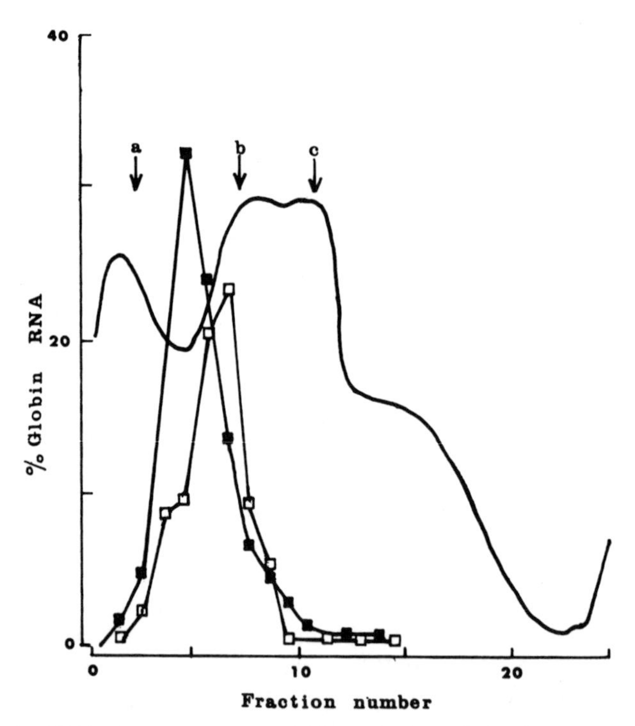

Figure 6. Globin RNA sequences in untreated Friend cell nuclear RNA. Nuclear RNA was isolated from citric acid-prepared nuclei by digestion with proteinase K in the presence of sodium dodecyl sarcosinate followed by centrifugation in CsCl (88). Cytoplasmic RNA was also extracted from parallel cultures of cells by using NP-40 lysis and phenol/chloroform extraction (13). RNA samples were centrifuged on 0–10% sucrose/99% DMSO gradients; fractions were collected and then assayed for globin RNA content by titration with globin cDNA (12). The results are then expressed as the proportion of the total globin RNA on the gradient which is present in each fraction. ——, absorbance at 280 nm; ▪, cytoplasmic RNA; □, nuclear RNA. The *arrows* a, b, and c indicate the positions of sedimentation of 4 S, 18 S, and 28 S RNA species.

sequences in the same nuclear RNA samples sediment with a modal sedimentation coefficient of about 35 S, which suggests that degradation of at least this high molecular weight RNA is not a serious problem. However, these experiments relate only to the most abundant globin RNA molecules present. More recent techniques have enabled newly synthesized globin gene transcripts to be detected in Friend cell nuclear RNA by molecular hybridization with globin cDNA elongated with oligo(dC), followed by recovery of hybridized RNA by affinity chromatography on poly(I)-Sephadex (109). These experiments have provided evidence of a globin mRNA precursor molecule sedimenting at about 15 S in 3-day DMSO-treated Friend cells. This agrees well with the size of the globin mRNA precursor in normal erythroid cells (mouse fetal liver) as determined by hybridizing pulse-labeled RNA with excess globin cDNA (110).

Heme

As in the normal reticulocyte (see ref. 111 for review), the amounts of heme and globin in Friend cells seem to be coordinated and little free heme accumulates (83). In fact, induction of the rate-limiting enzyme in the heme pathway (δ-aminolevulinic acid synthetase) is one of the early events induced by DMSO (112). This might suggest that control of globin synthesis by heme operates in Friend cells in a manner similar to that in normal erythroid cells, namely, via a hemin-controlled repressor (111). However, this is not the case. Unlike the situation in reticulocytes, protein synthesis in Friend cell lysates is not stimulated significantly by hemin; in fact, Friend cell lysates contain a protein that represses globin synthesis specifically in mixed Friend cell reticulocyte lysates and it is not affected by hemin (113). Furthermore, work has shown that the Friend cell repressor blocks peptide initiation at a point between the addition of methionyl-$tRNA^{fMet}$ to the ribosomal initiation complex and the NaF-sensitive reaction (114), whereas the hemin-sensitive repressor from reticulocytes affects the formation of the first initiation complex (111). As yet it is not known whether the situation regarding such repressors in Friend cells changes after DMSO treatment.

Nonglobin Proteins and Messenger RNAs

Other changes which occur during normal erythrocyte maturation have been detected during treatment of Friend cells with DMSO, for example, an increase in carbonic anhydrase (83) and alterations in the enzymes of purine metabolism (115). However, 2,3-diphosphoglyceric acid is apparently not induced (83) in Friend cells, although it is a normal erythrocyte function.

It appears that the profound ultrastructural changes characteristic of erythroid maturation in Friend cells (116, 117) may be accompanied by relatively minor changes in the overall distribution of proteins within the cell. Peterson and McConkey (118) used a two-dimensional system to separate more than 500 proteins from untreated and DMSO-treated Friend cells and found that only six nonhistone chromosomal proteins, two nucleoplasmic chromosomal proteins, and three cytoplasmic proteins change in abundance by more than 50% after

DMSO treatment. In addition, changes have been reported in the amounts of subfractions of histone F2a2 (119) and in another chromosomal protein (120). However, in view of the number of specific protein changes discussed elsewhere in this section, even this sophisticated method of analysis must be failing to resolve many Friend cell proteins. Essentially the same conclusion has been reached by Minty et al. in this laboratory (121) regarding changes in total poly(A)-containing mRNAs, as determined by molecular cross-hybridization studies with the use of cDNAs transcribed from the mRNAs themselves. With the exception of globin mRNA, the mRNA population as a whole seems to be surprisingly similar in untreated and DMSO-treated Friend cells. However, both these studies assume that most of the proteins or mRNAs isolated from DMSO-treated cells are produced by the differentiated cells and not by the minor fraction of cells remaining undifferentiated (15–20% of the population).

Membrane Changes

DMSO-treated Friend cells undergo a series of membrane changes characteristic of normal erythroid maturation. Immunological methods have been used to demonstrate the production of erythrocyte membrane proteins in mouse erythroleukemic cells differentiating intraperitoneally in vivo in perfusion chambers (76, 122, 123) or after DMSO treatment (76). Furthermore, the high molecular weight erythrocyte membrane protein, spectrin, accumulates (124), whereas H-2 antigens are lost (125) in Friend cells treated with DMSO. The level of spectrin in Friend cells increases from about 0.01% of the total proteins to a peak of about 0.25–0.3% after 3 days of treatment with DMSO and then falls somewhat after 4 days of treatment to about the level in the normal erythrocyte (0.1%) (126). The peak rate of spectrin accumulation occurs 1–2 days after DMSO treatment. Decreased membrane permeability (74) and increased membrane microviscosity after DMSO treatment (124, 125) are also relatively early events. Moreover, DMSO-treated Friend cells are more readily agglutinated by lectins (125, 127). This seems to be due to changes in mobility of lectin binding sites since the total number of sites remains roughly constant early after DMSO treatment and increases only 2–4-fold later (125). In fact, the changes in mobility of lectin binding sites and in membrane permeability seem to precede the change in membrane microviscosity as measured by emission anisotropy with the use of a fluorescent probe sensitive to the cholesterol to phospholipid ratio in the membrane (124). This suggests that the change in membrane permeability may be due to changes in the lipid-protein or protein-protein interactions in the membrane after DMSO treatment (124). Moreover, since a good correlation is found between membrane microviscosity and spectrin or hemoglobin levels in individual cells, these functions seem to be closely coordinated at the single cell level during Friend cell maturation (124).

Summary

Friend cells show evidence of erythroid maturation after treatment with DMSO, as judged by morphological criteria, the production of heme, globins, and globin mRNAs, erythrocyte membrane proteins, and other characteristic erythroid markers (Table 3). As far as can be determined, erythroid maturation is relatively normal, although minor aberrations have been reported. However, Friend cells are unresponsive to the normal hormone erythropoietin. Therefore, Friend cells may be a good model for studies of erythroid maturation but not for studies of how proliferation of erythroid precursor cells is regulated.

TIMING OF ERYTHROID MATURATION

Friend Cell Maturation as Stochastic Process

One interesting aspect of erythroid differentiation in Friend cells which has intrigued workers for many years is simply the timing of expression of erythroid markers and the manner in which individual cells become differentiated. Hemoglobin production in Friend cells does not begin to be detectable until cells have been exposed to DMSO for a period equivalent to one or more cell generation times (29, 128). After this latent period of exposure to DMSO, the fraction of cells producing hemoglobin subsequently is not reduced by removal of DMSO from the medium (100, 129). This has led to the concept that treatment of cells with inducer irreversibly commits the cells to differentiation and that the subsequent expression of erythroid functions follows without inducer. Although the exact timing of production of hemoglobin varies between cell lines and inducers, a considerable amount of evidence has now accumulated to support this basic hypothesis.

Although usually not more than 80–90% of cells in the population produce hemoglobin and differentiate terminally, this may be of no fundamental significance since one inducer (hexamethylene bis(acetamide)) is claimed to induce virtually the entire population (130). This proves that Friend cell differentiation

Table 3. Erythroid functions induced in Friend cells by DMSO

Erythroid functions
Hemoglobin pathway
Hemoglobin
Globin mRNAs
Heme
δ-Aminolevulinic acid synthetase
Membrane changes
Reduction in membrane permeability
Spectrin
Glycophorin
Increase in membrane agglutination by plant lectins
Increase in lectin binding sites (later)
Decrease in membrane fluidity
Others
Carbonic anhydrase
Nuclear condensation
Cessation of cell division

does not require a "feeder" population to support the differentiation of the rest of the population. Moreover, under suboptimal conditions, no correlation is found between the hemoglobin content of individual differentiated cells and the DMSO concentration used to induce them (131). In fact, the average cellular globin mRNA content correlates well with the percentage of cells containing hemoglobin, as determined either by microspectrometry or benzidine staining (131). These results are, therefore, consistent with the hypothesis that Friend cells are induced to differentiate in a stochastic or all-or-nothing manner, the probability of which is governed by the environmental conditions (131). A stochastic model has also been proposed for normal hematopoietic stem cell differentiation (132, 133).

This hypothesis has been supported by recent evidence (134) obtained by clonal analysis which shows that commitment of cells for differentiation limits the cell to four additional divisions, that heterogeneity exists in the cell population as committed cells appear (which is not due to genetic heterogeneity in the cells), and that a constant fraction of the cell population remains uncommitted to differentiation during each cell division cycle (134). Detailed mathematical analysis shows that these experimental observations are consistent with a stochastic model based on the assumption that the commitment process occurs only after a latent period and then ensues throughout the cell cycle with a probability characteristic for the given experimental conditions (134). The evidence suggests that this latent period or commitment process precedes or accompanies the increase in cytoplasmic globin mRNA (134). It will, therefore, be interesting to determine whether a latent period is also required before other early erythroid functions are expressed, for example, spectrin.

The length of the latent period (12–18 hr for DMSO induction) might not, in fact, be of any great biological significance, but rather simply represents the time required for establishing a critical concentration of inducer or other factor required for triggering differentiation. About 24–30 hr are apparently required for effective equilibration of Friend cells with DMSO (129), but this might differ widely for different inducers. For example, recent experiments show that clone B8 cells treated with butyric acid induce globin mRNA within one cell-doubling time (24 hr) (I. Pragnell, D. Jovin, and W. Ostertag, personal communication). However, interpretation of such results can be complex, because cell cycle analysis by cell sorting on the basis of DNA content reveals that 90% of butyric acid-treated B8 cells become arrested temporarily in G1 after 16 hr of treatment, whereas this does not occur in DMSO-treated cells (D. Jovin and W. Ostertag, personal communication).

Cell Cycle Dependence of Friend Cell Maturation

One of the critical assumptions in the stochastic model is that commitment to differentiation can occur at all stages throughout the cell cycle. Unfortunately, there is as yet no unequivocal answer to this question and the issue is still controversial. Levy and co-workers (129) have attempted to investigate directly whether DNA synthesis is required for induction of hemoglobin, using Friend cells synchronized by a double exposure to excess thymidine. Apparently, DMSO has to be in contact with the cells for 24–30 hr in order to achieve an effective intracellular concentration of DMSO. But, in addition, DMSO appears to be necessary during the S phase after release of the cell cycle block in order for hemoglobin to accumulate subsequently. These studies seem to imply that erythroid maturation in Friend cells requires some fundamental change in the state of the cell which is only accomplished in the presence of inducing agent at a critical part of the cell cycle.

This laboratory has approached the same question somewhat differently by single cell studies (135). Friend cells can be cloned in semi-solid medium and their proliferation arrested if necessary by addition of hydroxyurea, cytosine arabinoside, or FUdR or by isoleucine deprivation. Formation of hemoglobin can be detected in situ by benzidine staining. Experiments in this laboratory show that Friend cells arrested at the single cell stage in the presence of various inducers fail to accumulate hemoglobin, whereas doublet colonies in the same cultures frequently produce hemoglobin (Figure 7). Most of the single cells arrested by isoleucine deprivation and some of those arrested by hydroxyurea are still potentially functional since on reversal of the arresting conditions they proliferate and produce hemoglobin (Figure 7). Such studies can never exclude the possibility of unwanted side effects of the arresting conditions, but the results do suggest that single cells fail to accumulate hemoglobin due to cell cycle arrest per se. They, therefore, support the conclusion reached by Levy and co-workers (129).

In contrast, Leder and co-workers (136) have reported that butyric acid-induced production of hemoglobin by a different line of mouse erythroleukemic cells (T3C12) is unaffected by inhibition of DNA synthesis by hydroxyurea or cytosine arabinoside. These authors conclude that hemoglobin production is not dependent upon the capacity of the cells to undergo DNA synthesis. However, this interpretation is open to question on two grounds. First, hydroxyurea and cytosine arabinoside themselves have been reported to induce hemoglobin in T3C12 cells (76, 137), although this laboratory has not been able to confirm this with the use of our Friend cell lines. Secondly, in our experience, hemoglobin-containing single cells (clone T3C12) arrested by hydroxyurea in the presence of butyric acid are mainly binucleate, whereas mononucleate single cells stain for hemoglobin less frequently (unpublished results). Ikawa and co-workers (76) have also noticed a tendency for hemoglobinized T3C12 cells to be binucleate. Whether these binucleate cells arise by fusion of hemoglobinized cells or failure in cytokinesis after nuclear division is not known. If the latter explanation is correct, these results with the T3C12 line are not necessarily inconsistent with previous data because induction of hemoglobin may be associated with some event prior to nuclear division rather than with cytokinesis per se.

Clearly, further research in this important area requires the use of synchronous Friend cells obtained with minimal perturbation of the cells, for example,

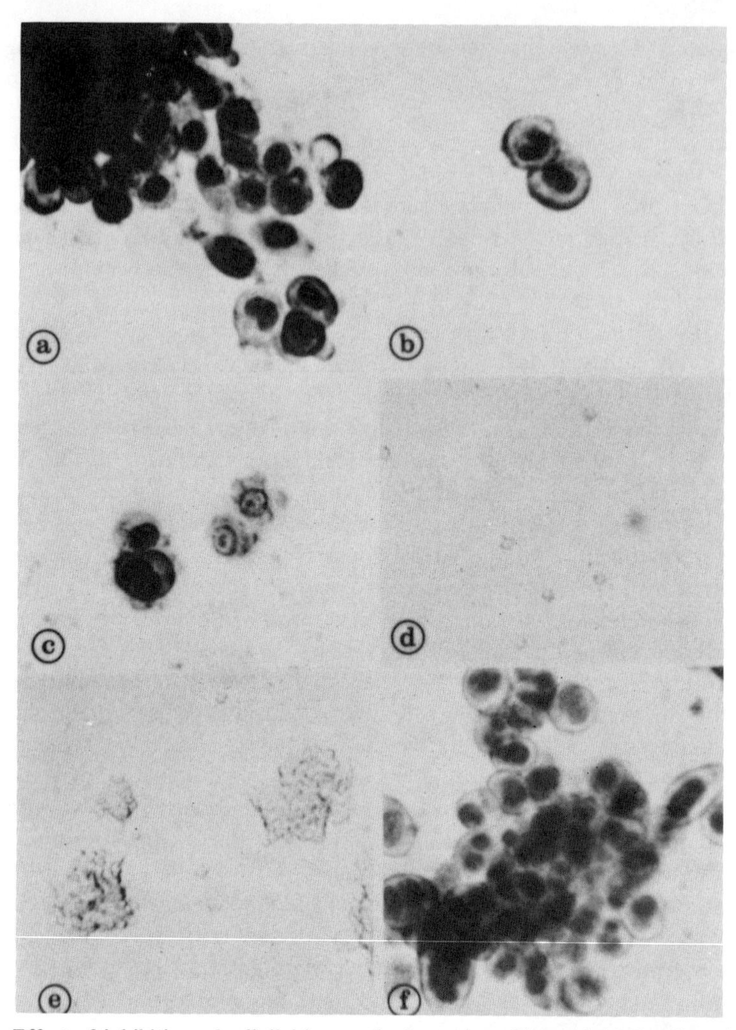

Figure 7. Effect of inhibition of cell division on the formation of hemoglobin by *N*-methylacetamide (NMA)-treated Friend cells cultured in medium containing dialyzed serum and lacking isoleucine. Cells were treated for 5 days with (*a*) isoleucine plus NMA (15 mM) or NMA alone (*b–d*). Subsequently, complete medium containing isoleucine and NMA was added to cultures (*b–d*) and then incubated for a further 5 days (results shown in *e* and *f*). Cultures *a–c* and *f* are stained with benzidine, whereas *d* and *e* are unstained. Plates *d* and *e* are at a lower magnification (× 16 objective) than the other plates.

with the use of variants which are temperature-sensitive for proliferation, by cell sorting on the basis of cell size or DNA content (124, 125), or possibly by selection of mitotic cells from adherent Friend cell variants (138) by the shake-off method. Such approaches would also permit an analysis as to whether in fact only part or the entire program of erythroid functions is dependent upon a prior cell cycle event.

Summary

Under favorable circumstances, virtually the entire population of Friend cells can be induced to undergo erythroid maturation. As a first approximation, Friend cells seem to differentiate stochastically, the probability of which depends upon the nature and concentration of inducing agent and other environmental factors. There is evidence that Friend cell differentiation is cell cycle-dependent, possibly requiring the presence of an inducing agent during DNA synthesis, but this question remains controversial.

MECHANISM OF ACTION OF INDUCING AGENTS

Inducing Agents

Ever since the discovery of the effect of DMSO on Friend cells in 1971 (29), the mechanism whereby DMSO releases the maturational block has been an intriguing topic of research. This question is important not only for a proper understanding of Friend cell differentiation per se; its elucidation may well shed light on the mechanism whereby transformation with Friend virus enables the committed erythroid stem cell to proliferate independent of the nature hormone, erythropoietin. Indeed, it has been claimed that prior treatment of Friend cells with DMSO renders them slightly sensitive to erythropoietin (139). Whatever the precise molecular mechanism, DMSO seems to act directly on the cell itself, since Friend cells adapted to growth in serum-free medium are partially inducible by DMSO (140). However, this effect of DMSO in the absence of serum seems to disappear on prolonged culture of the cells without serum.

In an attempt to discover how DMSO might be acting, several laboratories have screened a wide range of chemical agents (76, 130, 141–143, 145). This has produced a somewhat bewildering list of organic chemicals which induce hemoglobin in Friend cells. These are classified in Table 4, roughly in terms of their effectiveness as inducers. Polymethylene bis(acetamides) are a particularly interesting group of compounds since they are formed by linking two known good inducing groups (basically that of *N*-methylacetamide) via a carbon bridge. Hexamethylene bis(acetamide) (130) seems to be more effective than other inducers (for example DMSO, N-methylacetamide, dimethylacetamide, or butyric acid) in terms of the percentage of cells responding (99% rather than 40–90% for the other inducers), the average hemoglobin content per cell, and the extent of morphological differentiation.

Two reports (76, 137) have claimed that various inhibitors of DNA and RNA synthesis and other antimetabolites (e.g., *N*-dimethylrifampicin, bleomycin, cytosine arabinoside, mitomycin C, and cycloheximide) are also potent inducers of hemoglobin in Friend cells. These authors interpret these results to imply that inhibition of a combination of the synthetic pathways for RNA and DNA synthesis is necessary for induction of differentiation in Friend cells. However,

Table 4. Inducers of Friend cell differentiation

Very good inducers
Dimethylsulfoxide (29)
N-Methylacetamide (142, 143)
N-Methylpyrrolidone (142)
1-Methyl-2-piperidone (142)
Dimethylacetamide (142)
Hexamethylene bis(acetamide) (130)
Hypoxanthine (146)
Good inducers
Pyridine-*N*-oxide (142, 143)
Dimethylformamide (141–143)
Acetamide (142, 143)
Piperidone (142)
N-Methylformamide (142, 143)
Triethylene glycol (142)
Butyric acid (147, 148)
Hemin (149)
Propionamide (142)
2-Pyrrolidone (142)
Slight inducers or inactive compounds
Urea (142, 143)
Ethylene glycol (137, 143)
Pyridine (142, 143)
Acetone (142, 143)
Formamide (143)
Hexamethyl phosphoric triamide (142)
Ethylene carbonate (142)
Glycerol (76, 137, 143)
Various steroid hormones (137, 144)
Ethanol (137)
Cyclic AMP (137, 144)
Insulin (137)

interpretation of these results is problematical since the antimetabolites do not seem to inhibit the incorporation of precursors into DNA, RNA, or proteins and, therefore, may be acting by a totally unknown mechanism.

Various structural or physicochemical properties have been proposed as the common basis by which these chemical compounds induce Friend cell differentiation. Good inducers are generally of low molecular weight (60–150 daltons), are highly polar molecules, are often effective cryoprotective agents (143, 145), are capable of acting as Lewis' bases (150), and most partition in an octanol/water mixture in a similar manner (150). However, although these properties are characteristic of good inducers, there are exceptions and, therefore, they are not sufficient criteria for an inducer. Interestingly, although the optimal concentrations vary over a large range, that for any given inducer is about half its lethal concentration (142). Moreover, although the inducing effects of combinations of some agents are additive (at suboptimal concentrations of each agent), others may act synergistically (143).

More recently, other inducers have been discovered which are normal cellular components and are effective at much lower concentrations. These include purine derivatives (for example, hypoxanthine and certain analogs) (146), heme (149), and certain normal fatty acids (147, 148), particularly butyric acid. However, in no case is there any evidence that these natural compounds do in fact function physiologically. For example, purine derivatives can be effective without being incorporated into RNA or DNA, and their natural catabolites are not inducers (146). Similarly, natural catabolites of butyric acid do not act as inducers (147), and natural intermediates in the heme synthetic pathway (porphobilinogen and δ-aminolevulinic acid) are ineffective (149). Thus, these normal cellular constituents may be acting unphysiologically. In fact, heme is not a very effective inducer by itself, but it acts synergistically in combination with DMSO in a cell line which does not respond very well to DMSO alone (149). However, since hemoglobin induction by heme is faster than by DMSO, DMSO probably does not act simply by facilitating entry of heme into the cell (149).

Mechanism of Action

How these chemical inducers act is not clear. It has been speculated that they act by destabilizing DNA structure (142, 143) by analogy with the effect of DMSO and other DNA-denaturing reagents on the transcription of bacterial operons (151). But the relevance of these studies to the mechanism of action of DMSO on Friend cells is questionable, since they involve much higher concentrations of DMSO. An alternative and perhaps more plausible hypothesis is that inducers act by affecting the function of the plasma membrane. Three lines of evidence in particular support this hypothesis. First, a correlation has been established between the effects of DMSO and other cryoprotective agents on the phase transition temperature of artificial phospholipid membranes or microviscosity changes in Friend cell membranes and on the differentiation of Friend cells in vitro (124, 125, 152). Secondly, local anesthetics (which are known to interfere with membrane function) prevent induction of erythroid differentiation by a wide range of inducers (DMSO, acetamide, *N*-methylacetamide, dimethylacetamide, pyridine-*N*-oxide, 2-pyrrolidine, dimethylformamide, and hypoxanthine) without affecting cell proliferation (150). Finally, ouabain induces hemoglobin in Friend cells, to an extent dependent upon the K^+ concentration (153). This K^+ dependence is not a property of any other inducer tested. Since ouabain is known to specifically inhibit the membrane Na^+-K^+-Mg^{2+}-activated ATPase (EC 3.6.1.3.), this suggests that changes in the intracellular K^+ concentration may be involved in induction of hemoglobin by this agent.

An alternative approach in elucidating whether or not these diverse inducers act by a common mechanism involves the use of noninducible Friend cell variants. These can be selected by cloning or continuous growth of inducible cells in a particular inducing agent. In many instances the resistant Friend cell lines are stably noninducible by the inducing agent used for selection purposes. The lines can then be tested for induction by other inducers. In this way this laboratory has found three classes of DMSO-resistant cells: 1) those which do not respond to any inducer; 2) those which respond to all other inducers; and 3) those which respond to one or more, but not all, other inducers (see under "Genetic Studies"). Other workers have made similar observations: one DMSO-resistant line is inducible by purine analogs (146), others by butyric acid (90, 154), while another is responsive to a variety of other inducers (98). These collective results, therefore, indicate that the induction mechanism is not identical for all inducers.

Summary

A variety of chemical agents has been discovered which "induce" erythroid maturation in Friend cells. Some of these agents (e.g., purine derivatives, fatty acids, or heme) could be acting physiologically, but the evidence suggests this is not the case. It is not certain how the other chemical agents act, although alteration in membrane function is perhaps the most likely possibility. Genetic studies suggest that the agents do not all act by a single common mechanism.

GENETIC STUDIES

It has been argued in some detail in a previous section that DMSO-induced differentiation in Friend cells resembles closely the normal process of erythroid maturation. Since they grow continuously in culture under relatively well defined conditions, Friend cells offer the potential for genetic studies which at present are impossible with normal erythroid cells. The genetic studies which have been performed fall basically into three categories: 1) the isolation of noninducible variants; 2) characterization of the molecular basis of their noninducibility; and 3) studies of their dominance and complementation by cell fusion.

Isolation of Noninducible Variants

Irrespective of the time or conditions of treatment of Friend cells with inducers, usually only 80–90% of the cells produce hemoglobin and differentiate terminally. This finding, therefore, raises the question as to whether the cells which fail to differentiate represent a truly noninducible subpopulation. That this is not the case has been established by three facts. First, the DMSO response of subclones of a primary clone is relatively uniform; this excludes the existence of a substantial (i.e., 10–20%) noninducible subpopulation of cells (155, 156). Secondly, one inducer (hexamethylene bis(acetamide)) is claimed to induce virtually the entire population (i.e., ~99%) (130). Thirdly, the frequency of true noninducible Friend cells is less than 0.01%, as shown directly by cloning of Friend cells in 1.5–1.75% DMSO (see Table 5, refs. 154 and 157).

Table 5. Luria-Delbrück fluctuation test for formation of DMSO-resistant variants[a]

Experiment	A	B	C	D
Number of cells	10^5	3×10^5	10^5	3×10^5
Number of cultures	6	16	6	9
	replicate	parallel	replicate	parallel
Treatment	cloned in 1.5% DMSO		cloned in 1.75% DMSO	
Mean number (m) of DMSO-resistant clones	31	646	27	170
Variance (v) of distribution of number of DMSO-resistant clones	152	1.43×10^6	69	5.2×10^4
v/m	4.9	2,213	2.55	306

[a]A small clone of about 200 cells was isolated, disaggregated, and divided into 20 separate parallel cultures. These were grown up and 3×10^5 cells were cloned in 1.5% DMSO (experiment B). Cells (10^5) from replicate cultures of a single culture chosen at random were also cloned in 1.5% DMSO (experiment A). A similar experiment was performed with a separate primary clone, except that selection was in 1.75% DMSO (experiments C and D). Basically, experiments A and C measure the sampling error of the experiments, since all the cultures are identical until cloned in DMSO. Experiments B and D study the growth of parallel but independent cultures over about 14 cell generations prior to cloning DMSO. If DMSO-resistant cells arise during selection in DMSO, then all the parallel cultures should give identical frequencies of DMSO-resistant clones. If, however, non-inducible variants arise during culture, then the frequency of occurrence of DMSO-resistant clones will vary widely according to whether the non-inducible variant was formed early or late in the life of the culture. Analysis of the results shows that experiments B and D have a high ratio v/m, which is significant on a X^2 test at the 0.001 level, as compared to low values of v/m for experiments A and C. Thus, the results of the fluctuation test are consistent with a mutational origin of DMSO-resistant cells.

Growth or cloning of inducible Friend cells in DMSO has proved to be a useful method for obtaining noninducible cells in this laboratory (74, 90, 98, 100, 158). Studies in this laboratory (see Figure 8 for protocol) show clearly that noninducible cells are not produced by the DMSO treatment but arise during culture, as judged by the Luria-Delbrück fluctuation test (Table 5). Furthermore, mutagenesis with ICR 191 may increase the frequency of formation of DMSO-resistant cells, whereas mutagenesis with EMS seems not to be effective (157). These genetic studies are, therefore, consistent with a mutational origin for DMSO-resistant clones.

On the other hand, in our experience only 10–50% of DMSO-resistant clones remains stably noninducible on culture without DMSO as judged by subsequent recloning in DMSO. This apparent instability of some DMSO-resistant clones might be due to trivial reasons, for example, surviving inducible cells which by chance fail to differentiate during the first cloning in DMSO and, therefore, contaminate and subsequently overgrow the true DMSO-resistant cells. Alternatively, there may be a genuinely high reversion frequency of DMSO resistance.

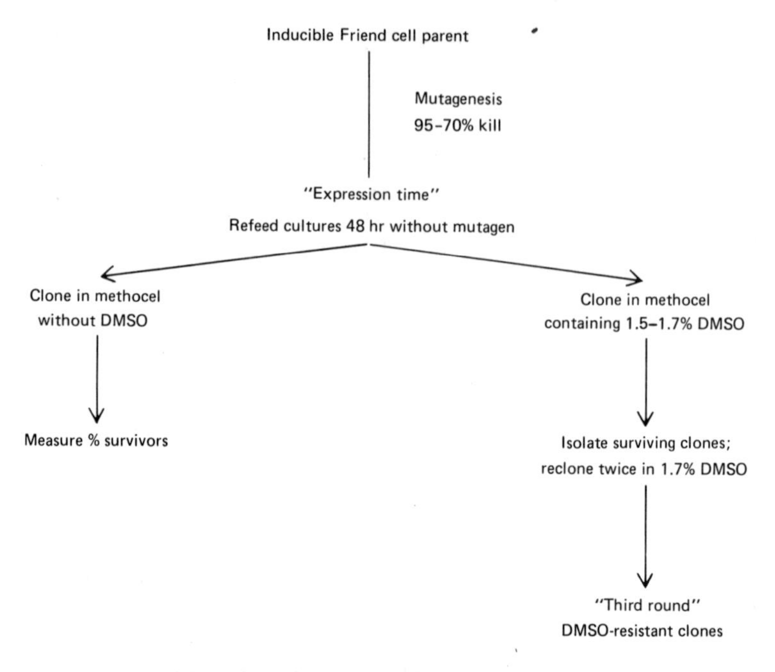

Figure 8. Isolation of DMSO-resistant variants.

Some (but not all) unstable DMSO-resistant clones have abnormally high chromosome numbers (157), and there is some evidence in other experimental systems to link phenotype instability with abnormally high chromosome numbers (159). In contrast, all our stable DMSO-resistant Friend cell clones are pseudo-diploid (37–40 chromosomes) like the parental inducible cells. Nevertheless, as might be expected, chromosomal changes are evident in DMSO-resistant clones (98, 155, 157).

Rovera and Meier (160) have also observed a high frequency of instability of DMSO-resistant clones as judged by the spontaneous formation of hemoglobin upon subsequent culture. Moreover, variant clones showing different extents of spontaneous differentiation change their phenotypes at high frequency under constant environmental conditions. For example, the frequency of change from high to low spontaneous differentiation is 10^{-1} variants per cell per generation as compared to 3×10^{-2} variants per cell per generation for the opposite phenotype change (160). These authors postulate a nonmutational (epigenetic) regulatory mechanism to explain such high rates of spontaneous phenotype change. But it is not at all clear how these frequent spontaneous phenotype changes are related to the generation of the true DMSO-resistant variants that occur at much lower frequency (154, 157), as discussed above.

Characterization of DMSO-resistant Variants

Effect of Other Inducers One of the interesting points to emerge from this author's studies is that some DMSO-resistant variants are not inducible by any of the other inducers, whereas others are inducible by some or all of the other inducers (Table 6). Other workers have obtained similar results (90, 98, 146, 154). This shows that all inducers do not operate by a common mechanism. This battery of noninducible variants should, therefore, be useful for elucidating the molecular parameters of induction, as discussed under "Erythroid Maturation in Friend Cells."

Molecular Characterization DMSO-resistant clones can be screened to determine the range of erythroid functions which are not expressed. An analysis of this kind should provide insights into which functions are tightly coupled and thus help to elucidate how the coordinate regulation of different functions is organized genetically.

None of the eight DMSO-resistant clones tested in this laboratory are inducible by DMSO for globin mRNA (Table 7), nor are two other clones studied by other workers (90, 98). Moreover, neither of the two clones examined in further detail induces globin RNA sequences in the nucleus (Table 7, ref. 158). These results could mean that induction of globin gene transcription is defective in these DMSO-resistant variants. Alternatively, globin RNA transcripts could be peculiarly unstable. It will be necessary to accomplish the difficult task of directly measuring the stability of primary globin RNA transcripts in the nucleus in order to distinguish between these possibilities.

Further work with the FW variant has shown that butyric acid (or to a lesser extent DMSO) induces globin mRNA and hemoglobin if heme is also present. Heme alone does not induce hemoglobin in the FW variant. These results suggest that induction of differentiation in the FW is ineffective due to a limitation in

Table 6. Hemoglobin production in DMSO-resistant variants treated with different inducers

Variant	DMSO	NMA[a]	NMP	HMBA	Heme	Heme + DMSO/BA
R707	–	–	–	–	–	–
TR28D	–	–	–	–	–	–
3BE1A5	–	–	–	–	–	
3BE3A6	–	–	–	–	–	
FW	–	–	–	–	–	+
TR25D	–	++	++	++	–	
3BE1A10	–	+	+	–	–	
3BE1A1	–	+	±	–	–	
3BE3B1	–	++	+	–	–	

[a]NMA, *N*-methylacetamide; NMP, *N*-methylpyrrolidone; HMBA, hexamethylene bis(acetamide); BA, butyric acid, –, <5% benzidine-positive cells; +, 15–30% benzidine-positive cells; ++, 30–60% benzidine-positive cells.

Table 7. Globin RNA and spectrin levels[a] in DMSO-resistant variants

Class of variant Hb response to			Globin RNA (ppm)			
DMSO	NMP[b]	Code	cRNA (ppm)	nRNA (ppm)	Spectrin	Terminal differentiation
+	+	FLC	500–1000	200–300	+	+
–	–	R707	5	6	–	–
		TR28D	15		–	–
		FW	30	20	+	–
–	+	TR25D	30		–	–
–	–	Lymphoma	<0.5		–	–

[a]All assays were performed on DMSO-treated cells unless stated otherwise. FLC is the parental inducible Friend cell line from which the DMSO-resistant variants were isolated. Globin mRNA levels were determined by titration of cytoplasmic RNA with globin cDNA. Since at least 80% of the α,β-globin cDNA is converted into hybrid by the RNA samples, both α- and β-globin mRNAs are present in low but significant amounts in the noninducible variants. Five other independently derived DMSO-resistant lines also fail to induce cytoplasmic globin mRNA. Spectrin was assayed at 3–4 days by the double absorption technique with the use of rabbit antispectrin antiserum (kindly donated by Dr. H. Eisen) and FITC-labeled goat antirabbit antiserum. The TR25D cell line induces both hemoglobin and spectrin with *N*-methylpyrrolidone.

[b]*N*-Methylpyrrolidone.

heme supply, but further work is necessary to show whether this interpretation is valid.

In general, it seems to be difficult to uncouple hemoglobin production from terminal differentiation. Friend cell variants which produce hemoglobin constitutively are either unstable or do so at a much lower level than is characteristic of induced Friend cells (Table 8, ref. 154). In fact, it is not certain that such variants are genuinely constitutive; the few cells which produce hemoglobin may represent a nondividing subpopulation within the proliferating pool of cells which themselves do not produce hemoglobin.

However, it has proven possible to obtain variants which are inducible for other erythroid functions despite the fact that they do not produce hemoglobin or differentiate terminally. Some of our variants have been screened in collaboration with Dr. H. Eisen to determine whether the erythrocyte membrane protein, spectrin, is induced. These studies show that spectrin is induced by DMSO in one variant (FW) which fails to induce hemoglobin or globin mRNA, whereas three others are noninducible for hemoglobin, globin mRNA, and spectrin. Dr. Eisen has performed a more extensive characterization of a range of his own DMSO-resistant variants and concludes that induction of spectrin, glycophorin, and a particularly characteristic chromosomal protein can be uncoupled from one another and from hemoglobin and terminal differentiation (Table 8, ref. 126).

Table 8. Induction of erythroid markers in DMSO-resistant Friend cell variants

Function assayed	Induction of erythroid functions by DMSO in different classes of variants[a]				
	I	II	III	IV	V
Terminal differentiation	–	–	–	–	–
Hemoglobin	–	–	–	–	slightly +
Spectrin	–	+	–	+	+
Glycophorin	–	+	–	–	+
Chromosomal protein	+	+	–	–	+
Lectin binding changes					+

[a]Unpublished results obtained by H. Eisen (personal communication).

Thus, some DMSO-resistant cells seem to be defective at a very early stage in the program of erythroid differentiation; they are noninducible by all inducers for all functions so far tested. On the other hand, some variants are clearly defective only in specific functions or groups of functions. Extension of this type of analysis should elucidate how individual functions are integrated and regulated within the overall program of differentiation.

Genetic Analysis by Cell Fusion Interactions of genes can readily be studied by fusing together two cells of interest, for example, by using Sendai virus. The nuclei of both parents in the resulting heterokaryon may undergo synchronous mitoses and thus produce a mononuclear hybrid cell in which both sets of parental genes can interact. Hybrid cells can be selected by employing conditions under which parental cells cannot survive, for example, the use of BUdR- and thioguanine-resistant parents coupled to the HAT selection medium devised by Littlefield (161). In this medium the metabolically defective parent cells are killed since they cannot incorporate either thymidine or hypoxanthine from the medium, nor can they synthesize pyrimidines or purines de novo due to the presence of aminopterin in the medium. Hybrid cells can survive this treatment due to complementation of the two recessive parental defects.

In intraspecific hybrids, chromosome loss is small and thus as a first approximation it can be assumed that loss of relevant structural or regulatory genes does not complicate interpretation of the results. Moreover, a series of noninducible Friend cell variants all derived from a common inducible parent should be syngeneic except for genetic differences responsible for the noninducible phenotype. Furthermore, since hybrids formed by fusing together two inducible Friend cells remain routinely inducible, cell fusion per se does not interfere with erythroid maturation.

This laboratory, therefore, used this approach to test the dominance of four of our DMSO-resistant clones, three of which are resistant to all inducers tested, whereas the other responds to *N*-methylpyrrolidone (Table 9). In hybrid clones between inducible Friend cells and two of the variants, extinction of hemoglobin

Table 9. Dominance analysis of Friend cell variants

Variant						
Status	Code	No. of hybrid clones tested	Hb	Globin mRNA	Nuclear globin RNA	Chromosome loss[a] (%)
I^{D+}	–	12	+			3
I^{D-}	R707	11	–	–	–	0, 0, 1, 3, 3, 5
		1	+			11
	FW	2	–	–		0, 9
		2	+	+		13, 17
	BE3A6	4	+			9
	BE1A5	4	+			4

[a]Each value for the percentage of chromosome loss refers to an independent hybrid clone. The values are calculated from the modal chromosome numbers of hybrid clones and parental cells.

is observed, except in clones in which there is evidence of slight chromosome segregation. In both cases induction of globin RNA in both nucleus and cytoplasm is also suppressed (157, 158). Thus, noninducibility in these two DMSO-resistant variants seems to operate in a *trans*-dominant manner, either at the level of globin gene transcription or by affecting the stability of the nuclear RNA transcripts. Other workers have reported analogous results; three groups have found that inducibility of both hemoglobin and globin mRNA is extinguished in hybrids between inducible Friend cells and fibroblast lines (131, 162–164) or poorly inducing Friend cell lines (131). However, our results with two other noninducible Friend cell variants show that defects in inducibility do not necessarily operate in a *trans*-dominant manner; in hybrids between inducible Friend cells and these variants, hemoglobin is partially or fully inducible, and there is no evidence for significant chromosome loss in the hybrid clones. Thus, noninducibility in these latter DMSO-resistant variants seems to be recessive or possibly *cis*-dominant. Which of these possibilities is in fact correct could be ascertained in principle by fusing the noninducible variants to a Friend cell inducible for an electrophoretically different hemoglobin. However, such a variant is not presently available. Apart from the problems and complications encountered with interspecific hybrids, this question could be answered in principle by fusing the noninducible Friend cell variants with a rat cell line in which rat hemoglobins are inducible by DMSO (165).

The evidence discussed above is all consistent with the hypothesis that gene expression in somatic cell hybrids is regulated at the level of gene transcription or at least in terms of accumulation of the relevant messenger RNA. Moreover, in the cell fusion experiments discussed above, different erythroid functions are regulated coordinately, in so far as this has been tested. However, cell hybrids formed between Friend cells and lymphoma cells of DBA/2 origin have proved to be interesting in this context (166). The relevant data are summarized in

Table 10. Erythroid maturation in Friend cell X lymphoma hybrid cells

Assay	Treatment (4-day)			
	Nil	1.5% DMSO	0.1 mM Heme	Heme + DMSO
Hemoglobin content[a]	–	–	+	++
Rate of globin chain synthesis (% total proteins)	⩽3	6	13	25
Globin mRNA content (molecules/cell)	1,700	9,000	13,000	31,000
Terminal differentiation	–	nil	slight	marked

[a]Estimated from the E^{540} extinction of cell lysates. Globin chain synthesis after a 5-min pulse of cells with labeled leucine is determined either by SDS-polyacrylamide gel electrophoresis or by CM-cellulose/urea chromatography. Globin mRNA levels are determined by titration of cytoplasmic RNA with globin cDNA, as described in the legend to Figure 5. The standard deviations of all the measurements are about 30% of the mean value. +: about 10 pg hemoglobin/cell; ++: about 15 pg hemoglobin/cell.

Table 10. In the Friend cell X lymphoma hybrid, induction of hemoglobin by DMSO is extinguished as is terminal differentiation, whereas globin mRNA is inducible; globin chains are slightly inducible by DMSO, although they do not form detectable amounts of hemoglobin as judged by benzidine staining. Moreover, pulse/chase experiments show that this failure to accumulate hemoglobin is not due to instability of globin chains. However, hemoglobin, globins, and globin mRNA are all inducible by heme (to 1.5–2 times the levels in DMSO-treated cells) or more efficiently by heme plus DMSO. Under all conditions, 80–90% of the globin mRNA molecules are present mainly in polysomes and are, therefore, presumably in the process of being translated.

Another interesting feature of these results is the finding that in the presence of heme plus DMSO the block in terminal differentiation is released and the hybrid cells differentiate at least until the orthochromatic stage. This illustrates how the ability of cells to produce hemoglobin and to differentiate terminally are tightly coupled.

Why heme is a better inducer of these hybrid cells than DMSO is not clear. Erythroid maturation in these hybrid cells may be limited physiologically due to deficient heme production, this defect being overcome by supplying heme exogenously to the cells. It would be necessary to define the defective enzymes in the heme pathway to prove this definitively. DMSO might well enhance the penetration of heme into the cells, which would explain why DMSO and heme together are particularly effective. However, it is also conceivable that heme is not acting physiologically, but rather as a chemical inducer perhaps acting on the cell membrane. The properties of the Friend cell plasma membrane could be modified by mixing with lymphoma membrane components with the result that heme is more effective than DMSO, although the reverse is true in Friend cells themselves.

Whatever the precise interpretation of these results, the overall effect is characteristic of this particular lymphoma line (L5178Y), since many independent hybrid clones from the same fusion experiment show very similar properties. However, hybrid cells formed by fusing a different lymphoma line (P388) with Friend cells are partially or fully inducible for hemoglobin (M. Allan and P. R. Harrison, unpublished results). Since the L5178Y cell line carries T-lymphocyte surface antigens, whereas the P388 line seems to be more immature and does not possess either T or B surface antigens (167), it is conceivable that expression of differentiated functions in a cell hybrid may be dependent upon the stage of differentiation of the parental cells. This hypothesis is being tested currently by other cell fusions involving cells at different stages of differentiation after commitment to the various hematopoietic pathways.

Finally, to conclude this section, it may be useful to discuss these genetic studies with Friend cells in the context of similar studies with other differentiating cells. In general, fusion of a differentiated cell with a cell not capable of differentiating in the same manner leads to extinction of the differentiated phenotype in the resulting hybrid, except if chromosome segregation is extensive, when re-expression of differentiated functions may occur (168–170). Our results show that cells which cannot differentiate do not necessarily have this dominant and negative effect on the genome of a differentiating cell, even in 1:1 hybrids and under conditions in which chromosome segregation is minimal. Interestingly, fusion of multipotential teratocarcinoma cells to thymocytes does not cause extinction of their ability to differentiate into many different tissue types (171). In other differentiating systems, continued expression of the differentiated phenotype is most commonly observed in 2:1 hybrids containing two genomes from the differentiated parent but a single genome from the nondifferentiating parent. In isolated cases, expression of differentiated markers characteristic of both parents is also observed (172–174). This has usually been interpreted in terms of activation of the normally "silent" genes from the nondifferentiating parent. However, since the mRNAs coding for their differentiated functions cannot be isolated, in no case has it been possible to demonstrate that such effects are regulated during gene transcription. Our own work with the Friend cell X lymphoma hybrid shows that interpretation of these types of experiments must take account of different levels of regulation and in particular of interactions between pathways responsible for different functions (e.g., heme and globins or hemoglobin and terminal differentiation in the Friend cell).

Summary

Noninducible Friend cell variants can be isolated by cloning inducible Friend cells in DMSO or other inducing agents. This method selects for cells which no longer differentiate terminally in the presence of DMSO. Fluctuation analysis studies are consistent with a mutational origin for these noninducible variants, but there is other evidence that Friend cells change phenotype more frequently than would be expected for true mutational changes. Molecular characterization of these variants has led to identification of different classes of noninducible variants: 1) those that are noninducible by all inducers for all erythroid functions; 2) those that are inducible for hemoglobin by some inducers, but not others; 3) those that are inducible by some inducers only in the presence of heme; and 4) those which are not inducible for late functions (e.g., hemoglobin), but are inducible for early functions (e.g., spectrin). Further progress with such genetic studies may be dependent upon devising other methods for selecting for mutants defective in a variety of specific erythroid functions. Noninducibility in these variants may be *trans*-dominant or recessive (or *cis*-dominant) as judged by cell fusion with the parental inducible Friend cell.

OVERALL CONCLUSIONS: CRITIQUE AND PERSPECTIVES

The wide interest of Friend virus-induced leukemia derives from the fact that its pathological, virological, and biological aspects are amenable to analysis by several experimental approaches.

As argued by Lilly (39), in the intact organism it can be usefully viewed as a model of a multigene disease, controlled by specific host loci which are becoming increasingly well understood. The relevance of Friend leukemia in mice to human leukemias is obviously a subject of some interest within this context, for similarities have been recognized with Di Gugliamo's erythroleukemia (78) or polycythemia vera (32, 58). Prchal and Axelrad (175), for example, have shown that bone marrow cells from polycythemia vera patients form erythroid colonies in vitro in the absence of erythropoietin (like Friend leukemia cells), whereas cells from normal patients or many other unrelated conditions require erythropoietin. However, whether these similarities are of fundamental significance or universal occurrence remains to be seen. Recently, Jasmin and co-workers (176) have argued that chronic granulocytic leukemia, polycythemia vera, and the erythroblastic component of erythroleukemia are compatible with the Friend leukemia physiopathological model, in contrast to the situation in acute leukemias. These authors suggest that the differentiated leukemias arise from an uncontrolled differentiation of a committed cell compartment, which stimulates proliferation of the stem cell compartment. Thus, these diseases are envisaged as being due to proliferation and accumulation of "subnormal" cells of shorter than normal life span, in a manner analogous to Friend leukemic cells. However, this assessment of the similarities between Friend leukemia and chronic human leukemias is not universally accepted. For example, the fact that the Philadelphia chromosome is present in cells of most hematopoietic lines in chronic granulocytic leukemia seems to argue that an uncommitted stem cell is primarily involved in this disease.

The interaction of Friend virus with erythroid precursor cells is also an important subject of research capable of exploitation by many approaches. In particular, the discovery that erythroid precursor cells can apparently be "transformed" in vitro to erythropoietin independence (59) opens up a new avenue of research. Moreover, since Friend virus can now be purified, the viral genome characterized and transcribed into a virtually complete DNA copy, this enables the techniques of molecular biology to be exploited to elucidate the integration and expression of viral genes and their interaction with the cellular genes involved in erythroid development. By fractionating the Friend virus genome by various means, it should be feasible in the foreseeable future to study the effects of specific viral genes. The development of assays for specific virus proteins should also facilitate this approach (177, 178). Such developments may help to elucidate how the Friend virus interacts specifically with the committed erythroid precursor cell and renders its subsequent differentiation independent of normal hormonal regulation.

Thirdly, the isolation of Friend cells in culture has clearly permitted a more detailed understanding of the process of erythroid maturation by the techniques of clonal analysis, molecular biology and biochemistry, and somatic cell genetics. Erythroid maturation in Friend cells seems to be relatively normal as judged by the appearance of a variety of erythrocyte markers. The isolation of Friend cell variants or mutants has perhaps been most useful in dissecting how the various functions are coordinated within a single differentiating cell type. How chemical "inducers" permit erythroid maturation in Friend cells is also a fruitful area for further research which is likely to benefit greatly from detailed membrane studies which make use of specific "noninducible" mutants. Such work may be relevant not only to the mechanism of action of the various inducers but also to how erythropoietin functions and how this function is impaired after Friend virus infection. The related issue—whether erythroid maturation in Friend cells is cell cycle-dependent—requires a definitive answer in order to assess how fundamentally Friend cell inducers alter the state of differentiation of the cell.

Despite the progress made with the use of the Friend cell system as a model of a differentiating cell, its relevance to the biology of erythroid development must be assessed critically in view of the following considerations. First, the available evidence strongly suggests that the Friend cell is derived from the committed erythroid cell precursor; it may, in fact, be equivalent to the proerythroblast in its stage of differentiation along the erythroid line. Therefore, Friend cell differentiation in culture represents the later stages of erythroid maturation, and, therefore, it is not directly relevant to the process of commitment to the erythroid line. Secondly, the Friend cell is transformed and malignant, and it is clearly abnormal in its lack of response to normal hormonal regulation. This means that the Friend cell system may not be relevant to the regulation of proliferation of normal, as opposed to leukemic, erythroid stem cells, although the evidence suggests that it is useful for studies of maturation of such precursors. Thus, it is questionable whether the Friend cell system has fundamentally widened the types of developmental processes which are amenable to experimental investigation. In fact, normal erythroid developmental systems, such as fetal liver, and the Friend cell system both relate to the same biological context, namely, the later stages of erythroid cell formation, but the latter system has permitted an analysis in greater depth than would perhaps otherwise have been possible.

Future advances concerning the molecular processes involved in hematopoietic cell commitment may, therefore, require alternative approaches. Techniques are now available for the growth and assay in vitro of multipotential stem cells (28) and various types of committed erythroid stem cells (21–23). Growth of multipotential stem cells requires the presence of feeder layers (28), and under these conditions the cells mature mainly into the granulocyte/mononuclear cell lineage (28). But it will be interesting to determine whether different environmental conditions can allow stem cells to mature in vitro along the erythroid line. An alternative approach may be to investigate conditions under which multipotential teratocarcinoma cells (see ref. 179 for review) can be induced to undergo erythroid differentiation, perhaps after interaction with other tissues present as feeder layers. Some type of precedent for such studies exists in the demonstration that teratocarcinoma cells can differentiate into keratinizing epithelium when grown on fibroblast feeder layers (180). Although such novel approaches are still somewhat speculative, they do offer the possibility of studying the process of commitment in vitro. Applications to these experimental systems of the combination of molecular and genetic approaches which have proved successful with the Friend cell in a more limited context may then allow greater insight into the molecular nature of commitment itself.

ACKNOWLEDGMENTS

The work owes much to contributions by my colleagues Nabeel Affara, Maggie Allan, David Conkie, John Paul, Ian Pragnell, and Tim Rutherford, and so far as the Friend virus work is concerned to the collaboration of Dr. W. Ostertag of the Max Planck Institute in Gottingen. The technical assistance of Janis Fleming, Kathy Gorski, Paul Hissey, Alistair McNab, and Jimmy Sommerville is also very much appreciated. I am grateful to John Paul for advice and encouragement during the period of this work.

REFERENCES

1. Harrison, P. R. (1976). Nature (Lond.) 262:353.
2. Denton, M. J., Spencer, N., and Arnstein, H. R. V. (1975). Biochem. J. 146:205.
3. Marchesi, S. L., Steers, E., Marchesi, V. T., and Tillack, T. W. (1970). Biochemistry 9:50.
4. Tomita, M., and Marchesi, V. T. (1975). Proc. Natl. Acad. Sci. U.S.A. 72:2964.

5. Klein, J. (1975). Biology of the Mouse Histocompatibility-2 complex, p. 620. Springer-Verlag, Heidelberg.
6. Arndt-Jovin, D. J., and Jovin, T. M. J. Supramol. Struct. (Suppl. 1), in press.
7. Furusawa, M., and Adachi, H. (1968). Exp. Cell Res. 50:497.
8. Verma, I. M., Temple, G. F., Fan, H., and Baltimore, D. (1972). Nature (New Biol.) 235:163.
9. Kacian, D. L., Spiegelman, S., Bank, A., Terada, M., Metafora, S., Dow, L., and Marks, P. A. (1972). Nature (New Biol.) 235:167.
10. Ross, J., Aviv, H., Scolnik, E., and Leder, P. (1972). Proc. Natl. Acad. Sci. U.S.A. 69:264.
11. Harrison, P. R., Birnie, G. D., Hell, A., Humphries, S., Young, B. D., and Paul, J. (1974). J. Mol. Biol. 84:539.
12. Young, B. D., Harrison, P. R., Gilmour, R. S., Birnie, G. D., Hell, A., Humphries, S., and Paul, J. (1974). J. Mol. Biol. 84:555.
13. Harrison, P. R., Conkie, D., Affara, N., and Paul, J. (1974). J. Cell Biol. 63:402.
14. Conkie, D., Kleiman, L., Harrison, P. R., and Paul, J. (1975). Exp. Cell Res. 93:315.
15. Terada, M., Cantor, L., Metafora, S., Rifkind, R. A., Bank, A., and Marks, P. A. (1972). Proc. Natl. Acad. Sci. U.S.A. 69:3575.
16. Ramirez, F., Gambino, R., Maniatis, G. M., Rifkind, R. A., Marks, P., and Bank, A. (1975). J. Biol. Chem. 250:6054.
17. McCulloch, E. A. (1970). *In* A. S. Gordon (ed.), Regulation of Hematopoiesis, Vol. 1, Chap. 7, pp. 132–59. Appleton-Century-Crofts, New York.
18. Lajtha, L. G., and Schofield, R. (1974). Differentiation 2:313.
19. Trentin, J. J. (1970). *In* A. S. Gordon (ed.), Regulation of Haematopoiesis, Vol. 1, pp. 161–186. Appleton-Century-Crofts, New York.
20. Wolf, W. S. (1974). Cell Tissue Kinet. 7:89.
21. Stephenson, J. A., Axelrad, A. A., McLeod, D. L., and Shreeve, M. (1971). Proc. Natl. Acad. Sci. U.S.A. 68:1542.
22. Iscove, N. W., Sieber, F., and Winterhalter, K. H. (1974). J. Cell. Physiol. 83:309.
23. Axelrad, A. A., McLeod, D. L., Shreeve, M. M., and Heath, D. S. (1973). *In* W. A. Robinson (ed.), Proceedings of the Second International Workshop on Hematopoiesis in Culture, pp. 226–237. DHEW Publication No. NIH 74-205.
24. Tambourin, P. E., and Wendling, E. (1975). Nature (Lond.) 256:320.
25. Miller, R. G., and Phillips, R. A. (1969). J. Cell Physiol. 73:191.
26. Harrison, P. R., Conkie, D., and Paul, J. (1973). *In* M. Balls and F. S. Billet (eds.), The Cell Cycle in Development and Differentiation, pp. 341–364. Cambridge University Press, London.
27. Cantor, L. W., Morris, A. J., Marks, P. A., and Rifkind, R. A. (1972). Proc. Natl. Acad. Sci. U.S.A. 69:1337.
28. Dexter, T. M., and Testa, M. G. (1976). Methods Cell Biol. 14:387.
29. Friend, C., Scher, W., Holland, J. G., and Sato, T. (1971). Proc. Natl. Acad. Sci. U.S.A. 68:378.
30. Friend, C. (1957). J. Exp. Med. 105:307.
31. Mirand, E. A. (1966). Natl. Cancer Inst. Monogr. 22:483.
32. Tambourin, P. E., Gallien-Lartigue, O., Wendling, F., and Huaulme, D. (1973). Br. J. Haematol. 24:511.

33. Smadja-Joffe, F., Jasmin, C., Malaise, E. P., and Bournoutian, C. (1973). Int. J. Cancer 11:300.
34. Smadja-Joffe, F., Klein, C., Kerdiles, C., Feinendegen, L., and Jasmin, C. (1976). Cell Tissue Kinet. 9:131.
35. Mirand, E. A. (1967). Science 156:832.
36. Mirand, E. A., Steeves, R. A., Lange, R. O., and Grace, J. T. (1968). Proc. Soc. Exp. Biol. Med. 128:844.
37. McGarry, M. P., and Mirand, E. A. (1973). Exp. Hematol. 1:174.
38. Sassa, S., Takaku, F., and Nakao, K. (1968). Blood 31:758.
39. Lilly, F. (1972). J. Natl. Cancer Inst. 49:927.
40. Steeves, R. A. (1975). J. Natl. Cancer Inst. 54:289.
41. Lilly, F., and Pincus, T. (1973). Adv. Cancer Res. 17:231.
42. Chirigos, M. A., Schwalk, E. D., and Scott, D. (1967). Cancer Res. 27:2249.
43. Mirand, E. A., Steeves, R. A., and Avila, L. (1968). Proc. Soc. Exp. Biol. Med. 127:900.
44. Steeves, R. A., Eckner, R. J., Bennett, N., Mirand, E. A., and Friedel, P. J. (1971). J. Natl. Cancer Inst. 46:1209.
45. Rowson, K. E., and Parr, I. B. (1970). Int. J. Cancer 5:96.
46. Fieldsteel, A. H., Kurahara, C., and Dawson, P. J. (1969). Nature (Lond.) 223:1274.
47. Fieldsteel, A. H., Dawson, P. J., and Kurahara, C. (1971). Int. J. Cancer 8:304.
48. Freedman, H. A., Lilly, F., and Steeves, R. A. (1974). Proc. Am. Assoc. Cancer Res. 15:131.
49. Bennett, M., Steeves, R. A., Cudkowicz, G., Mirand, E. A., and Russell, L. B. (1968). Science 162:564.
50. Steeves, R. A., Bennett, M., Mirand, E. A., and Cudkowicz, G. (1968). Nature (Lond.) 218:372.
51. Chirigos, M. A., and Marsh, R. W. (1966). Antimicrob. Agents Chemother. 6:489.
52. Odaka, T. (1969). Jap. J. Exp. Med. 39:99.
53. Levy, S. B., Rubenstein, C. B., and Tavassoli, M. (1976). J. Natl. Cancer Inst. 56:1189.
54. Levy, S. B., Rubenstein, C. B., and Friend, C. (1976). J. Natl. Cancer Inst. 56:1183.
55. Tambourin, P., and Wendling, F. (1971). Nature (New Biol.) 234:230.
56. Steeves, R. A., Mirand, E. A., Thomson, S., and Avila, L. (1969). Cancer Res. 29:1111.
57. Fredrickson, T., Tambourin, P., Jasmin, C., and Smajda, F. (1975). J. Natl. Cancer Inst. 55:443.
58. Liao, S. K., and Axelrad, A. A. (1975). Int. J. Cancer 15:467.
59. Clarke, B. J., Axelrad, A. A., Shreeve, M. M., and McLeod, D. L. (1975). Proc. Natl. Acad. Sci. U.S.A. 72:3556.
60. Wendling, F., Tambourin, P. E., and Jullien, P. (1972). Int. J. Cancer 9:554.
61. Okunewick, J. P., and Phillips, E. L. (1975). Blood 42:885.
62. Golde, D. W., Faille, A., Sullivan, A., and Friend, C. (1976). Cancer Res. 36:115.
63. Brown, W. M., and Axelrad, A. A. (1976). Int. J. Cancer 18:764.
64. Majumdar, S. K., and Bilenker, J. D. (1975). J. Nat. Cancer Inst. 54:503.
65. Matioli, G. (1973). J. Reticuloendothel. Soc. 14:380.

66. Rossi, G. B., Cudkowicz, G., and Friend, C. (1973). J. Natl. Cancer Inst. 50:249.
67. Thomson, S., and Axelrad, A. A. (1968). Cancer Res. 28:2105.
68. Axelrad, A. A., Cinader, B., Koh, S. W., and Van der Gaag, H. C. (1976). Cancer Res. 36:28.
69. Steeves, R. A., Mirand, E. A., and Thomson, S. (1970). *In* R. M. Dutcher (ed.), Comparative Leukemia Research, pp. 624–633. Karger, Basel.
70. Buffett, R. F., and Furth, J. (1959). Cancer Res. 19:1063.
71. Friend, C., and Haddad, J. R. (1960). J. Natl. Cancer Inst. 25:1279.
72. Dawson, P. J., Fieldsteel, A. H., and Bostick, W. I. (1963). Cancer Res. 23:349.
73. Ostertag, W., Melderis, H., Steinheider, G., Kluge, N., and Dube, S. (1972). Nature (New Biol.) 239:231.
74. Dube, S. K., Gaedicke, G., Kluge, N., Weimann, B. J., Melderis, H., Steinheider, G., Crozier, T., Beckmann, H., and Ostertag, W. (1974). *In* W. Nakaharo, T. Ono, T. Sugimoto, and H. Sugano (eds.), Differentiation and Control of Malignancy of Tumor Cells, pp. 99–135. University of Tokyo Press, Tokyo.
75. Ikawa, Y., and Sugano, H. (1966). Gann 57:641.
76. Ikawa, Y., Ross, J., Leder, P., Gielen, J., Packman, S., Ebert, P., Hayashi, K., and Sugano, H. (1974). *In* W. Nakahara, T. Ono, T. Sugimura, and H. Sugano (eds.), Differentiation and Control of Malignancy of Tumor Cells, pp. 515–547. University of Tokyo Press, Tokyo.
77. Metcalf, D., Furth, J., and Buffett, R. F. (1959). Cancer Res. 19:52.
78. Friend, C., Patuleia, M. C., and de Harven, E. (1966). Natl. Cancer Inst. Monogr. 22:505.
79. De Harven, E., and Friend, C. (1966). Natl. Cancer Inst. Monogr. 22:79.
80. Rossi, G. B., and Friend, C. (1967). Proc. Natl. Acad. Sci. U.S.A. 58:1373.
81. Patuleia, M., and Friend, C. (1967). Cancer Res. 27:726.
82. Rossi, G. B., and Friend, C. (1970). J. Cell Physiol. 76:159.
83. Kabat, D., Sherton, C. C., Evans, L. H., Bigley, R., and Koler, R. D. (1975). Cell 5:331.
84. Hankins, W. D., and Krantz, S. B. (1975). Nature (Lond.) 253:731.
85. Dube, S. K., Pragnell, I. B., Kluge, N., Gaedicke, G., Steinheider, G., and Ostertag, W. (1975). Proc. Natl. Acad. Sci. U.S.A. 72:1863.
86. Pragnell, I. B., Ostertag, W., Harrison, P. R., Williamson, R., and Paul, J. (1976). *In* N. Muller-Berat (ed.), Progress in Differentiation Research, pp. 501–511. North-Holland, Amsterdam.
87. Pragnell, I. B., Ostertag, W., Paul, J., and Williamson, R. W. J. Virol., in press.
88. Affara, N. A., and Young, B. D. (1976). MSE Scientific Instruments. Manor Royal, Crowley, Sheet No. A12/6/76.
89. Krieg, C. J., Ostertag, W., Pragnell, I., Swetly, P., Roesler, G., Clauss, V., and Weimann, B. J. Manuscript in preparation.
90. Ikawa, Y., Inoue, Y., Aida, M., Kameji, C., Shibata, C., and Sugano, H. (1976). Bibl. Haematol. 43:37.
91. Scher, W., Holland, J. G., and Friend, C. (1971). Blood 37:428.
92. Boyer, S. H., Wuu, K. D., Noyes, A. N., Young, R., Scher, W., Friend, C., Preisler, H. D., and Bank, A. (1972). Blood 40:823.
93. Schalekamp, M., Harrison, P. R., and Paul, J. (1975). J. Embryol. Exp. Morphol. 34:355.

94. Gilmour, R. S., Harrison, P. R., Windass, J. D., Affara, N. A., and Paul, J. (1974). Cell Differ. 3:9.
95. Harrison, P. R., Gilmour, R. S., Affara, N., Conkie, D., and Paul, J. (1974). Cell Differ. 3:23.
96. Ross, J., Ikawa, Y., and Leder, P. (1972). Proc. Natl. Acad. Sci. U.S.A. 69:3620.
97. Ross, J., Gielen, J., Packman, S., Ikawa, Y., and Leder, P. (1974). J. Mol. Biol. 87:697.
98. Ohta, Y., Tanaka, B., Terada, M., Miller, O. J., Bank, A., Marks, P. A., and Rifkind, R. A. (1976). Proc. Natl. Acad. Sci. U.S.A. 73:1232.
99. Preisler, H. D., Housman, D., Scher, W., and Friend, C. (1973). Proc. Natl. Acad. Sci. U.S.A. 70:2956.
100. Conkie, D., Affara, N., Harrison, P. R., Paul, J., and Jones, K. (1974). J. Cell Biol. 63:414.
101. Ostertag, W., Crozier, T., Kluge, N., Melderis, H., and Dube, S. (1973). Nature (New Biol.) 243:203.
102. Orkin, S. H., Swan, D., and Leder, P. (1975). J. Biol. Chem. 250:8753.
103. Lodish, H., and Small, B. (1976). Cell 7:59.
104. Aviv, H., Voloch, Z., Bastos, R., and Levy, S. (1976). Cell 8:495.
105. Williamson, R., Drewienkiewicz, C. E., and Paul, J. (1973). Nature (New Biol.) 241:66.
106. McNaughton, M., Freeman, K. B., and Bishop, J. O. (1974). Cell 1:117.
107. Imaizumi, T., Diggelmann, H., and Scherrer, K. (1973). Proc. Natl. Acad. Sci. U.S.A. 70:1122.
108. Spohr, G., Dettori, G., and Manzari, V. (1976). Cell 8:505.
109. Curtis, P. J., and Weissmann, C. (1977). J. Mol. Biol. 106:1061.
110. Ross, J. (1976). J. Mol. Biol. 106:403.
111. Hunt, T. (1976). Br. Med. Bull. 32:257.
112. Ebert, P. S., and Ikawa, Y. (1974). Proc. Soc. Exp. Biol. Med. 146:601.
113. Cimadevilla, J. M., and Hardesty, B. (1975). Biochem. Biophys. Res. Commun. 63:931.
114. Cimadevilla, J. M., Kramer, G., Pinphanichakarn, P., Konechi, P., and Hardesty, B. (1975). Arch. Biochem. Biophys. 171:145.
115. Reem, G. H., and Friend, C. (1975). Proc. Natl. Acad. Sci. U.S.A. 72:1630.
116. Sato, T., Friend, C., and de Harven, E. (1971). Cancer Res. 31:1402.
117. Darzynkiewicz, Z., Traganos, F., Sharpless, T., Friend, C., and Melamed, M. R. (1976). Exp. Cell Res. 99:301.
118. Peterson, J. L., and McConkey, E. H. (1976). J. Biol. Chem. 251:555.
119. Blankstein, L. A., and Levy, S. B. (1976). Nature (Lond.) 260:638.
120. Keppel, F., Allet, D., and Eisen, H. (1977). Proc. Natl. Acad. Sci. U.S.A., 74:653.
121. Minty, A., Birnie, G. B., and Paul, J. Manuscript in preparation.
122. Furasawa, M., Ikawa, Y., and Sugano, H. (1971). Proc. Jap. Acad. 47:220.
123. Ikawa, Y., Furusawa, M., and Sugano, H. (1973). Bibl. Haematol. 39:955.
124. Arndt-Jovin, D. H., Ostertag, W., Eisen, H., Klimek, F., and Jovin, T. M. (1976). J. Histochem. Cytochem. 24:332.
125. Arndt-Jovin, D. J., Ostertag, W., Eisen, H., and Jovin, T. M. (1976). *In* R. Neth, R. C. Gallo, K. Mannweiler, and W. C. Maloney (eds.), Modern Trends in Human Leukemia II, pp. 137–150. J. F. Lehmanns, Munich.
126. Eisen, H., Bach, R., and Emery, R. Proc. Natl. Acad. Sci. U.S.A., in press.

127. Eisen, H., Nasi, S., Georgotoulos, C., Arndt-Jovin, D., and Ostertag, W. (1977). Cell 10:689.
128. McClintock, P. R., and Papaconstantinou, J. (1974). Proc. Natl. Acad. Sci. U.S.A. 71:4551.
129. Levy, J., Terada, M., Rifkind, R. A., and Marks, P. A. (1975). Proc. Natl. Acad. Sci. U.S.A. 72:28.
130. Reuben, R. C., Wife, R. L., Breslow, R., Rifkind, R. A., and Marks, P. A. (1976). Proc. Natl. Acad. Sci. U.S.A. 73:862.
131. Orkin, S. H., Harosi, F. I., and Leder, P. (1975). Proc. Natl. Acad. Sci. U.S.A. 72:98.
132. Till, J. E., McCulloch, E. A., and Siminovitch, L. (1964). Proc. Natl. Acad. Sci. U.S.A. 51:29.
133. Korn, A. P., Henkelman, R. M., Ottensmeyer, F. P., and Till, J. E. (1973). Exp. Haematol. 1:362.
134. Gusella, J., Geller, R., Clarke, B., Wilks, V., and Housman, D. (1976). Cell 9:221.
135. Harrison, P. R., Conkie, D., Wood, A., and Yeoh, G. Manuscript in preparation.
136. Leder, A., Orkin, S., and Leder, P. (1975). Science 190:893.
137. Ebert, P. S. , Wars, I., and Buell, D. N. (1976). Cancer Res. 36:1809.
138. Demsey, A., and Grimley, P. M. (1976). Cancer Res. 36:384.
139. Preisler, H. D., and Giladi, M. (1974). Nature (Lond.) 251:645.
140. Kluge, N., Gaedicke, G., Steinheider, G., Dube, S., and Ostertag, W. (1974). Exp. Cell Res. 88:257.
141. Scher, W., Preisler, H. D., and Friend, C. (1973). J. Cell Physiol. 81:63.
142. Tanaka, M., Levy, J., Terada, M., Breslow, R., Rifkind, R. A., and Marks, P. A. (1975). Proc. Natl. Acad. Sci. U.S.A. 72:1003.
143. Preisler, H. D., and Lyman, G. (1975). Cell Differ. 4:179.
144. Gaedicke, G., Abedin, Z., Dube, S. K., Kluge, N., Neth, R., Steinheider, G., Weimann, B. J., and Ostertag, W. (1974). *In* R. Neth, R. Gallo, S. Spiegelman, and F. Stohlmann (eds.), Modern Trends in Human Leukemia, pp. 278–287. Grune and Stratton, New York.
145. Preisler, H. D., Christoff, G., and Taylor, E. (1976). Blood 47:363.
146. Gusella, J. F., and Housman, D. (1976). Cell 8:263.
147. Leder, A., and Leder, P. (1975). Cell 5:319.
148. Takahashi, E., Yamada, M., Saito, M., Kuboyama, M., and Ogasa, K. (1975). Gann 66:577.
149. Ross, J., and Sautner, D. (1976). Cell 8:513.
150. Bernstein, A., Boyd, A. S., Crichley, V., and Lamb, V. (1975). Presented at the Symposium on Biogenesis and Turnover of Membrane Molecules: Society of General Physiologists, September, 1975, Woods Hole, Massachusetts.
151. Nakanishi, S., Adhya, S., Gottesman, M., and Pastan, I. (1974). J. Biol. Chem. 249:4050.
152. Lyman, G. H., Preisler, H. D., and Papahadjopoulos, D. (1976). Nature (Lond.) 262:360.
153. Bernstein, A., Hunt, D. M., Crichley, V., and Mak, T. W. (1976). Cell 9:375.
154. Rovera, G., and Bonaiuto, J. (1976). Cancer Res. 36:4057.
155. Paul, J., and Hickey, I. (1974). Exp. Cell Res. 87:20.
156. Singer, D., Cooper, M., Maniatis, G. M., Marks, P. A., and Rifkind, R. A. (1974). Proc. Natl. Acad. Sci. U.S.A. 71:2668.

157. Harrison, P. R., Conkie, D., Rutherford, T., Affara, N., and Paul, J. Manuscript in preparation.
158. Harrison, P. R., Affara, N., Conkie, D., Rutherford, T., Sommerville, J., and Paul, J. (1976). *In* N. Muller-Berat (ed.), Progress in Differentiation Research, pp. 135–146. North Holland, Amsterdam.
159. Terzi, M., and Hawkins, T. S. C. (1975). Nature (Lond.) 253:361.
160. Rovera, G., and Meier, G. Manuscript in preparation.
161. Littlefield, J. S. (1966). Exp. Cell Res. 41:190.
162. Ruddle, F. H., and Kucherlapati, R. S. (1974). Sci. Am. 231:36.
163. Deisseroth, A., Burk, R., Picciano, D., Minna, J., French Anderson, W., and Nienhuis, A. (1975). Proc. Natl. Acad. Sci. U.S.A. 72:1102.
164. Deisseroth, A., Velez, R., Burk, R. D., Minna, J., French Anderson, W., and Nienhuis, A. (1976). Somatic Cell Genet. 2:373.
165. Kluge, N., Ostertag, W., Sugiyama, T., Jovin-Arndt, D., Steinheider, G., Furusawa, M., and Dube, S. K. (1976). Proc. Natl. Acad. Sci. U.S.A. 73:1237.
166. Harrison, P. R., Affara, N., McNab, A., and Paul, J. (1977). Exp. Cell Res., in press.
167. Shevach, E. M., Stobo, J. D., and Green, I. (1972). J. Immunol. 108:146.
168. Davis, F. M., and Adelberg, E. A. (1973). Bacteriol. Rev. 37:197.
169. Davidson, R. L. (1974). Annu. Rev. Genet. 8:195.
170. Davidson, R. L., and de la Cruz, F. (eds.). (1974). Somatic Cell Hybridisation. Raven Press, New York.
171. Miller, R. A., and Ruddle, F. H. (1976). Cell 9:45.
172. Colen, H. R., and Parkman, R. (1972). Science 176:1029.
173. Kao, F., and Puck, T. (1972). Proc. Natl. Acad. Sci. U.S.A. 69:3272.
174. Peterson, J., and Weiss, M. (1972). Proc. Natl. Acad. Sci. U.S.A. 69:571.
175. Prchal, J. F., and Axelrad, A. A. (1974). N. Engl. J. Med. 290:1382.
176. Jasmin, C., Smadja-Joffe, F., Klein, B., and Kerdiles-Lebousse, C. (1976). Cancer Res. 36:603.
177. Friedman, M., Lilly, F., and Nathenson, S. (1974). J. Virol. 14:1126.
178. Racevskis, J., and Koch, G. J. Virol., in press.
179. Martin, G. (1975). Cell 5:229.
180. Rheinwald, J. G., and Green, H. (1975). Cell 6:317.

International Review of Biochemistry
Biochemistry of Cell Differentiation II, Volume 15
Edited by J. Paul

7
Muscle Protein Synthesis and Its Control During the Differentiation of Skeletal Muscle Cells in Vitro

M. E. BUCKINGHAM
Institut Pasteur,
Paris, France

The work performed in this laboratory was supported by grants from the Fonds de Développement de la Recherche Scientifique et Technique, the Institut National de la Santé et de la Recherche Médicale, the Commissariat à l'Energie Atomique, the Ligue Nationale Française contre le Cancer, the Fondation pour la Recherche Médicale Française, and the Muscular Dystrophy Associations of America.

The study of molecular mechanisms underlying cell differentiation is facilitated if biochemical markers of the different states are available and if the process takes place when the cells are cultivated in vitro. These criteria have been mainly satisfied by systems undergoing terminal differentiation where a biochemically recognizable phenotype is expressed by cells already in an advanced developmental state. The flexibility remaining to the precursor cells is severely restricted. It may be possible, for example, to manipulate in tissue culture the development of chondrocytes or myoblasts (1), but usually the final phase of differentiation proceeds, either spontaneously or on triggering by a hormone, without any further element of choice.

The transition from dividing myoblasts to multinucleate muscle fibers or myotubes provides one of the most striking examples of terminal differentiation which can take place in vitro. The visible morphological phenomenon of cell fusion is accompanied by biochemical changes which result in the development of an excitable membrane, the assembly of the contractile apparatus, and an increase in the enzymes required for muscle metabolism. These processes take place spontaneously after one or more cycles of cell division, either in primary cultures or in established cell lines derived from fetal skeletal muscle (Figure 1) (for previous reviews, see refs. 2 and 3). In vivo, during embryological development, the formation of muscle fibers proceeds less synchronously over a much longer period of time, with biochemical differentiation frequently preceding incorporation of the myoblast into a fiber (see, for example, ref. 4). In tissue culture, morphological and biochemical differentiation are coordinated, although cell fusion may be arrested while muscle development proceeds normally, in medium containing low levels of calcium (5–10) or in the presence of cytochalasin B (11, 12). This clearly indicates, contrary to some previous proposals, that cell fusion is not a prerequisite for biochemical differentiation.

The synchrony with which the myoblast population differentiates is an important aspect of the in vitro system and is critical to any biochemical study of the transition from myoblast to myotube. The factors affecting the extent and synchrony of differentiation are at present ill defined. The role of cell division, with a postulated critical (or quantal) mitosis preceding differentiation (for reviews, see refs. 13 and 14), together with the timing of withdrawal from the G_1 phase of the cell cycle (e.g., ref. 15), has been discussed extensively. Empirically, initial plating densities and growth conditions certainly affect myogenesis. Inhibition of mitosis (16, 17) prevents myotube development, although possible side effects of the treatment cannot be ruled out. The absolute requirement for cell division in culture depends upon the system, since the developmental state of the mononucleated population apparently varies: chick myoblasts fuse without extensive cell growth (18), although this is not the case for most mammalian muscle cultures. Of particular note is the effect of the thymidine analog, bromodeoxyuridine (BUdR), which, when present during the last cell division (19–21), prevents subsequent differentiation. The effect of

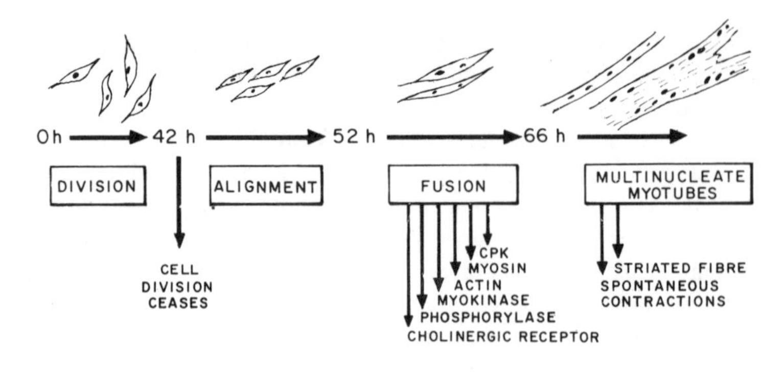

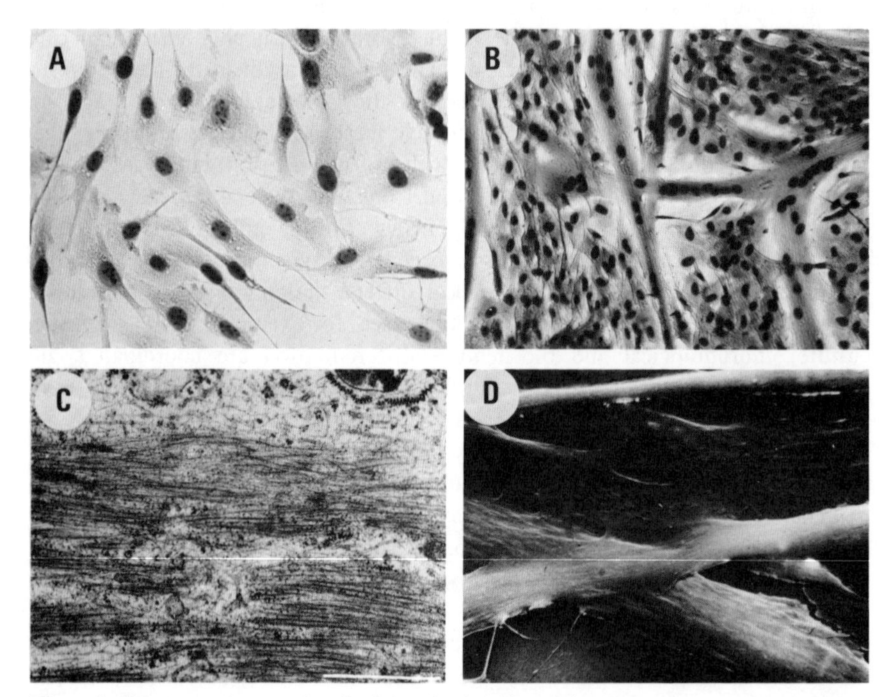

Figure 1. Primary cultures of cells from fetal calf muscle. The course of myogenesis is represented schematically. The photographs show dividing (*A*) and fused (*B*) cultures seen in phase contrast, fused cultures viewed by scanning electron microscopy (*D*), and transmission electron microscopy of a cross-section of a myotube showing the presence of contractile protein (*C*).

BUdR and its reversal by a further round of cell division provide some experimental support for the proposed significance of the mitotic cycle preceding differentiation.

The medium and especially the serum batch are important elements in determining the quality of myogenesis. A number of diffusible, fusion-promoting factors present in the culture medium, either derived from the embryonic extract (22) or possibly released by the cells themselves (18, 23), have been tentatively described. Insulin stimulates myotube formation (24), as does a factor isolated from crude extracts of brain or pituitary gland (25). These effects appear to be distinct from any influence on the rate of myoblast division, although in some cases they may be related to stimulation of collagen biosynthesis, since collagen promotes myotube formation (22). Inhibition of differentiation has been reported on addition of lectins (26), phospholipases (27), and of a low molecular weight factor from spinal cord cells (28). Many of these effects can be reversed, which results in relatively synchronous myotube formation, without any further cell division. Cyclic AMP also inhibits the progress of differentiation (29, 30), although this may be due primarily to an effect on myoblast growth (31). At fusion, decreased intracellular cyclic AMP levels and adenylate cyclase activities have been reported in rat cultures (32, 33), although in chick cells adenylate cyclase remains high (34). Immediately prior to cell fusion in chick myoblasts, a transitory increase in cellular cyclic AMP has been detected and is apparently a prerequisite for myotube formation (34). This is the first indication of a possible internal physiological event which may be important in "triggering" differentiation.

Of the actual cell culture systems available, the most extensively used are those of primary cultures of mammalian (rat, calf) or avian (chicken, quail) fetal skeletal muscle (for review of tissue culture methods, see ref. 3). In such cultures, contamination by fibroblasts and mature myoblasts affects the homogeneity and synchrony of myotube formation. Although it is possible to obtain pure populations of myotubes by eliminating the dividing cells with drugs such as cytosine arabinoside (35) or by sedimentation methods (36), the mononucleate population is always heterogeneous. In large scale primary cultures from fetal calf, 30–50% of the cells never fuses. The best system in this respect to have been reported is that of primary cultures from quail in which 85–90% of total nuclei enters myotubes (5).

One of the advantages of studying differentiation in muscle cells is that cell lines are also available. Diploid rat myoblast lines have been obtained by Yaffé from primary cultures of rat fetal skeletal muscle, with (L_5, L_6) or without (L_8) carcinogen treatment (2, 37). These lines provide a cloned population of muscle cells, although difficulties are encountered with the homogeneity of the population, which tends to differentiate more asynchronously than the primary cultures, and eventually even to lose the capacity to fuse after repeated transfers (37). In our laboratory certain clones show apparent uncoupling of biochemical markers of differentiation; otherwise normal clones of L_6, for example, never accumulate acetylcholine receptor and contain very low levels of creatine phosphokinase. Apart from spontaneously arising variants, the cell lines permit the selection of mutants, such as those already described which are α-amanitin-resistant (38), ouabain-resistant (39), or exhibit temperature-sensitive differentiation (40). They also provide the possibility of genetic manipulations and the investigation of myogenic expression in cell hybrids (41) and heterokaryons

(42). Continuous muscle cell lines have also been established from embryoid bodies of mouse teratocarcinomas. These are anaploid and differentiate normally (43).

A further culture system which has recently become available is that of myoblasts transformed by a temperature-sensitive virus (44, 45). At permissive temperatures for the virus, the cells are transformed and divide continuously in culture, whereas at the nonpermissive temperature the myoblasts differentiate normally. Apart from the intrinsic interest of the switch between viral and cellular control mechanisms, infection by a temperature-sensitive virus results in the establishment of a cell line which can be induced to differentiate in a synchronous manner upon change of temperature.

The primary cultures and lines which have been studied until now have been derived from normal fetal tissue. Potentially, muscle cell mutants such as that described for the nematode (46) or pathological states such as the myogenic defects of inherited dystrophy (47) are susceptible to investigation in tissue culture and may provide information about the biochemical coordination of muscle development.

Another aspect of muscle cell differentiation which has recently become accessible to biochemical study in tissue culture is that of the further development of muscle which takes place on innervation. Nervous contact produces important modifications in the excitable membrane, such as the localization and stabilization of the receptor under the synapse (for review, see ref. 48), and exerts a controlling influence on intracellular protein synthesis, modulating, for example, the type of myosin light chains produced (49). In vitro synapse formation has now been reported in a number of different nerve-muscle cell systems (50–54).

In the following discussion of muscle protein synthesis and its control during differentiation, only the initial stage of muscle development is considered, namely, the molecular changes which take place during the normally observed morphological transition from myoblast to myotube in vitro in primary cultures or muscle cell lines derived from fetal skeletal muscle.

MUSCLE PROTEINS AND THEIR SYNTHESIS

Differentiated myotubes are characterized by the accumulation of proteins required for muscular contraction. In other examples of terminal differentiation which have been the subject of extensive biochemical study, such as erythropoiesis (55), the differentiated state is characterized by the synthesis and accumulation of tissue-specific proteins, in this case globin. A protein such as the M isozyme of creatine phosphokinase is apparently unique to muscle, but the majority of muscle proteins is found in many different cell types (for review, see ref. 56). Indeed the development of contractile proteins, apparently essential for cell movement, may well have been of fundamental importance in the evolution of eukaryotes. There are, for example, indications from sequence data that some contractile proteins such as the 5,5′-dithiobis-(2-nitrobenzoic acid) (DTNB) light chain of myosin and troponin C may have evolved from a common precursor (57). The contractile apparatus of muscle represents, therefore, a highly specialized structure which is designed to permit organized movement of an organ or animal and which makes use of protein components whose universal distribution in small quantities permits cellular motility.

In fact, recent results on a number of contractile proteins suggest that families of very similar molecules, rather than single species, may be involved in different nonmuscle and muscle tissues. An example of this, which has been known for many years, is the variation in the myosin light chains seen in different types of muscle (58). Peptide analysis indicates that the different light chains have different primary sequences and, thus, that there is a set of closely related genes for these proteins (59). The number of genes which may exist for other muscle or nonmuscle contractile proteins is not yet clear. However, from the available information it would appear that during myogenesis there are qualitative changes in a number of contractile proteins and that these may reflect the expression of different genes.

In a differentiated myotube a large proportion of cellular protein consists of the structural components of the contractile apparatus, which may represent as much as 40% of total cell protein (60). However, if protein synthesis is examined, the difference between a dividing myoblast and a myotube is very much less striking. After exposure of the cells to labeled methionine for 1 or 2 hr, the majority of the polypeptides seen by two-dimensional gel analysis (61) is the same before and after differentiation (62). Quantitatively, the synthesis of myosin heavy chains, for example, increases (63) and, qualitatively, new forms of proteins such as actin appear (62), but there is no major burst of contractile protein synthesis at the expense of general cellular proteins. At the level of protein synthesis, therefore, the modifications taking place at differentiation are relatively minor. The changes occurring for individual proteins, and particularly the question of whether new gene products are involved, are discussed in the following paragraphs. The major contractile proteins (Figure 2) are considered first, followed by a brief account of changes in muscle enzymes and membrane proteins.

Contractile Proteins

Myosin Myosin, whether in muscle or nonmuscle tissue, is characterized by the capacity to bind reversibly to actin filaments and by the possession of an actin-activated ATPase activity. In addition, under appropriate conditions it undergoes self-assembly into characteristic filamentous structures. These properties vary according to the source of the myosin (56). The ATPase activity of fast and slow muscles, for example, is different, corresponding to the physiological requirements of the tissue and that of embryological myosin, although resembling fast muscle appears to have its own characteristic activity (64, 65). Myosins from nonmuscle sources, such as hepatocytes (66) or glial cells

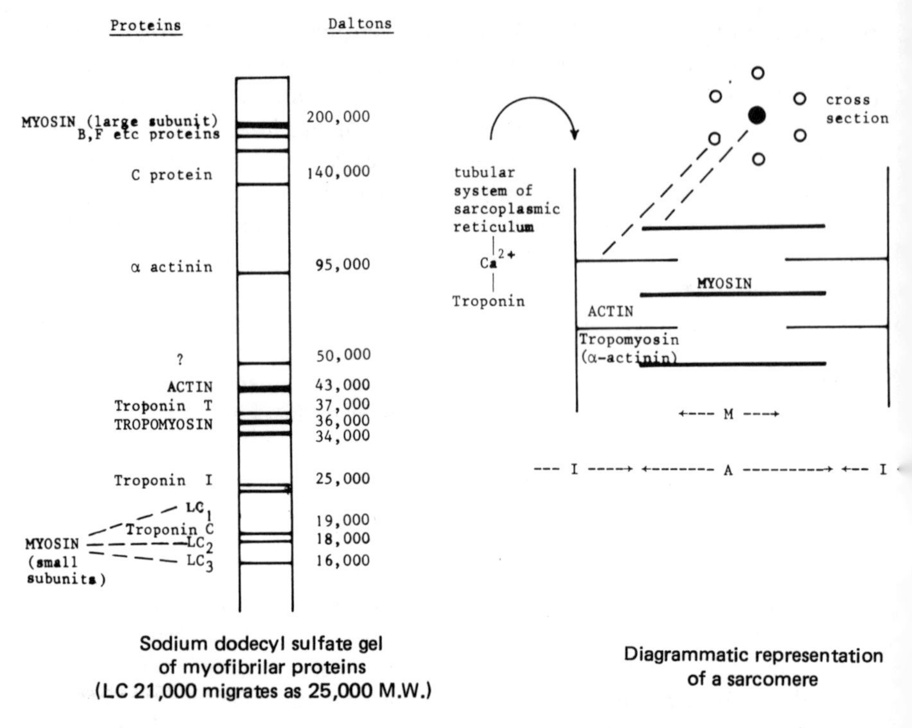

Figure 2. The contractile apparatus.

(67), also have different ATPase activities and differ, for example, in the degree to which they are activated by muscle actin. Smooth, skeletal, and platelet myosins also form different distinctive filamentous structures (68).

Different myosins can also be distinguished immunologically, although some confusion exists in the literature, partly because of probable minor contaminants such as the more highly antigenic C protein (69) in the antigen preparations used and partly because of differing sensitivities of different detection techniques (for discussion, see ref. 70). Immunofluorescent methods which are extensively used in cell culture work are more prone to interference from contaminants and are also more likely to detect minor cross-reactivity than the quantitative precipitation methods. Complement fixation tests between highly purified myosins from different cross-striated muscles indicate strict type specificity and less strict species specificity (70). Antibodies against platelet myosin (71) or mouse L cell myosin (72) do not cross-react with skeletal muscle myosin. By fluorescence methods, on the other hand, antibody against smooth muscle myosin does cross-react with myosin in nonmuscle cells (73). It is interesting, from the point of view of evolution, that, although their morphological and biochemical properties are very similar, myosins from myxomycetes, for example, give detectable immunological cross-reactions only with myosins from the same order and not with other closely related orders, nor with skeletal muscle myosin (74). During myogenesis, antibody against skeletal muscle myosin has been used to detect myosin in differentiated cells (75, 76). No apparent immunofluorescence was detected in undifferentiated myoblasts or fibroblasts in the population, although the threshold of detection in these experiments is an important factor, since differentiated muscle contains much greater quantities of myosin.

The enzymatic and immunological data on myosin from different sources suggest that a family of closely related molecules may exist. However, it is from studies on the polypeptide components of the molecule and, ultimately, from their sequence analysis that precise information on possible gene diversity can be obtained. Structurally, myosin has a molecular weight of about 450,000 and is composed of two large subunits of about 200,000 daltons and a number (2–4) of small subunits ranging in molecular weight from 15,000 to 30,000. Available information on qualitative aspects of the large and small subunits of myosin in adult and embryological muscle and in nonmuscle tissue, together with information on their appearance during muscle cell differentiation in vitro, is discussed in the next two sections.

Large Subunit of Myosin Because of the technical difficulties encountered with a polypeptide of the size of the large subunit of myosin, only limited sequence data are available. Comparison of a methylhistidine-containing sequence in adult rabbit skeletal myosin with the corresponding sequence in cardiac myosin shows that, out of 13 residues, 3 are different. In the equivalent peptide from fetal myosin, two conservative amino acid substitutions are found (77). There is, therefore, unambiguous evidence for at least three genes for the large subunit of myosin. Other evidence comes from enzymatic or chemical degradation studies. Different rates of tryptic digestion and different peptide products are seen on treatment of fast and slow cardiac and embryonic myosins from rabbit (64). Comparison of the staining pattern of crystal structures formed by light meromyosins (part of the molecule derived from the large subunits) from these different tissues would also suggest that the myosin heavy chains are different (64). Analysis of the chemical cleavage products of myosin heavy chains from different tissues of the adult chicken suggests that there are at least five different types, namely, skeletal, cardiac and smooth muscle, platelet, and brain myosins. Some cells, such as fibroblasts, apparently contain a mixture of platelet and brain-type heavy chains (78). It is, in fact, not entirely clear whether or not different populations of heavy chains are present in any given muscle type, although results, for example, on the number of cysteine-containing tryptic peptides in skeletal muscle might argue for homogeneity of the large subunits (79).

During myogenesis there is immunological evidence for the appearance of myosin which interacts with antibody against skeletal muscle myosin (76). More recent experiments with the use of antibody against light meromyosin (which does not contain the small subunits) have correlated the in situ binding of the antibody with the first appearance of interdigiting thick and thin filaments in chick muscle cultures (80). Since antibodies against nonmuscle myosins used under similar conditions react with the corresponding nonmuscle cell, it is suggested that the lack of cross-reactivity with myoblasts in these experiments is not due to quantitative differences, but rather to the fact that dividing myoblasts contain a different type of myosin. A similar conclusion is reached from experiments on cultures of quail muscle cells, in which antibody against adult skeletal muscle does not form an immunoprecipitate with newly synthesized myosin in dividing myoblasts, but precipitates myosin from fused cultures (81). It should be noted that the antibodies used are against the heavy chains of adult skeletal muscle myosin. Fetal myosin apparently contains slightly different heavy chains, and, although there is no evidence on the type of muscle heavy chain synthesized in culture, the data on light chains suggest that the myosin present in cultured myotubes probably reflects qualitatively that in embryonic muscle (see next section).

The quantitative aspects of the increase in myosin heavy chains during myogenesis in vitro have been extensively documented, since the large size of the polypeptide has resulted in its use as a convenient marker of differentiation. Results obtained by the direct analysis of cytoplasmic proteins on gels should be treated with caution, since other high molecular weight polypeptides present in detergent-treated extracts, such as the recently described LETS protein (MW 230,000) (82) or filamin (MW 250,000) (83), may have confused estimations of low levels of myosin synthesis. Precipitation at low ionic strength has been used as a method of partial purification and removes most other high molecular weight species. One of the difficulties experienced in comparing different quantitative results in the literature for many of the proteins which have been followed during differentiation arises because of the different terms of reference used in expressing the results. Rates of myosin synthesis, for example, have been expressed per Petri dish (8, 84, 85), per milligram of total protein (10), as percentage of total protein synthesis (10, 60, 86), or per nucleus or quantity of DNA (5, 63, 81, 85). In an attempt to obtain an absolute value independent of other parameters, such as variations in total protein content per myotube nucleus after fusion or an increase in protein and DNA due to continued division of contaminating cells, it is probably most satisfactory to express results per fiber nucleus (5, 81, 85).

Most estimations of myosin heavy chain synthesis have demonstrated a 7–15-fold increase at the time of fusion (8, 10, 60, 63, 84–86). It has been shown that no substantial change takes place in the size of the precursor amino acid pool during differentiation (5, 85) and, therefore, that these measurements probably represent a real change in the rate of synthesis. Since the value obtained is expressed as a function of the total culture, the proportion of cells which differentiate affects the result (see ref. 88). In primary cultures of quail muscle, 85–90% of nuclei enters myotubes (cf. other systems' 50–70%) and a 50-fold increase in the rate of myosin synthesis has been reported (5). It was calculated that this value corresponds to a rate of synthesis of the myosin heavy chain of 25,000–30,000 molecules per fiber nucleus per minute and that this figure remains fairly constant, the onset of the increased rate of myosin synthesis corresponding to the onset of fusion (5). If corresponding calculations are made on the number of molecules synthesized per fiber nucleus per minute from the data presented for differentiating chick cultures (85), a figure which is 5 times lower is obtained. This is in keeping with the difference in the increase in myosin synthesis reported by Emerson and Beckner (5), compared with that found in other systems. It is not evident why results expressed as synthetic rates per fiber nucleus should differ to such an extent, since parameters such as the extent of fusion do not directly affect the estimation. However, one explanation may be that maximal rates of myosin synthesis per fused nucleus did not reach a plateau in the chick cultures under discussion, whereas in the quail system a constant rate of myosin synthesis was immediately attained on fusion. This is not the case in many systems in which a lag in the increase in myosin synthesis has been found after fusion (63, 84, 85), although it should be noted that, by using immunofluorescent techniques, nondividing mononucleate cells (postmitotic myoblasts) have been reported to contain muscle myosin before their integration into fibers (76). It seems probable that the precise correlation between increased myosin synthesis and cell fusion depends upon the culture conditions (81); the uncoupling experiments described previously indicate that there is no obligatory relation.

In the quail culture system, synthesis of the large subunit of myosin accounts for about 10% of total protein synthesis in differentiated cultures (5), whereas in the other systems this figure varies from 2–7% (10, 60, 63, 88), depending upon the homogeneity of the culture (88). Estimations of the basal level of myosin synthesis in undifferentiated cultures depend upon the extent of contamination by differentiated cells, but figures of 0.5–2% of total protein synthesis have been found (7, 10, 60, 88) and as low as 0.2% in quail cultures (5). The relative rates of actin and myosin synthesis change at differentiation from a myosin to actin molar ratio of 0.9 in chick myoblasts to 2.0 in fused cultures or 3.1 in purified myotubes; this has been cited as further evidence for the involvement of different gene products (87, 88; see also below). The steady state level of the two proteins undergoes a similar change, suggesting that their synthesis and degradation are coordinated (87). The accumulation of myosin has been measured directly (8, 10, 12, 60, 86) or as an increase in muscle-type ATPase activity (87). At fusion in L_6 cell lines, for example, there is a 15–20-fold increase in accumulated protein, compared with a 7-fold increase in the rate of myosin synthesis, indicating a decreased rate of myosin turnover in myotubes (86) and, indeed, an approximately 2-fold increase in the half-life of myosin has been reported in primary chick cultures (89).

Small Subunits of Myosin It is well established that different light chains are present in different types of muscle and that there is a correlation between the function of the muscle, its ATPase activity, and its type of light chain

(90–92). This is illustrated by cross-innervation experiments in which a change in the physiological activity of the muscle is accompanied by a change in the small subunits of myosin (49). Myosin from fast skeletal muscles contains four light chains associated with the two large subunits of the molecule (93): two distinct chains (21,000 and 16,000 daltons), released on alkali treatment, which are of regulatory importance in the myosin ATPase, and two chains of similar size (18,000 daltons) (one of which is susceptible to secondary modification by phosphorylation (94)) which are released on DTNB treatment and which have been shown in other muscles to play a role in calcium binding (95). In fact, the stoichiometry is not strictly adhered to, and it is probable that within any one muscle there is a heterogeneous population of myosin molecules containing different combinations of small subunits (93, 96). As summarized in Table I, fast skeletal muscle myosin contains four types of light chain, slow skeletal muscle three, and cardiac myosin two (96). Peptide and, in some cases, complete sequence analyses have been carried out on the light chains. Considerable homologies have been found, for example among the alkali class, suggesting a common evolutionary origin (95). In particular, the sequences of the alkali light chains of fast skeletal muscle have a common COOH terminal sequence of 141 residues out of the 149 residues of the smaller of the two subunits (97). Very few changes are observed among the different chains between species, indicating a high degree of conservation (96).

Embryonic skeletal muscle contains two major light chains which have the same migration properties on one-dimensional gels as the two larger subunits of fast muscle myosin (98, 99). Trace amounts of a third chain are also seen (99). Other properties of fetal myosin more resemble those of slow muscle myosin (65, 99). As already discussed under "Large Subunit of Myosin," it has been shown that the fetal heavy chain contains a distinct sequence (77). The question of whether fetal light chains are identical with adult subunits or are the products of different genes awaits peptide and sequence analysis. During postnatal development, different muscle types mature at different times and, depending upon the age of the young animal, varying combinations of light chains are seen in the muscle (100). Indeed, in embryonic muscle, the similarity between fetal light chains and those of fast muscle may reflect the predominance of this muscle type. The soleus muscle dissected from the embryo at birth contains the two light chains characteristic of adult slow muscle, in addition to the fast light chains (100). Thus, it seems probable that in the embryo as well different types of muscle fiber may be distinguishable (99, 100). In tissue culture, primary cultures (62, 80, 88, 89, 101), or cell lines (101), the light chain pattern seen on differentiation reflects that seen in the embryo, by one- (80, 88, 99, 101) or two-dimensional (62, 101) gel analysis. The absence of innervation in vitro does not apparently affect the early expression of light chains. Prefusion primary cultures synthesize two types of light chains which correspond to those described for certain nonmuscle myosins, including that of fibroblasts (80, 88), and which are distinct from muscle light chains. The myosin synthesized in these cultures also has the characteristic ATPase activity of fibroblast myosin. It is assumed that myoblasts also synthesize these chains, although fibroblast contamination (10–30%) (88) of the myoblast cultures may obscure a clear demonstration of this. In similar experiments on rat cell lines, the detection techniques were not sufficiently sensitive to answer this question (101). In fused cultures and in embryonic muscle, trace amounts of a third light chain (MW circa 16,000) have been found, but in experiments on purified cultures (80, 88) and in fused rat cell lines (101) this chain was absent; it is apparently contributed by contaminating fibroblasts. If the myoblasts synthesize myosin light chains of the fibroblast type, it would appear, therefore, that this synthesis is suppressed when the embryonic muscle phenotype begins to be expressed. A precise quantitative estimate of the synthesis of the myosin light chains during differentiation has not been made. They represent about 10% of the myosin molecule and, therefore, on the basis of heavy chain synthesis, would represent about 0.5% of total protein synthesis in differentiated cultures. This corresponds to the figure quoted in a preliminary report on calf primary cultures (102).

Actin Actin and myosin constitute the major structural components of the sliding filaments which provide the basis for muscular contraction. Actin is a simpler molecule than myosin and consists of a single polypeptide of about 43,000 daltons. It is characterized functionally by its interactions with myosin, resulting in the activation of the Mg^{2+}-dependent myosin ATPase. It also interacts with tropomyosin and troponin. Structurally, actin polymerizes from a globular monomer into a double helical filamentous polymer and also forms characteristic "arrowhead" complexes, of regular repeat distance, with heavy meromyosin. Actin, like myosin, contains a number of rare methylated amino acids, in particular *N*-methylhistidine. A further property, recently described, which has proved useful in its characterization and purification, is the interaction of actin with DNAse I (103). In the last few years a protein with these properties has been found as a relatively major constituent of many different cells and has been described as cytoplasmic actin (for review, see ref. 56). It is present in both a soluble and a filamentous form (104). Together with the microfilaments of tubulin (200 Å) and the intermediate sized filaments (100 Å), actin filaments (60 Å) are a major structural component of the cell (105). They are apparently important in cellular anatomy, forming a cytoskeleton as visualized by immunofluorescence (106). Filamentous actin and its potential depolymerization to a soluble form (104, 107) participate in a number of phenomena which affect cell structure and function, such as cell motility (108), cytoplasmic streaming (109), cell attachment or anchorage (110, 111), and cell division (112).

All actins are extremely similar in function and in amino acid composition, minor differences being attributable to possible impurities (56); this, together with the apparent identity of tryptic peptides from muscle and nonmuscle actins from different sources, led to the idea that actin was a single, highly conserved, universal protein (113). More recently, differences in tryptic peptides have been

Table 1. Numbers of probable genes for major contractile proteins[a]

Tissue	Myosin			Actin[b]	Tropomyosin[c]	Troponin[d]
	Whole molecule[e]	Heavy chains[f]	Light chains (LC)[g]			
Fast skeletal muscle (FSM)	FSM	FSM (seq)	FLC_1 21,000 (seq) FLC_2 (2) 18,000 (seq) FLC_3 (16,000 (seq)	α (seq)	α (seq) β (seq) (also α', β' ?)	Tn T Tn I^F (Tn I^S) Tn C
Slow skeletal muscle (SSM)	SSM	SSM	SLC_1 (2) 22,000 (seq) SLC_2 19,000	α	α β	Tn T (Tn I^F) Tn I^S Tn C
Embryonic skeletal muscle (SM)	(EM)SSM	EM (seq)	As FLC and/or SLC	α	α β	Tn T Tn I^F (Tn I^S) Tn C
Smooth muscle (SM)	SM	SM	$SMLC_1$ 20,000 $SMLC_2$ 17,000	?	α^{SM} β^{SM} ?	

Cardiac muscle (CM)	CM	CM (seq)	CLC_1 22,000 (seq) CLC_2 19,000 (seq)	α^C (seq)	α	Tn T Tn I^C Tn C
Nonmuscle tissues (NM)	NM	NM (2)	$NMLC_1$ 20,000 $NMLC_2$ 16,000	β,γ (seq)	30,000	
Myoblast	NM	NM ?	NM	β,γ	?	
Myotube	EM/FSM ?	EM ?	EM	$\underline{\alpha}$ β,γ	α $\underline{\beta}$	?
Possible number of structural genes		7	13	4	6/7	5

[a]The summary of results presented here is based on references and data referred to in the text and principally concerns mammalian and avian proteins. Indications of distinct molecules are derived from analyses by immunology (i), enzymology (e), physical methods (p), gel electrophoresis (ge), classical peptide mapping (pm), and enzymatic (ed) and chemical (cd) degradation studies. Where sequence data provide conclusive evidence for different structural genes, this is indicated (seq). Molecular weights are given in daltons. Where different components are present, the major one is underlined; minor species are shown in parentheses.

[b]MW 43,000; g, e, pm.

[c]MW 34,000 (α) and 36,000 (β); ge, pm, i.

[d]MW 37,000 (Tn T), 24,000 (Tn I), and 19,000 (Tn C); ge, e, cd.

[e]MW 450,000; i, e, p.

[f]MW 200,000; i, ed, cd, p.

[g]MW 16,000–22,000; i, e, ge, pm.

demonstrated (114, 115), and sequence data on some of the CnBr peptides have shown unequivocally that nonmuscle actin from human blood platelets has a threonine residue at position 129 where cardiac and skeletal muscle actins have valine (116). A preliminary communication further describes a one amino acid substitution between bovine brain and cardiac and skeletal muscle actins (117). Thus, it would appear that there are at least three genes for actin from different muscle and nonmuscle origins. In the presence of urea and sodium dodecyl sulfate, different forms of actin can be demonstrated by gel electrophoresis of chick brain and skeletal muscle actins (118) or by isoelectric focusing on acrylamide gels (61) of actins from rat or calf muscle, kidney, and a number of different types of tissue culture cells. A form of actin found only in muscle and designated α-actin is seen, together with a nonmuscle form which, by isoelectric focusing, is separable into two components, β- and γ-actins. Peptide analyses show several differences between α- and β + γ-actins (36, 119) and one or two minor differences between β- and γ-actins (36). Pulse-labeling experiments on cells in culture have revealed two further short-lived components described as δ- and ε-actins which may represent unstable modified forms of β- and γ-actins (36). It is argued that the difference between α and β + γ does not result from a secondary modification of the actins since α-actin is only synthesized in cell-free translation systems by messenger RNA preparations from muscle tissue or fused muscle cultures (36, 62, 119). In vitro, both β- and γ-actins are synthesized, and, furthermore, the difference in isoelectric point between β- and γ-actins is too small (0.03 unit) to be accounted for by most secondary modifications (36, 62). Circumstantial evidence would suggest, therefore, that β- and γ-actins may be different molecules, although a clear demonstration of this awaits sequencing data. The function of the two nonmuscle forms of actin remains unknown. They appear to have a similar intracellular distribution and both to be present in different cell types (36).

During muscle development in vivo, increasing quantities of muscle actin are present, and in adult muscle the β and γ forms are not readily detectable (36, 62). During myogenesis in vitro, dividing myoblasts contain β- and γ-actins (36, 62), and actin filaments have been demonstrated microscopically in these cells (36, 120). Synthesis of actin represents a significant part of total protein synthesis, estimated to be as much as 5–7% by two-dimensional gel analysis in primary cultures before differentiation. At fusion, in this system, the proportion of actin synthesis even decreases slightly (5%), but α-actin now represents 30–50% of that synthesized (102). In chick primary cultures a 10-fold increase in actin synthesis has been described at fusion (121), although the increase relative to total protein synthesis has not been calculated. It is also possible that the glycerination procedure used on the cells prior to characterization of actin (121) may have resulted in the loss of soluble nonfilamentous cytoplasmic actin in prefusion cells. In general (29, 62, 88), the increase in actin synthesis and in the quantity of this protein present in differentiated myotubes is less striking than in the case of myosin, since actin is a more major component of nonmuscle cells. Qualitatively, however, there is a clear difference in the type of actin synthesized in myotubes, and differentiation is characterized by the appearance of α-actin, the exclusively muscle form of the protein (36, 62, 117). Examination of purified myotubes indicates that β- and γ-actins continue to be synthesized after differentiation (36). Whether the α-actin of myotubes in culture or of embryonic tissue is in fact identical with adult skeletal muscle actin remains to be established by sequence analysis.

Tropomyosin Tropomyosin is a major regulatory protein of muscle, acting together with troponin on the calcium ion-dependent interaction of myosin and actin (122). It is closely associated with actin in the thin filaments. In vitro, tropomyosin forms a repeating paracrystalline structure of characteristic periodicity (123). The protein is a dimer and, in muscle, two types of subunits can be distinguished electrophoretically: α-tropomyosin (MW 34,000) and β-tropomyosin (MW 36,000) (124). Different muscle types have different proportions of the two forms, fast muscle, for example, having an α to β ratio of 3.8:1, whereas slow muscle has a ratio of 1:1 (125). The two tropomyosins have very similar amino acid compositions (124) and at least half of their tyrosine peptides in common (126). The sequence of the COOH-terminal half of rabbit skeletal muscle tropomyosin suggests that it may in fact consist of at least four very similar polypeptides (127). Analysis of chymotryptic peptides from different types of muscle indicates that calf cardiac and skeletal tropomyosins are apparently identical, whereas smooth muscle is slightly different (123).

A nonmuscle form of tropomyosin has been isolated from many cell types and shown to have very similar functional and structural properties. It is apparently a single polypeptide of about 30,000 molecular weight (123, 128, 129), although values of 35,000 have been reported for tropomyosins from some cultured cells (130). Peptide analysis of tropomyosins from calf tissues of different embryological origin, such as brain (ectodermal), pancreas (endodermal), and platelet (mesodermal), which have identical molecular weights (30,000) and functional properties, shows that their peptide maps after chymotryptic digestion are also identical (123). Comparison with the peptide map of tropomyosin from skeletal or cardiac muscle shows that the molecules are very similar; 21 out of 31 peptides are common to both types. Similarly, muscle and nonmuscle tropomyosins show very little species variation (123). It has been suggested that the tropomyosins have arisen by gene duplication from a precursor sequence; some support for this comes from sequence analysis of α-tropomyosin, which has a repeating sequence of 42 amino acids (131). Variations in the sequence of different tropomyosins are all located in the NH_2-terminal (127). In nonmuscle cells, tropomyosin is also associated with actin and by immunofluorescent techniques can be shown to be located in a periodic manner along the actin filaments (130).

During muscle development a change takes place in the realtive proportions of α- and β-tropomyosin. In fetal rabbit, 70% of the tropomyosin in skeletal muscle is of the β type, and this is reduced to 20% after birth (132). This observation is in keeping with the analogy between the activity of embryonic muscle and slow muscle (64, 98). Skeletal muscle from chick embryos contains a tropomyosin-troponin complex of activity very similar to that of the adult, but with β-tropomyosin again as the predominant tropomyosin type (133). Similarly, a study of the proteins present in embryonic chick muscle during development has shown that β-tropomyosin is the major component, whereas α-tropomyosin predominates in the adult (118). Despite the fact that it is one of the major muscle proteins, very few studies have been reported on tropomyosin during myogenesis in vitro. Fluorescent antibodies against skeletal muscle tropomyosin did not cross-react with dividing myoblasts, nor with nonmuscle cell types (76), although this latter finding is contrary to the results reported in the preceding paragraph (130). As discussed for myosin, the greater quantities of contractile proteins in differentiated myotubes make them relatively more easily detectable by immunofluorescence. In fact, given the similarities in sequence between different tropomyosins, immunological cross-reactivity is not surprising and probably varies with the antibody preparation. The synthesis of tropomyosin during myogenesis has been examined in primary calf cultures. Two-dimensional gel analysis of pulse-labeled protein permits the identification of α- and β-tropomyosin and shows an increase in their synthesis, particularly of the β form, in myotubes (62). At fusion, α- and β-tropomyosins represented about 0.8% of total protein synthesis. There was an apparent diminution (0.1–0.025%) in the synthesis of nonmuscle tropomyosins (102); synthesis of muscle tropomyosins prior to fusion was presumably due to contamination by differentiated cells. In rat cell lines, synthesis of α- and β-tropomyosins was clearly detectable in fused cultures, but not in undifferentiated myoblasts (101). The nonmuscle form of tropomyosin was not identified in these reports, and the question of its level of synthesis in myoblasts and continued synthesis in myotubes remains open.

Troponin Nervous stimulation of the muscle cell membrane results in an influx of Ca^{2+} ions into the sarcoplasmic reticulum. In vertebrate striated muscle they bind to troponin, which by the intermediary of tropomyosin interacts with the actin filaments, thus activating muscle contraction (122). In smooth muscle (134) and in some invertebrate muscles (95), troponin is absent, and contraction is regulated by the direct binding of Ca^{2+} ions to one of the regulatory light chains of myosin (95). In this respect, and also in that troponin has not been reported in nonmuscle cells, the protein apparently differs from myosin, actin, and tropomyosin in being less universally associated with contraction. There are three troponin polypeptides present in equimolar amounts, Tn T (37,000 daltons), Tn I (24,000 daltons), and Tn C (19,000 daltons), each with a defined role in the Ca^{2+} regulatory mechanism (135). Three different forms have been described for troponin I, one which predominates in slow, one in fast, and one in cardiac muscle (136, 137). In cross-innervation experiments the relative proportions of the two polypeptides of troponin I change according to the type of activity induced in the muscle. Although it forms a complex with troponin, tropomyosin does not undergo an equivalent change in the proportion of α and β chains on cross-innervation (136). In embryonic muscle, on the contrary, although many of its properties resemble slow muscle and the β form of tropomyosin is predominant, the troponin components are characteristic of adult fast muscle (138). In this and the preceding example, troponin would appear to be under separate control from tropomyosin. Its behavior is more analogous to that of the light chains of myosin. No information is at present available on troponin synthesis during myogenesis in vitro. Since it is a major component of striated muscle and is apparently specific to this class of muscle, it is a potentially useful marker of the regulatory development of the contractile apparatus.

Other Protein Components of Contractile Apparatus Myosin, actin, tropomyosin, and troponin are the four major components of vertebrate skeletal muscle, being present in molar ratios of 1:7:1:1 (139). In addition to these proteins, a number of minor protein components of myofibrils have been described (140). Of these, α-actinin would appear to have the universal distribution characteristic of myosin, actin, and tropomyosin. It is isolated as a protein complex, sedimenting at 6 S, and is composed of polypeptides of 95,000 molecular weight. Immunofluorescent studies of muscle fibers have shown that α-actinin is associated with actin and is localized particularly at the Z-line (141). In nonmuscle cells it is closely associated with the actin filaments (141, 142). In skeletal muscle a number of other high molecular weight proteins have been identified, such as the C protein (140,000 daltons) and B and F proteins (150,000–180,000 daltons), all of which copurify with myosin (69, 140). Further high molecular weight proteins have been identified as M-line components, e.g., M-line protein (MW circa 165,000) (140, 143). It is probable that such minor myofibrillar proteins may be important in the structural organization of the sarcomere, and their role in the development of the contractile apparatus during myogenesis is of potential interest. One such protein component of unknown function, which is apparently myofibrillar in origin, has been noted during differentiation in vitro. In primary cultures from calf muscle, two-dimensional gel analysis of newly synthesized proteins shows a pronounced spot at about 50,000 daltons which appears at differentiation (62). It represents about 0.45% of total protein synthesis at this time (102).

Summary: Contractile Proteins From the preceding discussion of the contractile proteins, it can be seen that families of similar genes exist for these proteins (see Table I) and that muscle-specific forms equivalent to those found in the corresponding embryological tissue probably begin to be synthesized at differentiation (Table 2). Current evidence would suggest that the synthesis of nonmuscle forms of actin (36, 102) and possibly of tropomyosin (102) persists in myotubes, whereas that of the light chains characteristic of nonmuscle myosin is not detectable after differentiation in purified cultures (89). Preliminary data on the quantitative determination of nonmuscle protein synthesis would suggest that synthesis of these proteins is reduced after fusion (102). The coordination of muscle protein synthesis and the assembly of the proteins into a contractile

Table 2. Qualitative and quantitative changes in muscle proteins during myogenesis in vitro[a]

Protein	Qualitative changes at differentiation	Quantitative changes at differentiation: Total protein synthesis (%)	Quantitative changes at differentiation: Fold increase
Contractile proteins			
Myosin heavy chains		2–7 10 (quail system)	× 10 synthesis/μg of DNA (quail × 50–100) × 20 accumulated protein
Myosin light chains	Fibroblast type → embryonic muscle 20,000 → 21,000 16,000 → 18,000 (16,000)	~0.5	
Actin	β, γ → α, β, γ	~5 (α-actin 2.5)	Little change
Tropomyosin	Nonmuscle (+α (+β)) → muscle β + α (+ nonmuscle)	~0.8	
Membrane proteins			
Acetylcholine receptor	Present in significant amounts of myoblasts before differentiation?	0.005–0.05	× 15 in toxin-binding/Petri dish (× 400 . . . in cell line L_6)

Acetylcholinesterase	In myotubes present in multiple forms as in vivo		X 10 in activity/mg of protein
Ca^{2+} ATPase (sarcoplasmic reticulum)	Present in myoblasts before differentiation?	1	X 10 in activity/mg of protein
LETS protein	Present on undifferentiated cells		Several-fold increase on surface
Enzymes	General conversion to muscle forms		Fold increase in activity/mg of total protein
Creatine phosphokinase	BB → MB + MM		X 6–10
Aldolase	AC forms → mainly A forms		X 4–10
Phosphorylase	B → M		X 5–6
Glycogen synthetase			X 3–4
Phosphofructokinase	Conversion to mainly PFK_1		X 2–4
Phosphorylase kinase			X 4
Myokinase			X 4–6

[a]Summarizes data discussed in the text. In general, average figures for primary cultures or cell lines are used.

unit are poorly understood. It is clear that they are not coded for by polycistronic messages but are synthesized on separate polysomes (145). These appear by electron microscopy to be intimately associated with the filaments of the contractile apparatus in differentiating myotubes (12), and nascent chains may be immediately integrated (for a review of myofibrillar assembly, see ref. 145).

Before discussing the question of messenger RNA synthesis and utilization which underlies the problem of coordinate expression of the muscle phenotype, two other categories of muscle protein are mentioned briefly, namely, muscle enzymes and the proteins of the excitable membrane. Most work on messenger transcription and translation during myogenesis has been concerned with the contractile components, and it is for this reason that they have been discussed at greater length.

Enzymes of Muscle Metabolism

Muscle tissue contains a pattern of metabolic enzymes adapted to its particular physiological requirements. They are characterized quantitatively by relatively high activities and qualitatively by specific isozymic forms, usually composed of subunits found in different proportions in other tissues.

During muscle cell differentiation in vitro, increases have been reported in the specific activities of the following enzymes known to increase during muscle development in vivo: phosphorylase (X 5–6) (146–148), glycogen synthetase (X 3–4) (147), phosphorylase kinase (X 4) (147), myokinase (X 4–8) (86, 146, 149), phosphofructokinase (X 2) (148), aldolase (X 4) (9, 150), and creatine phosphokinase (X 6–10) (7, 9, 12, 86, 149–152). The results listed have in most cases been expressed per milligram of cell protein. On this basis there is a 3–10-fold increase in the muscle enzymes examined both in primary cultures from various sources and in rat cell lines. The latter tend to have rather lower initial levels and a slower rate of increase in activity, reflecting their greater initial homogeneity of undifferentiated cell type and lower synchrony of fusion (146). In most cases the increase in enzyme activity occurs at the same time as other parameters of differentiation; reported lags in the increase in activity after fusion (151) may result from the method of expressing the results (150) or from differences in the culture conditions. One report on the L_6 line indicates that enzymes such as myokinase or creatine phosphokinase may increase at high cell densities irrespective of myotube formation (149). As discussed previously, biochemical differentiation is not necessarily associated with cell fusion. However, it should also be borne in mind that enzyme activities depend upon the metabolic state of the cell and that some tissue culture conditions may result in altered metabolic activities. It is perhaps surprising that muscle cells in culture do mimic muscle tissue with respect to the enzymes mentioned. Under standard conditions of growth and differentiation, the increased enzyme activity is apparently located in the myotubes, as clearly shown, for example, by histological staining in the case of phosphorylase (146, 153). Experiments with inhibitors of protein synthesis suggest that the increased activity is probably due to an increase in the de novo synthesis of the enzyme (146). The extent to which increased enzyme levels at differentiation may be regarded as markers of muscle varies: a high level of creatine phosphokinase is characteristic of brain and muscle, and phosphorylase of liver, brain, and muscle, whereas myokinase is high in many types of tissue.

Qualitative changes which may be regarded as markers of differentiation have been investigated for a number of enzymes known to have muscle isozymes. Phosphorylase has five isozymic forms which can be distinguished electrophoretically, and it has been shown that, while the brain form is present in myoblasts, the muscle form is predominant after fusion (148). Similarly, phosphofructokinase, which has four forms, is principally present in myotubes as the muscle enzyme PFK_1 (148). Aldolase is a tetrameric protein composed of three subunits, A, B, and C, of which B is not found in muscle and A is the predominant subunit found in adult muscle. During myoblast differentiation the AC combinations of subunits which exist initially are replaced by isozymic forms principally composed of A subunits (9, 150, 151). Creatine phosphokinase is the enzyme which has been most extensively used as a marker of muscle cell differentiation. This is due to the fact that it has a form which is composed of polypeptides specific to muscle tissue. The enzyme is a dimer existing as the BB nonmuscle form in most cells and as the MM isozyme in muscle. The physiological function of the M isozyme is poorly understood. About 5–10% of creatine phosphokinase is associated with the M-line of the contractile apparatus and may thus fulfill a special role, but most of the enzyme remains free in the cytoplasm. There is a suggestion that a proportion of the A form of aldolase may also be associated specifically with the contractile apparatus (154). During myogenesis in vitro, the M isozyme appears in myotubes (153), and MM creatine phosphokinase rapidly becomes the predominant form of the enzyme (148, 150, 153). The actual timing of the increase in M rather than B subunits may depend upon culture conditions. Some authors have reported an immediate increase in the M form, corresponding to the increase in enzyme activity (101). In other cases, the MB form seems to increase initially (150), whereas an initial increase in exclusively B subunits, only followed by the appearance of the M isozyme, has been reported (155). In muscle cells derived from teratoma embryoid bodies, cell fusion is accompanied by an increase in the B form of the enzyme. The myotubes subsequently degenerate, and the M form of creatine phosphokinase never appears (156; cf. 43). With respect to this enzyme, it should be noted that the assay is liable to interference from myokinase, and, furthermore, that the latter enzyme migrates in the same position as the MM form of creatine phosphokinase during the electrophoresis procedure employed. Particularly with certain clones of the L_6 cell lines which have high levels of myokinase, interference by this enzyme presents a problem (157).

In conclusion, on differentiation of muscle cells in vitro, a number of enzymes undergo quantitative and qualitative changes to isozymic activities characteristic of differentiated muscle. The onset of increased activity is correlated with other biochemical changes and is probably due to de novo protein synthesis. The subunit composition of the enzyme also changes at differentiation to form the combination of polypeptides found in muscle tissue. In the case of creatine phosphokinase the muscle isozyme is composed of polypeptides which are unique to muscle tissue, and for this reason it has been used frequently as a marker of differentiation.

Membrane Proteins

The development of an excitable membrane is an important aspect of muscle cell differentiation. Two major proteins are associated with the reception of neuromuscular impulses, namely, the acetylcholine receptor and acetylcholinesterase. Both of these proteins have been studied during myogenesis in vitro, and the metabolism of the receptor, in particular, has been examined in some detail. For the present, most information about the synthesis of defined membrane components during myogenesis is confined to these two proteins. However, before discussing them in more detail, it is worth mentioning briefly the ATPase of the sarcoplasmic reticulum and the recently described LETS protein, since both are of potential interest in the study of myogenesis.

The ATPase active in Ca^{2+} ion transport has a molecular weight of 100,000 on denaturing gels and is the major protein component of the sarcoplasmic reticulum (158, 159). The formation of "T" junctions characteristic of this membrane structure can be seen in tissue culture (12, 145), and the Ca^{2+} binding ATPase is a potential marker for the development of membrane-mediated control of contraction. There is one report of the appearance of this enzyme during myogenesis in vitro (160), indicating that its level increases 10-fold due to de novo synthesis at the time of myotube formation.

Another defined membrane component that has been examined during myogenesis is the LETS protein (230,000 daltons), which is found on the exterior surface of most cell membranes and which is absent in transformed cells. Iodination of the cell surface proteins of muscle cells indicates that the LETS protein increases several fold at fusion (82). This may be related to the entry of the cells into a G_0 phase of the cell cycle (82) or to the process of cell fusion itself (see ref. 161 for morphological and electrical description), about which there is little biochemical information, or, alternatively, this may indicate that this protein is important in the development of the excitable membrane. Because LETS is apparently concerned in cell/cell contact, it may play a role in fusion, and, subsequently, in nerve/myotube interaction.

Acetylcholine Receptor The acetylcholine receptors which have been most extensively studied are those purified from the electric organs of certain fish, since these contain very high concentrations of the protein (e.g., 15,000 mol/μm^3). Identification and purification of the receptor have been facilitated by the development of assay systems which use affinity labels, in particular, labeled toxins, which bind at the active site (for reviews, see refs. 162 and 163). The acetylcholine receptor from the electric organ of *Torpedo* is a macromolecule of 300,000–400,000 daltons composed of five or six subunits. The receptor from *Electrophorus* is apparently smaller, circa 230,000 daltons. The nature of the polypeptide components is still controversial, but it would appear that the component active in binding receptor has a molecular weight of about 40,000. Other subunits of different size which have been reported may be products of aggregation or degradation (163). In muscle cultures from fetal calf tissue the use of toxin affinity columns has permitted the purification of labeled receptor which sediments at 9 S and which, when analyzed on two-dimensional gels, is composed of subunits of 40,000 molecular weight, focusing in three spots, possibly as a result of partial modification or glycosylation of the molecules (164, 102).

Acetylcholine receptor appears on the myotube cell membrane at fusion (164–169). Like other markers of biochemical differentiation, its appearance is not necessarily linked to this event, and equivalent quantities of receptor can be detected on the membranes of mononucleated myoblasts grown in medium containing a low concentration of Ca^{2+} ions to inhibit fusion (7). Experiments with inhibitors of protein synthesis had suggested that de novo synthesis of the receptor occurred at fusion (166, 170), and this has since been confirmed by incorporation studies with heavy isotopes (171) and by radioactive labeling of purified receptor (164). From the latter experiments it can be calculated that synthesis of receptor in differentiating myotubes of calf primary cultures represents 0.005–0.05% of total protein synthesis (cf. myosin circa 5%). In this primary culture system the rate of receptor synthesis increases at least 15-fold from the onset of fusion (48). In these experiments, contrary to other reports (169), there was no evidence of a cellular pool of toxin-binding material; newly synthesized material appeared rapidly on the cell membrane (48). Accumulated receptor, measured by toxin binding on the cell surface, increased at least 100-fold in this system (102). Similarly, in muscle cell lines, a 400-fold increase in toxin-binding capacity per milligram of protein has been reported (166). It is not clear whether basal levels of toxin-binding activity observed in myoblasts (166) or in nonmuscle cells are significant. It should be noted that, although the acetylcholine receptor is often regarded as unique to muscle and nerve, a protein with similar properties has been found in the membrane of erythrocytes, and the possibility exists that closely similar molecules may be implicated in membrane-mediated effects on cell rigidity (172).

A further aspect of the appearance of the acetylcholine receptor at fusion is its type of distribution on the surface of the myotube. In innervated muscle a stable form of the receptor is localized under the synapse (162). In culture the newly synthesized receptor is distributed uniformly over the surface of the myotube at a density of about 10^3 molecules of receptor/μm^2, which corresponds to about 2% of membrane protein (167). Patches of receptor subsequently form (168, 173), and these eventually disappear, so that "old" myotubes contain very little receptor (173). The half-life of the molecule has been measured either directly by pulse-chase experiments on labeled receptor (174) or indirectly by measuring the loss of labeled toxin from cultured myotubes (170). In both cases a half-life of about 17 hr has been found, which corresponds approximately to that reported for extrasynaptic receptor in vivo (175). Receptor localized under the synapse has a half-life of several days (175). It remains to be proven that this is the same molecule as that found prior to innervation. However, it seems possible that acetylcholine receptor of short half-life (17 hr) which is initially distributed on the myotube surface becomes concentrated in patches at about the time when innervation normally occurs and that nervous contact results in stabilization of the molecule. A corresponding reduction in its rate of synthesis results in the disappearance of extrasynaptic receptor and, in noninnervated myotubes in culture, the loss of receptor from the patches. Denervated muscle appears to undergo the reversal of this process with the reappearance of the less stable form of extrasynaptic receptor (176).

Changes in the localization and stability of acetylcholine receptor under the influence of the nerve may eventually be susceptible to study in tissue culture, as mentioned previously. It is clear from the experiments discussed that, although the synthesis of receptor represents a very small proportion of total protein synthesis, it is nonetheless possible with the technique of affinity chromatography to study its metabolism during myogenesis.

Acetylcholinesterase Acetylcholinesterase, like the acetylcholine receptor, has been purified from the electric organs of *Torpedo* and *Electrophorus*, in which it can be shown to be a complex molecule composed of multiples (approximately 9 S, 14 S, and 18 S) of a basic tetrameric structure, containing apparently identical subunits of 80,000 daltons (177, 178). Similar multiple forms are found in muscle tissue. In adult muscle the high molecular weight form predominates and is mostly localized under the synapse. In embryonic muscle, however, all three forms are present and are localized within the cells, along the sarcoplasmic reticulum and around the cell periphery (177). In fact, the enzyme is probably not, strictly speaking, a membrane protein, although its function is closely related to that of the excitable membrane. By analogy with the molecule of the electric organ, it seems probable that different forms of the enzyme represent different subunit combinations. However, it is not entirely clear whether the sarcoplasmic activity, for example, which appears to have different isozymic properties, is derived from the same basic molecular structure or not.

During myogenesis in vitro a large increase in acetylcholinesterase activity (about 10-fold when expressed per milligram of protein) is detected at fusion (179–181). In unfused cultures enzyme activity is barely measurable, but in situ staining indicates activity in some mononucleated cells (180). These may be postmitotic myoblasts rather than dividing, clearly undifferentiated cells, since these observations have been made in primary chick cultures, in which a relatively high proportion of myoblasts are ready to fuse without further cell division (18, 19). Fibroblasts were found to be negative to the staining test (180). The enzyme in muscle cells is present in the multiple forms seen in fetal muscle and has similar drug sensitivity (179). Studies with inhibitors of protein synthesis suggest that it is synthesized de novo and has a rapid rate of turnover (181). In situ staining indicates a distribution within and around the perimeter of the myotubes, whereas on older myotubes concentrated patches of the enzyme are seen, analogous to those found with the acetylcholine receptor (182). In addition, there is a continuous process of release of the enzyme into the culture medium (179).

The appearance of acetylcholinesterase during myogenesis, therefore, resembles in many respects that of the acetylcholine receptor. Circumstantial evidence suggests that the esterase as well is synthesized de novo at fusion, and it appears to undergo similar changes in localization during myotube development. It has, in fact, been suggested that the genes coding for the two proteins are under similar regulatory controls, since in certain neuroblastoma cell mutants they seem to display coordinate variation (183). A further mechanistic point of some interest is the suggestion that the de novo synthesis of acetylcholinesterase may be susceptible to regulation by acetylcholine (184).

Summary

It is clear from the preceding sections that myogenesis is characterized by qualitative as well as quantitative changes in the types of protein present. These changes are summarized in Table 2. In the case of the muscle enzymes, characteristic combinations of subunits, which are not necessarily themselves unique to muscle, appear at differentiation, creatine phosphokinase being an example of an enzyme which does have muscle-specific polypeptide chains. The proteins associated with the development of an excitable membrane are present in quantity in fused myotubes. It is not yet clear whether they exist as minor components in other types of membrane, nor is it known how many gene products a protein such as the acetylcholine receptor represents. It is possible that families of receptor molecules may exist, with slightly different isozymes present in the excitable membranes of muscle or nerve, for example. The contractile proteins are a clear example of the existence of families of very similar molecules in which different members are expressed in different types of muscle and, for many of the proteins, in nonmuscle cells as well. Evidence on the continuation of the expression of nonmuscle contractile proteins after differentiation is confused. Certainly, in muscle tissue, very little if any nonmuscle actin is detectable. It is possible that there is a gradual phasing out of the synthesis of nonmuscle proteins of this type.

The question of coordination in the synthesis and assembly of the different protein components (enzymes, membrane, and contractile proteins) during myogenesis remains to be explored. Uncoupling agents, such as those mentioned in the introduction, result in complete biochemical expression in the absence of cell fusion. It is possible that reagents may be found that will produce uncoupling of the different biochemical phenomena. Some information may also come from the muscle cell mutants mentioned in the introduction. However, the mechanisms governing coordinate expression of the muscle phenotype, and also those responsible for the triggering of this expression, are best approached by the study of muscle-specific proteins at the level of the synthesis and utilization of their messenger RNA and, ultimately, of their gene organization. In the next section the approaches which have been adopted in studying muscle messenger RNA metabolism are discussed.

CONTROL OF MUSCLE PROTEIN SYNTHESIS DURING DIFFERENTIATION

During muscle cell differentiation there is direct evidence from labeling studies for the de novo synthesis of myosin, actin, tropomyosin, the acetylcholine receptor, and the Ca^{2+}-activated ATPase of the sarcoplasmic reticulum (see previous section). Indirect evidence that the appearance of proteins such as acetylcholinesterase and the M form of creatine phosphokinase is also a result of new protein synthesis comes from experiments with the inhibitor cycloheximide (2, 84, 180). In order to understand the mechanisms controlling muscle protein synthesis, it is necessary to investigate the metabolism of muscle messenger RNAs. The first question that arises is whether these RNAs are newly synthesized at differentiation or whether they are pre-existent in the cytoplasm of undifferentiated muscle cells. Having established whether biochemical differentiation for a given muscle protein is controlled at the level of gene transcription or by post-transcriptional regulation, the next step is to elucidate this mechanism. Further questions then concern the coordinate expression of different markers, and, finally, the chromosomal organization of genes for the families of different muscle and related nonmuscle proteins. Most experimental work on muscle messenger RNA (mRNA) has been directed toward resolving the first basic question and investigating some possible mechanisms of post-transcriptional control. The approaches which have been adopted will be discussed in the following sections.

Transcriptional or Post-transcriptional Control

Experiments with Actinomycin D Actinomycin D has been used to inhibit transcription prior to differentiation, and the appearance of muscle proteins has then been compared in control and treated cultures. These experiments are limited by the toxicity of the drug which prevents its use on cultures for more than a few hours. Interpretation of the results is complicated by possible side effects of actinomycin on translation (185, 186). Apparent stimulation of existing protein synthesis is one such side effect, which may be due to the inhibition of unstable RNA species with an as yet undefined role in protein synthesis (187) or simply to the increased availability of translational components resulting from the decrease in cellular mRNA on inhibition of its synthesis (188), which results in an "overshoot" type of phenomenon (189).

The addition of actinomycin D at a sufficiently high concentration (2 μg/ml) to inhibit mRNA as well as ribosomal RNA (rRNA) synthesis did not prevent the initial increase in myosin synthesis and creatine phosphokinase activity when added to primary cultures from fetal rat muscle for a 6-hr period immediately preceding differentiation (84). Indeed, although the final level was reduced, the initial rise in myosin synthesis or enzyme activity occurred earlier in treated cultures. Similar effects were also observed on cell fusion. Addition of actinomycin D at different times just before and during differentiation resulted in progressively less reduction in the final level of fusion and creatine phosphokinase activity the later the drug was added (84). It should be noted that reduction in the final levels attained may be simply due to the toxic effects of more prolonged exposure to the drug. With primary chick cultures, exposure to actinomycin D at 0.05 μg/ml for 8 hr prior to fusion did not prevent fiber formation, but resulted in a reduction of the initial rate of myosin synthesis to 27% of the control level. Addition of the drug at this concentration (0.05 μg/ml) after the onset of differentiation had no effect on the continuing increase in myosin synthesis (190). It should be noted that in experiments with rat cultures it has been demonstrated that mRNA synthesis is inhibited. This is not the case with the experiments on chick cultures, in which, particularly with the low doses of actinomycin D used, mRNA synthesis may continue, and the results should, therefore, be treated with some reservations.

These results might suggest that mRNAs for the proteins examined have already accumulated, or partially accumulated in the case of chick cultures in which the time scale is more contracted, in the 6–8-hr period preceding differentiation. However, since myosin and creatine phosphokinase are synthesized in undifferentiated muscle cells, the observed increase in the presence of actinomycin D may be due to a superinduction type of effect on existing synthesis. The possibility is rendered less likely by the demonstration that the creatine phosphokinase activity seen in treated cultures of rat muscle is a result of an increase in the muscle-specific form of the enzyme. No increase occurs in the B form of the enzyme present in undifferentiated cultures (191). Some reservations remain about these experiments, since de novo synthesis of the enzyme has not been demonstrated directly and since myokinase may interfere with the test for detecting the presence of the M form of creatine phosphokinase.

In Vitro Translation of Muscle Cell RNA Experiments with the use of inhibitors of RNA synthesis can provide only rather circumstantial evidence about control mechanisms. In order to obtain direct information it is necessary to look at the messenger RNAs present in the cell. This eventually means isolating and characterizing individual messengers for specific muscle proteins. However, a more immediately accessible approach is to look at the translational products synthesized in vitro by extracts of total mRNA from muscle cells and hence obtain information about the presence of different mRNA species. Actin, tropomyosin, and myosin have been identified among the products synthesized by muscle mRNA in vitro.

Actin Actin tends to be a major species synthesized by cellular mRNAs in vitro (see, for example, the results on actin synthesis directed by total mRNA from neuroblastoma cells, in ref. 192). Synthesis of actin by cytoplasmic mRNA preparations from muscle cells has been demonstrated for total polyadenylated RNA from chick primary cultures in the wheat germ system (121), for total RNA from calf primary cultures in the Schrier and Staehelin (193) in vitro system (62), and for RNA isolated from embryonic chick muscle in the wheat germ system (118, 194). In cultures of chick muscle, changes in the proportion of actin synthesized in vitro compared with total protein synthesis (1% in myoblasts, 8% in fused cultures) corresponded to the increase in actin synthesis measured per Petri dish in vivo (about $\times$ 9) (121). In cultures prevented from differentiating by the addition of BUdR or in the low Ca^{2+} medium used by these authors, both in vivo and in vitro actin synthesis remained at the level found in undifferentiated myoblasts. Actin, in these experiments, was characterized by peptide analyses. A quantitative feature of the results should be pointed out. The increase in actin synthesis in vivo on differentiation is higher than that described elsewhere (see under "Actin"), although the relative increase is probably less than that reported per Petri dish if expressed as a percentage of total protein synthesis, since cell number and hence protein synthesis per plate probably increase during the period in culture. Thus, the capacity of cellular RNA to synthesize actin in vitro may actually increase at differentiation by a relatively greater amount than that required for the increase observed in vivo. However, quantitative comparisons of this kind are difficult since the efficiency of the translation system is ill defined. As discussed previously, different forms of actin can be identified in muscle cells pre- and postfusion by isoelectric focusing (62). Two-dimensional gel analysis was carried out on the products synthesized in the Schreier and Staehelin in vitro system by total cytoplasmic RNA extracted from calf muscle cultures before and after differentiation. RNA from prefusion cultures synthesized β- and γ-actins with only trace amounts of α-actin, whereas RNA from fused cultures synthesized α-actin as a major form. The relative quantity of γ- to β-actin synthesized was also increased. The results on the synthesis of the muscle form of actin (α-actin) by pre- and postfusion RNA in vitro correspond to the qualitative observations on the type of actin synthesized before and after differentiation in vivo (62). Similar observations have been made of a parallel change in the type of actin synthesized in vivo and in vitro in developing chick muscle tissue (118).

α- and β-Tropomyosins The in vitro synthesis of α- and β-tropomyosins has also been examined by two-dimensional gel analysis, which permits the separation of these two forms (62). In primary cultures of calf muscle, α- and some β-tropomyosin are synthesized in detectable amounts before fusion, probably in contaminating differentiated cells or myotubes, since at least in the steady state the isozymes are only detected in muscle. The synthesis of nonmuscle tropomyosin has also been tentatively described in this system. After fusion there is an approximately 3-fold increase in the synthesis of α- and β-tropomyosins. In vitro protein synthesis in the Schreier and Staehelin system directed by total RNA from pre- and postfusion cells shows a rather different situation. The synthesis of β-tropomyosin predominates in prefusion extracts, whereas in vivo α-tropomyosin is relatively more strongly labeled. Postfusion, β-tropomyosin continues to be synthesized in greater amounts in vitro, which corresponds to the situation seen in vivo in well fused cultures (62, 102). It would appear, therefore, from these experiments that messenger RNA for β-tropomyosin may be present in quantity in prefusion cells and that it is not translated to a corresponding extent in vivo until after fusion, although differing efficiencies of the translation system for the two isozymes are also a consideration (see discussion at the end of the section).

Myosin: Light Chains The translation of myosin light chains has been investigated in RNA extracts from primary calf cultures (62, 102) and from rat primary cultures and cell lines (101, 191). In vitro synthesis of the two embryonic muscle subunits–$LC_1{}'$ + LC_1 and LC_2, identified by two-dimensional electrophoresis (focusing × sodium dodecyl sulfate)–takes place in significant amounts only with RNA extracts from postfusion calf cultures (102). Similarly, in rat cell lines, substantial amounts of LC_1 and LC_2 are synthesized by poly(A)-containing RNA from postfusion cultures, but not from prefusion cultures. Again, this corresponds to the situation in vivo. Identification of the products synthesized in the wheat germ cell-free system used in these experiments was based on the capacity of the labeled light chains to combine with the large subunit of myosin and by their migration on one- and two-dimensional gels. RNA extracted from differentiated rat primary cultures directed the synthesis of LC_1, LC_2, and a third light chain, LC_3. Polyadenylated RNA from primary cultures of rat fibroblasts apparently synthesized none of these subunits. LC_3 is present in the thigh muscle of adult rats, but is only synthesized to a minor extent in differentiated rat primary cultures (101). It would appear, therefore, that the in vitro synthesis of this light chain does not totally reflect the in vivo situation in the rat culture system. In calf cultures, LC_3 was not a notable species either in vivo or in vitro (102).

The timing of light chain synthesis has been examined during the differentiation of primary rat cultures. An initial report (191) suggested that the corresponding messenger RNAs were present before significant light chain synthesis occurred in vivo. However, more detailed characterization of the light chains indicated that, although their synthesis may be detectable in vitro a few hours before differentiation, their major synthesis in vitro, as in vivo, takes place later (101).

Myosin: Heavy Chain The myosin heavy chain is not satisfactorily translated in the two in vitro systems used in the experiments which have been discussed. Furthermore, the two-dimensional gel system which is convenient for identifying cell-free products cannot be used for the large subunit of myosin since it does not focus under these conditions (62). The in vitro synthesis of myosin heavy chains by fractions of mRNA from embryonic muscle and, to a lesser extent, from muscle cultures has been described in several reports and is discussed in the next section. Experiments with RNA from tissue culture cells have been carried out in the reticulocyte lysate, and the myosin product has been characterized and quantitatively determined by immunoprecipitation (195). It has been shown in this way that total RNA extracted from the cytoplasm of differentiated chick cultures synthesizes appreciable quantities of myosin heavy chain (195). RNA from prefusion cultures synthesizes only relatively small quantities of precipitable myosin. It is difficult to determine whether this RNA comes from contaminating differentiated cells or myotubes in the prefusion cultures, or whether low levels of myosin mRNA are present in undifferentiated cells. It is also possible that the antibody cross-reacts with nonmuscle myosins, in which case the low levels of synthesis detected in vitro before differentiation may be due to this. In a previous report (196) the same authors did not detect any myosin heavy chain synthesis in vitro with RNA extracted from the polysomes of prefusion cells. This might suggest, given the low levels of synthesis found with total cytoplasmic RNA from these cells (195), that the myosin mRNA is present in a nonpolysomal form before fusion. However, the detection methods with the use of myosin antibody in the experiments with total RNA (195) are more sensitive, and low levels of myosin synthesis directed by polysomal RNA may have been obscured by the background of other proteins in this region of the gel (196). Under conditions in which biochemical differentiation is inhibited, no detectable mRNA coding for the heavy chain is found in the polysomal RNA fraction in these experiments (196).

Discussion of In Vitro Experiments The results on the translation of messenger RNA extracted from muscle cells at different stages of development indicate that translatable mRNAs for differentiated proteins are not present in quantity in undifferentiated cells. In general a close correlation exists between the proteins synthesized in vitro and those synthesized in the corresponding intact cells. Differences are seen in the case of β-tropomyosin, for example, or possibly of LC_3 in rat primary cultures, but this may simply reflect the greater efficiency of the in vitro system for these particular messengers. The proteins are detectable in vivo and, therefore, small quantities of their messengers are certainly present. They may be more efficiently translated in vitro than those of other proteins, either due to an absolute difference in translatability or to a difference in the particular in vitro system used. This is the case, for example, with the messengers for α- and β-globin, which are translated with different efficiencies (197); a similar phenomenon is seen in vivo when cells are exposed to hypertonic medium (189). Quantitative interpretation of the results requires the determination of the efficiencies of translation of each mRNA. Until this becomes possible with purified muscle mRNA species, interpretation of reported in vitro/in vivo differences is difficult. Related to the question of translational efficiency is the fact that this approach only detects translatable messenger RNA. The possibility remains that muscle mRNAs are present in undifferentiated cells in a modified form which renders them untranslatable (see under "Cytoplasmic Distribution of Messenger RNA: Messenger RNP Particles").

Discrete Species of Muscle Messenger RNA: Messenger for Large Subunit of Myosin Most attempts to isolate a specific mRNA from muscle have concentrated on the messenger for the large subunit of myosin on the grounds that the size of its polypeptide is such that its mRNA should be larger than the bulk of cell messengers. Isolation procedures have been based on the criterion of size, and characterization has been carried out by the analysis of products synthesized in vitro. A crude fraction of RNA coding for actin has also been isolated from ribonucleoprotein particles (8–20 S) of embryonic chick muscle (194). However, no further purification has yet been reported.

As early as 1969, a fraction of RNA sedimenting at 26 S was isolated from the heavy polysome fraction of embryonic chick muscle and shown to direct the synthesis of the large subunit of myosin in vitro (198). Identification of the radioactive product synthesized in a homologous cell-free system was based on its copurification with carrier myosin on separation by DEAE chromatography and gel electrophoresis and on antibody precipitation (198). Subsequently, the same laboratory demonstrated the synthesis of myosin in a partially fractionated in vitro system containing muscle initiation factors and chick erythroblast ribosomes. After separation of the [^{35}S]methionine-labeled myosin product, a tryptic digest did not reveal any additional labeled peptides when the fingerprint was compared with that of cold carrier myosin (199). The material which copurified with carrier myosin in these experiments represented 10–40% of the total radioactivity incorporated in the in vitro system. Similarly, RNA isolated from the heavy polysomes of chick embryonic muscle and further separated from ribosomal RNA by millipore binding and cellulose chromatography has been shown to synthesize myosin in a homologous in vitro system. When subjected to peptide analysis after cleavage with cyanogen bromide, this material was shown to contain peptides identical with those of the myosin heavy chain (200); it can be calculated that this represented about 40% of the total labeled protein. The remaining material may represent uncompleted myosin chains or other polypeptides. These authors also demonstrated the synthesis of myosin in a heterologous in vitro system (reticulocyte lysate), in which the problem of endogenous myosin synthesis is avoided, but the full peptide characterization of the product was not carried out (201). Cytoplasmic RNA migrating in the 26 S region of a sucrose gradient has been isolated from fused muscle cell cultures and also has been shown to synthesize myosin heavy chains in a reticulocyte lysate. The product was separated by precipitation with myosin antibody in the presence of carrier myosin and characterized by its migration on sodium dodecyl sulfate-acrylamide gels (195). Quantitative determination of the proportion of myosin heavy chain synthesized by the fraction of added muscle RNA is complicated by the high level of endogenous synthesis in the lysate.

The characteristics of the RNA fractions which have been shown to synthesize myosin are as follows. They have been isolated from chick embryonic muscle tissue, either from the heavy polysomes (198, 201) or, more recently, from a ribonucleoprotein fraction sedimenting between 70–90 S (202), and from muscle cells in culture from ribonucleoprotein particles (30–40 S) released on puromycin treatment of polysomes (196) or from total cytoplasmic RNA (195). The RNA migrates at 26 S on sucrose gradients and at 30–32 S relative to 28 S rRNA on polyacrylamide (203) or formamide-polyacrylamide gels (201), although the latter may depend upon the degree of denaturation of the RNA, since similar fractions from muscle cells apparently migrate at 26 S (195).

There has been some controversy as to whether the RNA contains a tract of poly(A). It has been separated from rRNA by chromatography on Sepharose 4B (203), by millipore filtration followed by separation on cellulose (200), or by its capacity to bind to oligo(dT)-cellulose (204). Partial retention of myosin heavy chain RNA on oligo(dT)-cellulose has been reported for extracts of total cytoplasmic RNA from cultured cells (195), whereas that released from polysomes on puromycin treatment was not bound under the conditions used (196). Gel electrophoresis of the poly(A) tract after labeling with tritiated borohydride and digestion with pancreatic RNAse indicated that it contained about 170 nucleotides compared with marker poly(A) sequences (201). It should be noted, however, that the borohydride technique used is now known to be equally applicable to the capped 5′ end of messengers (205). These experiments, therefore, do not necessarily demonstrate a tract of poly(A) at the 3′ end. Assuming that the messenger preparation codes mainly for the myosin large subunit and that all the molecules contain poly(A), the authors calculate that there are additional untranslated sequences representing about 5% of the molecule (201).

The difficulty with this type of calculation is that it assumes that the messenger preparation is homogeneous. The criterion of a single band on gel electrophoresis is unsatisfactory since it may result simply from taking a sufficiently narrow cut of the original RNA. The demonstration of myosin synthesis in vitro is also inadequate, and, where an attempt at quantitative determination is possible, it can be shown that myosin is probably not the only product synthesized by the RNA preparations used (see previous discussion). Indeed, it is clear that other cellular proteins (e.g., LETS), and hence messengers, probably exist in the size range selected. Estimations of mRNA purity based on in vitro protein synthesis assume that all possible messenger components are equally translated, and quantitative determination for large messengers in particular is rendered difficult by their relatively inefficient translation in many in vitro systems. Further difficulties arise from premature chain termination and from endogenous protein synthesis, which in homologous systems may include myosin and which in heterologous systems also complicates the estimation of other components in the mRNA preparation. Once the major in vitro product has been identified as myosin, a more satisfactory way of determining the homogeneity of the messenger preparation is to make the complementary DNA (cDNA) copy and analyze the kinetics of the back-hybridization reaction (206). In theory, any apparent heterogeneity may be due to two possible causes, either contaminants (i.e., mRNAs of similar size) not coding for myosin or distinct mRNA molecules coding for different isozymes of myosin. These situations should be distinguishable by analyzing the products of in vitro translation. Back-hybridization has been carried out in the case of 26 S RNA purified from ribonucleoprotein particles (70–90 S) of embryonic chick muscle (207, 208). The $R_0t_{1/2}$ value of the back-hybridization reaction is 2.0 ± 1.4 (208). This value is higher than expected for a single messenger species of this size under the hybridization conditions used and suggests that the fraction is heterogeneous (for discussion, see under "Direct Measurement of Steady State Populations of Messenger RNA in Muscle Cells").

In conclusion, therefore, it has been shown that a fraction of muscle RNA sedimenting at 26 S will code for the large subunit of myosin, but the homogeneity of this material remains uncertain. It is very important to obtain a pure RNA species in order to characterize the messenger molecule itself, and also in order to use its cDNA copy for hybridization studies during myogenesis. It is evident that in the case of other muscle mRNAs, and probably of the mRNA for the heavy subunit of myosin as well, selective isolation procedures are necessary. Precipitation of polysomes by specific antibodies (209) has not yet been exploited for the purification of mRNAs coding for muscle protein. It is unlikely, given the relatively small proportion of total mRNA which codes for any muscle-specific protein, that this technique will result in pure preparations of mRNA. However, it may be useful as a preliminary selection step. It is with the development of genetic engineering techniques (210) as a means of obtaining bacterial clones which contain plasmids bearing a homogeneous cDNA fragment that the possibility of obtaining pure muscle mRNAs from a heterogeneous population of mRNAs (or their cDNAs) becomes feasible.

Synthesis and Stability of Messenger RNA During Myogenesis

Pulse-labeled RNA Even without isolating specific mRNA molecules, some information about mRNA metabolism during differentiation can be obtained from labeling studies on the total RNA population. Pulse-labeled cytoplasmic RNA has been examined during myogenesis in primary cultures from fetal calf muscle (63, 211). Cells were labeled with [^{3}H]uridine for 2 hr, and total cytoplasmic RNA was extracted and separated on sucrose gradients. Poly(A)-containing messenger RNA was detected in the gradient fractions by its binding to poly(U) filters. In cultures before and after fusion, most mRNA species (poly(A)$^+$) were distributed in the 10–20 S region of the gradient. However, in cells at all stages, a pronounced peak of pulse-labeled RNA sedimented at 26 S. Further material could be distinguished from the bulk of the RNA at 23 S, 30 S, and 32 S. Peaks of pulse-labeled RNA of about the same S value have also been described in rat muscle cell lines (212). The material sedimenting at 26 S (32 S relative to 28 S rRNA on nondenaturing acrylamide gels) potentially contained mRNA for the myosin heavy chain (see previous section). Circumstantial evidence that this was the case came from experiments on other cell types which suggested that a pronounced peak of pulse-labeled RNA at 26 S was not a universal phenomenon in cell lines, but appeared to be characteristic of cells with a capacity for synthesizing myosin in quantity (41, 213). Examination of the polysomal distribution of mRNA indicated that, whereas most pulse-labeled mRNAs were distributed between polysomal (circa 60%) and nonpolysomal (circa 40%) cytoplasmic forms, the bulk of the 26 S RNA was nonpolysomal prior to differentiation and was found in the heavy polysomes at the time when myosin synthesis increased in the cultures. Direct attempts to demonstrate identity between pre- and postfusion 26 S RNA material by hybridization were, in fact, unsatisfactory because of the high R_0t values obtained with the crude mRNA preparations used (211).

Quantitatively, the rate of rRNA synthesis in primary calf cultures is reduced approximately 10-fold at the stage of cell alignment/fusion, whereas poly(A)-containing RNA synthesis falls to about 60% of the level in dividing cells (63). A decrease in RNA synthesis has also been reported in chick cultures, although this was apparently due to a reduction in rRNA synthesis rather than in messenger RNA (214, 215). DNA-RNA hybridization studies also indicated that the relative incorporation of radioactive precursors into RNA hybridizing with nonrepetitive compared to repetitive sequences (including rRNA) was greater after fusion (216). Further evidence of a decrease in RNA synthesis comes from earlier studies which showed that there is a reduction in extractable RNA polymerase activity during myogenesis (217).

Pulse-labeling studies have thus shown that the overall pattern of mRNA synthesis in muscle cultures is very similar before and after differentiation, and indeed this might be expected from the general similarity in the patterns of protein synthesis (see under "Muscle Proteins and Their Synthesis" and "In Vitro Translation of Muscle Cell RNA"). There is a reduction in RNA synthesis at fusion, principally of rRNA as is generally observed in stationary compared with exponentially growing cultures. Messengers specific for muscle proteins have not been identified, but there is indirect evidence to suggest that part of the pulse-labeled RNA sedimenting at 26 S may code for the large subunit of muscle myosin. This RNA is synthesized in dividing cultures, but apparently not translated.

Messenger RNA Stability Pulse-chase experiments were carried out in order to estimate messenger half-lives and especially to determine whether 26 S RNA accumulated before differentiation (63, 211). It was found that most poly(A)-containing RNA, including 26 S RNA, had a half-life of about 10 hr in dividing cultures. In the mainly stationary phase, when alignment of myoblasts begins to take place prior to fusion, major changes in messenger stability occurred. The bulk of mRNA then had a half-life of about 20–25 hr, whereas certain fractions, revealed by chase experiments, had longer half-lives (>30 hr). The RNA sedimenting at 26 S had an estimated half-life of about 50 hr. Pulse-chase experiments are subject to a number of artifacts such as changing pool sizes or changing efficiencies of the chase technique. An attempt was made to control these various parameters, but the half-lives quoted are probably only of relative significance. It is noteworthy in this context that a class of small RNAs with a shorter half-life than the bulk of cell mRNA ($t_{1/2} < 6$ hr) before differentiation continues to be synthesized and to have a similar short half-life after fusion.

It can be concluded from the experiments on primary calf cultures that classes of RNA with different stabilities are present at all times and that the majority of poly(A)-containing RNA is more stable (× 2–5) in differentiating cultures. The increase in mRNA half-life is probably an underestimate because of heterogeneity in the cell population; dividing fibroblasts, for example, continue to contribute to the estimation after fusion. The chase experiments revealed a number of species, including 26 S RNA, which are initially present as stable RNAs in a predominantly nonpolysomal form and can be chased into the polysomes during differentiation. This might suggest that post-transcriptional control of both mRNA stability and its availability for translation may be important in muscle cell development. It remains to be demonstrated whether stability changes actually occur in RNA species already being synthesized, as implied (63) in the case of 26 S RNA, for example. The interpretation of the labeling experiments is limited by the fact that only poly(A)-containing RNA was examined and by the absence of clear identification of muscle-specific mRNAs. The first point is considered in a later section; the second, as already discussed, remains a major limitation of existing results on the system.

The possibility that stable mRNAs for muscle proteins may be present immediately before and during differentiation was also suggested by the experiments with actinomycin D, which have already been discussed (190, 191). An earlier attempt to estimate the half-life of myosin mRNA during differentiation by looking at the decay of myosin synthesizing activity in the presence of actinomycin D resulted in shorter estimated values (circa 6 hr), although longer than those estimated for total protein synthesis (circa 15 hr) (212). These values are probably less reliable than those obtained by direct pulse-chase techniques since actinomycin is known to have secondary effects on mRNA stability (185, 218). Estimations of mRNA stability in differentiating muscle cells have also been obtained by looking at in vitro protein synthesis directed by RNA from rat cell lines treated with actinomycin D (results not yet published in detail, but see ref. 191). Total cytoplasmic polyadenylated RNA was extracted from the cultures at different times after application of the drug and translated in the wheat germ system. The synthesis of proteins such as tubulin, actin, and the light chains of myosin was estimated by densitometric scanning of autoradiographs after one-dimensional gel electrophoresis of the in vitro products. Identification by this means is obviously approximate (see under "Large Subunit of

Myosin"), and experiments with actinomycin D are subject to artifacts; however, half-lives of longer than 20 hr were estimated for the mRNAs coding for these proteins (191), providing a further indication that messengers for muscle proteins may be relatively stable in differentiating cultures.

Populations of RNA of different stabilities have been reported in other cell types, both in dividing (219, 220) and stationary phase (221) cultures. Two components, one with a short half-life (⩽6 hr) and one with a longer half-life (circa 24 hr), have been described in dividing HeLa cells, for example (219). In these experiments and in similar results obtained with insect cells (220), the longer lived components were smaller sized RNA molecules, in contrast to the situation in muscle cultures. A relationship between the half-life of the majority of cell messengers and the cell generation time has been demonstrated in a number of systems. In HeLa cells (219), for example, they are similar, whereas in other mammalian cultured cells the half-life has been reported as approximately half the cell generation time (222). This latter situation would appear to pertain to muscle cultures (cell generation time ~ 18–20 hr, $t_{1/2}$ most mRNA ~ 10 hr). In stationary phase cells, the bulk of the mRNA is apparently of longer half-life; for example, in resting lymphocytes, the more stable class of mRNA has a half-life which exceeds 24 hr (221). Messengers coding for the specific proteins of differentiated cells have been reported to be relatively very stable, with half-lives of several days (see, for example, globin (223), ovalbumen (224), zymogen (225)). In maturing reticulocytes, messengers of different stabilities are found (226, 227). The potential role of messenger stability in modulating the extent of synthesis of differentiated proteins has been discussed, for example, in the context of zymogen synthesis during silk moth development (225) and of globin synthesis during erythropoiesis (226, 227).

During muscle cell differentiation it is difficult to distinguish between phenomena which may simply reflect the state of confluency and those actually related to myogenesis. This is particularly true for mammalian cultures which fuse at high cell densities. The problem arises mainly when bulk mRNA is examined. In the case of mRNA species which are apparently specifically involved in the synthesis of differentiated proteins, the situation is clearer. In conclusion, the pulse-chase data which have been discussed provide suggestive evidence of the importance of differences in mRNA stability for mRNA coding for the large subunit of myosin and for certain other mRNA species also apparently not engaged in the polysomes before differentiation.

Direct Measurement of Steady State Populations of Messenger RNA in Muscle Cells

Hybridization with Poly(U) Pulse-labeling studies provide information on mRNA metabolism during myogenesis. Some idea of the steady state levels of mRNA can be obtained from estimating the content of poly(A)-containing RNA by hybridization with radioactive poly(U) (228). With the use of this technique, cytoplasmic RNA, separated on sucrose gradients, was analyzed before and after fusion in primary calf cultures (213, 229). No major differences were observed in the population of poly(A)-containing RNA, which represented 3–3.5% of total cytoplasmic RNA. During differentiation, rather more large messengers (⩾18 S) were present relative to the total RNA, and a shoulder of RNA in the 26 S region of the gradient was detectable. However, this material was not sufficiently well resolved from the rest of the RNA for questions about quantity and accumulation to be answered. In general, the steady state levels observed corresponded approximately to those expected from the pulse-chase experiments, in which mRNA synthesis was less (circa 60%) after fusion, but many mRNAs were more stable (circa × 2–3).

Complementary DNA Hybridization with Total Messenger RNA The experiments which are likely to yield most information about steady state populations of mRNA are those which use complementary DNA-RNA hybridization techniques. Employing this approach, it is possible to draw general conclusions about the classes of RNA and their relative abundance without purifying specific RNAs (230). Experiments with the use of total polysomal mRNA (poly(A)$^+$) have been carried out with myoblast cell lines derived from a mouse teratocarcinoma (231). From an analysis of the kinetics of the back-hybridization reaction between excess polysomal RNA (poly(A)$^+$) and the corresponding cDNA from differentiated and undifferentiated cultures, it was estimated that about 11,500 different sequences per cell were present in the polysomes of dividing cells (231) (cf. reported values of about 10,000 (HeLa), 9,000 (3T6, stationary phase), 8,000 (3T6 log phase) (232)) and that about 30 represented abundant species (about 1,500 copies/cell), 800 intermediate sequences, and the rest (approximately 11,000) were present as five or fewer copies per cell. In differentiated cultures, about 8,000 different sequences were detected in polysomal RNA, of which about 12 were abundant (circa 2,500 copies), about 2,250 were present at intermediate frequencies, and the rest (circa 6,000) were present as five copies or fewer. Thus, there is a reduction in RNA complexity during myogenesis.

Cross-hybridization experiments with heterologous cDNA and RNA give further information on the extent to which dividing myoblasts and differentiated cultures have sequences in common. These experiments (231) on teratoma-derived muscle cell lines led to the conclusion that sequences present in myoblast cDNA are also present in differentiated cells as a higher abundance class, and that about 6% of myoblast cDNA and 12% of the cDNA of differentiated cells are unique to the respective stages of development. These experiments do not give any information about the numbers of sequences involved in each case. Presumably, most of the sequences specific to myoblasts are present in the low abundance class, given the relatively higher complexity of undifferentiated cells. However, it is not clear in which abundance class the 12% of differentiated cDNA is found, although from the kinetics of hybridization the sequences which fail to hybridize appear to originate from the more abundant species and, therefore, may represent a small number of sequences. These results refer at present only to polysomal poly(A)-containing RNA. In view of the pulse-chase experiments, it will be of interest to see whether additional RNA species are detected in the cytoplasm in a nonpolysomal form before differentiation. It is also important to see whether the mRNA populations of primary muscle cultures have similar characteristics since teratoma-derived muscle lines have a number of peculiarities (e.g., they are anaploid (43)).

DNA-RNA Hybridization Some information on RNA populations in primary chick cultures has been derived from hybridization experiments in which pulse-labeled total cellular RNA was hybridized to excess total DNA (216). The kinetics of hybridization were analyzed in order to make a distinction between the proportion of repetitive and nonrepetitive sequences represented in pre- and postfusion RNA. From this type of data it could be concluded that relatively more nonrepetitive sequences were transcribed after fusion. Since total RNA and DNA were used, repetitive sequences in this case were presumably mainly ribosomal.

Experiments which give information about mRNA transcribed from repetitive sequences have been performed on a nonfusing rat myoblast cell line (233). Messenger RNA (poly(A)$^+$, polysomal), both pulse-labeled and labeled by iodination in the steady state, was hybridized with excess DNA. From an analysis of the C_ot curves, it was concluded that about 20% of the mRNA was transcribed from repetitive sequences. This material alone annealed with reiterated DNA when the latter was separated from total DNA, suggesting that it was not covalently linked to the bulk of mRNA. The explanation for these findings and the possible role of such sequences are not clear. In sea urchin gastrula (234), it was estimated that less than 3% of polysomal mRNA contained repeated sequences. In teratoma myoblasts, mRNA-containing reiterated sequences do not represent a significant fraction of polysomal mRNA (poly(A)$^+$) (231).

A further type of experiment which is complementary to the cDNA approach and which also potentially provides information about mRNA sequences has been performed by hybridizing excess total RNA extracted from primary chick muscle to labeled "nonrepetitive" DNA isolated by isopycnic centrifugation from the chick genome (235). It was found that the proportion of this DNA which hybridized with the total RNA increased from 7–8% in myoblasts to about 10% at differentiation, suggesting that more transcripts were present after fusion. Because of the difficulty of obtaining highly radioactive DNA (*DNA), it is not clear whether these experiments were carried out in sufficient RNA excess to detect single copy RNA species. Since total RNA has been used, nuclear RNA, which is several times more complex (230) than cytoplasmic RNA, will make a relatively large contribution to the results. Again, future studies on the different compartments of cellular RNA during myogenesis will be of interest. The results of these experiments on *DNA-RNA hybridization, although more limited than the cDNA analysis in the information obtained, have the advantage of providing data on the changes in the absolute number of sequences during myogenesis.

Discussion of Hybridization Experiments with Populations of RNA Hybridization experiments with populations of RNA suffer from a number of disadvantages. From a practical point of view, in the unique *DNA-RNA experiments, it is difficult to prepare DNA which is sufficiently radioactive so that the quantities of cellular RNA available are enough to drive the reaction to C_0t values where less abundant RNA molecules are detectable. Furthermore, fractionation of the DNA into classes of different sequence repetition may not be totally clear-cut. It is easier to obtain very radioactive cDNA and thus to hybridize under conditions in which single copy messengers begin to be detected. Theoretically, the cDNA reflects the mRNA population and, therefore, provides more refined information about abundance classes, whereas hybridization with radioactive DNA provides a more accurate estimation of the total complexity of the RNA sequences present, especially for few or single copy mRNAs. In fact, in the cDNA experiments, only poly(A)-containing RNA is examined (see under "Messenger RNA Molecule: Poly(A)"), and it is assumed that all molecules are copied with equal efficiency by the reverse transcriptase so that the cDNAs obtained faithfully reflect the messenger RNA population. This can be controlled to some extent by looking at the saturation curve for hybridization at lower RNA to cDNA ratios (231).

Minor variations are introduced in both types of experiment by differences in size and sequence repetition between messengers. A further limitation lies in the lack of functional identification of messengers, although two-dimensional gel analysis (61) of products synthesized in vitro by total RNA fractions has sufficient resolving power to identify about 500–1,000 species, and assuming comparable translational efficiencies in vitro and in vivo should give some indication of which are the more abundant mRNA species present (see under "In Vitro Translation of Muscle Cell RNA"). It should be noted that, in the experiments described, the contribution of contaminating nondifferentiated cells, e.g., fibroblasts in the primary cultures, has not been assessed, and, more important, the effect of the stationary phase, rather than the differentiated state, in inducing changes in RNA populations should be considered.

Hybridization studies with the use of a cDNA synthesized from a purified mRNA avoid many of the difficulties in interpretation encountered with such experiments when a global cDNA is employed. The more precise information which can be obtained by studying a single species provides both a control and a complement to the general view resulting from studies with populations of RNA. Because of the lack of purified muscle mRNA preparations (see under "Discrete Species of Muscle Messenger RNA: Messenger for Large Subunit of Myosin"), very few such studies have been attempted.

Complementary DNA Hybridization with 26 S RNA Fraction In chick cultures, hybridization studies have been reported with the use of a cDNA synthesized from an mRNA fraction coding for the large subunit of myosin (207, 208). The principal question here is the purity of the cDNA probe. From the back-hybridization kinetics between the mRNA and its cDNA, it should be possible to estimate the homogeneity of the mRNA preparation (206). However, this is difficult from the published data since the $R_0t_{1/2}$ value for a purified mRNA of about the same size (e.g., Mengo virus RNA), which could be used as a standard for the hybridization conditions employed, is not given. The $R_0t_{1/2}$ value for the back hybridization of Mengo RNA (7,200 nucleotides; cf. mysoin about 6,500 (201)) was found experimentally to be about 4–5 times that of globin (236). Globin mRNA under hybridization conditions resembling those used for the myosin cDNA (i.e., formamide, etc.) has a $R_0t_{1/2}$ value of 4×10^{-3} (237); thus, an expected value for myosin would be about 2×10^{-2}. The quoted value (i.e., 2) is, in fact, two orders of magnitude higher than this, suggesting a high degree of complexity. The back-hybridization data would suggest, therefore, that the cDNA probe is heterogeneous, probably due to contaminants, either cytoplasmic or nuclear. There is a possibility that there are many similar isozymes of the myosin heavy chain of muscle (see under "Large Subunit of Myosin"), and this would then increase the apparent complexity of the cDNA.

An indication of the specificity of the probe for muscle comes from its $R_0t_{1/2}$ value with fibroblast RNA (including rRNA), which is about 8-fold higher (1.075×10^3) than that with RNA from fused cultures. Comparison of the $R_0t_{1/2}$ values for RNA from cultures during myogenesis showed that the number of copies of RNA hybridizing with the probe increased from about 500 per nucleus in myoblasts immediately after plating to 4,500 copies at 24 hr and 12,000 copies at 48 hr. At the time when myosin synthesis is maximal, about 2,000 copies are present per myotube nucleus. This latter figure corresponds to that calculated in quail cultures in order to account for the rate of myosin synthesis observed (5). It is concluded (207, 208) that myosin mRNA is present in chick myoblast cultures prior to fusion in greater quantities than required for the observed level of myosin synthesis and that it accumulates prior to differentiation. The cellular origin of the materials used was as follows: the cDNA was synthesized from RNA coding for myosin present in a nonpolysomal cytoplasmic fraction (ribonucleotprotein, see below); the cellular RNA contained all polysomal and large nonpolysomal RNAs. As far as the large subunit of myosin is concerned, therefore, the results are on total cytoplasmic RNA. One of the difficulties in this type of experiment is to obtain enough RNA from cultured cells to be in sufficient mRNA excess with respect to the cDNA during hybridization. It is not clear that this was the case in these experiments since the hybridization was not taken to more than 50–70% saturation. Another general problem is that of the heterogeneity of the cell population. In the results discussed here, a correction has been applied for fibroblast contamination. The contribution of fibroblasts and also of postmitotic myoblasts in the prefusion population (particularly in chick cultures) and of nonfusing myoblasts in the postfusion population to any mRNA data is difficult to assess. The use of in situ hybridization may help to clarify this problem.

Summary It is evident from the preceding sections that there is not yet a clear answer to the basic question of whether messenger RNAs for muscle proteins are present in the cytoplasm of undifferentiated myoblasts. The lack of pure muscle-specific mRNAs and the corresponding cDNA probes has meant that the available information is derived from more indirect approaches. The results which have been presented are summarized in Table 3. The conclusions should be treated with some reservations, for the reasons which have been discussed.

Possible Regulatory Mechanisms During Myogenesis

The synthesis of muscle proteins during the differentiation of muscle cells may be regulated by changes in the synthesis, nuclear processing, cytoplasmic processing, and translation of the corresponding mRNA molecules. In the preceding section the detection of cytoplasmic mRNAs has been discussed. Since the development of muscle-specific cDNA probes still presents difficulties, investigation of muscle transcripts in nuclear RNA or in the transcription products from isolated chromatin is not yet feasible. Indications of changes in nuclear activity during myogenesis come from observations of the activity of nuclear enzymes. RNA polymerase activity decreases (217), although this may be primarily related to the decrease in cell division. Nuclear protein kinase activity increases, as does the extent of phosphorylation of nonhistone proteins (238), although this too may not be directly related to the process of differentiation since the increase depends upon the time in culture and is not seen in vivo. Because it is more accessible to study, interest has principally centered on possible cytoplasmic regulation of mRNA expression in muscle cells. Control of mRNA stability and translation are major facets of cytoplasmic mRNA regulation, and there is some evidence to suggest that both may be implicated in muscle differentiation (see previous sections). Factors which may be important in one or both of these phenomena are discussed in the following sections.

Messenger RNA Molecule: Poly(A) Like other eukaryotic mRNAs (239), those coding for muscle proteins appear to be monocistronic (239–243); the large and small subunits of myosin, for example, are synthesized off different sized polysomes. Nonpolysomal as well as polysomal mRNA, when isolated under denaturing conditions, is apparently translatable in vitro (194, 202), suggesting that the coding sequence, when nonpolysomal, does not undergo cytoplasmic modifications. In order to obtain more direct information about muscle mRNA sequences, it is necessary to carry out fingerprint and sequence analyses which obviously depend upon the availability of purified molecules. "Capping" of mRNAs may modify their function, and although the role of cap is unclear, two different types of 5′-methylation have been described for cytoplasmic mRNAs (205). Since most cellular messengers appear to be capped, the same might be expected for muscle cell mRNAs, but the presence and characteristics of this modification have not yet been investigated in the mRNA popula-

Table 3. Summary of experiments which have attempted to determine whether muscle mRNAs are present in undifferentiated myoblasts

Method of detection	Muscle mRNA	Dividing myoblasts	Prefusion/ alignment	Fusion
Actinomycin D: observations on protein synthesis	Myosin heavy chain	–	+	+
	Creatine phosphokinase M	–	+	+
In vitro protein synthesis with total mRNA: detection of products on one- or two-dimensional gels	Actin	β, γ		α, β, γ
	Myosin light chains	–		$LC_1' + LC_1, LC_2, (LC_3)$
	Myosin heavy chain (total RNA)	(+) but significant?		+
	Myosin heavy chain (polysomal)	–		+
	Tropomyosin	$\beta\ (\alpha)$		β, α
Pulse-labeling	26 S fraction (? myosin heavy chain)	+ (RNP, $t_{1/2}$ ~ 10 hr)	+ (RNP, $t_{1/2}$ ~ 55 hr)	+ (RNP-polysomes, $t_{1/2}$ ~ 55 hr)
	Other mRNAs		RNP	RNP-polysomes
*cDNA-RNA hybridization	Total polysomal mRNA	Circa 6% cDNA is specific to this stage		Circa 12% cDNA is specific to this stage
		Reduction in complexity →		
		Proportion of less abundant species becomes more abundant		
	cDNA from heavy RNP fraction (? myosin mRNA)-RNA from polysomes and large RNPs	Copies/cell: 500 → 4,500 → 12,000 →		Circa 2,000 copies/cell
Unique *DNA-RNA hybridization	Total cell RNA	7–8% unique DNA transcribed		10% unique DNA transcribed

tion during myogenesis. The presence of poly(A) at the 3′-terminal of the molecule is another feature of many eukaryotic messengers (239). Possible functions of the poly(A) sequence remain controversial, but it has been proposed that it may affect mRNA translation or stability (for review, see ref. 244). The proportion of cytoplasmic mRNA which contains poly(A) and the length of the sequence associated with different classes of mRNA have been examined during muscle differentiation in culture.

A number of different experiments suggest that most mRNAs in muscle cells contain poly(A). This was the conclusion reached from the results of [^{3}H] poly(U) hybridization across polysome gradients from fused rat primary cultures, in which a standard length of poly(A) per messenger (see below) was assumed (245). In vitro protein synthesis in the Schreier and Staehelin system (193) is not inhibited by rRNA, and it is, therefore, possible to compare the products synthesized by total cytoplasmic RNA and RNA which has been passed over oligo(dT)-cellulose. From one-dimensional gel analysis of the products synthesized by total cytoplasmic RNA, RNA enriched for poly(A)$^+$ sequences, and RNA not retained on oligo(dT)-cellulose, the proportion of mRNA which is poly(A)$^+$ was investigated in primary cultures from calf muscle (213, 246). The only class of messenger for which there was no enrichment in the poly(A)$^+$ fraction was that coding for the histones. Some actin and trace amounts of other peptides were also synthesized by the RNA not retained on oligo(dT)-cellulose. This is not necessarily poly(A)$^+$, but may contain shorter tracts of poly(A) ($<$20 nucleotides). Questions of the translatability in vitro, efficiency of binding to the column, and possible degradation of poly(A) during the isolation procedure affect the quantitative interpretation of these results, but they suggest that most translatable mRNA in calf muscle cells contains poly(A). Another indication of this comes from studies of nonpolysomal pulse-labeled mRNA during myogenesis in primary calf cultures (247). Messenger RNA in the form of ribonucleoprotein particles (RNPs) can be separated from ribosomal material by density gradient centrifugation on metrizamide (see next section). Most pulse-labeled mRNA ($>$80%) in this fraction was bound to poly(U) filters. In dividing myoblasts, the fraction which did not bind was rather higher (circa 25%), possibly due to the presence of more histone mRNA. Pulse-labeled 26 S RNA isolated in this way was also bound to poly(U) filters (92–95%) (see also reports on the poly(A) content of 26 S messenger preparations under "Discrete Species of Muscle Messenger RNA: Messenger for Large Subunit of Myosin").

Estimations of poly(A) content in such experiments are restricted to nonpolysomal mRNA and are limited by the efficiency of poly(A):poly(U) binding on the filters; however, they indicate that the bulk of pulse-labeled mRNA in this fraction is poly(A)$^+$. A further indication that no major change in poly(A) content occurs during muscle cell differentiation comes from hybridization studies with [^{3}H]poly(U) on total cytoplasmic RNA and polysomal/nonpolysomal RNA (229). The question of whether a proportion of muscle mRNA does not contain poly(A) and whether this proportion changes during myogenesis is of great importance in the interpretation of many of the experiments discussed in the first part of the chapter, e.g., studies of pulse-labeling, cDNA hybridization, and in vitro translation in certain systems. It can be concluded from the preceding discussion that there is not a major class of poly(A)$^-$ RNA in muscle cells, but the possibility that minor messenger components important in differentiation are poly(A)$^-$cannot be excluded.

The overall mRNA content of poly(A) does not change during myogenesis, but the length of the poly(A) sequence may undergo alterations. This was investigated by [^{3}H]poly(U) hybridization of isolated poly(A) sequences in different size classes of mRNA from primary cultures of calf muscle (229). Within the limits of sensitivity of the experiments, no correlation between the cytoplasmic localization (i.e., polysomal or not) and stability of messenger RNA could be detected. The length of poly(A) associated with different size classes of mRNA remained constant throughout the period in culture. A correlation was, however, found between the size of the mRNA (6–30 S) and its length of poly(A) (20–170 nucleotides) (229). A similar observation had been made for mRNA extracted from polysomes of embryonic chick muscle of different sizes (201). This phenomenon may be of general significance in the metabolism of mRNA, but it would not appear that poly(A) plays a specific role in the regulation of messenger stability or translation during myogenesis.

Cytoplasmic Distribution of Messenger RNA-Messenger Ribonucleoprotein Particles Pulse-labeling experiments on primary cultures of calf myoblasts indicated that mRNA (poly(A)$^+$) was distributed between the supernatant (0–80 S) and polysomal regions of a gradient in an approximate ratio of 40:60 and that this ratio remained about the same in pulse-chase experiments (63). Estimation of the distribution in the steady state ([^{3}H]poly(U) hybridization) gave a similar result (229). The overall distribution varied very little during differentiation, although the pulse-labeled RNA sedimenting at 26 S in dividing myoblasts was predominantly nonpolysomal. In the period immediately preceding fusion, RNA of this size, together with certain other species revealed after a chase, was also nonpolysomal and could be chased into the polysomes during differentiation. Pulse-chase experiments of this type (63, 211), therefore, indicate the potential importance of the nonpolysomal compartment in sequestering mRNAs during myogenesis, although the results should be treated with some caution since the mRNA species involved have not been clearly identified. Experiments on the in vitro translation of RNA isolated from puromycin-dissociated polysomal particles indicated that myosin mRNA was present in the polysomes in significant amounts only after fusion (196). In contrast, it has been claimed on the basis of polysome run-off experiments and one-dimensional gel analysis of the products that there is a difference of less than a factor of two between the quantity of myosin mRNA present in polysomes from primary chick cultures before and after fusion (248). This is contrary to most reports on the extent of in vivo myosin synthesis, including those described here in which a 7-fold increase in myosin synthesis is seen at fusion. It is suggested that myosin mRNA is associated with polysomes before fusion, but is not translated efficiently in vivo.

Since the messenger is apparently in the heavy polysomes, it would appear that elongation, not initiation, is limiting. However, before the run-off experiments can be taken as clear evidence for this phenomenon, further characterization of the in vitro product as muscle myosin is required.

Characterization of the nonpolysomal form of pulse-labeled mRNA in muscle cells by its RNA-protein content and by its buoyant density on CsCl (249) or metrizamide (247) suggests that the particles have properties characteristic of free cytoplasmic mRNPs (for review, see ref. 250). In the case of 26 S RNA, the messenger is found in the 80–100 S region of a polysome gradient; it was only after EDTA treatment to disrupt ribosomal and polysomal structures that it could be demonstrated that the RNA was in fact not associated with these structures, but was contained in an mRNP particle comigrating in this region of the gradient (63, 211). The 80–100 S region of polysome gradients has been used as a source of mRNP particles for the preparation of myosin mRNA. It has been shown that RNA isolated in this way from embryonic chick muscle will direct the synthesis of the large subunit of myosin in vitro (202, 204). Correspondingly, actin is synthesized in vitro by RNA extracted from 16–40 S RNP particles present in postribosomal supernatants of the same tissue (194).

The protein components associated with free RNP particles are at present ill defined. The difficulty has been to isolate particles free of any possible spurious cytoplasmic contamination, which, given the lack of any functional criteria, is difficult to control. Two selective techniques for their separation are available, namely, binding to poly(U)-Sepharose (251) or density gradient centrifugation on metrizamide (247). Most available information on free RNPs concerns particles isolated from sucrose gradients without any selective step and with the consequent additional problem of contamination by enzyme complexes, etc., cosedimenting in the same region. mRNPs isolated from embryonic chick muscle, from the region of a sucrose gradient where mRNA activity for actin is found, contain a major polypeptide of 44,000 molecular weight, together with other minor species (194). These proteins are larger than most of those associated with rRNA. A similar observation has been made in a preliminary report on mRNP proteins in mRNP particles separated on metrizamide from muscle cells in culture (252). The potential regulatory role of proteins associated with mRNA molecules, either in influencing their susceptibility to degradation or their translatability, is of importance in considering post-transcriptional control of protein synthesis. The only protein to have been clearly described is that associated with the poly(A) sequence of mRNA (253). A second protein has also been identified in polysomally released mRNPs (254). These two components seem to be fairly universally associated with mRNA in polysomes (250). The possible diversity of proteins in nonpolysomal RNPs remains to be determined.

Proposed Regulatory Components

Translational Control RNA In addition to the possible role of mRNP proteins in the cytoplasmic regulation of mRNA activity during myogenesis, it has been proposed (202, 204, 255) that a small RNA component of the particles, named translational control RNA (tcRNA), may play a controlling role.

An RNA-like component which influenced in vitro protein translation, apparently at the level of initiation (256), was first identified in dialysates during the preparation of initiation factors (257). The material apparently conferred some specificity on translation since a factor isolated from the dialysate of muscle initiation factors inhibited globin translation, and, correspondingly, inhibition of myosin synthesis was observed by the dialysate of reticulocyte factors (257). A small RNA-like molecule from the dialysate of reticulocyte factors has been reported to stimulate globin synthesis in a reticulocyte in vitro system (258). The problem with these observations is the ill determined nature of the stimulatory or inhibitory components and the possibility of effects due to materials such as spermidine (259) or mRNA degradation products. Furthermore, doubt has been raised as to whether the activity shows mRNA specificity (260).

Studies of tcRNA have been taken further in preparations from embryonic muscle, and two subfractions have been described, one isolated from dialysates of polysomes, the other from mRNPs (202, 204, 255). Both materials are RNA-like and contain stretches of poly(U) (258), which represent about 50% of mRNP-tcRNA and rather less of the smaller polysomal tcRNA. The mRNP-derived material inhibits the translation of mRNP-derived mRNA, while having no effect on polysomal mRNA. Polysomally derived tcRNA has no such inhibitory effect, but stimulates the translation of polysomal RNA. From studies on the RNAse resistance of mRNA-tcRNA mixtures, it has been suggested that nuclease-resistant complexes (possibly circular forms of the tcRNA-mRNA in mRNPs) exit, in which poly(A)-poly(U) hybrids are formed between the two molecules. The translation and RNAse sensitivity of de-adenylated mRNA are unaffected by tcRNA. Both of these effects of tcRNA show some degree of specificity for the mRNA from the same origin; thus, for example, tcRNA from the mRNPs containing myosin heavy chain messenger activity is more inhibitory on the translation of this preparation than on that coding for smaller sized proteins (204). The situation is not entirely clear, however, since in an earlier report myosin tcRNA did inhibit globin synthesis (256; cf. 204). It is suggested that there may be sequence homologies between tcRNA and mRNA from the same RNP, and a model has been proposed for the role of tcRNA in the regulation of mRNA availability based on the circularization of mRNA in mRNPs. Such a proposal, derived principally from the results of experiments with muscle components, is obviously attractive as a post-transcriptional control mechanism during myogenesis. However, the current difficulty with the type of experiments described is the precise nature of the material used. Until the active component is clearly characterized as a small RNA molecule, preferably with sequence data, it is difficult to assess the full significance of tcRNA.

Initiation Factors Much of the original argument for the existence of protein initiation factors with specificity for particular messengers was based on comparative studies between reticulocyte and muscle systems. In 1969 the first report appeared of a muscle factor requirement for the translation of muscle mRNAs in a reticulocyte lysate and, correspondingly, for reticulocyte factors for the translation of globin mRNA in an in vitro system derived from muscle (261). The specificity apparently lay in the fraction responsible for mRNA binding to the ribosome (262, 263), subsequently identified as IF_3 (199). The requirement for homologous factors was not absolute, at least for globin synthesis, but they greatly increased the efficiency of translation (264). That such an effect was messenger-dependent rather than tissue-dependent was shown in experiments on myosin and myoglobin synthesis in preparations from red and white leg muscle of chicken. It was shown that myoglobin synthesis, characteristic of red muscle in vivo, required initiation factors from red muscle, whereas myosin synthesis, which proceeds in vivo in both tissues, took place in vitro when factor preparations from either tissue were present (265). The possible regulatory importance of initiation factors during development was deduced from the observation that the appearance in extracts of red muscle of an IF_3 fraction which permits myoglobin synthesis in vitro coincides with the beginning of myoglobin synthesis in this muscle in vivo (266). Crude IF_3 preparations from red muscle have been separated into different fractions, one of which permits myosin synthesis and the other, myoglobin synthesis in vitro (257). It should be pointed out that the mRNA for myoglobin is apparently only present in translatable amounts in vitro (266) when myoglobin synthesis is initiated in vivo, suggesting that the quantity of mRNA present, rather than (or as well as) factors permitting its translation, is probably of regulatory importance.

The suggestion from the experiments described is that for the efficient translation of a given muscle mRNA there is a qualitative requirement for the corresponding initiation factors. Experiments (195, 196, 200) which demonstrate the synthesis of myosin heavy chains when the corresponding mRNA is added to a reticulocyte lysate, without the addition of muscle initiation factors, demonstrate that the requirement is not absolute. Quantitative comparison of the extent of protein synthesis with globin mRNA or muscle mRNAs and partially purified preparations of reticulocyte or muscle initiation factors in the Schreier and Staehelin (193) in vitro system suggest quantitative, but not qualitative, differences between preparations; in general, reticulocyte factors were more active with both globin and muscle messenger preparations (267). A major difficulty in the experiments described, on the apparent specificity of initiation factors, is that purified fractions of proteins have not been used and the exact point in the mechanism of protein synthesis, as it is now understood, at which specificity intervenes has not been precisely defined (for discussion, see ref. 48). Where specificity has been thought to exist in other systems, for example, with E.M.C. RNA, the factor concerned has subsequently been identified as a general component of protein synthesis (268). The question of tcRNA, its existence, and its proposed specificity, as already discussed, raises similar problems of purification and definition.

Discussion of Translational Specificity The current trend in thinking about eukaryotic protein synthesis is not to favor the presence of components which are messenger-specific. However, although there may be no absolute specificity, it appears that different messengers are translated with different efficiencies (see under "Discussion of In Vitro Experiments"). This is illustrated by the difference in efficiency of translation between the α- and β-globin messengers (197) or by experiments in which changes in the cellular environment result in a change in the rates of protein synthesis, seen, for example, as a preferential inhibition of host rather than viral mRNA translation (269) or as the differential inhibition of the synthesis of immunoglobulin light and heavy chains and nonimmunoglobulin polypeptides in plasmocytoma cells (189). Observations of this kind are in keeping with the suggestion that different messengers have inherently different translational efficiencies perhaps due to differences in their initiation sequences, possible sequences homologous with rRNA (270) or tertiary structure. It is possible that a general regulatory control may be exercised on the expression of differentiated mRNAs, such as those coding for muscle proteins, in comparison with other cell messengers, on the basis of their reduced ability to form functional mRNA-ribosome initiation complexes, for example. Efficient synthesis of muscle proteins would then require that some previously limiting component of protein synthesis ceased to be so. Whether, in addition, factors which are specific for individual messengers or classes of messengers operate at the level of translation remains an open question. Indeed, the translational capacity of the cell may not be limiting in any way under normal conditions, and protein synthesis during differentiation simply may be dependent upon the availability of mRNA.

Experiments Which Provide General Indication of Possible Regulatory Mechanisms: Cell Manipulations A general indication of the level at which differentiation is regulated, and eventually of the type of mechanism involved, can be expected to come from cellular manipulations such as cell enucleation, cell reconstitution, and somatic cell hybridization with muscle cells at different stages of development. Very few such experiments have been carried out on muscle cells. A hybrid formed between an L_6 myoblast and a fibroblast did not differentiate and did not synthesize a peak of pulse-labeled 26 S RNA, a result in keeping with other observations on the suppression of "luxury" functions in hybrids (42). Heterokaryons between chick erythroblasts and rat (L_6) myoblasts were subsequently incorporated into myotubes by the normal process of muscle cell fusion with other rat myoblasts. Although chick surface antigens were synthesized by the reactivated erythroblast nucleus, no chick muscle myosin was detected (271). Myotubes formed by fusion between chick and rat myoblasts synthesized both types of myosin. The erythroblast result suggests that either the chick genes for muscle functions are not accessible for reactivation or that the timing of heterokaryon formation with respect to the myoblasts' own differentiation is important and that in these experiments the myoblasts had already undergone the process of activation of their own muscle genes. Further experiments of this kind, together with reconstitution experiments with myoblasts at different stages of development, would be of interest in distinguishing the roles of nuclear and cytoplasmic components during muscle cell differentiation. In order to interpret such experiments, whether between normal and variant myoblasts (48) or between myoblasts and other cell types, it is necessary to have specific probes which can be used on cell extracts or in situ. Antibodies against specific muscle proteins are available, but, in order to investigate gene expression in heterokaryons, hybrids, and reconstituted cells in more detail, cDNA probes against specific muscle mRNAs are required.

CONCLUDING COMMENTS AND SPECULATIONS

In the preceding sections a number of aspects of protein synthesis during myogenesis were discussed. The cell system itself has been considered briefly, but most attention has been directed toward the following questions: the numbers of genes coding for muscle proteins, qualitative and quantitative changes in protein synthesis on muscle cell differentiation, the possibility that mRNAs coding for muscle-specific proteins are already present in the cytoplasm of dividing myoblasts, and some tentative investigations of elements which may be important in regulating mRNA stability or utilization during differentiation. These aspects have been chosen because there are experimental data available on which to base a discussion.

Cellular Considerations

From a practical point of view, a clear-cut interpretation of many results is hampered by heterogeneity in the cell system, due to both the presence of different cell types (e.g., fibroblasts in the case of primary cultures) and to variously differentiated muscle cell types (e.g., partially differentiated cells in the myoblast population or undifferentiated cells in the cultures of myotubes). A further difficulty arises in assessing the contribution of the stationary state to general changes observed at differentiation. To some extent the situation can be clarified by isolating purified myotubes and by controlling the contribution of the stationary phase with the use of agents which prevent differentiation. Heterogeneity in the myoblast population due to mononucleated cells at different stages of development is more difficult to circumvent. The synchrony with which differentiation takes place is another limiting aspect of the system at present. It should become more susceptible to control once the physiological regulation of myogenesis in vitro is better understood.

An important theoretical point about the myogenic systems which are currently studied concerns the degree to which the myoblast population is committed to a certain program of differentiation. Among the mononucleate muscle cells there are probably a minority which are at an earlier stage of development, but the majority undergo terminal differentiation in vitro to form muscle fibers. The proteins synthesized apparently resemble those in embryonic muscle. It is not clear whether myoblasts from fetal or neonatal muscle, which is already distinguishable as fast or slow, will synthesize the muscle proteins corresponding to the tissue of origin or whether they will synthesize the contractile proteins characteristic of embryological muscle at an earlier stage of development. The difference between different types of skeletal muscle is primarily due to the kind of innervation which the fiber receives. The question is, therefore, whether mononucleated myoblasts in a muscle are influenced by its nervous control. One might predict that myoblasts will differentiate in vitro to form muscle fibers characteristic of early embryological tissue and that subsequent differentiation will take place only if the cultures are innervated in vitro by the appropriate nerve. Such experiments remain to be explored. In addition, further information is required about the proteins in early embryological muscle. At present, sequence data on the heavy chain suggest a distinctive type of myosin, but it is not yet clear, for example, what happens to the light chains at early stages of development in vivo.

Innervation of muscle cultures represents a further stage of terminal differentiation, and future biochemical studies on the in vivo system will be of interest. Nervous control of light chain synthesis, for example, raises interesting problems of regulation. The study of earlier stages of muscle cell differentiation in vitro remains less accessible. One possible approach is to study the growth and development of early mesodermal tissue in culture, although it is uncertain to what extent this will differentiate in vitro. The quantitative difficulties which arise in any biochemical study may be overcome if it proves possible to transform precursor muscle cells with temperature sensitive viruses, as described for myoblasts. This then provides a means of amplifying the material before studying its development at permissive temperatures. The study of lines derived from teratoma embryoid bodies which will differentiate spontaneously in vitro to form a range of differentiated cell types, including muscle (272), also provides potential material with which to study earlier stages of muscle cell development.

Contractile Proteins

One general idea which emerges from the discussion of contractile proteins (see under "Muscle Proteins and Their Synthesis") is that families of these proteins exist and that characteristic members are present in different types of muscle and nonmuscle tissues. At present the information, particularly about nonmuscle tissues, is limited. Chemical degradation studies suggest that more than one type of myosin heavy chain may exist in different nonmuscle tissues; it remains to be seen whether sequence analysis will reveal differences in myosin light chains. At present the data on sequence diversity are limited, but it is possible that as more tissues become documented a much greater range of heterogeneity in nonmuscle contractile proteins will become evident. Should minor differences in the proteins associated with cell motility be found in cells at earlier stages of development, then the study of differentiation during early myogenesis would immediately become more accessible to molecular investigation since biochemical markers for more primitive stages of differentiation would be available.

Related to the potential developmental interest of contractile proteins is the study of their evolution and of the organization of their genes. It seems probable that the different actins, for example, arose from gene duplication, accompanied by minor mutational divergence. It is notable that the selective pressure on the sequence of this protein must have been considerable, since only one or two minor amino acid changes are observed. The same is true to a lesser extent with other known contractile sequences. Not only does it seem likely that the existing families of contractile proteins each arose from a common ancestor, but also, as mentioned under "Muscle Proteins and Their Synthesis," sequence data would suggest that different families may have evolved from common sequences. It is possible that a common contractile protein(s) existed originally. In this context the similarities between actin and bacterial elongation factor Tu are intriguing (273). Even in bacteria this protein (Tu) seems to have several roles. It is, for example, also a subunit of Qβ-replicase (274). As far as the organization of genes for the contractile proteins is concerned, two extreme situations can be envisaged, namely, that all the genes for a given family of proteins are associated or that all the genes concerned with the contractile phenotype of a particular muscle or nonmuscle cell are grouped together. Naïvely, association could take place by localization of the genes on the same chromosome, or appropriate juxtapositioning of these elements may be achieved at interphase without any necessity for sequential organization of contractile genes on the chromosome. It is clear, for example, from genetic studies that the α- and β-globin genes are not found on the same chromosome (275, 276). A further question concerns the number of gene copies that exist for a protein such as actin, which represents a significant proportion of total protein synthesis (~6%) in most cell types. Specialized proteins like the globins are only present as one or two gene copies (230); this may also be the case for α-actin. However, more universally distributed proteins such as β- and γ-actin or tubulin may be present in multiple copies. The answers to this and other questions of gene organization await the development of cDNA probes to specific muscle proteins.

Selectivity of Transcription During Differentiation

A fundamental question about eukaryotic differentiation concerns possible selective transcription of different genes according to the developmental stage of the organism. One extreme view would be that all genes are transcribed in all tissues and their expression entirely depends upon post-transcriptional regulation. Indeed, one report claims that a few copies of globin genes are found in the nuclei of tissues which do not synthesize globin (277). However, the finding of a few copies of the gene, if significant, may simply reflect some "leakiness" of the repression system. It is clear that in many cases, such as the hormonal stimulation of ovalbumen synthesis (224), changes take place at the level of gene transcription. During development, studies have been initiated on differences between populations of RNA in different cell types. The first reports concern polysomal mRNA and suggest that there is a reduction in the number of genes expressed in the gastrula, compared with the embryo, and that mRNA in differentiated tissues in general is of lower complexity that that in the embryo or in reproductive tissue (234). Studies on primitive (multipotential) teratoma cells, compared with a muscle cell line from the same origin, suggest that the polysomal mRNA of the latter has a higher complexity and that this is reduced on terminal differentiation (231). Experiments of this kind on nonpolysomal cytoplasmic RNA and nuclear RNA will be necessary before the extent of transcriptional selection during differentiation becomes clear. It remains to be seen whether there is a reduction or an amplification in the number of structural genes transcribed during early development.

Cytoplasmic Messenger RNA During Terminal Differentiation

During the terminal stage of myogenesis there is little change in the synthesis of those proteins which are readily detectable (500–1,000). The synthesis of proteins such as the histones, which are clearly associated with cell division, is reduced, as is the synthesis of certain nonmuscle contractile proteins, whereas that of the muscle proteins becomes a significant fraction of total protein synthesis. Such observations concern the most abundant mRNA species. Analysis of the mRNA population indicates that messengers coding for muscle proteins are not accumulated in dividing myoblasts. They may, however, be present as relatively unstable nonpolysomal cytoplasmic species, as suggested by the labeling data. Hybridization studies indicate that certain mRNA species which are in the least abundant class before differentiation become abundant in differentiated cultures. Accumulation of cytoplasmic mRNAs coding for muscle proteins takes place at differentiation, and some results suggest that this event may precede the increase in muscle protein synthesis (see under "Control of Muscle Protein Synthesis During Differentiation").

These tentative conclusions about terminal myogenesis are derived from rather circumstantial experiments. The case of globin synthesis during erythropoiesis (278) or of crystallin synthesis during the terminal differentiation of lens cells (279) has been examined more directly with the use of cDNA probes for the corresponding mRNAs. In both cases, accumulation of mRNA for globin or δ-crystallin occurs at the same time as the onset of synthesis of the corresponding protein. Immature erythroid cells, for example, do not contain a pool of untranslated globin mRNA. Whether a lag is detected between the accumulation of globin mRNA and the major increase in globin synthesis depends upon the methods used; for example, cytochemical staining of hemoglobin suggested that the protein was only synthesized later (280), whereas labeling of globin (278) or of crystallin in the lens system (279) suggests that messenger and protein synthesis are coordinated. Whether or not small quantities of specific mRNAs are found in the undifferentiated cells also depends upon the sensitivity of the detection methods. In the case of globin and crystallin they are probably not present. In other systems, such as that of keratin synthesis

(281) or the differentiation of auxin-treated pea epicotyls (282), mRNA accumulation appears to precede the synthesis of differentiated protein. However, here again the sensitivity of the detection methods should be questioned. Most systems in which differentiation is under hormonal control provide clear examples of transcriptional regulation of gene expression (e.g., ovalbumen (224) or vitellogenin (283)), although there is some evidence that, once stimulated, subsequent protein synthesis may be regulated post-transcriptionally. Vitellogenin mRNA, for example, would appear to be present in a nonpolysomal form in the liver of animals which had been previously stimulated to produce the protein, but which have ceased to synthesize it (284). There are many examples of systems in which mRNA is present in the cytoplasm, but apparently is not associated with polysomes. This is the case for histone mRNAs in oocytes of sea urchins (285) or in the oocytes of silk moths, in which the mRNA is stored as messenger ribonucleoprotein (286). Even β-galactosidase in *Escherichia coli* has been shown to be present in an apparently nonpolysomal form under certain conditions (287).

The factors determining mRNA utilization and hence protein synthesis have been discussed in the case of myogenesis (see under "Possible Regulatory Mechanisms During Myogenesis"). The simplest proposition is that the quantity of mRNA present determines the extent of synthesis of the corresponding protein, hence the significance in a highly differentiated cell of the accumulation of mRNAs coding for differentiated proteins. However, it is clear that different mRNAs may be translated with varying degrees of efficiency (see under "Control of Muscle Protein Synthesis During Differentiation"), the globin system providing a clear example of this (197, 288). Another alternative is that mRNA is more or less available for translation and may even be removed from the translational cycle and retained in a storage form, either because of modifications in the translational machinery for this mRNA or because of specific masking of the mRNA.

Myogenesis as Example of Differentiation

It seems probable that mRNA concentration, different efficiencies of translation, and specific masking may all play a role in protein synthesis during development, as indeed may transcriptional and post-transcriptional control in the nucleus (289).

It is unlikely that all developmental systems conform to the same regulatory scheme, and, indeed, the terminal differentiation of muscle appears to differ from a number of other examples (e.g., globin and crystallin) in that it represents a less dramatic example of specialization at the level of protein synthesis. Structurally and functionally, a differentiated muscle fiber is obviously very highly specialized; yet in cultured myotubes myosin represents only about 5% of total protein synthesis, and the acetylcholine receptor only 0.05%, compared with, for example, globin synthesis, which is 50–70% of total protein synthesis in a reticulocyte. This difference may be related to the conditions of cell culture, but it would nonetheless appear that the metabolism of muscle is less limited than that of a reticulocyte or lens cell. Furthermore, in a myotube, the types of protein synthesized represent slightly different members of families of universal proteins. These features of muscle cell differentiation, together with its flexibility in undergoing subsequent changes on innervation, might suggest that it is a terminally differentiating system in which the regulation of phenotypic expression may be more analogous to that during earlier stages of development. Such differences between myogenesis and other examples of terminal differentiation add to its intrinsic interest as a system in which to study differentiation.

ACKNOWLEDGMENTS

I am grateful to François Gros for his support and encouragement. I should especially like to thank the following of my colleagues for helpful discussion and comments on the manuscript: N. Affara, M. Fiszman, R. Whalen, and W. Wright.

REFERENCES

1. Schubert, D., and Lacorbière, M. (1976). Proc. Natl. Acad. Sci. U.S.A. 73:1989.
2. Yaffé, D. (1969). Curr. Top. Dev. Biol. 4:37.
3. Hauschka, S. D. (1972). *In* G. H. Rothblat and V. J. Cristofalo (eds.), Growth, Nutrition and Metabolism of Cells in Culture, Vol. II, p. 67. Academic Press, New York.
4. Stockdale, F. E., and Holtzer, H. (1961). Exp. Cell Res. 61:508.
5. Emerson, C. P., and Beckner, S. K. (1975). J. Mol. Biol. 93:431.
6. Adamo, S., Zani, B., Siracusa, G., and Molinaro, M. (1976). Cell Differ. 5:53.
7. Merlie, J. P., and Gros, F. (1976). Exp. Cell Res. 97:406.
8. Moss, P. S., and Strohman, R. C. (1976). Dev. Biol. 48:431.
9. Turner, D. C., Gmür, R., Siegrist, M., Burckhardt, E., and Eppenberger, H. M. (1976). Dev. Biol. 48:258.
10. Vertel, B. M., and Fischman, D. A. (1976). Dev. Biol. 48:438.
11. Sanger, J. W., Holtzer, S., and Holtzer, H. (1971). Nature (New Biol.) 229:121.
12. Wahrmann, J. P., Drugeon, G., Delain, E., and Delain, D. (1976). Biochimie 58:551.
13. Holtzer, H., and Sanger, J. W. (1970). Proceedings of the International Scientific Conference of the Muscular Dystrophy Associations of America, Cleveland, Ohio. Excerpta Medica International Congress Series No. 240, p. 122.
14. Holtzer, H., Weintraub, H., Mayne, R., and Mochan, B. (1972). Curr. Top. Dev. Biol. 7:229.
15. Buckley, P. A., and Konigsberg, I. R. (1974). Dev. Biol. 37:193.
16. Bischoff, R., and Holtzer, H. (1967). J. Cell Biol. 36:111.
17. O'Neill, M. C., and Stockdale, F. E. (1972). Dev. Biol. 29:410.
18. Doering, J. L., and Fischman, D. A. (1974). Dev. Biol. 36:225.
19. Stockdale, F., Okazaki, K., Nameroff, M., and Holtzer, H. (1964). Science 146:533.

20. Coleman, J. R., Coleman, A. W., and Hartline, E. T. H. (1969). Dev. Biol. 19:527.
21. Bischoff, R., and Holtzer, H. (1970). J. Cell Biol. 44:134.
22. De La Haba, G., Komali, H. M., and Tiede, D. M. (1975). Proc. Natl. Acad. Sci. U.S.A. 72:2729.
23. Konigsberg, I. R. (1971). Dev. Biol. 26:133.
24. Mandel, J. L., and Pearson, M. L. (1974). Nature (Lond.) 251:618.
25. Gospodarowicz, D., Weseman, J., and Moran, J. (1975). Nature (Lond.) 256:216.
26. Den, H., Malinzak, D. A., Keaton, H. J., and Rosenberg, A. (1975). J. Cell Biol. 67:826.
27. Nameroff, M., Trotter, J. A., Keller, J. M., and Munar, E. (1973). J. Cell Biol. 58:107.
28. Kagen, L. J., Collins, K., Roberts, L., and Butt, A. (1976). Dev. Biol. 48:25.
29. Wahrmann, J. P., Winand, R., and Luzzati, D. (1973). Nature (New Biol.) 245:112.
30. Zalin, R. J. (1973). Exp. Cell Res. 78:152.
31. Epstein, C. J., De Asua, L. J., and Rozengurt, E. (1975). J. Cell. Physiol. 86:83.
32. Reporter, M. (1972). Biochem. Biophys. Res. Commun. 48:598.
33. Wahrmann, J. P., Winand, R., and Luzzati, D. (1973). Biochem. Biophys. Res. Commun. 52:576.
34. Zalin, R. J., and Montague, W. (1974). Cell 2:103.
35. Holtzer, H., Rubenstein, N., Chi, J., Dienstman, S., and Biehl, J. (1973). *In* A. Milhorat (ed.), Exploratory Concepts in Muscular Dystrophy, p. 3. Excerpta Medica, Amsterdam.
36. Garrels, J. I., and Gibson, W. (1976). Cell 9:793.
37. Richter, C., and Yaffé, D. (1970). Dev. Biol. 23:1.
38. Somers, D. G., Pearson, M. L., and Ingles, C. J. (1975). J. Biol. Chem. 250:4825.
39. Luzzati, D. (1974). Biochimie 56:1567.
40. Loomis, W. F., Jr., Wahrmann, J. P., and Luzzati, D. (1973). Proc. Natl. Acad. Sci. U.S.A. 70:425.
41. Buckingham, M. E., Cohen, A., Gros, F., Luzzati, D., Charmot, D., and Drugeon, G. (1974). Biochimie 56:1571.
42. Carlsson, S. A., Ringertz, N. R., and Savage, R. E. (1974). Exp. Cell Res. 84:255.
43. Boon, T., Buckingham, M. E., Dexter, D. L., Jakob, H., and Jacob, F. (1974). Anal. Microbiol. 125B:13.
44. Holtzer, H., Biehl, J., Yeoh, G., Meganathan, R., and Kaji, A. (1975). Proc. Natl. Acad. Sci. U.S.A. 72:4051.
45. Fiszman, M. Y., and Fuchs, P. (1975). Nature (Lond.) 254:429.
46. Epstein, H. F., and Thomson, J. N. (1974). Nature (Lond.) 250:579.
47. Linkhart, T. A., Yee, G. W., Nieberg, P. S., and Wilson, B. W. (1976). Dev. Biol. 48:447.
48. Merlie, J. P., Buckingham, M. E., and Whalen, R. G. Curr. Top. Dev. Biol., in press.
49. Streter, F. A., Gergely, J., Salmons, S., and Romanul, F. (1973). Nature (New Biol.) 241:17.
50. Harris, A. J., Heinemann, S., Schubert, D., and Tarakis, H. (1971). Nature (Lond.) 231:296.

51. Kidokoro, Y., and Heinemann, S. (1974). Nature (Lond.) 252:593.
52. Crain, S. M., and Peterson, E. R. (1974). Ann. N.Y. Acad. Sci. 228:6.
53. Nurse, C. A., and O'Lague, P. H. (1975). Proc. Natl. Acad. Sci. U.S.A. 72:1955.
54. Nelson, P., Christian, C., and Niremberg, M. (1976). Proc. Natl. Acad. Sci. U.S.A. 73:123.
55. Marks, P. A., Rifkind, R. A., and Bank, A. (1974). *In* J. Paul (ed.), MTP International Review of Science, Biochemistry Series I, Vol. 9, p. 129. University Park Press, Baltimore.
56. Pollard, T. D., and Weihing, R. R. (1974). C.R.C. Crit. Rev. Biochem. 2:1.
57. Collins, J. H. (1976). Nature (Lond.) 259:699.
58. Sarkar, S., Streter, F. A., and Gergely, J. (1971). Proc. Natl. Acad. Sci. U.S.A. 68:946.
59. Weeds, A. G., and Frank, G. (1973). Cold Spring Harbor Symp. Quant. Biol. 37:9.
60. Luzzati, D., and Drugeon, G. (1972). Biochimie 54:1157.
61. O'Farrell, P. H. (1975). J. Biol. Chem. 250:4007.
62. Whalen, R. G., Butler-Browne, G. S., and Gros, F. (1976). Proc. Natl. Acad. Sci. U.S.A. 73:2018.
63. Buckingham, M. E., Cohen, A., and Gros, F. (1976). J. Mol. Biol. 103:611.
64. Stracher, A., Balint, M., and Gergely, J. (1975). Dev. Biol. 46:317.
65. Dow, J., and Stracher, A. (1971). Biochemistry 10:1316.
66. Brandon, D. L. (1976). Eur. J. Biochem. 65:139.
67. Ash, J. F. (1975). J. Biol. Chem. 250:3560.
68. Pollard, T. D. (1975). J. Cell Biol. 67:93.
69. Offer, G., Moos, C., and Starr, R. (1973). J. Mol. Biol. 74:653.
70. Bruggmann, S., and Jenny, E. (1975). Biochim. Biophys. Acta 412:39.
71. Pollard, T., Thomas, S., and Niederman, R. (1974). Anal. Biochem. 60:258.
72. Willingham, M., Ostlund, R., and Pasten, I. (1974). Proc. Natl. Acad. Sci. U.S.A. 71:4144.
73. Weber, K., and Groeschel-Stewart, U. (1974). Proc. Natl. Acad. Sci. U.S.A. 71:4561.
74. Kessler, D., Nachmias, V. T., and Loewy, A. G. (1976). J. Cell Biol. 69:393.
75. Coleman, J. R., and Coleman, A. W. (1968). J. Cell. Physiol. (Suppl. 1) 72:19.
76. Holtzer, H., Sanger, J. W., Ishikawa, H., and Strahs, K. (1973). Cold Spring Harbor Symp. Quant. Biol. 37:549.
77. Huszar, G. (1972). Nature (New Biol.) 240:260.
78. Burridge, K., and Bray, D. (1975). J. Mol. Biol. 99:1.
79. Kimura, M., and Kielley, W. W. (1966). Biochem. Z. 345:188.
80. Chi, J. C., Fellini, S. A., and Holtzer, H. (1975). Proc. Natl. Acad. Sci. U.S.A. 72:4999.
81. Emerson, C. P. Proceedings of the International Scientific Conference of the Muscular Dystrophy Associations of America, Durango, Colorado. Excerpta Medica, in press.
82. Hynes, R. O., Martin, G. S., Shearer, M., Critchley, D. R., and Epstein, C. J. (1976). Dev. Biol. 48:35.
83. Kuan Wang, J., Ash, J. F., and Singer, S. J. (1975). Proc. Natl. Acad. Sci. U.S.A. 72:4483.

84. Yaffé, D., and Dym, H. (1973). Cold Spring Harbor Symp. Quant. Biol. 37:543.
85. Paterson, B., and Strohman, R. C. (1972). Dev. Biol. 29:113.
86. Schubert, D., Tarikas, H., Humphries, S., Heinemann, S., and Patrick, J. (1973). Dev. Biol. 33:18.
87. Rubinstein, N. A., Chi, J. C. H., and Holtzer, H. (1974). Biochem. Biophys. Res. Commun. 57:438.
88. Chi, J. C. H., Rubinstein, N., Strahs, K., and Holtzer, H. (1975). J. Cell Biol. 67:523.
89. Holtzer, H., Rubinstein, N., Dienstman, S., Chi, J. C. H., Biehl, J., and Somlyo, A. (1975). Biochimie 56:1575.
90. Sarkar, S., Streter, F. A., and Gergely, J. (1971). Proc. Natl. Acad. Sci. U.S.A. 68:946.
91. Lowey, S., and Risby, D. (1971). Nature (Lond.) 234:81.
92. Streter, F. A., Sarkar, S., and Gergely, J. (1972). Nature (New Biol.) 239:124.
93. Holt, J. C., and Lowey, S. (1975). Biochemistry 14:4600.
94. Perrie, W. T., Smillie, L. B., and Perry, S. V. (1973). Biochem. J. 135:151.
95. Kendrick-Jones, J., Szentkiralyi, E. M., and Szent-Györgyi, A. G. (1976). J. Mol. Biol. 104:747.
96. Weeds, A. G. (1976). Eur. J. Biochem. 66:157.
97. Frank, G., and Weeds, A. G. (1974). Eur. J. Biochem. 44:317.
98. Dow, J., and Stracher, A. (1971). Proc. Natl. Acad. Sci. U.S.A. 68:1107.
99. Streter, F., Holtzer, S., Gergely, J., and Holtzer, H. (1972). J. Cell Biol. 55:586.
100. Pelloni-Mueller, G., Ermini, M., and Jenny, E. (1976) FEBS Lett. 67:68.
101. Yaffé, D., Yablonka, Z., and Kessler, G. Proceedings of the International Scientific Conference of the Muscular Dystrophy Associations of America, Durango, Colorado. Excerpta Medica, in press.
102. Whalen, R. G., Buckingham, M. E., Goto, S., Merlie, J. P., and Gros, F. Proceedings of the International Scientific Conference of the Muscular Dystrophy Associations of America, Durango, Colorado. Excerpta Medica, in press.
103. Lazarides, E., and Lindberg, U. (1974). Proc. Natl. Acad. Sci. U.S.A. 71:4742.
104. Tilney, L. G., and Detmers, P. (1975). J. Cell Biol. 66:508.
105. Goldman, R. D., and Knipe, D. M. (1972). Cold Spring Harbor Symp. Quant. Biol. 37:523.
106. Lazarides, E., and Weber, K. (1974). Proc. Natl. Acad. Sci. U.S.A. 71:2268.
107. Norberg, R., Lidman, K., and Fagraeus, A. (1975). Cell 6:507.
108. Huxley, H. E. (1973). Nature (Lond.) 243:445.
109. Kersey, Y. M., Hepler, P. K., Palevitz, B. A., and Wessels, N. K. (1976). Proc. Natl. Acad. Sci. U.S.A. 73:165.
110. Pollack, R., Osborn, M., and Weber, K. (1975). Proc. Natl. Acad. Sci. U.S.A. 72:994.
111. Pollack, R., and Rifkin, D. (1975). Cell 6:495.
112. Sanger, J. W. (1975). Proc. Natl. Acad. Sci. U.S.A. 72:1913.
113. Bray, D. (1973). Cold Spring Harbor Symp. Quant. Biol. 37:567.
114. Gruenstein, E., and Rich, A. (1975). Biochem. Biophys. Res. Commun. 64:472.
115. Gruenstein, E., Rich, A., and Weihling, R. R. (1975). J. Cell Biol. 64:223.

116. Elzinga, M., Maron, B. J., and Adelstein, R. S. (1976). Science 191:94.
117. Lu, R., and Elzinga, M. (1976). Fed. Proc. 35:1359.
118. Storti, R. V., Coen, D. M., and Rich, A. (1976). Cell 8:521.
119. Storti, R. V., and Rich, A. (1976). Proc. Natl. Acad. Sci. U.S.A. 73:2346.
120. Ishikawa, H., Bishoff, R., and Holtzer, H. (1969). J. Cell Biol. 43:281.
121. Paterson, B., Roberts, B. E., and Yaffé, D. (1974). Proc. Natl. Acad. Sci. U.S.A. 71:4467.
122. Ebashi, S., and Endo, M. (1968). Prog. Biophys. Mol. Biol. 18:123.
123. Fine, R. E., and Blitz, A. L. (1975). J. Mol. Biol. 95:447.
124. Cummins, P., and Perry, S. V. (1973). Biochem. J. 133:765.
125. Cummins, P., and Perry, S. V. (1974). Biochem. J. 141:43.
126. Johnson, L. (1974). Biochim. Biophys. Acta 371:219.
127. Hodges, R. S., Sodek, J., Smillie, B., and Jarasek, L. (1973). Cold Spring Harbor Symp. Quant. Biol. 37:299.
128. Cohen, I., and Cohen, C. (1972). J. Mol. Biol. 68:383.
129. Fine, R. E., Blitz, A. C., Hitchcock, S. E., and Kaminer, B. (1973). Nature (New Biol.) 245:182.
130. Lazarides, E. (1975). J. Cell Biol. 65:549.
131. McLachlan, A. D., Stewart, M., and Smillie, L. B. (1975). J. Mol. Biol. 98:281.
132. Amphlett, G. W., Syska, H., and Perry, S. V. (1976). FEBS Lett. 63:22.
133. Roy, R. K., Potter, J. D., and Sarkar, S. (1976). Biochem. Biophys. Res. Commun. 70:28.
134. Frederiksen, D. W. (1976). Proc. Natl. Acad. Sci. U.S.A. 73:2706.
135. Perry, S. V. (1974). *In* A. T. Milhorat (ed.), Exploratory Concepts in Muscular Dystrophy II, p. 319. Excerpta Medica, Amsterdam.
136. Amphlett, G. W., Perry, S. V., Syska, H., Brown, M. D., and Urbura, G. (1975). Nature (Lond.) 257:602.
137. Syska, H., Perry, S. V., and Trayer, I. P. (1974). FEBS Lett. 40:253.
138. Amphlett, G. W., Syska, H., and Perry, S. V. (1976). FEBS Lett. 63:22.
139. Potter, J. D. (1974). Arch. Biochem. Biophys. 162:436.
140. Etlinger, J. D., and Fischman, D. A. (1973). Cold Spring Harbor Symp. Quant. Biol. 37:511.
141. Lazarides, E., and Burridge, K. (1975). Proc. Natl. Acad. Sci. U.S.A. 71:4742.
142. Lazarides, E. (1976). J. Cell Biol. 68:202.
143. Masaki, T., and Takaiti, O. (1972). J. Biochem. (Tokyo) 71:355.
144. Sarkar, S., and Cooke, P. H. (1970). Biochem. Biophys. Res. Commun. 41:918.
145. Fischman, D. A. (1970). Curr. Top. Dev. Biol. 5:235.
146. Shainberg, A., Yagil, G., and Yaffé, D. (1971). Dev. Biol. 25:1.
147. Wahrmann, J. P., Gros, F., and Luzzati, D. (1973). Biochimie 55:457.
148. Delain, D., Meienhofer, M. C., Proux, D., and Schapira, F. (1973). Differentiation 1:344.
149. Tarikas, H., and Schubert, D. (1974). Proc. Natl. Acad. Sci. U.S.A. 71:2377.
150. Turner, D. C., Maier, V., and Eppenberger, H. M. (1974). Dev. Biol. 37:63.
151. Morris, G. E., and Cole, R. J. (1972). Exp. Cell Res. 75:191.
152. Zalin, R. (1972). Biochem. J. 130:79.
153. Turner, D. C., Gmür, R., Lebherz, H. G., Siegrist, M., Wallimann, T., and Eppenberger, H. M. (1976). Dev. Biol. 48:284.
154. Turner, D. C., and Eppenberger, H. M. (1973). Enzyme 15:224.

155. Morris, G. E., Piper, M., and Cole, R. J. (1976). Nature (Lond.) 263:76.
156. Gearhart, J. D., and Mintz, B. (1975). Cell 6:61.
157. Buckingham, M. E., Cohen, A., and Gros, F. Manuscript in preparation.
158. Meissner, G., Conner, G. E., and Fleischer, S. (1973). Biochim. Biophys. Acta 298(2):246.
159. Margreth, A., Salviati, G., Salviati, S., and Dalla Libera, L. (1975). *In* E. Carafoli (ed.), Calcium Transport in Contraction and Secretion, p. 383. North Holland, Amsterdam.
160. Holland, D. L., and MacLennan, D. H. (1976). J. Biol. Chem. 251:2030.
161. Rash, J. E., and Fambrough, D. (1973). Dev. Biol. 30:166.
162. Changeux, J. P. (1975). *In* L. L. Iverson, S. D. Iverson, and S. H. Snyder (eds.), Handbook of Psychopharmacology, Vol. 6, p. 235. Plenum Press, New York.
163. Briley, M. S., and Changeux, J. P. *In* R. J. Bradley (ed.), International Review of Neurology, Vol. 19. Academic Press, New York, in press.
164. Merlie, J. P., Sobel, A., Changeux, J. P., and Gros, F. (1975). Proc. Natl. Acad. Sci. U.S.A. 72:4028.
165. Fambrough, D., and Rash, J. E. (1971). Dev. Biol. 26:55.
166. Patrick, J., Heinemann, S. F., Lindstrom, J., Schubert, D., and Steinbach, J. H. (1972). Proc. Natl. Acad. Sci. U.S.A. 69:2762.
167. Hartzell, H. C., and Fambrough, D. M. (1973). Dev. Biol. 30:153.
168. Sytkowski, A. J., Vogel, Z., and Nirenberg, M. (1973). Proc. Natl. Acad. Sci. U.S.A. 70:270.
169. Patterson, B., and Prives, J. (1973). J. Cell Biol. 59:241.
170. Devreotes, P., and Fambrough, D. M. (1975). J. Cell Biol. 65:335.
171. Devreotes, P. N., and Fambrough, D. M. (1976). Proc. Natl. Acad. Sci. U.S.A. 73:161.
172. Huestis, W. H., and McConnell, H. M. (1974). Biochem. Biophys. Res. Commun. 57:726.
173. Prives, J., Silman, I., and Amsterdam, A. (1976). Cell 7:543.
174. Merlie, J. P., Changeux, J. P., and Gros, F. (1976). Nature (Lond.) 264:74.
175. Berg, D. K., and Hall, Z. W. (1974). Science 184:473.
176. Brockes, J. P., and Hall, Z. W. (1975). Proc. Natl. Acad. Sci. U.S.A. 72:1368.
177. Rieger, F., Bon, S., Massoulié, J., Cartaud, J., Picard, B., and Benda, P. (1976). Eur. J. Biochem. 68:513.
178. Bon, S., Huet, M., Lemonnier, M., Rieger, F., and Massoulié, J. (1976). Eur. J. Biochem. 68:523.
179. Wilson, B. W., Nieberg, P. S., Walker, C. R., Linkhart, T. A., and Fry, D. M. (1973). Dev. Biol. 33:285.
180. Fluck, R. A., and Strohman, R. C. (1973). Dev. Biol. 33:417.
181. Wilson, B. W., and Walker, L. R. (1974). Proc. Natl. Acad. Sci. U.S.A. 71:3194.
182. Harvey, A. L., and Dryden, W. F. (1974). Differentiation 2:237.
183. Simantov, R., and Sachs, L. (1973). Proc. Natl. Acad. Sci. U.S.A. 70: 2902.
184. Oh, T. H., and Johnson, D. D. (1972). Exp. Neurol. 37:360.
185. Singer, R. H., and Penman, S. (1972). Nature (Lond.) 240:100.
186. Murphy, L., and Attardi, G. (1973). Proc. Natl. Acad. Sci. U.S.A. 70:115.
187. Tomkins, G. M., Gelehrter, T. D., Granner, D., Martin, D., Jr., Samuels, H. H., and Thompson, E. B. (1969). Science 166:1474.
188. Schimke, R. T. (1974). *In* J. Paul (ed.), MTP International Review of

Science, Biochemistry Series I, Vol. 9, p. 183. University Park Press, Baltimore.
189. Nuss, D. L., and Koch, G. (1976). J. Mol. Biol. 102:601.
190. Molinaro, M., Zani, B., Martinozzi, M., and Monesi, V. (1974). Exp. Cell Res. 88:402.
191. Yaffé, D., Yablonka, Z., Kessler, G., and Dym, H. (1975). *In* G. Bernardi and F. Gros (eds.), Proceedings of the Tenth FEBS Meeting, Vol. 38, p. 313. North Holland, Amsterdam.
192. Gozes, I., Schmitt, H., and Littauer, U. Z. (1975). Proc. Natl. Acad. Sci. U.S.A. 72:701.
193. Schreier, M. H., and Staehelin, T. (1973). J. Mol. Biol. 73:329.
194. Bag, J., and Sarkar, S. (1975). Biochemistry 14:3800.
195. Strohman, R. C., Moss, P. S., Micou-Eastwood, J., Spector, D., Przybyla, A., and Paterson, B. (1977). Cell 10:265.
196. Przybyla, A., and Strohman, R. C. (1974). Proc. Natl. Acad. Sci. U.S.A. 71:662.
197. Lodish, H. F. (1974). Nature (Lond.) 251:385.
198. Heywood, S. M., and Nwagwu, M. (1969). Biochemistry 8:3839.
199. Rourke, A. W., and Heywood, S. M. (1972). Biochemistry 11:2061.
200. Sarkar, S., Mukherjee, S. P., Sutton, A., Mondal, H., and Chen, V. (1973). Prep. Biochem. 3(6):583.
201. Mondal, H., Sutton, A., Chen, V. J., and Sarkar, S. (1974). Biochem. Biophys. Res. Commun. 56:988.
202. Heywood, S. M., Kennedy, D. S., and Bester, A. J. (1975). FEBS Lett. 53:69.
203. Morris, G. E., Buzash, E., Rourke, A., Tepperman, I. C., Thompson, W., and Heywood, S. M. (1973). Cold Spring Harbor Symp. Quant. Biol. 37:535.
204. Heywood, S. M., Kennedy, D. S., and Bester, A. J. (1975). Eur. J. Biochem. 58:587.
205. Perry, R. P. (1976). Annu. Rev. Biochem. 45:605.
206. Birnstiel, M. L., Sells, B. H., and Purdom, I. F. (1972). J. Mol. Biol. 63:21.
207. Robbins, J., and Heywood, S. M. (1976). Biochem. Biophys. Res. Commun. 68:918.
208. Robbins, J., and Heywood, S. M. Proc. Natl. Acad. Sci. U.S.A., in press.
209. Palacios, R., Sullivan, D., Summers, M. N., Kiely, M. L., and Schimke, R. T. (1973). J. Biol. Chem. 248:540.
210. Rabbitts, T. H. (1976). Nature (Lond.) 260:221.
211. Buckingham, M. E., Caput, D., Cohen, A., Whalen, R. G., and Gros, F. (1974). Proc. Natl. Acad. Sci. U.S.A. 71:1466.
212. Luzzati, D., and Drugeon, G. (1972). Biochimie 54:1157.
213. Buckingham, M. E., Goto, S., Whalen, R. G., and Gros, F. (1975). *In* G. Bernardi and F. Gros (eds.), Proceedings of the Tenth FEBS Meeting, Vol. 38, p. 325. North Holland, Amsterdam.
214. Man, N., and Cole, R. J. (1972). Exp. Cell Res. 72:429.
215. Clissold, P., and Cole, R. J. (1973). Exp. Cell Res. 80:159.
216. Man, N. T., and Cole, R. J. (1974). Exp. Cell Res. 83:328.
217. Marchock, A. C., and Wolff, J. A. (1968). Biochim. Biophys. Acta 155:378.
218. Greenberg, J. R. (1972). Nature (Lond.) 240:102.
219. Singer, R. H., and Penman, S. (1973). J. Mol. Biol. 78:321.
220. Spadling, A., Hui, H., and Penman, S. (1975). Cell 4:131.

221. Berger, S. L., and Cooper, H. L. (1975). Proc. Natl. Acad. Sci. U.S.A. 72:3873.
222. Sensky, T. E., Haines, M. E., and Rees, K. R. (1975). Biochim. Biophys. Acta 407:430.
223. Hunt, J. A. (1974). Biochem. J. 138:487.
224. Palmiter, R. D. (1975). Cell 4:189.
225. Kafatos, F. C. (1972). Curr. Top. Dev. Biol. 7:125.
226. Lodish, H. F., and Small, B. (1976). Cell 7:59.
227. Aviv, H., Voloch, Z., Bastos, R., and Levy, S. (1976). Cell 8:495.
228. Rosbash, M., and Ford, P. J. (1974). J. Mol. Biol. 85:87.
229. Goto, S., Buckingham, M. E., and Gros, F. In press.
230. Lewin, B. (1975). Cell 4:77.
231. Affara, N., Jacquet, M., Gros, F., Jakob, E., and Jacob, F. Manuscript in preparation.
232. Williams, J. G., and Penman, S. (1975). Cell 6:197.
233. Campo, M. S., and Bishop, J. O. (1974). J. Mol. Biol. 90:649.
234. Galan, G. A., Britten, R. J., and Davidson, E. H. (1974). Cell 2:9.
235. Colbert, D. A., Edwards, K., and Coleman, J. R. (1976). Differentiation 5:91.
236. Jacquet, M., Caput, D., Falcoff, E., Falcoff, R., and Gros, F. (1977). Biochimie 59:189.
237. Birnie, G. D., Macphail, E., Young, B. D., Getz, M. J., and Paul, J. (1974). Cell Differ. 3:221.
238. Man, N. T., Morris, G. E., and Cole, R. J. (1975). Dev. Biol. 47:81.
239. Brawerman, G. (1974). Annu. Rev. Biochem. 43:621.
240. Heywood, S. M., Dowben, R. M., and Rich, A. (1967). Proc. Natl. Acad. Sci. U.S.A. 57:1002.
241. Heywood, S. M., and Rich, A. (1968). Proc. Natl. Acad. Sci. U.S.A. 59:590.
242. Sarkar, S., and Cooke, P. H. (1970). Biochem. Biophys. Res. Commun. 41:918.
243. Low, R. B., Vournakis, J. N., and Rich, A. (1971). Biochemistry 10:1813.
244. Brawerman, G. (1976). Prog. Nucleic Acid Res. Mol. Biol. 17:118.
245. Kaufman, S. J., and Gross, K. W. (1974). Biochim. Biophys. Acta 353:133.
246. Whalen, R. G., and Gros, F. (1976). Manuscript in preparation.
247. Buckingham, M. E., and Gros, F. (1975). FEBS Lett. 53:355.
248. Young, R. B., Goll, D. E., and Stromer, M. H. (1975). Dev. Biol. 47:123.
249. Spirin, A. S. (1969). Eur. J. Biochem. 10:20.
250. Williamson, R. (1973). FEBS Lett. 37:1.
251. Lindberg, U., and Sundquist, B. (1974). J. Mol. Biol. 86:451.
252. Buckingham, M. E., and Gros, F. (1975). *In* D. Rickwood (ed.), Proceedings of a Colloquium on the Use of Iodinated Density Gradient Media for Biological Separations, p. 71. Information Retrieval.
253. Kwan, S. W., and Brawerman, G. (1972). Proc. Natl. Acad. Sci. U.S.A. 69:3247.
254. Blobel, G. (1973). Proc. Natl. Acad. Sci. U.S.A. 70:924.
255. Bester, A. J., Kennedy, D. S., and Heywood, S. M. (1975). Proc. Natl. Acad. Sci. U.S.A. 72:1523.
256. Kennedy, D. S., Bester, A. J., and Heywood, S. M. (1974). Biochem. Biophys. Res. Commun. 61:415.
257. Heywood, S. M., Kennedy, D. S., and Bester, A. J. (1974). Proc. Natl. Acad. Sci. U.S.A. 71:2428.

258. Bogdanovsky, D., Hermann, W., and Schapira, G. (1973). Biochem. Biophys. Res. Commun. 54:25.
259. Atkins, J. F., Lewis, J. B., Anderson, C. W., and Gestland, R. F. (1975). J. Biol. Chem. 250:5688.
260. Salden, M., and Bloemendal, H. (1976). Biochem. Biophys. Res. Commun. 68:157.
261. Heywood, S. M. (1969). Cold Spring Harbor Symp. Quant. Biol. 34:799.
262. Heywood, S. M. (1970). Nature (Lond.) 255:696.
263. Heywood, S. M., and Thompson, W. C. (1971). Biochem. Biophys. Res. Commun. 43:470.
264. Heywood, S. M. (1970). Proc. Natl. Acad. Sci. U.S.A. 67:1782.
265. Thompson, W. C., Buzash, E. A., and Heywood, S. M. (1973). Biochemistry 12:4559.
266. Heywood, S. M., and Kennedy, D. S. (1974). Dev. Biol. 38:390.
267. Whalen, R. G., and Gros, F. (1977). Biochim. Biophys. Acta 475:393.
268. Marcker, K. A., Blair, G. E., Dahl, H. H. M., and Lelong, J. C. (1975). *In* F. Chapeville and M. Grunberg-Manago (eds.), Proceedings of the Tenth FEBS Meeting, Vol. 39, p. 297. North Holland, Amsterdam.
269. England, J. M., Howett, M. K., and Tan, K. B. (1975). J. Virol. 16:1101.
270. Steitz, J. A., and Jakes, K. (1975). Proc. Natl. Acad. Sci. U.S.A. 72:4734.
271. Ege, T., Krondahl, U., and Ringertz, N. R. (1974). Exp. Cell Res. 88:428.
272. Nicolas, J. F., Dubois, P., Jakob, H., Gaillard, J., and Jacob, F. (1976). Ann. Microbiol. 126A:1.
273. Beck, B., Arscott, P. G., and Jacobson, A. Proc. Natl. Acad. Sci. U.S.A., in press.
274. Weissmann, C., Billeter, M. A., Goodman, H. M., Hindley, J., and Weber, H. (1973). Ann. Rev. Biochem. 42:303.
275. Lehmann, H., and Huntsman, R. G. (1966). Man's Haemoglobins. North Holland, Amsterdam.
276. Deisseroth, A., Velez, R., and Nienhuis, A. W. Science, in press.
277. Humphries, S., Windass, J., and Williamson, R. (1976). Cell 7:267.
278. Harrison, P. R. (1976). Nature (Lond.) 262:353.
279. Milstone, L. M., Zelenka, P., and Piatigorsky, J. (1976). Dev. Biol. 48:197.
280. Ramirez, F., Giambino, R., Maniatis, G. M., Rifking, R. A., Marks, P. A., and Bank, A. (1975). J. Biol. Chem. 250:6054.
281. Desvaux-Chabrol, J. (1976). Biochimie 58:563.
282. Verma, D. P. S., MacLachlan, G. A., Byrne, H., and Ewings, D. (1975). J. Biol. Chem. 250:1019.
283. Shapiro, D. J., Baker, H. J., and Stitt, D. T. (1976). J. Biol. Chem. 251:3105.
284. Mullinix, K. P., Wetekam, W., Deeley, R. G., Gordon, J. I., Meyers, M., Kent, K. A., and Goldberger, R. F. (1976). Proc. Natl. Acad. Sci. U.S.A. 73:1442.
285. Lifton, R. P., and Kedes, L. H. (1976). Dev. Biol. 48:47.
286. Paglia, L. M., Kastern, W. H., and Berry, S. J. (1976). Dev. Biol. 51:182.
287. Ennis, H. L., and Kievitt, K. D. (1976). J. Biol. Chem. 251:2854.
288. Orkin, S. H., Swan, D., and Leder, P. (1975). J. Biol. Chem. 250:8753.
289. Scherrer, K. (1974). *In* A. Kohn and A. Statkay (eds.), Control of Gene Expression, p. 169. Plenum Publishing Corporation, New York.

International Review of Biochemistry
Biochemistry of Cell Differentiation II, Volume 15
Edited by J. Paul

8
Teratocarcinoma Cells as a Model for Mammalian Development

B. L. M. HOGAN
Imperial Cancer Research Fund,
Mill Hill Laboratories,
London, England

Teratomas are tumors arising in the testis or ovary or in early embryos which have been implanted into extrauterine sites. As befits the derivation of their name from the Greek "teraton"—meaning monster or wondrous spectacle—they consist of a chaotic mixture of adult and embryonic tissues, varying in the maturity of their differentiation, but easily recognizable histologically as skin, nerve, muscle, cartilage, gut, trophoblast, yolk sac, etc. Interspersed among these tissues are nests of undifferentiated "embryonal" cells, which are the pluripotent stem cells of the tumors. Rather different criteria have been used by clinicians and experimental biologists to classify tumors into teratomas and teratocarcinomas. A clinician calls a tumor a teratocarcinoma if, in addition to differentiated tissues, it contains undifferentiated embryonal cells and is likely to metastasize to other sites. A rather different definition is used by experimental biologists. To them a tumor is a teratocarcinoma so long as it contains embryonal cells, grows progressively, and is transplantable to syngeneic animals, whereas a tumor is a teratoma if the pool of stem cells has been exhausted by differentiation or necrosis and the tumor grows slowly (if at all) and is not transplantable. The latter terminology will be used here.

It is only recently that the full potential of teratocarcinomas as experimental material for studying biochemically the processes of determination and differentiation in mammalian cells has been recognized. This has followed from the development of techniques for culturing large numbers of the pluripotent stem cells and triggering their differentiation in vitro, together with the discovery of several biochemical and immunological similarities between teratocarcinoma cells and cells of the normal early embryo. But probably the most decisive factor in establishing teratocarcinomas as legitimate material for developmental biologists to study has been the beautiful experiments of Mintz and Illmensee (1) and Papioannou et al. (2), in which mouse teratocarcinoma cells have been integrated into the inner cell mass of 4-day embryos and have given rise in some cases to apparently completely normal chimeric adults with derivatives of teratocarcinoma cells incorporated into many different tissues, including the germ line. These experiments show that when placed in the environment of the early embryo some teratocarcinoma cells, at least, are able to participate in normal morphogenesis and differentiation, which in turn means that their biochemical properties are, or can be, very similar to those of normal embryos.

The aims of this review are, firstly, to explore the analogies which can be made between teratocarcinoma cells and cells of the normal early embryo and, secondly, to discuss critically the argument that the way in which teratocarcinoma cells differentiate in vitro closely resembles both morphologically and biochemically the behavior of normal embryos. By necessity most of the evidence comes from work with teratocarcinomas of mice, but some reference has been made to work with human cells. Although the histology and natural history of in vivo tumors are discussed to some extent under "Histogenesis of Teratomas in Vivo," readers seeking detailed accounts are referred to earlier reviews (3–5). Recent in vitro experiments with cultured teratocarcinoma cells have been reviewed by Martin (6) and much useful background material, as well as new experimental data, is to be found in the proceedings of a meeting held at the Roche Institute, New Jersey, in May, 1975, published as a book entitled *Teratomas and Differentiation* (7).

SUMMARY OF EARLY DEVELOPMENT OF MOUSE EMBRYO

In order that the reader may fully understand the analogies which are made between teratocarcinoma cells and those of the normal embryo, it is necessary at this stage to summarize briefly the key events in early mouse embryogenesis. No attempt is made to provide a comprehensive account, and fuller details are available in several excellent articles and reviews (8, 9).

Three days after fertilization, the embryo is a ball or morula of 32 cells (Figure 1). At around this time the outer layer of cells becomes morphologically different from the rest and is known as the trophectoderm. The outside cells pump water inward to make a fluid-filled cavity or blastocele, while the enclosed cells are displaced to one side to form a disk or inner cell mass (ICM) (Figure 1*b*). It is not known exactly when the outer cells become restricted, or "determined," in their fate and irreversibly committed to form trophectoderm, but the process seems to occur around the 16–32-cell stage. Prior to this, at the 8-cell stage, there is good evidence that all the cells of the morula are equivalent in their developmental potential (10, 11). According to the so-called "inside-outside" hypothesis, the determination of the late morula is related to the position or microenvironment in which the cells find themselves at the time when the fluid pumping activity starts; labeled cells on the outside nearly always colonize the trophectoderm, whereas inside cells are mostly found to have formed part of the inner cell mass (12). If cells of the 8-cell morula are separated so that they are prevented from becoming part of a ball of cells with an inside and outside,

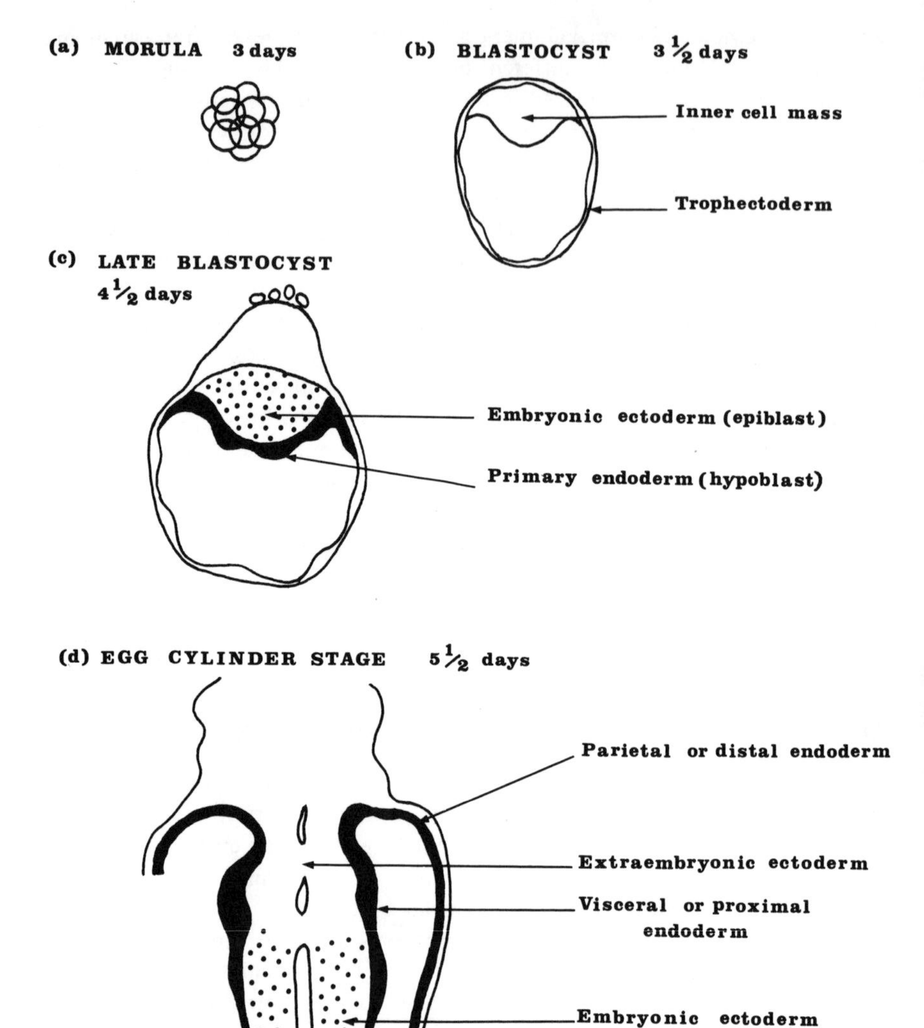

Figure 1. Schematic representation of the early development of the mouse embryo. The diagrams are not drawn to scale.

then each one, or both of its descendants, develops morphologically and biochemically into a trophectoderm cell (13).

A detailled study has been made by Ducibella et al. (14) of the intercellular junctions in late morulae and early blastocysts with the use of lanthanum to see if molecules can penetrate from the outside into the spaces between the cells. In

16–32-cell morulae, a continuous network of tight junctions between the outer layer of cells provides an effective seal against the penetration of the lanthanum, but this is not true of the tight junctions seen in earlier morulae. The inside cells of late morulae and early blastocysts are connected to each other and to outside cells through gap junctions. It is not known whether the appearance of the apical network of tight junctions and the determination of the trophectoderm are causally related, but the network could allow the buildup of certain molecules inside the morula, and these might influence the differentiation of the inner cells.

At 3½ days the ICM consists of about 15 cells, and during the next 24 hr a second determination event takes place (Figure 1*c*) so that the outer layer of cells in contact with the blastocele fluid becomes committed to form primary endoderm, or hypoblast, while the enclosed cells go on to form what is known as epiblast or, misleadingly, as embryonic ectoderm (in fact, it is thought that this group of cells gives rise to all the tissues of the fetus, including endoderm-derived tissues such as gut (see under "Origin from Embryonic Ectoderm"). In contrast to the epiblast the outer endoderm cells have an extensive endoplasmic reticulum with many ribosomes, intracellular vacuoles, microvilli on the outer surface, and a certain amount of basement membrane material on their inner surface. Intercellular junctions in the ICM of early-to-late blastocysts have been studied by Nadijcka and Hillman (15). They found tight junctions between like and unlike cells at all stages, but it is not known whether those between endoderm cells form an effective barrier to lanthanum and other large molecules.

There is some evidence that inside-outside cell position may be an important factor in determining the fate of cells in the ICM. For example, the ICM may be isolated from 3½-day blastocysts manually (16) or by the technique of immunosurgery, which involves killing the outer trophectoderm cells by incubating the blastocysts successively in rabbit anti-mouse spleen antiserum and complement (17). (The inner cells are not killed because the antiserum cannot penetrate between the trophectoderm cells.) During the next 24 hr the isolated ICM forms a continuous layer of endoderm cells over its entire surface, not just on the side which was exposed to the blastocele (see Figure 3). If two or more of these isolated 3½-day ICMs are fused together, then only a single layer of endoderm cells develops. This shows that, when exposed to "outside" conditions, cells of the ICM do not form trophoblast, but instead differentiate into endoderm. The fact that only a single monolayer of endoderm cells develops whether the aggregate is smaller than normal (due to damage during manipulation) or larger (due to fusion between ICMs) makes it unlikely that there is sorting out or migration of predetermined but cytologically undifferentiated endoderm cells to the outside, although this possibility has not been formally disproved. Unlike the separated 8-cell morulae, it is not known whether individual cells isolated from the 3½-day ICM will all differentiate into endoderm.

During subsequent stages of development (Figure 1*d*) the egg cylinder becomes elaborated into the fetus proper and two subclasses of extraembryonic endoderm become distinguishable—namely, parietal and visceral endoderm. Cells of the parietal endoderm adjacent to the trophoblast secrete large amounts of basement membrane to form what is known as Reichert's membrane (see under "Basement Membrane Material"). The visceral endoderm cells form part of the visceral yolk sac, which is the first hematopoietic tissue of the embryo. During midgestation, when the parietal yolk sac has broken down, it is the main route through which proteins are actively absorbed from the maternal blood into the fetal circulation. The visceral endoderm cells have enormous numbers of microvilli, are actively pinocytotic and highly vacuolated, and are separated by a thin basement membrane from underlying mesodermal cells, in which islands of hematopoietic tissue are found.

HISTOGENESIS OF TERATOMAS IN VIVO

The description of tumors growing in vivo given here is not meant to provide more than a background to understanding the origin and in vitro behavior of teratocarcinoma cells. Fuller details can be found in other reviews (3, 18).

Spontaneous Testicular Teratomas in Mice and Humans

The development of teratocarcinomas in the testes of the 129/Sv strain of mice has been studied carefully by Stevens (18, 19). Up to 10% of the males of this strain have congenital testicular teratomas, and the incidence can be as high as 30% in the subline 129/ter Sv (20). The tumors are first seen as unorganized clusters of undifferentiated cells during the 12th day of gestation within the tubules of the fetal testis. In slightly older embryos they are seen to have formed two distinct and separate kinds of epithelial arrangements of cells, each surrounding a cavity; these epithelia may be closely apposed over a limited region or separated by mesodermal cells. One kind of epithelial arrangement resembles the embryonic ectoderm of 6-day mouse embryos (see above and Figure 1*d*); the other resembles endoderm, but since this does not surround the ectodermal cells it cannot be directly compared with the primary endoderm layer of normal embryos (see above and Figure 1*c*). In older animals some of the ectodermal vesicles have developed into neuroepithelium, whereas the endoderm vesicles have matured into ciliated and secretory gutlike epithelium. Later still, up to 15 different tissues may be found, including skin; neural tissue; cartilage; bone with marrow; skeletal, cardiac, and smooth muscle; connective tissue; and respiratory and alimentary epithelium. However, in thousands of samples, no liver, kidney, or gonad has been seen and trophoblast is also very rare. Interspersed among the mature tissue are nests of undifferentiated cells, but by the time the mouse is about 30 days old these have undergone differentiation and the tumor becomes a teratoma. In a small number of cases the undifferentiated embryonal cells persist, and the tumor grows progressively and is transplantable. Proof that the embryonal cells are the stem cells of the tumor was obtained by Kleinsmith and Pierce (21), who showed that 10% of all isolated single embryonal cells (from an ascites tumor) injected intraperitoneally gave rise to tumors with a whole range of differentiated tissues.

When transferred to the peritoneum, testicular teratocarcinomas multiply as solid tumors on the mesentery and as numerous floating aggregates known as embryoid bodies. The simplest of these consist of an inner core of undifferentiated cells surrounded by a monolayer of cells resembling the primary endoderm of 4½–5 day embryos. More complex, cystic embryoid bodies are also found, and these contain visceral and parietal endoderm, neuroepithelium, connective tissue, and hematopoietic cells. Embryoid bodies have also been observed to form in solid tumors transplanted subcutaneously, but never in primary tumors.

During the serial transplantation of testicular teratocarcinomas, changes may sometimes occur in their ability to differentiate (22). At one extreme the tumor may differentiate completely and become benign. Alternatively, the tumor may contain progressively less differentiated tissue, with embryonal cells predominating; the ultimate result of this progression may be a tumor consisting of embryonal cells alone. Finally, the tumor may come to consist predominantly of one kind of differentiated tissue, such as muscle, nerve, or, most commonly, parietal endoderm (23). Embryonal cells may be present, but these are gradually lost as the differentiated cells, which have probably undergone malignant transformation, overgrow the other tissues. There is little evidence that embryonal cells ever become restricted in the sense of giving rise consistently to only one kind of differentiated tissue.

The commonest form of testicular teratocarcinoma in man is not congenital, as in 129/Sv mice, but arises later in life, with a peak incidence at around 30 years of age. The tumors frequently metastasize, producing secondary tumors containing either a wide range of differentiated tissues or only highly malignant embryonal carcinoma or trophoblast (choriocarcinoma) cells. Trophoblast, which secretes chorionic gonadotropin, is also frequently found in the primary tumors (24, 25). Closely apposed vesicles of ectoderm and endoderm similar to those in fetal 129 mouse testis have been observed in human teratocarcinomas (25).

Spontaneous Ovarian Teratomas in Mice and Humans

In mice, spontaneous ovarian teratomas are very rare except in the inbred LT/Sv strain, in which the incidence is about 50%. In these mice, tumors arise from parthenogenetically activated ova which develop almost normally through morula and blastocyst stages to the egg cylinder stage (equivalent to about 6½ days of normal development). They then become disorganized, and undifferentiated cells migrate into the surrounding tissue (18). Most tumors soon become completely differentiated, but about 20% retain undifferentiated embryonal cells and can be transplanted, when they behave very much like their testicular-derived counterparts.

Human ovarian teratomas are rare and are usually benign. The differentiated tissues are often quite well organized, and it is in these tumors that fingers, teeth, and clumps of hair have been described (24).

Tumors Derived from Embryos Implanted into Extrauterine Sites

In general, it is possible to obtain teratocarcinomas experimentally by implanting early mouse embryos beneath the testis or kidney capsule of syngeneic hosts. However, a number of qualifications to this statement must be made (26). Firstly, in certain strains of mice, in particular C57BL and AKR, if the graft is successful the embryonic cells soon undergo differentiation, and only mature teratomas are found at the site of implantation. This is a function of the host, since C57BL embryos implanted into C57BL × C3H hybrids, for example, do develop with a reasonable frequency into teratocarcinomas retaining large numbers of undifferentiated cells. Secondly, in permissive strains, the age of the embryo when grafted influences the frequency with which teratocarcinomas, as opposed to mature teratomas, are obtained. Up to the age of 6–7 days there is a progressive increase in the frequency of teratocarcinomas (to about 50% of all successful grafts), but it falls off sharply after that. Younger embryos seem to develop normally to the 6–7-day stage and then become disorganized (27), as described for LT/Sv ovarian tumors; they then either differentiate into teratomas or form transplantable teratocarcinomas which behave very much like transplantable testicular tumors.

The most extensively studied of all teratocarcinomas, the OTT 6050 tumor, was derived by Stevens in 1967 from a 6-day 129/Sv $Sl^J\ C\ P$ embryo implanted into the testis (28). The primary tumor contained a range of differentiated cells as well as embryonal cells, and spread up the lymph vessels to the left kidney. At the tenth transplant generation part of the tumor was injected into the peritoneum, where it formed simple and cystic embryoid bodies. Several sublines of the ascites tumor were established, e.g., OTT 6050A and OTT 6050B.

In conclusion, although their original histogenesis is rather different, testicular-, ovarian-, or embryo-derived teratocarcinomas in mice all behave in very much the same way when transplanted, and all will form embryoid bodies. The secondary solid tumors contain embryonal cells and usually a wide range of differentiated tissues, occasionally including cells morphologically like trophoblast (29–31). However, it is not possible to deduce from serial sections of solid tumors the exact sequence in which different tissues appeared and how this relates to normal embryogenesis, in which, for example, trophoblast appears before endoderm, which precedes mesoderm, etc. (see under "Summary of Early Development of Mouse Embryo").

Early in normal embryogenesis the family tree of cells derived from the zygote is partitioned into groups or blocks which come to have the same appearance (i.e., to express the same sets of genes). This happens quite regularly, and the cells always have the same spatial relationship to each other, which has allowed embryologists to classify blocks of cells into classes, or germ layers, called endoderm, ectoderm, and mesoderm. However, it is fairly common in adults for one class of cell to express, for obscure reasons, a set of genes usually associated with another germ layer. For example, in humans patches of cells in the stomach or pancreas (endoderm) may synthesize keratin (usually associated with ectoderm); this condition is known as squamous metaplasia (32). In teratomas, in which there is no set pattern or regularity to cell lineages and spatial relationships, the classification of cells into germ layers has less meaning. Nevertheless, for sheer convenience, the terms "ectoderm," "endoderm," and "mesoderm" are used here in describing the differentiation of teratocarcinoma cells in the subsequent sections.

THEORIES ABOUT ORIGIN OF TERATOCARCINOMA CELLS

As has been seen, pluripotent teratocarcinoma cells can arise either in the gonads or in ectopic embryos. However, in neither case is it clear exactly which cells are the precursors of the embryonal stem cells or whether their conversion involves a genetic event, such as a mutation or chromosome rearrangement, or release from environmental controls by disruption of normal cell interactions, or both (see under "Integration of Teratocarcinoma Cells into Normal Mouse Embryos").

There are two main theories about the origin of teratocarcinoma cells: they arise from germ cells or they arise from cells in the embryonic ectoderm (or epiblast) (see Figure 1*d*). As shall be seen, both theories may be to a certain extent correct, but until the rather gray area surrounding the relation of germ cells to embryonic ectoderm is cleared up the problem is likely to remain somewhat confused.

Origin from Primordial Germ Cells

The germ cells of mammals undergo an extraordinary migration before they take up their position in the germinal ridges. In mouse embryos this has been followed by staining sections histochemically for alkaline phosphatase, an enzyme which is particularly active in germ cells (see below). A few positive cells can be detected in the primitive streak mesoderm of the 7½-day embryo (8). (The primitive streak is formed at about 7 days when cells of the embryonic ectoderm migrate between the epiblast and the endoderm, forming an intermediate layer of mesoderm.) The alkaline phosphatase-positive cells first migrate to the visceral yolk sac endoderm and later along the hind gut to the genital ridges.

There is good evidence that in mice testicular teratocarcinomas arise from germ cells. About 80% of the genital ridges of 11½–12½-day male (but not female) 129/Sv mouse embryos give rise to terato(carcino)mas when grafted into the adult testis. However, if the genital ridges are grafted from 129 mice homozygous for the Sl mutation, which blocks the early migration and proliferation of primordial germ cells, then no tumors are produced (33). What is not clear is the mechanism by which a male germ cell is converted into a teratocarcinoma stem cell and why this normally happens only in the 129 strain of mice and not in others.

During spermatogenesis a diploid ($2n$) spermatogonium gives rise to a $4n$ primary spermatocyte. In the first meiotic division this gives two $2n$ secondary spermatocytes (one essentially XX and the other YY), each homozygous for most genetic loci at which the animal is heterozygous, except where crossing over has taken place. The second meiotic division gives four haploid spermatids. Theoretically a teratocarcinoma stem cell could arise from either a spermatogonium, a secondary spermatocyte in which the second meiotic division has been suppressed, or a haploid spermatid which has doubled its chromosomes. The last two possibilities would be eliminated if at least some testicular teratocarcinoma cells could be shown to contain both an X and a Y chromosome. Although some studies have been made on the karyotype of testicular teratocarcinoma cells (34, 35), until recently techniques for identifying chromosomes have not really been accurate enough. However, G. G. Magrane (personal communication) has now shown by G-banding that in a nonclonal line of cultured human testicular teratocarcinoma cells some cells have two X chromosomes and two Y chromosomes, while others have one X and no Y. These results are compatible with the stem cell arising from a spermatogonium.

Additional evidence which has been cited in favor of the derivation of embryonal carcinoma cells from germ cells is the high level of alkaline phosphatase activity in the two cell types and their ultrastructural similarities (36). However, as discussed by Damjanov and Solter (37), embryonal cells (whether testicular-, ovarian-, or embryo-derived) are morphologically very similar to embryonic ectoderm cells, which also have high alkaline phosphatase levels. The histogenesis of testicular teratocarcinomas outlined under "Histogenesis of Teratomas in Vivo" precludes the possibility that a male germ cell gives rise to a teratocarcinoma stem cell by behaving like a fertilized zygote and developing through normal embryo stages to yield embryonic ectoderm cells.

As far as ovarian teratomas are concerned, in the LT strain of mice there is little doubt that the tumors are derived from parthenogenetically activated eggs which develop to the equivalent of the 7-day embryo and then become disorganized (see under "Summary of Early Development of Mouse Embryo"). The immediate precursor to the embryonal cells could, therefore, be an embryonic ectoderm cell.

In human females there is evidence that a teratoma arises from a secondary oocyte parthenogenetically activated following fusion of its two postmeiotic haploid nuclei. Studies on the morphology of the chromosomes of several human teratomas with the use of sensitive cytochemical banding techniques showed that at all sites near the centromeres, where the chromosomes from normal tissues were heteromorphic, those of the tumor were homomorphic. The tumors were also mainly homozygous for electrophoretic variants of enzymes where the normal tissues were heterozygous, but they were heterozygous for some markers, as expected if crossing over had taken place (38).

Origin from Embryonic Ectoderm

It was seen earlier that embryos implanted into ectopic sites or arising parthenogenetically in the ovary developed to about the 7-day stage before becoming disorganized.

In normal 7-day embryos there is good evidence that at least some of the cells of the embryonic ectoderm or epiblast are pluripotent and can give rise to many different tissue types. Škreb et al. (39) found that when they implanted the isolated embryonic ectoderm from 8–8½-day rat embryos (equivalent to about 6½–7-day mouse embryos) into the kidney capsule the grafts developed into teratomas. The tumors contained a wide range of tissues, including nerve (ectoderm), gut (endoderm), cartilage, and muscle (mesoderm). B. L. M. Hogan (unpublished results) has cultured embryonic ectoderm isolated immunosurgically from mouse embryos developing in vitro and has been able to obtain many different cell types, including keratinized squamous epithelium, nerve, beating muscle, cartilage, and epithelium resembling endoderm. In neither case is it clear what sort of morphogenetic movements or cell interactions occurs during the early growth of the explant and how this compares with the process of "gastrulation" in the intact embryo where mesoderm, and possibly endoderm, cells migrate from the ectoderm layer. Similarly, it is not clear whether all the cells of the early embryonic ectoderm are pluripotent or whether this property is confined to a small subpopulation, for example, the precursors of the germ cells.

What seems to be clear is that, by 8 days in the mouse, pluripotent cells in the ectoderm either are no longer present or are not convertible to teratocarcinoma cells, since transplantable tumors cannot be obtained from embryos of this age.

GROWTH AND DIFFERENTIATION OF TERATOCARCINOMA CELLS IN VITRO

In the years following the establishment of transplantable teratocarcinomas in mice, several laboratories have used the tumors to isolate cells which will grow and differentiate in tissue culture. In this section the subject of in vitro culture is not treated historically, nor is every line established in every laboratory described. Instead, this section emphasizes the most recent methods for maintaining cell lines, so that, when required, they can be triggered to differentiate rapidly and in a predictable sequence. This property of teratocarcinoma cells is obviously the most useful for studying biochemically the control of cell differentiation.

Isolation of Clonal Lines of Teratocarcinoma Cells

In general, two different methods have been used to establish tissue culture lines of mouse teratocarcinoma cells. The first method ensures that the cells are grown and maintained on monolayers of feeder fibroblasts right from the start, and these lines seem to retain, over long periods, their ability to differentiate. The second method initially selects cells which will grow in the absence of feeder layers; although this has many technical advantages, after some time the lines appear to differentiate less readily in culture.

In the first method (31, 40–42), either intact or disassociated embryoid bodies from ascites tumors or minced and collagenase-treated solid tumor tissue are plated out on a monolayer of feeder fibroblasts. These are usually made from x-irradiated or mitomycin C-treated mouse fibroblasts, although there has been no systematic survey of the relative ability of cells from other species or tissues to act as feeder layers. After several days, many different kinds of differentiated cells can be seen in the cultures, and also nests of tightly packed embryonal cells, distinguishable by their undifferentiated morphology, small size, indistinct cell boundaries, relative paucity of cytoplasm, and large, dense nucleoli. The general morphology of these embryonal cells is shown in Figure 2. On repeated subculture of the mixed population (usually with a mixture of EDTA and trypsin in phosphate-buffered saline solution) embryonal cells predominate in the culture. Alternatively, embryonal cell colonies can be isolated manually from the original culture and dispersed into dishes containing feeder cells.

In the second method (30, 43–46), intact or dissociated embryoid bodies are allowed to attach to plastic tissue culture dishes, preferably with a coating of gelatin, and after several days many different kinds of cells grow out, including nests of embryonal cells with the morphology described above. On repeated subculture these cells may overgrow the others. Although no definite rules can be formulated, successful establishment of cultures of predominantly embryonal cells from these mixed populations seems to depend on several factors, in particular, the line of ascites tumor chosen, but also the density at which the cells are initially subcultured and the use of gelatin-coated dishes. During the early subcultures, differentiated cells in the mixed population may in fact act as feeders for the embryonal cells.

Once more or less homogeneous cultures of embryonal cells have been obtained, several methods have been used to isolate clones from single cells. The most reliable method is to pick individual cells in a finely drawn out Pasteur pipette and to seed these into small tissue culture dishes or wells containing feeder fibroblasts (31, 41, 48). When small colonies become visible they are trypsinized and seeded out into larger dishes containing feeder layers. The efficiency of cloning by this method is only 15–30%, but the reason for this low value is not clear; it does not seem to be due to damage to the cells during isolation because in one study 80% of the cells were shown to divide once or twice before dying (31).

Another method used for cloning is to seed the embryonal cells out at very low density onto a layer of feeder fibroblasts and to pick individual colonies, which grow up; this procedure is repeated several times. This increases the probability that the final clone has been derived from a single cell.

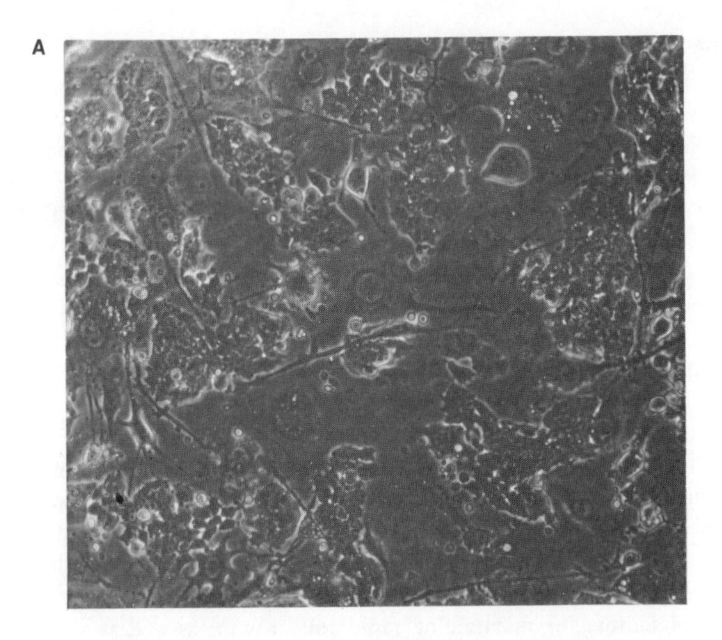

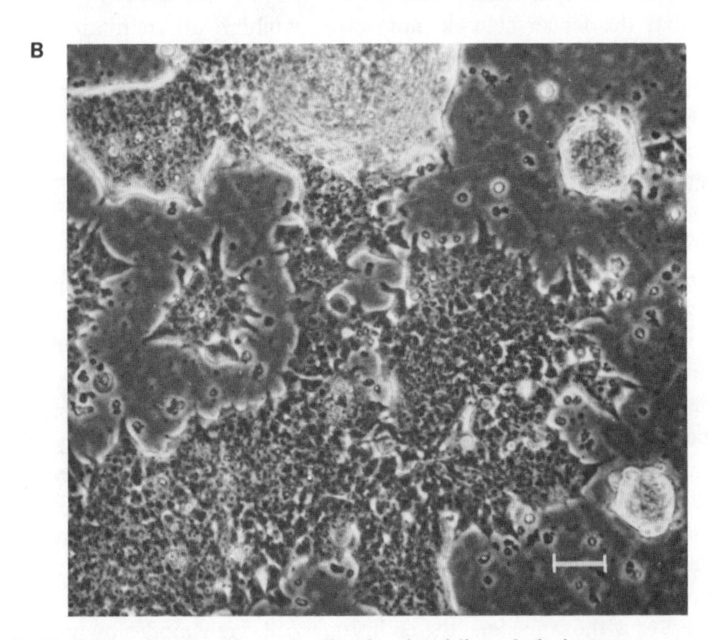

Figure 2. Undifferentiated embryonal cells of a clonal line of pluripotent teratocarcinoma cells derived from the OTT 6050 ascites tumor. *A*, tumor cells growing on a layer of mouse feeder fibroblasts; *B*, tumor cells growing on a gelatin-coated tissue culture plastic Petri dish. *Bar* is 100 μm.

Routine Maintenance of Undifferentiated Embryonal Cells

Stock cultures of embryonal cells remain as homogeneous populations of undifferentiated cells so long as they are subcultured in the exponential phase of growth before they become confluent. In general, the population doubling time is about 20 hr, but the cells grow slightly faster as they become adapted to growing in culture. The medium most often used is Dulbecco's modified Eagle's medium, and either calf or fetal bovine serum is added in concentrations from 5–20%, depending on the cell lines. Apart from the feeder dependence of some lines (see below), no particular growth requirements of the cells have been studied. It would be interesting to know what range of hormone receptors the embryonal cells express.

Cell lines which have been isolated and maintained on feeder fibroblasts can often be adapted to grow on gelatin-coated tissue culture dishes (31, 47), although this involves an initial selection period in which there is much differentiation and cell death before a population of embryonal cells grows up. Some lines, in particular those isolated from recently induced primary tumors, do not seem to survive this selection period and cannot be grown without feeders (31, 42). The precise function of the feeder layer is not clear; it may provide nutritional factors for maintaining rapid cell growth, as well as a substrate on which the cells can spread out, rather than aggregate into clumps. It is important to note that the feeder cells do not actually inhibit differentiation, as a wide range of cell types can be obtained under certain conditions from colonies of embryonal cells growing on feeder layers (see under "Differentiation by Routes Other Than Formation of Floating Embryoid Bodies").

The morphology of the feeder-dependent and independent embryonal cells is very similar (Figure 2).

Differentiation of Teratocarcinoma Cells in Culture

Formation of Embryoid Bodies During early studies on the growth and differentiation of teratocarcinoma cells in culture, people noticed that aggregates of cells would detach into the medium and develop into multilayered structures resembling the embryoid bodies found in ascites fluid (40, 43). However, Martin and Evans (41) were the first to recognize that whole cultures of embryonal cells could be induced, in a controlled way, to differentiate into embryoid bodies rapidly and fairly synchronously, and that this behavior could be exploited to study determination and differentiation biochemically. Martin and Evans made their discovery because the new clones which they had isolated would grow well only on layers of feeder fibroblasts. When homogeneous cultures of the embryonal cells on feeder layers were subcultured at low density into tissue culture dishes without feeders, the embryonal cells grew more slowly for a while; after two further subcultures very few cells were found attached to the dishes and these sparse populations died out (42). The reason for this was that the embryonal cells did not attach very firmly to the tissue culture dish and began to form clumps which could be dislodged with gentle pipetting. After about a day, the aggregates in suspension, which remained quite healthy, had clearly differentiated into simple embryoid bodies, with an outer layer of endoderm cells surrounding an inner core of undifferentiated cells. Following this initial observation, the process of making floating aggregates from mass cultures of embryonal cells on feeder layers was simplified by directly subculturing the cells into bacteriological Petri dishes to which they could not adhere. All the clumps of embryonal cells remained in suspension and developed into simple embryoid bodies in 12–36 hr. So long as these aggregates were kept in suspension they did not develop further, but several kinds of differentiated cells grew up when they were allowed to reattach to tissue culture dishes (see under "Acetylcholinesterase, Creatine Phosphokinase, and Myokinase").

The clones which gave the results described above were isolated by Martin and Evans from a stock culture of the SIKR teratocarcinoma line which had been in culture for many generations following its establishment from a 3-day embryo-derived tumor (30). Later clones, isolated from earlier frozen stocks of the SIKR parent line, or from new primary cultures of the solid tumor, behaved in a rather different way. The embryonal cells grew very poorly indeed when subcultured in the absence of feeder fibroblasts and rapidly formed aggregates which detached and developed into simple embryoid bodies. Over the next few days these embryoid bodies became more complex and formed a fluid-filled cyst on one side; on histological examination they were shown to contain neuroepithelium and mesodermal derivatives as well as visceral and parietal endoderm.

On the basis of their observations, Martin and Evans have classified teratocarcinoma cells in culture into three types (42). Type I are so-called "nullipotent" cells, which do not differentiate at all. Type II are pluripotent cells, which form only simple embryoid bodies when kept in suspension, and type III form simple embryoid bodies which subsequently become cystic. The distinctions between the three cell types are far from absolute, and type III cells may gradually come to behave like types II and I with increasing time in culture. This is particularly true if freshly isolated clones originally maintained on feeder layers and behaving like type III cells are adapted to grow in the absence of feeder cells on gelatin-coated dishes. After a few months, instead of forming cystic embryoid bodies when put into suspension, they form only simple ones, and then more slowly and with decreasing efficiency (47). Some nullipotent cell lines have arisen from cells originally grown in the absence of feeder cells (44), and all will grow rapidly on untreated tissue culture dishes. Obviously, the best cells to use for biochemical studies on differentiation are freshly isolated clones maintained on feeder layers as homogeneous populations of embryonal cells, which can be triggered to form embryoid bodies by subculturing the cells into bacteriological Petri dishes. Unfortunately the presence of the feeder cells often interferes with biochemical assays, and for this reason lines adapted to grow on gelatin-coated dishes have to be used, although they differentiate less well.

Since the properties of lines gradually change in culture, it is difficult to be precise about the time course and synchrony of development of embryoid bodies when embryonal cell aggregates are placed in suspension, but the description which follows will refer to a good type III culture. After 24–36 hr in suspension the aggregates of embryonal cells have developed a clearly defined outer monolayer of cells (Figure 3*A*) which in both the light and electron microscope resembles the primary endoderm layer of normal 4½-day embryos. The simple embryoid bodies formed from the teratocarcinoma cells also bear a striking resemblance to the structures formed when inner cell masses isolated immunosurgically (see under "Summary of Early Development of Mouse Embryo") from normal 3½–4-day blastocysts are cultured in vitro (Figure 3*b*) (17 and B. L. M. Hogan, unpublished observations). In both cases the outer cells have the following features, some of which are illustrated in Figure 4: microvilli on the outer surface, a well developed endoplasmic reticulum with numerous ribosomes, cytoplasmic vacuoles, lateral tight junctions between the cells, and a basement membrane. The material of the basement membrane resembles morphologically material present in the swollen cisternae of the endoplasmic reticulum of the endoderm cells (see under "Basement Membrane Material" for further details on the chemistry and biosynthesis of this basement membrane material). In the cultured normal embryos, the basement membrane never seems to become very thick, but in teratocarcinoma embryoid bodies harvested after several days in suspension the membrane is often very thick indeed (Figure 5), and whole groups of cells which resemble morphologically the well differentiated parietal endoderm cells of normal embryos can be seen embedded in masses of basement membrane material (see Figure 8).

After 2–3 days in culture, those simple embryoid bodies which are going to become more complex develop a fluid-filled cyst on one side (Figures 6–8). Over the next 10 days, before the whole structure becomes necrotic, this cyst can reach a diameter of about 5 mm. In sections, the cells forming the wall of the cyst resemble cells of the visceral yolk sac of normal midgestation embryos in utero (49). They have numerous cytoplasmic vacuoles and microvilli on the outer surface and are separated from an underlying mesoderm layer by a thin basement membrane. About 50% of normal mouse blastocysts attached to a substratum (50) and most isolated inner cell masses allowed to remain in suspension (B. L. M. Hogan, unpublished observations) also develop such fluid-filled cysts and visceral endoderm cells in culture.

Cystic embryoid bodies derived from teratocarcinoma cells show only a few other histologically distinguishable tissues, besides visceral and parietal endoderm cells, however long they are kept in suspension. Tubes of cells resembling neuroepithelium are often seen, together with sheets of cells which can be loosely called "mesenchyme." Nests of cells synthesizing red pigment, which is presumably hemoglobin, have been seen in about 20% of cystic embryoid bodies developed in vitro from a clonal line of embryonal cells from the OTT 6050 tumor (B. L. M. Hogan, unpublished observations), as well as in other embryoid bodies in vitro (51). It is not known why embryoid bodies in suspension fail to show a wider range of differentiated tissues, but, if they are allowed to attach to plastic tissue culture dishes, many different cell types appear over the course of

A

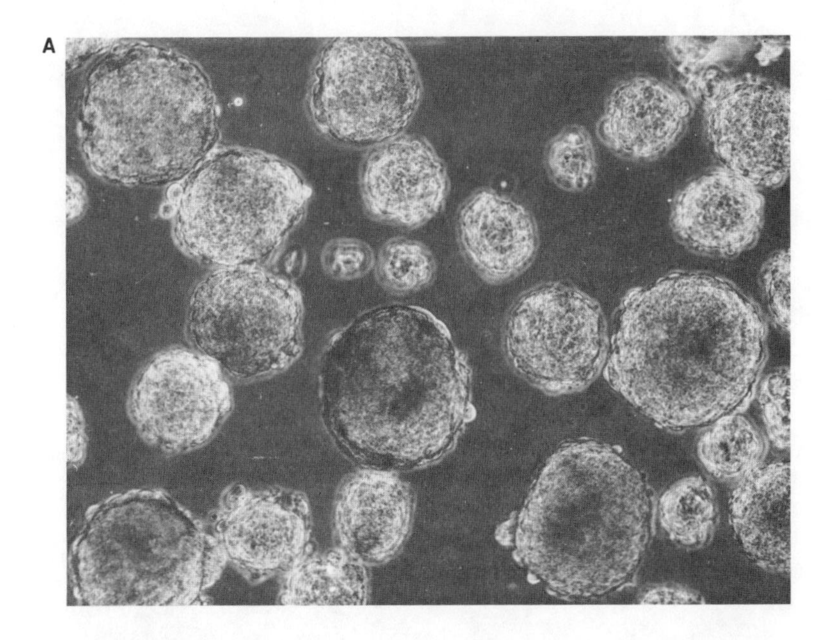

B

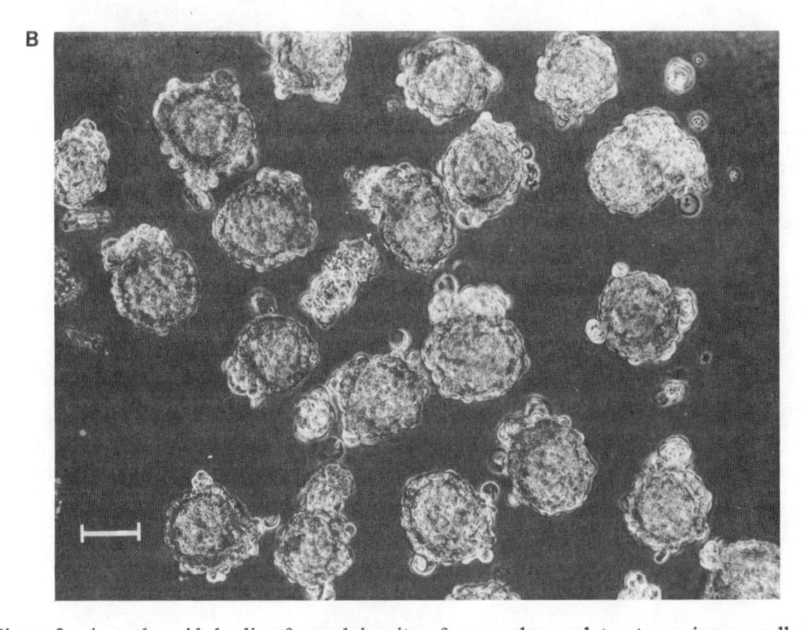

Figure 3. *A*, embryoid bodies formed in vitro from embryonal teratocarcinoma cells incubated in suspension in a bacteriological Petri dish for 3 days; *B*, embryoid bodies from inner cell masses isolated immunosurgically from C3H mouse blastocysts and incubated for 2 days. *Bar* is 100 μm.

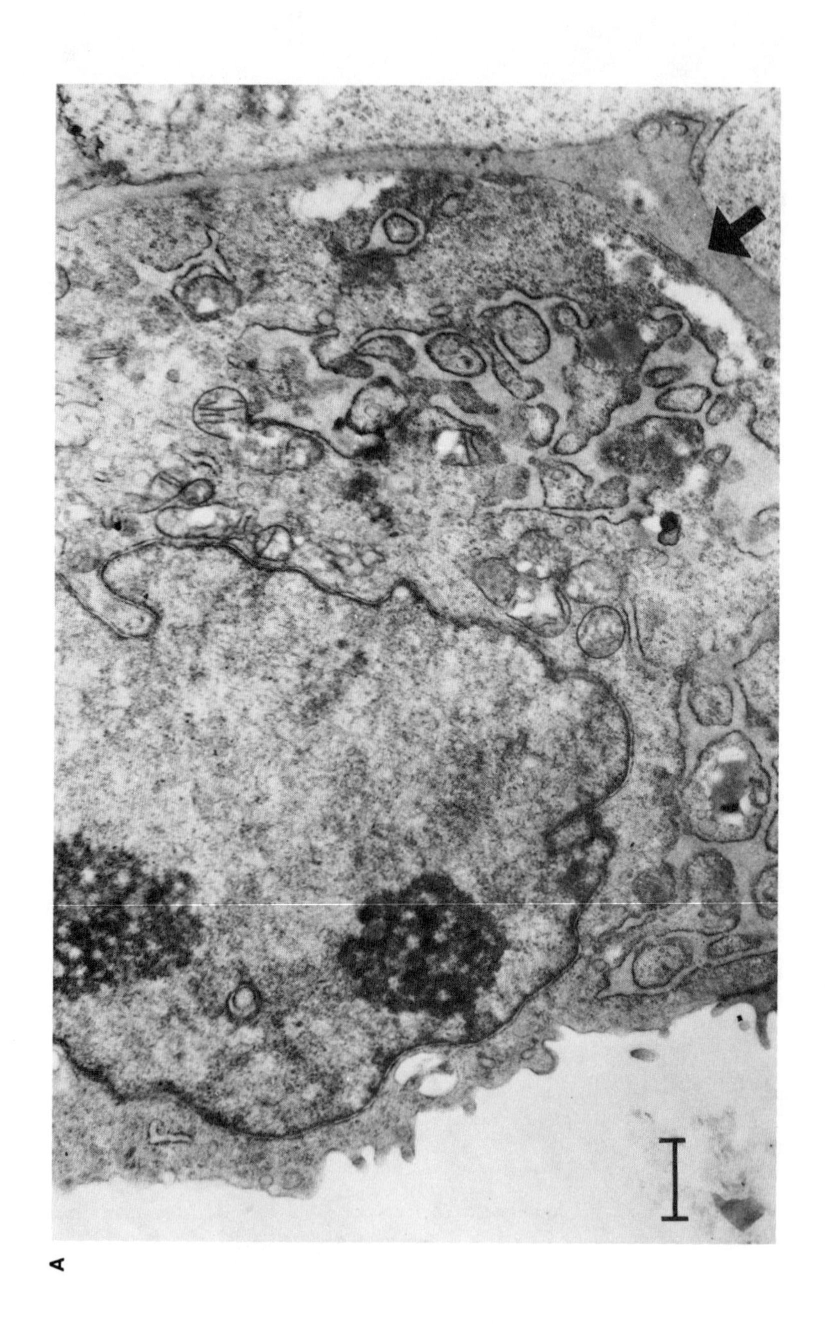

A

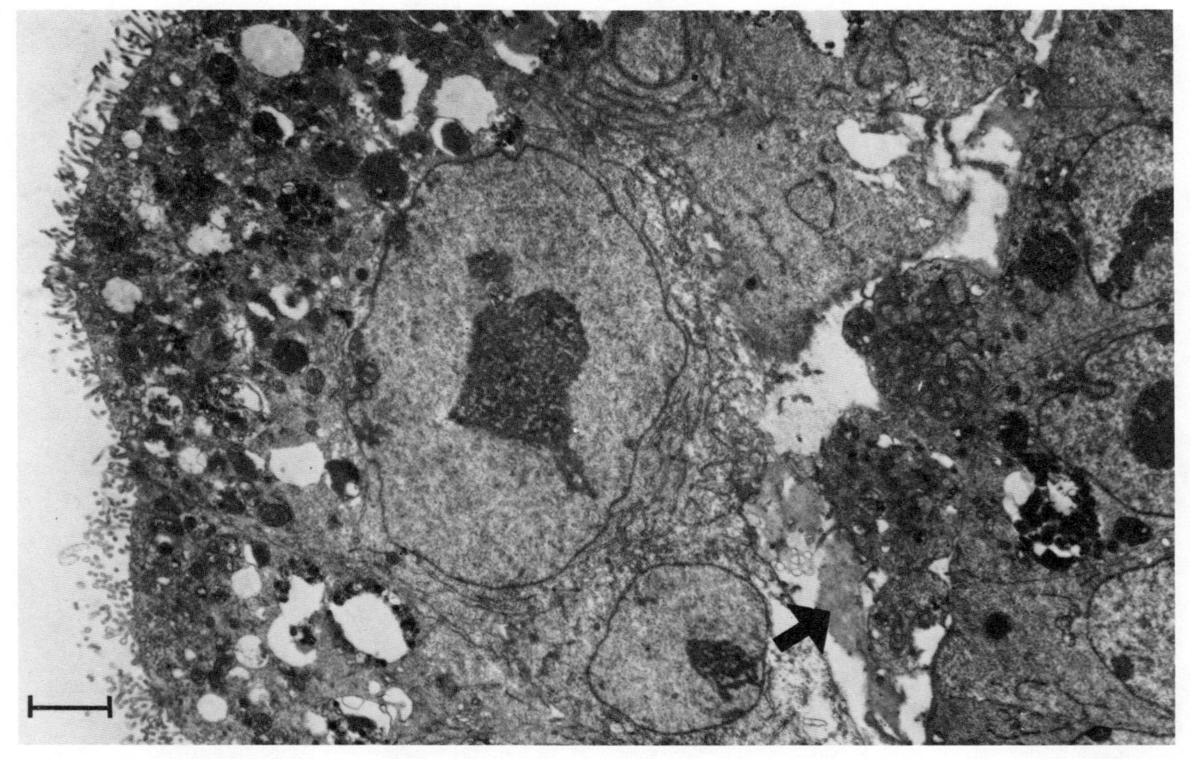

Figure 4. *A*, section through endoderm cells on the outside of an embryoid body formed 18 hr after placing aggregates of embryonal teratocarcinoma cells in suspension in a bacteriological Petri dish. *B*, section through endoderm cells on the outside of an embryoid body formed from an inner cell mass isolated from a normal C3H blastocyst and incubated in vitro for 78 hr. *Arrows* point to basement membrane material secreted by the endoderm cells. *Bar* is 1 μm in *A* and 5 μm in *B*.

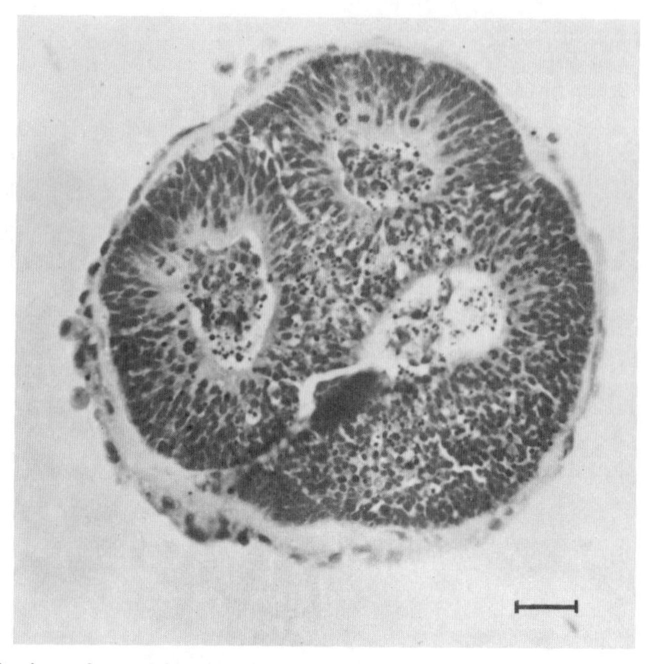

Figure 5. Section of an embryoid body formed from an aggregate of embryonal cells incubated in vitro for 6 days. Note the thick basement membrane between the outer layer of endoderm cells, some of which are sloughing off, and the inner cells which have become organized into neuroectodermal-like tubes. All sections stained with hematoxylin and eosin. *Bar* is 100 μm.

several weeks, including beating muscle, nerve, pigment, keratinized squamous epithelium, and glandular tissue (52). This is described further under "Acetylcholinesterase, Creatine Phosphokinase, and Myokinase." Although this later appearance of beating muscle and nerve, for example, is rather dramatic, the most exploitable event from the point of view of studying biochemically the control of cell differentiation seems to be the very first appearance, on the outside of aggregates of embryonal cells, of cells resembling the primary endoderm of 4½-day embryos. There are, however, a number of somewhat problematic features about this differentiation. Firstly, the aggregates consistently form an outer layer of endoderm cells rather than trophectoderm, even though transplantable teratocarcinomas growing in mice can apparently give rise, albeit rarely, to tissue morphologically resembling trophoblast. The fact that the first thing to be formed is an endoderm layer suggests that, in culture, embryonal cells cloned from embryo-derived tumors behave like the inner cell masses of 3½-day blastocysts. However, as seen under "Histogenesis of Teratomas in Vivo," normal embryos implanted into the testis or kidney capsule always develop to the 6½–7-day stage before becoming disorganized and giving rise to teratocarcinomas. It has been argued that in these cases the cells which give rise to the tumor are derived from the embryonic ectoderm or epiblast. However, groups of ectodermal cells isolated from normal cultured embryos similar to those illustrated in Figure 3*B* by stripping off the endoderm layer do not remake an endoderm layer (B. L. M. Hogan, unpublished observations). For the moment, the paradox that teratocarcinoma cells seem to arise from embryonic ectoderm and yet behave in culture like cells of the inner cell mass must remain unresolved.

Although the population of embryonal cells from which the floating aggregates are formed appears to be morphologically homogeneous, it has still not been formally proven that the rapid appearance of simple embryoid bodies is the result of no more than the "sorting out" of two subpopulations of cells generated earlier; indeed, as will be discussed under "Surface Antigens as Markers of Differentiation," two laboratories have shown that cultures of morphologically homogeneous pluripotent embryonal cells are heterogeneous for a surface antigen. However, it seems more likely that, in both aggregates of embryonal cells and in normal inner cell masses, the outer cells differentiate into endoderm in response to some signal from their microenvironment and position relative to the inside cells, as predicted by the inside-outside hypothesis of cell determination described under "Summary of Early Development of Mouse Embryo." The question of whether sorting out does occur could conceivably be resolved by labeling cells with antibodies or other markers and observing their fate in sections or with time lapse photography. Certainly, many more studies

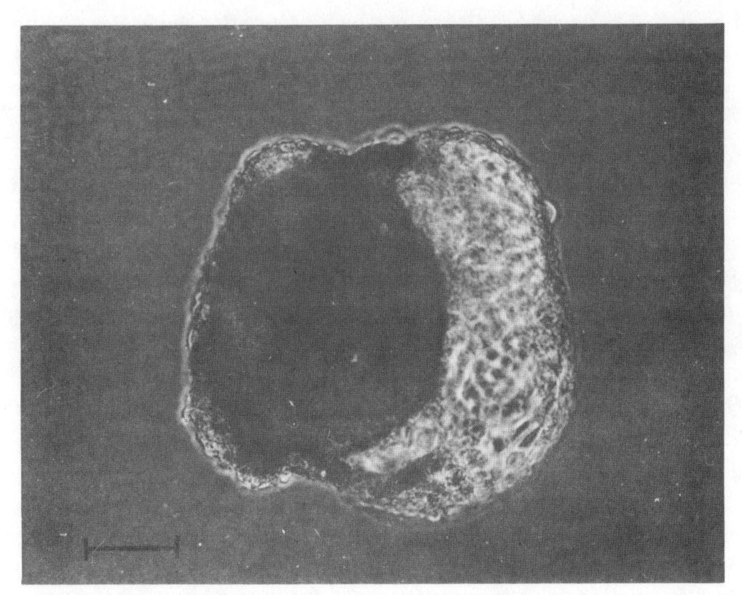

Figure 6. Cystic embryoid body formed as described for Figure 5 after 6 days in culture in a bacteriological Petri dish. Phase contrast microscopy. *Bar* is 100 μm.

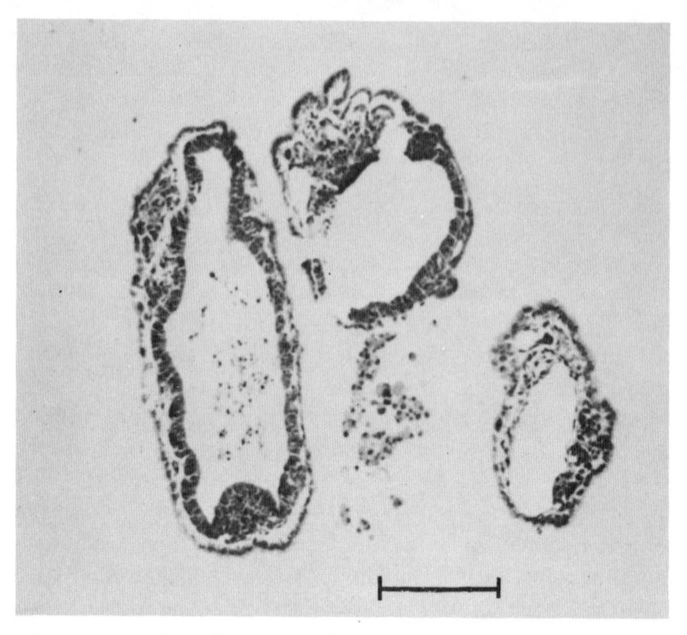

Figure 7. Section of three cystic embryoid bodies formed after 8 days incubation in vitro, showing visceral and parietal endoderm, ectoderm, and mesoderm. *Bar* is 500 μm.

need to be made on very early events which occur after aggregates of embryonal cells are placed in suspension. For example, although tight junctions clearly exist between endoderm cells after 24 hr, it is not known how soon they appear and whether they prevent the penetration of large molecules and so might serve to separate the inside cells from the surrounding medium.

Many other questions about the early determination and differentiation of embryonal cells to form endoderm remain to be answered. For example, if the endoderm layer is stripped off a simple embryoid body soon after it has been formed, will the inside cells give rise to a new layer, or have some of them become restricted in their developmental potential? It would also be interesting to know how soon and under what circumstances endoderm cells become distinguishable as two subpopulations, visceral and parietal.

Differentiation by Routes Other Than Formation of Floating Embryoid Bodies In the preceding section considerable emphasis was given to the fairly rapid and synchronous development of embryoid bodies from floating aggregates of embryonal cells. However, formation of floating embryoid bodies is by no means the only route through which teratocarcinoma cells can differentiate. Small, isolated colonies of embryonal cells on feeder layers will, unless subcultured, differentiate into many different cell types; in this case, however, the rounded colonies probably first form endoderm on their upper surface and behave much like embryoid bodies which have become reattached to a plastic

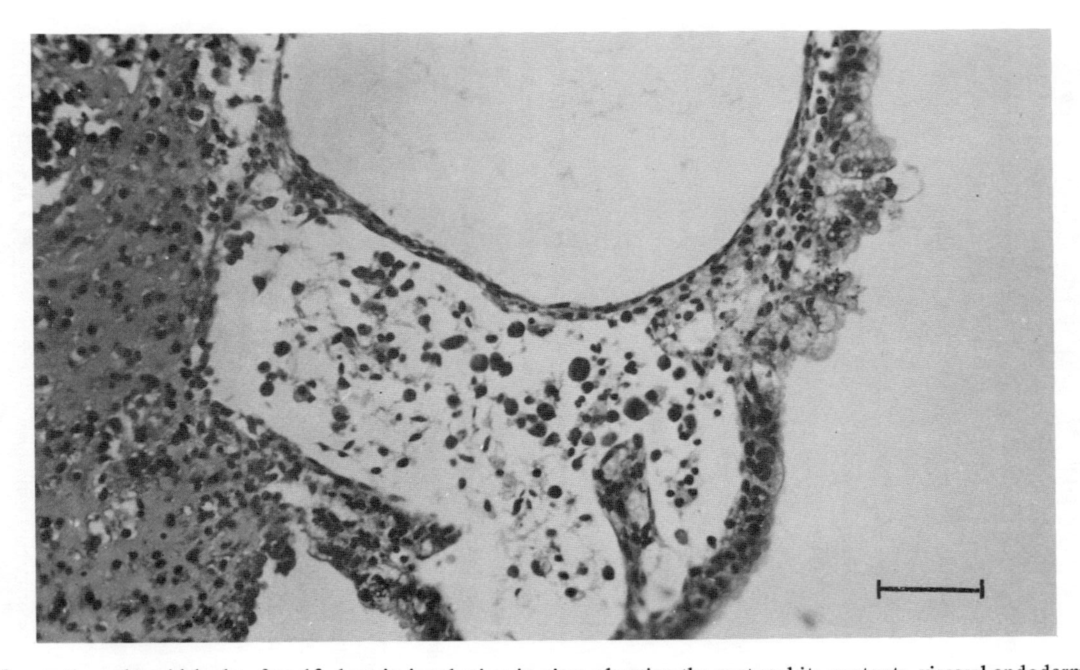

Figure 8. Section of a cystic embryoid body after 13 days in incubation in vitro, showing the cyst and its contents, visceral endoderm (*right*), a group of hemopoietic cells (*center*), and a patch of parietal endoderm cells embedded in basement membrane material (*left*). *Bar* is 200 μm.

surface. In contrast, cultures of embryonal cells which have been grown for some time in the absence of feeder cells only very slowly form embryoid bodies from floating aggregates, if at all. However, these cultures do differentiate in plastic dishes if they are allowed to become very dense, provided that the medium is replaced daily, and scattered patches of differentiated cells appear after several days, by which time there has been massive lysis of undifferentiated cells (45, 46, 53). Although the first patches observed are usually of endoderm and nerve, followed by a whole range of cells such as keratinized epithelium, pigment, muscle, etc., it is not known whether the tissues always arise in a particular sequence.

It should be recalled that spontaneous testicular teratocarcinomas in 129 mice arise from clusters of undifferentiated cells which do not go through an embryoid-body-like configuration in which endoderm cells completely surround ectoderm-like cells, and the exact sequence of their early differentiation is also not known.

The question of whether or not teratocarcinoma cells differentiate in a regular sequence is a crucial one. On the one hand, their differentiation may proceed as a cascade of binary choices between alternative states of determination, with each decision being separated by at least one round of cell division (54). Alternatively, the cells may be able to make multiple, rather than binary, choices between many different states, the particular choice depending on which state is more stable under the culture conditions used. The two models make different predictions about the behavior of teratocarcinoma cells in culture. For example, the number of steps between embryonal cell and myoblast could be much smaller on the multiple choice rather than on the binary cascade model if environmental conditions favoring the stability of the myoblast state could be found. The models of binary versus multiple choice can, of course, be applied to normal embryogenesis, but it may be easier to test them with the use of teratocarcinoma cells in culture which are released from the constraints of making a viable organism and yet, when put back into the embryo, are able to participate in normal morphogenesis.

BIOCHEMICAL MARKERS OF DIFFERENTIATION

As seen in the previous section, considerable weight has been placed upon morphological criteria for judging the differentiation of teratocarcinoma cells in vitro. However, these are not very informative on the molecular mechanisms behind the differentiation, particularly when there is embryological evidence that cells may be committed to a particular fate some time before they show any gross morphological changes. This section reviews some of the studies which are in progress on changes in the levels of enzymes and other proteins in teratocarcinoma cells during differentiation in vitro. Almost nothing is known about how the genes for these proteins are controlled or about changes in such things as DNA-binding proteins or DNA methylation patterns during differentiation.

Immunological studies using surface antigens as markers of differentiation will be discussed under "Surface Antigens as Markers of Differentiation."

Alkaline Phosphatase

The physiological function of this enzyme (APase EC 3.1.3.1.), which is a rather nonspecific phosphatase with a pH optimum in vitro of around 9–10, is not known (55). It is a glycoprotein located in the plasma membrane of some, but not all, adult tissues and is present in high specific activity in cells of early embryos, including morulae, ICM, and embryonic ectoderm, as well as in primordial germ cells (44, 56). There are several isoenzymes of APase; in the mouse these are not well resolved electrophoretically, but differ in heat stability, inhibition by L-phenylalanine, and chromatographic properties.

Bernstine et al. (44) have studied the APase of several in vivo and in vitro teratocarcinoma lines. High activity was found in all embryonal cells, whether they came from nullipotential lines which did not differentiate or from sublines which formed simple or cystic embryoid bodies intraperitoneally, and the isoenzyme was similar to that from placenta and kidney (and different from liver and intestine), as judged by heat inactivation and inhibition by L-phenylalanine (44). However, embryonal cell APase can be distinguished from both placenta and kidney enzymes by chromatography on DEAE-cellulose (56), but the full significance of these results is not yet clear.

When sections of both simple and cystic embryoid bodies were stained histochemically for APase, an interesting distribution of enzyme activity was found; it was high in the undifferentiated embryonal cells, but low in the surrounding endoderm (44). This distribution was also observed by Martin and Evans in simple embryoid bodies formed in vitro (41). It, therefore, appears that as embryonal cells differentiate into endoderm they switch off APase activity. APase activity is also high in the inner cell mass and embryonic ectoderm of blastocysts and low in primary endoderm cells (37) so that the enzyme appears to be switched off during normal embryogenesis and the behavior of teratocarcinoma cells mimics this process. However, many questions need to be answered about the control of APase activity both in normal embryos and in teratocarcinomas. For example, how does the disappearance of the enzyme correlate with morphological changes in the cells, does it still occur in the absence of RNA synthesis, and is the residual APase activity in endoderm cells the same isoenzyme as in embryonal or ICM cells?

In an attempt to gain information about the control of APase activity, Bernstine and Ephrussi have followed its behavior in hybrids between embryonal carcinoma cells and mouse neuroblastoma cells (in which enzyme activity is low) (56). Surprisingly, in three separate hybrid clones, they found a significantly higher specific activity of APase than in the parental teratocarcinoma cell line, taken at the same phase of the growth cycle. Preliminary studies suggest that the same isoenzyme form is expressed in the hybrids as in the embryonal cells, but further work is needed to clarify these results and the additional finding that hybrids between embryonal cells and rat hepatoma express only the low level characteristic of the latter cell line.

Basement Membrane Material

Extensive studies have been made on the chemistry, immunology, and biosynthesis of the basement membrane (BM), known as Reichert's membrane, which in normal rodent embryos is secreted by parietal endoderm cells adjacent to the trophectoderm (see under "Summary of Early Development of Mouse Embryo" and Figure 1*d*). This amorphous or finely fibrillar material, consisting of both glycoprotein and collagen components, is immunologically and chemically very similar to basement membranes made in adult animals by epithelial cells in kidney, lens, and testis, for example. In the early mouse embryo only the parietal yolk sac endoderm secretes large quantities of BM material, whereas the visceral endoderm and the endoderm layer adjacent to the embryonic ectoderm normally make only a very small amount (57). However, as has been seen, isolated inner cell masses in culture develop an outer endoderm layer, part or all of which secretes significant amounts of BM material, and in this respect the behavior of teratocarcinoma cell clumps in suspension and normal isolated inner cell masses is very similar (see under "Growth and Differentiation of Teratocarcinoma Cells in Vitro" and Figure 4 *A* and *B*). If it can be shown rigorously that the BM material made by endoderm cells in embryoid bodies and in isolated inner cell masses is chemically identical with Reichert's membrane, then the genes coding for the constituents of this BM (see below) must be switched on, or expressed to a far greater extent, when teratocarcinoma cells or inner cell mass cells differentiate into endoderm. Not only would it be interesting to know what factors control and coordinate the synthesis of BM material by endoderm cells in general, but also why parietal endoderm makes a lot and visceral endoderm only a little.

The pioneering work on Reichert's membrane material was done by Pierce and his collaborators (23). Pierce hit upon the idea of exploiting the ascites form of a 129 mouse testicular teratocarcinoma which had become restricted to forming only parietal endoderm cells in vivo or in vitro (23). These tumors (known as parietal yolk sac carcinomas) secrete large amounts of an amorphous hyaline material similar in morphology and chemical composition to Reichert's membrane. Antibodies were made in rabbits against this parietal yolk sac material, and, after absorbing against spleen, they specifically reacted with embryonic Reichert's membrane and the basement membranes of most adult epithelia, but not with connective tissue or blood vessels (58). With the use of immunological techniques, Pierce et al. were also able to show that the BM material was accumulated intracellularly in the endoplasmic reticulum of parietal endoderm cells (58).

More recent studies on the synthesis of Reichert's membrane in the 14-day rat embryo with the use of [^{14}C]proline and [^{3}H]glucosamine as tracers have shown that both the collagen and the glycoprotein component of the basement membrane are made and secreted by parietal yolk sac cells (59). The collagen component is in fact synthesized and secreted as procollagen, which, in contrast to procollagen in connective tissue, is apparently not cleaved to collagen, which may account for the absence of closely packed, striated fibers in the membrane (60). The collagen of basement membranes also has a different primary structure and is hence coded for by a different gene than collagen in connective tissue or cartilage; it consists of three identical $\alpha 1$(IV) chains wound in a triple helix and is known as type IV collagen. In addition, more of the proline and lysine residues are hydroxylated, and there is a higher carbohydrate content than in other collagens (61, 62). Obviously, the synthesis and secretion of Reichert's membrane material require the coordinated expression of a large number of genes. For example, the enzymes for the hydroxylation of proline and lysine and the glucosylation and glycosylation of proteins need to be switched on, in addition to those for type IV collagen and noncollagen protein. Exactly how many genes are involved and how they are controlled during endoderm differentiation remain problems for the future.

α-Fetoprotein

α-Fetoprotein (AFP) is a serum protein with a molecular weight of approximately 65–70,000 daltons. It contains traces of carbohydrate, which is normally synthesized only by the visceral yolk sac and liver of mammalian embryos, and is not present in the blood of healthy adults. Antibodies have been made to purified AFPs from several species; anti-human AFP serum cross-reacts with AFP from other species, including rodents (63). The physiological function of the protein is not known, although some studies show that it has strong estrogen-binding activity (64).

The synthesis of AFP by visceral yolk sac cells of midgestation rodent embryos is fairly well documented (65), but very little is known about the onset of its synthesis early in development and whether, for example, it is made by primary endoderm or only later by the visceral (proximal) endoderm subpopulation. Engelhardt et al. (66) quote preliminary studies which suggest that small amounts of AFP are present in some cells of the proximal endoderm of 6–6½-day mouse embryos, but not in distal endoderm, mesoderm, and ectoderm. The same authors have made a careful study with the use of specific indirect immunofluorescence techniques of the distribution of AFP in cells of the OTT 6050 teratocarcinoma line grown in vivo as solid tumors or as embryoid bodies. In solid tumors, AFP was localized not only in tubules and sheets resembling visceral endoderm, but also in some cells of less easily identifiable epithelia and, to a small extent, in the squamous epithelium of keratin pearls. In simple and complex embryoid bodies, AFP was found in the outer endoderm layer, but not necessarily in the cytoplasm of all of the cells, and in the fluid of the cysts, along with transferrin, albumin, and trace amounts of γ-globulin.

Careful studies are in progress on the synthesis of AFP during the in vitro differentiation of several teratocarcinoma cell lines (E. Adamson and M. J. Evans, personal communication). The undifferentiated embryonal cells do not make the protein, but in some cases it is synthesized and secreted several days after placing aggregates in suspension so that they form embryoid bodies. Although these observations are still preliminary, they strongly suggest that the gene for AFP is activated some time during the differentiation of teratocarcinoma cells into endoderm, but it remains to be seen whether it occurs early in this process or during the elaboration of the parietal endoderm subclass.

Acetylcholinesterase, Creatine Phosphokinase, and Myokinase

When simple or cystic embryoid bodies present in ascites fluid or formed in suspension in vitro are allowed to attach to a plastic tissue culture surface, they give rise to a wide variety of differentiated tissues. Even with the use of a cloned line, the precise sequence in which these differentiated cells appear varies from one embryoid body to another and from one culture to another (B. L. M. Hogan, personal observations) (52). In general, however, the first cells to grow out after the bodies have attached are endoderm cells; fibroblast-like cells, neurons, muscle, and pigment cells appear later. The neurons always grow out over sheets of other cells and do not seem to adhere easily to bare plastic. Nests of undifferentiated embryonal cells also tend not to spread over the plastic surface. Dense clumps of cells always mark the original site of the embryoid body, and in these clumps cartilage and keratin pearls may be found.

Two groups have studied the appearance of the enzymes acetylcholinesterase (AChE) and creatine phosphokinase (CPK) in cultures of embryoid bodies on plastic surfaces, but have interpreted their findings rather differently.

In adults, AChE and CPK are found predominantly in neural tissue and in muscle. CPK exists in two isoenzyme forms which are separated on the basis of their electrophoretic mobility. Brain and nerve cells express only the B form, whereas adult muscle expresses only the M form. However, in fetal muscle CPK is present initially as the B form, and there is a transition to the M form, which can be seen both in vivo and in culture, after the mononucleate myoblasts fuse into multinucleate fibers (67, 68).

By using embryoid bodies of the OTT 6050A ascites subline, Levine et al. (69) found that acetylcholinesterase activity (AChE units/mg of protein) was initially very low, but increased dramatically 5–6 days after attachment of the cells to plastic and continued to rise even after 40 days in culture. Creatine phosphokinase specific activity was also initially very low, but showed a similar sharp rise 5–6 days after the embryoid bodies were allowed to attach. Embryoid bodies kept in suspension in the same medium showed no increase in either enzyme activity, and a nullipotential line of embryonal cells had virtually none of these enzymes at all. Levine et al. found only the B form of CPK in their cultures, and, since there were many neural-like cells present at the time when AChE and CPK activities were high (70), they concluded that the appearance of these enzymes was biochemical evidence of the differentiation of nerve cells.

Gearhart and Mintz (71) also observed low AChE activity in unattached OTT 6050 embryoid bodies and a large increase several days after these were allowed to attach to a plastic surface. Maximal activity was reached after 3–4 weeks in culture, and the AChE was histochemically localized in uninucleate and multinucleate muscle cells, although some of the latter fibers were negative. In a more recent study (72) the authors have reported that CPK activity also rises in cultures of attached embryoid bodies, reaching a maximum after about 3 weeks and then declining. Only the B form of the enzyme was detectable throughout, and the muscle-specific enzyme myokinase remained low. These two discrepancies are attributed to "aberrant late myogenesis" in the cultures, reflected in the paucity of striations and the presence of virus particles in the fibers.

There are several possible reasons for the different interpretations of the two groups. For example, the culture conditions may favor either the growth of nerve or muscle cells, or the sublines of OTT 6050 embryoid bodies may have somehow diverged in their behavior. Whatever the explanation, the enzyme activities measured are rather late markers of the differentiation of cells already determined to be either nerve or muscle, and they tell us nothing about the more interesting problem of how the embryonal cells give rise to cells with such different developmental commitments.

Chromatin Proteins

A preliminary study has been made of the histone and nonhistone chromatin proteins isolated from nuclei of cultured embryonal cells of the OTT 6050 line, as compared with similar proteins isolated from mature tissues of 129/Sv mice (73). However, the separation and characterization of the proteins have not yet been taken far enough to yield any meaningful results.

SURFACE ANTIGENS AS MARKERS OF DIFFERENTIATION

There is good reason to believe that there are considerable changes during early embryonic development in the kinds of proteins synthesized by cells and inserted into their plasma membranes. For example, some surface proteins may be used to control specific cell-cell recognitions, allowing mutually adhesive cells to sort themselves out from a mixed population and to form distinct sheets, tubes, or blocks of tissues. These "cell recognition" proteins may only be expressed for a limited period of time, when crucial morphogenetic events are taking place, or they may become modified later by association with other proteins or carbohydrate groups. Other surface proteins may act as specific receptors for growth-promoting or differentiation-inducing agents (e.g., peptides, proteins, nucleic acids, or steroids) produced by one tissue to act on either adjacent or distant populations of cells. Still other membrane proteins may be used by cells in conjunction with their cytoplasmic microtubules and microfilaments to maintain a specific shape or attachment to extracellular material. Finally, the embryo has to protect itself against immunological rejection by the mother. Evidence suggests that cells of the trophectoderm which come into direct contact with maternal tissues do not express the major histocompatibility antigens most effective in eliciting circulating antibodies and graft rejection (H-2 in mouse, HLA in man) (74, 75), and other mechanisms may also be used to reduce the antigenicity of the early embryo.

There are many sensitive and well established immunological techniques for studying surface antigens, and these are now being exploited to study changes in the membrane components (proteins and glycoproteins) of both normal mouse embryos and teratocarcinoma cells as they differentiate. Antisera against the surface antigens of these cells can be raised by injecting whole cells either into animals of another species (for example, rabbits) or into adult mice of the same inbred strain. The first method produces antibodies to a whole range of proteins, some of which may be shared by embryonic and adult mouse cells, so that the antiserum must first be absorbed with many different adult tissues before it is specific for antigens present on the embryonic cells. Another possible drawback to this method is that the sequence of some surface proteins crucial for development may have been strictly conserved during the evolution of vertebrates, so that they may be only weakly antigenic, even in heterospecific immunization. The second method is based upon the assumption that surface proteins on embryonic cells act as autoantigens in adults if the proteins are no longer being expressed at the time when the immune system which distinguishes "self" from "not self" is established. Obviously, antisera raised in syngeneic animals are initially much more specific than those raised in other species, but careful absorption controls must still be done. Finally, there are now many inbred strains of mice which differ genetically at only the H-2 complex, or subregions within it, so that quite specific antisera are available for studying the expression of the H-2 genes during development.

Several laboratories have now raised antisera against the undifferentiated embryonal cells of mouse teratocarcinoma lines. In the first series of experiments, carried out in collaboration between François Jacob's laboratory in the Pasteur Institute and Dorothea Bennett's laboratory in Cornell University Medical College, x-irradiated cells of the nullipotent F9 in vitro subline, originally derived from the OTT 6050 tumor, were injected into male 129/Sv-CP mice every 2 weeks, and antisera were collected several days after the final immunization (76). Antisera prepared in this way showed much greater complement-dependent cytotoxicity toward F9 cells than did nonimmune sera, even at dilutions of 1:3,000, and 95% of the cells were killed. In contrast, none of the cells in several clonal lines of differentiated cells derived from the OTT 6050 tumor were affected. All of the cells in an independently isolated nullipotent line were killed, but this was true for only 40–60% of the morphologically undifferentiated cells in exponentially growing cultures of two pluripotent teratocarcinoma lines (53, 76). The reason for this immunological heterogeneity in an apparently homogeneous population of embryonal cells is not known, but it has been observed by others using different antisera and techniques (see below). Several established mouse cell lines and mouse tissues were then tested for their ability to be killed by anti-F9 serum or to absorb out cytotoxic activity toward F9 cells. All cells tested were negative except for testicular cells and sperm; the latter could absorb out all of the activity against F9 cells. Finally, anti-F9 serum reacted with early embryos from the 2–8-cell morulae stage as judged by indirect staining with peroxidase-coupled sheep antimouse immunoglobulin antibodies. These results led Artzt et al. (77) to the conclusion that one, or possibly several, antigen(s) expressed on the surface of nullipotent embryonal cells is also synthesized by cells of the early morula and by sperm. Since this surface component(s) is immunogenic in syngeneic animals, the authors argued that it must appear and then disappear early in development (or only be expressed on cells, such as sperm, in immunologically privileged sites); this suggested that the antigen(s) plays a role in cell interactions during a specific and transient phase of early development.

Later experiments from the two laboratories have to some extent strengthened the theory outlined above, while leaving several problems unresolved. One set of experiments (77) tested the ability of sperm from several inbred strains of mice, differing only at one or two loci within the T/t complex, to absorb out cytotoxic activity against F9 cells from anti-F9 serum. The T/t complex, about which relatively little is known at present, consists of at least six complementation groups, each with several allelic forms which can be found in populations of wild mice. Homozygous mutations at each locus are lethal and block embryonic development at specific stages; for example, t^{12}/t^{12} mutants homozygous at the t^{12} locus die when they are morulae, whereas t^{w5}/t^{w5} mutants homozygous at the t^{w5} locus reach the egg cylinder stage, but the embryonic ectoderm then becomes necrotic, followed by the rest of the embryo. It has been suggested that the genes of the T/t complex code for surface proteins involved in cell recognition and morphogenesis during specific transient stages of development, but they must also have other functions since they are expressed on mature sperm cells (for reviews see 78–80). In the absorption experiments, it was found that twice as many sperm were needed from ($+^{t12}/t^{12}$) animals heterozygous at the t^{12} locus as from ($+^{t12}/+^{t12}$) homozygous animals to remove the same amount of cytotoxic activity from anti-F9 antiserum. Heterozygosity at two other loci (T, t^{w1}) had no such effect. This suggested that at least one of the antigens on F9 cells is the product of the wild type allele of the t^{12} locus. This correlated nicely with the observation that anti-F9 serum apparently reacted most strongly with normal 8–16-cell morulae (76) and the fact that t^{12}/t^{12} mutant embryos die as morulae. However, more recent studies have shown that anti-F9 serum also reacts with cell membranes of both the trophectoderm and inner cell mass of early and late blastocysts (81). Technical problems make it difficult to look for F9 antigen in slightly older embryos which have just implanted into the uterus.

As mentioned above, the Pasteur group has found that 40–60% of the cells in an exponentially growing population of pluripotent embryonal cells reacts with anti-F9 antiserum. In contrast, none of the cells reacts with an antiserum specific for the major histocompatibility complex H-2^b as judged by indirect immunofluorescence (53, 82). However, if the culture is allowed to reach confluence, differentiated cells gradually appear in the culture, and some of these are now H-2^b positive and F9 antigen negative (53). At present, it is not known which are the first cell types to express H-2 antigens in differentiating cultures; this is a problem of considerable interest. In spite of the fact that many studies have been made on the time of appearance of H-2 antigens in normal mouse embryos, the question is still very much unresolved (for reviews see 83–85). In brief, most techniques used for detecting H-2 antigens (e.g., direct and indirect immunofluorescence with anti-H-2 antisera, graft rejection, cytotoxic T cell lysis) do not give a positive result with morulae, blastocysts, or postimplantation embryos until at about 7 days of development. However, one group using a sensitive immunoperoxidase labeling technique does find low levels of H-2 antigens on early blastocysts and increased levels on the inner cell mass (but not the trophectoderm) of blastocysts cultured in vitro (74). All these observations are complicated by the fact that the different methods may detect either products of different subregions of the H-2 complex or different parts of the same product of one locus. Obviously, much more work needs to be done to clarify the situation and to provide evidence for or against the hypothesis (78–80) that the T and the H-2 complexes, which are both closely linked on chromosome 17 in the mouse, code for structurally analogous (86) sets of surface proteins which are used in a mutually exclusive way for cell-cell recognition during early embryogenesis and adult life, respectively.

Returning to the consideration of teratocarcinoma cells, another group, working in University College, London, has produced an antiteratocarcinoma antiserum by injecting x-irradiated pluripotent SIKR cells into syngeneic male 129/Sv Sl^{J}CP mice (87). The cells injected were both embryonal and fibroblast-like in morphology, but the antiserum reacted only with the embryonal cells, as judged by direct and indirect fluorescence assays, the serum being noncytotoxic. The so-called “C antigen” detected by the antiserum of Stern et al. (87) has many similarities with the F9 antigen of the Pasteur group; for example, it is expressed on $>80\%$ of nullipotent, and approximately 50% of pluripotent, embryonal cells in culture, and is absent from all differentiated cell lines derived from the parental teratocarcinoma. The C antigen is also absent from the outer endoderm cells of simple embryoid bodies formed in vitro from aggregates of SIKR embryonal cells. However, the C antigen does seem to be present on cells in brain and kidney in adult mice, since these tissues absorb out antiembryonal cell activity. The University College group also found that H-2 antigens were not expressed by embryonal cells or endoderm cells on the outside of embryoid bodies as judged by the failure of these cells to absorb cytotoxic activity against H-2^b thymocytes or lymph node cells from anti-H-2^b antiserum. Finally, the antigen Thy-1 was shown to be absent from both the pluripotent embryonal cells and endodermal cells.

Edidin and coworkers (84, 88–90) have taken a rather different approach to the study of surface antigens on teratocarcinoma cells and early embryos. They injected rabbits with adjuvant plus live cells cultured from the poorly differentiating tumor 402 AX, derived from a spontaneous 129 testicular teratocarcinoma, and obtained, after absorption with normal 129 tissues, an antiserum which reacts by indirect immunofluorescence with the original teratocarcinoma cells, with many other mouse tumor cells (e.g., melanoma and hepatoma), with established cultured cell lines (e.g., an L cell derivative), and with cell lines (50) grown from normal mouse blastocysts in culture (84, 88–90). The whole serum may be subdivided into fractions I, II, and III by absorption with different tumor lines; the antigen, or family of antigens defined by fraction I, is present on eggs, morulae, and the inner cell mass (but not the trophectoderm) of cultured blastocysts, and it has been speculated that it represents a precursor to H-2 antigens. However, as in the other studies discussed in this section, much more work needs to be done before the picture is clearly defined.

VIRUS INFECTION OF CULTURED TERATOCARCINOMA CELLS

There is some evidence that early embryonic cells differ from adult tissues in their ability to support the replication of certain exogenous viruses, but the difficulty of obtaining enough material has meant that very little is known about exactly when and how the cells change during development with respect to the way in which they handle virus infection. The availability of lines of teratocarcinoma cells which differentiate in vitro has now opened up the possibility of exploiting the enormous amount which is known about the molecular biology of some virus-cell interactions to study the biochemical changes which take place in pluripotent cells as they become committed to different developmental pathways.

Embryo cells also differ from adult tissues in the way in which they handle genetically inherited endogenous viruses. Depending on the stage of development, endogenous virus may be either partially expressed or completely replicated, and studies with teratocarcinoma cells may help us to understand not only how cells control integrated viral genes, but also their own genetic material during development. Since most is known about the replication of exogenous and endogenous viruses in embryos and teratocarcinoma cells of mice, only work with this species will be discussed, but obviously the results have important implications for human embryogenesis and viral-induced malformations.

Polyoma and SV40

Polyoma (Py) and simian virus 40 (SV40) are simple, double-stranded, circular DNA papovaviruses, isolated originally from mice and monkeys, respectively. They each contain about 6,000 base pairs of DNA and replicate in the nucleus of the host cell, where one or more of the DNA molecules becomes covalently integrated into the chromosomes (91).

The interaction of Py with mouse cells has been most extensively studied in secondary or long-term cultures of fibroblasts derived from approximately 13-day mouse embryos (e.g., the 3T3 line). In these cultures, Py replicates lytically; many new virus particles are released and the cells die. Very rarely, however, a cell is transformed by Py; in this case, complete virus particles are not made, but the cell does express a virus-coded nuclear antigen (T antigen) and is altered in many properties such as adhesiveness, saturation density, and morphology. Cell transformation by Py is also seen when the virus is injected into newborn (but not adult) mice, and tumors appear in many different tissues and cell types (hence "poly"oma). The interaction of Py with very early mouse embryos has only been studied at an ultrastructural level by Biczysko et al. (92). They found that intact virus was not attached to or taken up by 2-cell embryos, but it was absorbed by pinocytosis into trophoblast cells of 4-day blastocysts. These blastocysts could be cultured in vitro for up to 5 days and developed to the same very limited extent as the controls; evidence was seen of lytic infection of endodermal cells. Egg cylinders dissected from 8-day embryos also absorbed virus particles, this time through the endoderm cells, and during later cultivation in vitro lytic infection of endoderm, hematopoietic tissue, and neural cells occurred. Unfortunately, these studies do not provide unequivocal evidence for or against the lytic replication of Py in the embryonic ectoderm, or epiblast, which, as was seen under "Origin from Embryonic Ectoderm," is probably a pluripotent tissue from which teratocarcinoma cells can arise.

As far as the effect of Py on teratocarcinoma cells in culture is concerned, two different results have been obtained. With the use of a pluripotent cell line derived from the OTT 6050 tumor, Lehman and his colleagues (93, 94) reported that Py does not replicate lytically in the undifferentiated embryonal cells and that these cells do not express T antigen or viral coat protein. However, two lines of differentiated cells (parietal yolk sac and mesenchymal spindle cells) derived from the parental line do support lytic infection of Py. Similar results have been obtained with undifferentiated embryonal cells of another OTT 6050-derived pluripotent teratocarcinoma line by Kelly and Boccara (95). On the other hand, N. Teich (personal communication) has observed lytic infection by Py of a nullipotent SIKR-derived teratocarcinoma line which does not differentiate. At the moment, it is not clear whether the two different results are due to a genuine change in embryonal cells when they become nullipotent or to some difference in experimental design between the laboratories.

Unlike Py, SV40 does not lytically infect embryo-derived secondary or established mouse fibroblast cultures, but it causes a small fraction of the cells to become transformed. They show many altered growth properties and express T antigen in the nucleus, but they do not make complete virus particles. However, intact SV40 virus can be "rescued" after fusion of a transformed cell with a permissive cell which supports lytic infection. The interaction of SV40 virus with early mouse embryos has been studied at an ultrastructural level by Biczysko et al. (92); uptake of virus particles by endocytosis was seen in 2-cell embryos, in the trophectoderm of 4-day blastocysts, and in the endoderm of 8-day egg cylinders. Although the blastocysts and egg cylinders grew normally in culture, evidence for production of complete new virus particles was seen in the nucleus of a few cells of the endoderm, hematopoietic tissue, and neuroepithelium. Jaenisch and Mintz (96) injected naked SV40 DNA molecules into the blastocele of 3-day mouse blastocysts and found that these developed normally in utero and gave rise to healthy mice which had no signs of tumors after 1 year. However, when DNA from various tissues of these mice was hybridized to labeled SV40 DNA, evidence was obtained for integration of viral DNA sequences into the host chromosomes, although the possibility of rescuing complete virus by fusion of host and permissive cells was not tested.

Lehman and his collaborators (93, 94) have reported that after adding SV40 virus to cultures of pluripotent teratocarcinoma cells the undifferentiated embryonal cells were negative for T antigen expression, whereas most of the differentiated cells were positive. The apparent inability of SV40 to elicit T antigen expression in embryonal cells has been observed by others (95); the block is not due to lack of uptake or uncoating of the virus, and fusion of T antigen negative cells with permissive cells did not yield intact virus (93). In addition to the differentiated cells in mixed cultures, lines of parietal yolk sac and mesenchymal spindle cells also showed T antigen expression and altered growth properties following SV40 infection (93, 94). More recently, Speers and Lehman (97) have reported that treating cultures of embryonal cells with BUdR for several days prior to SV40 infection increases the proportion of cells positive for T antigen expression relative to controls. The treatment with BUdR causes embryonal cells to become flattened and to have other morphological changes at an ultrastructural level; it is these cells which are T antigen positive. However, until more is known about biochemical markers of differentiation in the BUdR-treated cells, the significance of these results is not clear.

Adenovirus Type 2

Adenoviruses are much larger and more complex than the papovaviruses, containing about 7 times as much DNA in a linear, double-stranded molecule. Like the papovaviruses, they replicate in the nucleus (91). In teratocarcinoma cultures, adenovirus type 2 appears to infect both embryonal and differentiated cells and seems to elicit T and capsid antigen synthesis and production of infectious virus, although the yield of virus is much less than from HeLa cells (95).

Minute Virus of Mice

The parvovirus called minute virus of mice (MVM) is a small, linear, single-stranded DNA virus which is unable to induce DNA synthesis in its host and can, therefore, only lytically infect cells which are proliferating in vivo or in culture. It is known that parvoviruses lytically infect tissues of early and late gestation rodent embryos in utero, including cells in the placenta and brain, but infection of normal late neonatal or adult animals gives rise to no obvious disease even though many tissues are actively proliferating (e.g., bone marrow, skin, and gut) (for review see 98). MVM lytically infects fibroblast cultures derived from midgestation 129 mouse embryos (P. Tattersall, personal communication).

With the use of OTT 6050-derived teratocarcinoma cells, Miller et al. (99) have found that rapidly proliferating pluripotent or nullipotent embryonal cells will not support lytic replication of MVM and that the block is at some stage before production of viral protein. Similar results with pluripotent embryonal cells have been obtained by P. Tattersall (personal communication). Of the various differentiated cell lines derived from the parental teratocarcinoma line or from normal blastocysts, parietal endoderm is not lytically infected (99) (Tattersall, personal communication), whereas several fibroblast-like lines are. Hybrids between nonsusceptible embryonal cells and susceptible Friend leukemic cells are resistant to lytic infection (99).

C-type RNA Viruses

The natural history and molecular biology of C-type RNA viruses are very complex topics (91), so only a brief description can be given here of the interesting work being carried out on the interaction of these viruses with early mouse embryo and teratocarcinoma cells. All mouse strains that have been studied, and probably all wild mice, have the genetic information for C-type RNA viruses integrated into their genomes in the form of one or more DNA copies which are transmitted vertically through the germ line. These endogenous RNA viruses fall into two groups: xenotropic, in which the intact virus particles once released cannot reinfect mouse cells but can infect cells of several other species, and ecotropic, in which the virus particles can reinfect mouse cells. There are many homologies between the DNA sequence and the viral-coded antigens of the two groups and morphologically they are identical, so that, before the development of molecular probes such as labeled DNA and antisera specific for xenotropic or ecotropic viruses, studies were unable to distinguish between them. Some inbred mouse strains have endogenous viruses of both classes, whereas others, of which the 129 strain is an example, contain endogenous xenotropic, but, as far as is known, no ecotropic virus (N. Teich, personal communication). How the xenotropic viruses, which are unable to infect mouse cells, originally came to be integrated ubiquitously into the mouse genome is completely unknown.

Whether or not endogenous xenotropic or ecotropic C-type virus DNA is expressed as complete, infectious virus particles depends on many factors. For example, in the AKR strain of mice infectious virus is detectable in 15–18-day embryos and reaches high titers in certain tissues quite soon after birth, whereas in other strains infectious virus may normally appear only late in life or, as in the case of the 129 strain, not at all. However, in all strains so far tested ecotropic or xenotropic virus antigens or both are expressed in cells of early midgestation embryos, even if infectious virus is not made. This observation has led to the speculation that the C-type viral genome contains, or may be generated de novo from, genes normally expressed during embryogenesis and differentiation (100, 101).

Since the endogenous xenotropic virus of 129 mice is not normally expressed as infectious virus, teratocarcinoma cells from this strain are unlikely to provide much information about how mouse cells decide to switch on the complete replication of endogenous virus during development. However, experiments are in progress to study the interaction of 129-derived SIKR pluripotent teratocarcinoma cells with exogenous Moloney murine leukemia virus (Moloney MLV). This is a typical C-type virus, with a 2 × 35S RNA genome, plus several species of host cell transfer RNA and reverse transcriptase enzyme, enclosed within a complex glycoprotein and lipid capsid and envelope. N. Teich (manuscript in preparation) has shown that undifferentiated embryonal cells can take up the virus and integrate the genetic information in the form of a DNA copy into their chromosomes, but do not support the replication of new infectious virus particles. When embryonal cells which were exposed to MLV were allowed to differentiate into a wide variety of tissues, there was no production of infectious MLV. On the other hand, some cells (probably fibroblasts) in the differentiated cultures could be productively infected by exogenous MLV, although a much larger yield of virus was obtained from fibroblasts cultured from normal 14-day 129 mouse embryos. The finding that Moloney MLV can integrate into 129 embryonal cells accords well with the observations by Jaenisch and collaborators (102) that, when 4–8-cell 129 × BALB/c hybrid embryos were exposed to Moloney MLV and subsequently transferred to foster mothers, they developed into mice carrying the viral genome in every tissue examined, as well as in the germ line (103). In some of these animals, infectious Moloney MLV could be isolated from the serum and from tumor tissue. It, therefore, remains to be seen why infectious MLV is not produced when 129 teratocarcinoma cells which have been exposed to the virus differentiate in culture.

INTEGRATION OF TERATOCARCINOMA CELLS INTO NORMAL MOUSE EMBRYOS

The first attempt to integrate teratocarcinoma cells into the normal mouse embryos was made by Brinster (104, 105). He microinjected 4-day blastocysts of random-bred Swiss albino mice with 2–4 cells from completely dissociated simple embryoid bodies of the OTT 6050 ascites tumor, which was originally derived from a 6-day 129/Sv Sl^J CP (agouti, black) embryo (see under "Tumors Derived from Embryos Implanted into Extrauterine Sites"). The blastocysts were returned to pseudopregnant foster mothers, and the resulting 60 offspring were examined for pigmentation of the coat and eyes and for survival of skin grafts from 129 donors. One male had small patches of black agouti hair in its albino coat, and there was a significant increase in the survival time of 129 skin grafts onto the experimental animals compared with the controls.

Similar but much more detailed experiments were later carried out in the laboratory of Mintz and colleagues (1, 106). Groups of about five embryonal cells, obtained from the isolated cores of simple embryoid bodies of the OTT 6050 ascites tumor, were injected into 4-day blastocysts of the C57 BL/6-b/b (nonagouti, brown) strain, and the embryos were then transferred to pseudopregnant females. Of the 183 blastocysts in foster mothers allowed to go to term, 48 developed fully, and some of these mice have been analyzed for a whole range of biochemical and morphological markers. One male had an almost entirely black agouti coat, with only small stripes of brown, mainly agouti, hair. This showed that both melanocytes (originating in the neural crest) and mesodermal components of the hair follicle were derived from teratocarcinoma cells since the "agouti" phenotype (a subterminal band of pheomelanin pigment on the hair shaft) is controlled by cells in the hair follicles and not by the melanocytes. The coat of this animal also showed the phenotype produced by the steel (Sl^J) gene, and most of the red and white cells in the circulating blood were derived from the teratocarcinoma cells since they contained the glucose phosphate isomerase (GPI) isoenzyme and hemoglobin characteristic of the 129 strain. The same animal was also tested for immunoglobulin allotypes at the Ig-1 and Ig-4 loci and for allelic variants of the major urinary protein complex (produced in the liver). The results showed that most of the plasma cells and hepatocytes were of the 129 genotype. Even more impressive is the fact that, in mating tests with C57 BL/6-b/b females, this remarkable male mouse (appropriately named Teritom) has transferred 129 genes to several offspring, so that teratocarcinoma cells must have been incorporated into its germ line. A number of other animals proved to be clearly mosaic for 129 cells in some tissue (for example, thymus, kidney, and skeletal muscle), but not in others. Finally, at the time of publication of the results, none of the surviving animals had developed tumors.

These beautifully designed experiments show unambiguously that, placed in the environment of the embryo, embryonal cells from in vivo maintained embryoid bodies of the OTT 6050 teratocarcinoma can participate in normal morphogenesis and differentiate into a wide range of apparently normal tissues, including some such as kidney, liver, thymus, and testis which are never found in solid tumors. However, several details need to be clarified before the full significance of these results can be evaluated. In particular, it would be extremely interesting to have a detailed analysis by chromosome banding techniques of the karyotype of the embryonal core cells of the OTT 6050 tumor line used in these experiments. Dunn and Stevens (107) originally reported that cells of the OTT 6050 tumors had a normal XY karyotype of 40 chromosomes; this was also claimed by Mintz (1, 106) for the core cells of the embryoid bodies used in her chimera experiments. However, the techniques used were not precise enough to clearly distinguish different chromosomes.

More detailed experiments carried out in Oxford (C. Graham, personal communication) (31) and in the Pasteur Institute, Paris, have shown that embryonal cells in OTT 6050 ascites embryoid bodies and in cultured lines derived from them are lacking the Y chromosome and are trisomic for chromosome 8, with several other minor rearrangements. Since the OTT 6050 transplantable tumor was divided into several sublines soon after it was established, it is possible that one of these sublines lost the Y chromosome and underwent other karyotypic changes before being distributed to laboratories in Europe and elsewhere. It would not be too surprising if banding studies on the chromosomes of embryonal cells of the OTT 6050 embryoid bodies used in Mintz's laboratory did reveal certain rearrangements which nevertheless do not interfere with the cells' capacity to participate in normal embryogenesis, since in people suffering from Downs syndrome trisomy for chromosome 21 produces only very subtle changes in overall development.

Independently, another set of experiments was designed to study the integration of teratocarcinoma cells into normal embryos in the laboratory of Richard Gardner in Oxford (2). In contrast to those of Brinster and Mintz, these experiments used three teratocarcinoma lines which had been adapted to grow in vitro—a pluripotent SIKR line established from the OTT 5568 129/Sv mouse embryo-derived tumor (30) and two new lines established from separate C3H embryos implanted into the kidney capsule of syngeneic mice. The SIKR line had been in culture for about 45 generations before being used in the experiment, whereas the two C3H lines, C86 and C17, had been cultured for a much shorter time. C17 cells injected into adult mice gave well differentiated teratocarcinomas, but C86 cells produced poorly differentiated tumors consisting mainly of embryonal cells and neuroepithelium (31). None of these in vitro cell lines has a normal karyotype; the SIKR line, although having a modal chromosome number of 40, is XO, and C86 and C17 are XX, but have a number of chromosomal rearrangements (31). Clumps of 20–40 embryonal cells were injected into 3½–4-day blastocysts of two mouse strains differing from 129 at several loci, and the embryos were then transferred to pseudopregnant foster mothers. Midgestation embryos or full-term offspring were checked for external chimerism in the coat and eyes (as judged by the presence of pigmented melanocytes and agouti hairs) and internal chimerism in various tissues (as judged by glucose 6-phosphate isoenzyme analysis). The yield of chimeric offspring was very low; of 121 mice born, only 11 were seen to contain cells derived from the original teratocarcinoma cells. The one chimeric mouse which contained progeny of SIKR teratocarcinoma cells was an apparently completely normal male when it was killed for analysis of internal tissues at 6 weeks of age. It was clearly mosaic in several tissues, including coat, blood, heart, skeletal muscle, and gonads, but not in the germ line, which is hardly surprising since the karyotype of the injected cells was XO. All six chimeric offspring which had been injected as blastocysts with C86 cells developed rapidly growing tumors a few days after birth. The tumors were poorly differentiated teratocarcinomas, consisting mainly of embryonal cells, presumably of C86 origin. GPI analysis suggested that progeny of the injected teratocarcinoma cells had become incorporated into several normal tissues, but the presence of numerous small tumor metastases or nests of embryonal cells could not be eliminated. The remaining four chimeric mice were derived from blastocysts injected with C17 cells. One developed two subcutaneous teratocarcinomas and died at 7 days, but the others appeared completely normal, with good evidence of external chimerism. Test breeding showed no sign of C17 cells in the germ line. However, after about 9 weeks, two of these mice developed muscle tissue tumors which histologically resembled fibrosarcomas rather than teratocarcinomas and seemed to have been derived from C17 cells (M. J. Evans, personal communication).

The results of the experiments carried out in all three laboratories on the integration of teratocarcinoma cells into normal blastocysts can be interpreted in two completely different ways, depending on how one views the origin of teratocarcinomas. If one believes that the growth of the teratocarcinoma simply represents the natural inherent ability of embryonic ectoderm cells to multiply and migrate unrestrainedly when moved away from their normal controlling environment in the 6-day egg cylinder, then it is hardly surprising that these cells can be reintroduced to normal morphogenesis by being returned to the blastocyst. If this view is correct, then formation and reversal of the teratocarcinoma phenotype may have absolutely no bearing on the nature of the phenotype of the typical malignant cell in the common cancers arising in adult life. Although this explanation of the behavior of teratocarcinoma cells would ascribe the occasional appearance of teratocarcinoma tumors in the chimeric mice to the presence of embryonal cell "rests," it really cannot readily accommodate a high frequency of appearance of tumors from differentiated cells. The other way of looking at the experiments is to say that the rapid proliferation of embryonic ectoderm cells that occurs when the embryos are transplanted into ectopic sites allows mutation and selection to occur so that, in a minority of cases, a malignant teratocarcinoma cell arises; this teratocarcinoma cell owes its phenotype to the expression of predominantly embryonic genes, some of which have undergone mutation. When these cells are put back into the entraining environment of the blastocyst, many, if not all, of these genes become switched off and the malignant phenotype becomes masked. Unlike the previous hypothesis, this explanation can easily accommodate the formation of tumors from differentiated cells since these are assumed to retain the original mutations in covert form. Clearly, the experimental evidence available at present does not allow one to distinguish between the unorthodox, epigenetic explanation of malignancy, which argues for the complete reversal of the malignant phenotype, and the more conventional explanation in terms of irreversible mutation and selection.

Note added in proof:

SURFACE ANTIGENS AS MARKERS OF DIFFERENTIATION

There is now good evidence against the hypothesis that the F9 antigen is the product of the wild type t^{w32} (t^{12}) locus. Kemler et al. (108) have used indirect immunofluoresence to study the distribution of the F9 antigen on the surface of embryos obtained from crosses between a variety of t/t^x x $+/t^x$ heterozygotes; F9 antigen was present on all embryos derived from crosses between t^{w18} or T heterozygotes, but was absent from a proportion of embryos derived from both

t^{w32} and t^{w5} heterozygotes. In other words, F9 antigen appears to segregate as expected for the product of both the t^{w32} and t^{w5} loci. Other studies have shown that F9 antigen can be detected on embryonic ectoderm cells until days 8–9 of gestation (109).

ACKNOWLEDGMENT

I should like to thank Drs. J. Cairns, C. Graham, M. J. Evans, and M. Monk for stimulating and critical discussions, Ms. G. McArthur for technical assistance in the work cited from my own laboratory, and Ms. B. Marriott for her efficiency and patience in typing the manuscript.

REFERENCES

1. Mintz, B., and Illmensee, K. (1975). Proc. Natl. Acad. Sci. U.S.A. 72:3585.
2. Papioannou, V. E., McBurney, M. W., Gardner, R. L., and Evans, M. J. (1975). Nature (Lond.) 258:70.
3. Stevens, L. C. (1967). Adv. Morphog. 6:1.
4. Pierce, G. B. (1967). Curr. Top. Dev. Biol. 2:223.
5. Damjanov, I., and Solter, D. (1974). Curr. Top. Pathol. 59:69.
6. Martin, G. R. (1975). Cell 5:229.
7. Sherman, M. I., and Solter, D. (eds.). (1975). Teratomas and Differentiation. Academic Press, New York.
8. Symposium on Embryogenesis in Mammals, London (1975). Ciba Found. Symp. 40 (New Series).
9. M. Balls and A. E. Wild (eds.). (1975). The Early Development of Mammals. 2nd Symp. Brit. Soc. Dev. Biol. Cambridge University Press, Cambridge.
10. Tarkowski, A. K., and Wroblewska, J. (1967). J. Embryol. Exp. Morphol. 18:155.
11. Gardner, R. L., and Rossant, J. (1975). Ciba Found. Symp. 40:5 (New Series).
12. Hillman, N., Sherman, M. I., and Graham, C. (1972). J. Embryol. Exp. Morphol. 28:263.
13. Sherman, M. I. (1975). *In* M. Balls and A. E. Wild (eds.), The Early Development of Mammals, pp. 145–166. Cambridge University Press, Cambridge.
14. Ducibella, T., Albertini, D. F., Anderson, E., and Biggers, J. D. (1975). Dev. Biol. 45:231.
15. Nadijcka, M., and Hillman, N. (1974). J. Embryol. Exp. Morphol. 32:675.
16. Rossant, J. (1975). J. Embryol. Exp. Morphol. 33:991.
17. Solter, D., and Knowles, B. B. (1975). Proc. Natl. Acad. Sci. U.S.A. 72:5099.
18. Stevens, L. C. (1975). *In* M. I. Sherman and D. Solter (eds.), Teratomas and Differentiation, pp. 17–32. Academic Press, New York.
19. Stevens, L. C. (1959). J. Natl. Cancer Inst. 23:1249.
20. Stevens, L. C. (1973). J. Natl. Cancer Inst. 50:235.
21. Kleinsmith, L. J., and Pierce, G. B. (1964). Cancer Res. 24:1544.
22. Stevens, L. C. (1958). J. Natl. Cancer Inst. 20:1257.
23. Pierce, G. B., and Dixon, F. J. (1959). Cancer 12:584.

24. Willis, R. A. (1960). Pathology of Tumours, Ed. 3. Butterworths, London.
25. Dixon, F. J., and Moore, R. A. (1953). Cancer 6:427.
26. Solter, D., Adams, N., Damjanov, I., and Koprowski, H. (1975). *In* M. I. Sherman and D. Solter (eds.), Teratomas and Differentiation, pp. 139–166. Academic Press, New York.
27. Stevens, L. C. (1968). J. Embryol. Exp. Morphol. 20:329.
28. Stevens, L. C. (1970). Dev. Biol. 21:364.
29. Pierce, G. B., Dixon, F. J., and Verney, E. L. (1960). Lab. Invest. 9:583.
30. Evans, M. J. (1972). J. Embryol. Exp. Morphol. 28:163.
31. McBurney, M. W. (1977). J. Cell Physiol. 89:441.
32. Willis, R. A. (1962). The Borderland of Embryology and Pathology, Ed. 2. Butterworths, London.
33. Stevens, L. C. (1976). J. Natl. Cancer Inst. 38:549.
34. Stevens, L. C., and Bunker, M. C. (1964). J. Natl. Cancer Inst. 33:65.
35. Martineau, M. (1969). J. Pathol. 99:271.
36. Pierce, G. B., and Beals, T. F. (1964). Cancer Res. 24:1553.
37. Damjanov, I., and Solter, D. (1975). *In* M. I. Sherman and D. Solter (eds.), Teratomas and Differentiation, pp. 209–220. Academic Press, New York.
38. Linder, D., Kaiser-McCaw, B., and Hecht, F. (1975). N. Engl. J. Med. 292:63.
39. Škreb, N., Švajgar, A. and Levak-Švajger, B. (1975). Ciba Found. Symp. 40:27 (New Series).
40. Kahan, B. W., and Ephrussi, B. (1970). J. Natl. Cancer Inst. 44:1015.
41. Martin, G. R., and Evans, M. J. (1975). Proc. Natl. Acad. Sci. U.S.A. 72:1441.
42. Martin, G. R., and Evans, M. J. (1975). *In* M. I. Sherman and D. Solter (eds.), Teratomas and Differentiation, pp. 169–187. Academic Press, New York.
43. Rosenthal, M. D., Wishnow, R. M., and Sato, G. H. (1970). J. Natl. Cancer Inst. 44:1001.
44. Bernstine, E. G., Hooper, M. L., Grandchamp, S., and Ephrussi, B. (1973). Proc. Natl. Acad. Sci. U.S.A. 70:3899.
45. Jakob, H., Boon, T., Gaillard, J., Nicolas, J.-F., and Jacob, F. (1973). Ann. Microbiol. (Inst. Pasteur) 124B:269.
46. Lehman, J. M., Speers, W. C., Swartzendruber, D. E., and Pierce, G. B. (1974). J. Cell Physiol. 84:13.
47. Hogan, B. L. M. (1976). Nature (Lond.) 263:136.
48. Jami, J., and Ritz, E. (1974). J. Natl. Cancer Inst. 52:1547.
49. Padykula, H. A., Deren, J. J., and Hastings-Wilson, T. (1966). Dev. Biol. 13:311.
50. Sherman, M. I. (1975). Cell 5:343.
51. Hsu, Y.-C., and Baskar, J. (1974). J. Natl. Cancer Inst. 53:177.
52. Martin, G. R., and Evans, M. J. (1975). Cell 6:467.
53. Nicolas, J. F., Dubois, P., Jakob, H., Gaillard, J., and Jacob, F. (1975). Ann. Microbiol. (Inst. Pasteur) 126A:3.
54. Kauffman, S. (1975). Ciba Found. Symp. 29:201 (New Series).
55. Fernley, H. N. (1971). *In* P. D. Boyer (ed.), The Enzymes, pp. 417–430. Academic Press, New York.
56. Bernstine, E. G., and Ephrussi, B. (1975). *In* M. I. Sherman and D. Solter (eds.), Teratomas and Differentiation, pp. 271–287. Academic Press, New York.
57. Pierce, G. B. (1966). Dev. Biol. 13:231.

58. Pierce, G. B., Beals, T. F., Sriram, J., and Midgley, A. R. (1964). Am. J. Pathol. 45:929.
59. Clark, C. C., Tomichek, E. A., Koszalka, T. R., Minor, R. R., and Kefalides, N. A. (1975). J. Biol. Chem. 250:5259.
60. Minor, R. R., Clark, C. C., Strause, E. L., Koszalka, T. R., Brent, R. L., and Kefalides, N. A. (1976). J. Biol. Chem. 251:1789.
61. Fietzek, P. P., and Kühn, K. (1976). Int. Rev. Connect. Tissue Res. 7:1.
62. Kefalides, N. A. (1976). Int. Rev. Connect. Tissue Res. 6:63.
63. Abelev, G. I. (1971). Adv. Cancer Res. 14:295.
64. Uriel, J., Bouillon, D., Aussel, C., and Dupiers, M. (1976). Proc. Natl. Acad. Sci. U.S.A. 73:1452.
65. Gitlin, D., and Boesman, M. (1967). J. Clin. Invest. 46:1010.
66. Engelhardt, N. V., Poltorania, V. S., and Yazova, A. K. (1973). Int. J. Cancer 11:448.
67. Morris, G. E., Cooke, A., and Cole, R. J. (1972). Exp. Cell Res. 74:582.
68. Morris, G. E., Piper, M., and Cole, R. (1976). Nature (Lond.) 263:76.
69. Levine, A. J., Torosian, M., Sarokhan, A. J., and Teresky, A. K. (1974). J. Cell Physiol. 84:311.
70. Teresky, A. K., Marsden, M., Kuff, E. L., and Levine, A. J. (1974). J. Cell Physiol. 84:319.
71. Gearhart, J. D., and Mintz, B. (1974). Proc. Natl. Acad. Sci. U.S.A. 71:1734.
72. Gearhart, J. D., and Mintz, B. (1975). Cell 6:61.
73. Jami, J., and Loeb, J. E. (1975). *In* M. I. Sherman and D. Solter (eds.), Teratomas and Differentiation, pp. 221–233. Academic Press, New York.
74. Searle, R. F., Sellens, M. H., Elson, J., Jenkinson, E. J., and Billington, W. D. (1976). J. Exp. Med. 143:348.
75. Goodfellow, P. N., Barnstable, C. J., Bodmer, W. F., Snary, D., and Crumpton, M. J. (1976). Transplantation 12:597.
76. Artzt, K., Dubois, P., Bennett, D., Condamine, H., Babinet, C., and Jacob, F. (1973). Proc. Natl. Acad. Sci. U.S.A. 70:2988.
77. Artzt, K., Bennett, D., and Jacob, F. (1974). Proc. Natl. Acad. Sci. U.S.A. 71:811.
78. Bennett, D. (1975). Cell 6:441.
79. Artzt, K., and Bennett, D. (1975). Nature (Lond.) 256:545.
80. Bennett, D. (1975). *In* M. Balls and A. E. Wild (eds.), The Early Development of Mammals, pp. 207–218. Cambridge University Press, Cambridge.
81. Babinet, C., Condamine, H., Fellous, M., Gachelin, G., Kemler, R., and Jacob, F. (1975). *In* M. I. Sherman and D. Solter (eds.), Teratomas and Differentiation, pp. 101–107. Academic Press, New York.
82. Artzt, K., and Jacob, F. (1974). Transplantation 17:632.
83. Billington, W. D., and Jenkinson, E. J. (1975). *In* M. Balls and A. E. Wild (eds.), The Early Development of Mammals, pp. 219–232. Cambridge University Press, Cambridge.
84. Edidin, M. (1976). Ciba Found. Symp. 40:177. (New Series).
85. Muggleton-Harris, A., and Johnson, M. H. (1976). J. Embryol. Exp. Morphol. 35:59.
86. Vitetta, E. S., Artzt, K., Bennett, D., Boyse, E. A., and Jacob, F. (1975). Proc. Natl. Acad. Sci. U.S.A. 72:3215.
87. Stern, P. L., Martin, G. R., and Evans, M. J. (1975). Cell 6:455.
88. Gooding, L. R., and Edidin, M. (1974). J. Exp. Med. 140:61.

89. Edidin, M., and Gooding, L. (1975). *In* M. I. Sherman and D. Solter (eds.), Teratomas and Differentiation, pp. 109–121. Academic Press, New York.
90. Gooding, L. R., Hsu, Y.-C., and Edidin, M. (1976). Dev. Biol. 49:479.
91. Tooze, J. (1973). The Molecular Biology of Tumor Viruses. Cold Spring Harbor, New York.
92. Biczysko, W., Solter, D., Pienkowski, M., and Koprowski, H. (1973). J. Natl. Cancer Inst. 51:1945.
93. Swartzendruber, D. E., and Lehman, J. M. (1975). J. Cell Physiol. 85:179.
94. Lehman, J. M., Klein, I. B., and Hackenberg, R. M. (1975). *In* M. I. Sherman and D. Solter (eds.), Teratomas and Differentiation, pp. 289–301. Academic Press, New York.
95. Kelly, F., and Boccara, M. (1976). Nature (Lond.) 262:4019.
96. Jaenisch, R., and Mintz, B. (1974). Proc. Natl. Acad. Sci. U.S.A. 71:1250.
97. Speers, W. C., and Lehman, J. M. (1976). J. Cell Physiol. 88:297.
98. Siegel, G. (1976). *In* S. Gard and C. Hallaner (eds.), Virology Monographs, Vol. 15. Springer-Verlag, Wien.
99. Miller, R. A., Ward, D. C., and Ruddle, F. H. J. Cell Physiol., in press.
100. Huebner, R. J., Kellof, G. J., Sarma, P. S., Lane, W. T., Turner, H. C., Gilden, R. V., Oroszlan, S., Meier, H., Myers, D. D., and Peters, R. L. (1970). Proc. Natl. Acad. Sci. U.S.A. 67:366.
101. Temin, H. M. (1971). J. Natl. Cancer Inst. 46:3.
102. Jaenisch, R., Fan, H., and Cooker, B. (1975). Proc. Natl. Acad. Sci. U.S.A. 72:4008.
103. Jaenisch, R. (1976). Proc. Natl. Acad. Sci. U.S.A. 73:1260.
104. Brinster, R. L. (1974). J. Exp. Med. 140:1049.
105. Brinster, R. L. (1975). *In* M. I. Sherman and D. Solter (eds.), Teratomas and Differentiation, pp. 51–58. Academic Press, New York.
106. Mintz, B., Illmensee, K., and Gearhart, J. D. (1975). *In* M. I. Sherman and D. Solter (eds.), Teratomas and Differentiation, pp. 59–82. Academic Press, New York.
107. Dunn, G. R., and Stevens, L. C. (1970). J. Natl. Cancer Inst. 44:99.
108. Kemler, R., Babinet, C., Condamine, H., Gachelin, G., Guenet, J. L., and Jacob, F. (1976). Proc. Natl. Acad. Sci. U.S.A. 73:4080.
109. Jacob, F. (1977). Immunological Rev. 33:3.

International Review of Biochemistry
Biochemistry of Cell Differentiation II, Volume 15
Edited by J. Paul

9
Structural Organization and Transcription of the Genome of *Dictyostelium discoideum*

R. A. FIRTEL[1]

Department of Biology,
University of California at San Diego,
La Jolla, California, U.S.A.

AND

A. JACOBSON[2]

Department of Microbiology,
University of Massachusetts Medical School,
Worcester, Massachusetts, U.S.A.

---

[1] Recipient of grants from The American Cancer Society, the National Science Foundation, and the University of California's Cancer Research Coordination Committee.

[2] Recipient of a grant and a Faculty Research Award from The American Cancer Society.

DICTYOSTELIUM DISCOIDEUM AND ITS DEVELOPMENTAL BIOLOGY

For good reasons, biologists tend to concentrate their efforts on two kinds of organisms—the "unipurpose" organism and the "all-purpose" organism. The former class includes those organisms (or cell types) which facilitate the study of a very specific event, e.g., cells which accumulate extraordinary amounts of a single protein. The latter class includes those organisms which have a large number of useful attributes, all of which facilitate studies of diverse biological problems.

The cellular slime mold, *Dictyostelium discoideum*, is rapidly becoming a member of the all-purpose class. In 1956, Sussman (1) noted that only five laboratories in the entire world were studying the biology of the slime molds. Today it is estimated that the number has increased at least 50-fold. In 1956, most efforts were concentrated on an understanding of aggregation and morphogenesis. Those studies continue today. In addition, intensive work is being pursued in studies of slime mold cell surfaces, gene structure, RNA, protein and carbohydrate metabolism, photo- and chemotaxis, sexual and parasexual genetics, spore formation and germination, and other aspects of the developmental biology of the organism (for reviews see refs. 2–9). This explosion of research has been facilitated by the organism's attributes, which include the following:

1) *D. discoideum* is a simple eukaryote which undergoes a rigorously defined developmental sequence (2, 8).
2) The organism can be cultivated in large quantities in association with bacteria, in an axenic medium, or in a new, defined medium (2, 8, 10).
3) The organism is haploid. Clones can be isolated on bacterial lawns, mutants are easy to generate, and sexual and parasexual genetics has been established (11, 12).
4) The nuclear genome is small, relative to that of metazoans, and messenger RNA (mRNA) metabolism is correspondingly simplified (13–17).

At the risk of oversimplifying, it can be said that, collectively, slime mold biologists are interested in understanding the regulation of the organism's life cycle, which is depicted in Figure 1. In the center of the figure is the morphogenetic sequence which has commanded the most attention. It indicates that amoebae grow vegetatively in the presence of a food source, but, when starved, aggregate into multicellular structures. The chemotactic agent for aggregation is cyclic AMP (18). Aggregates form pseudoplasmodia (slugs), structures which show biochemically differentiated regions (2–7). Slugs are also phototactic and chemotactic and are capable of migrating. Ultimately, the pseudoplasmodium forms a "Mexican hat." Within the center of this structure, "prestalk" cells vacuolate and extend lengthwise, forming a stalk with a spore mass at its tip. Eighty per cent of the cells originally in the aggregate differentiate into spores. These have the capacity to germinate in the presence of a food source, liberating individual amoebae. The presence or absence of numerous factors affects and regulates this developmental sequence or, alternatively, these factors have their presence or absence regulated by the developmental sequence. These factors include food, heat, light, moisture, specific ions, small molecules, and macromolecules (2, 7). The last category has attracted the attention of developmental biologists working on many systems, all of whom are intent on knowing what the changes in gene expression are during the development of an organism and what regulates these changes. In this chapter, the known developmental changes in slime mold RNA and protein synthesis are summarized and the current status of research on the structure and transcription of the slime mold genome is discussed at length.

Developmental Changes in Protein Synthesis

Experiments by Sussman and co-workers on the changes in specific activity of certain enzyme during slime mold morphogenesis first focused attention on the usefulness of this organism for studies on the regulation of gene expression. This work was previously reviewed in Volume 1 of this series by Newell (9) and is only summarized here. The initial studies focused on enzymes of carbohydrate metabolism, but since that time the list has spread to a variety of enzymes (2) (Figure 2). In a typical experiment, extracts are prepared from several developmental stages and assayed for specific activity of a given enzyme. In a typical

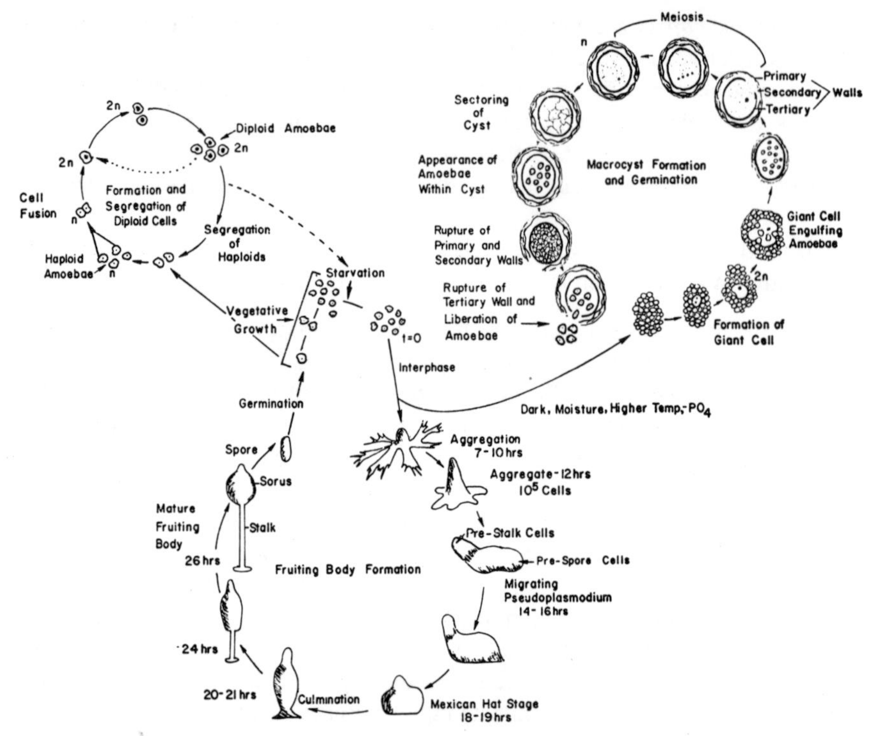

Figure 1. Stages of the slime mold life cycle. *Center*, the commonly studied aspects of the slime mold life cycle: vegetative growth of haploid amoebae and fruiting body formation. Fruiting body formation occurs when starved amoebae are placed on an air-water interface in the presence of visible light. Fruiting body formation culminates with the differentiation of approximately 80% of the cells into spores which, under appropriate conditions, germinate and liberate amoebae, which can then enter vegetative growth. *Upper right*, the formation and germination of macrocysts. Entry into this cycle is favored by starved amoebae in the dark, in the presence of excess moisture or in the absence of phosphate. Two amoebae fuse to form a diploid giant cell. The eventual meiotic divisions that yield four haploid nuclei are not observable cytologically, but are included in the figure because their existence is implied from genetic experiments. Amoebae liberated in macrocyst germination can commence vegetative growth or aggregation. *Upper left*, the formation and segregation of diploid cells, the cycle that is exploited for parasexual genetics. Diploid cells can grow, as such, vegetatively and also can enter fruiting body formation. In the latter case, diploid spores are found (see refs. 2, 3, and 7 for details).

result, a given enzyme activity is virtually absent for several hours of development, then rises dramatically, reaches a characteristic level, and then falls off. Timing of appearance and disappearance, as well as level of activity, is reproducible and well coordinated with the morphogenetic sequence. Moreover, appearance of enzyme activity is generally inhibited by prior treatment of cells with drugs which inhibit RNA and protein synthesis (19, 20, 21). This implication of de novo protein synthesis has been definitively confirmed for two enzymes (22, 23) and for a carbohydrate-binding protein (24).

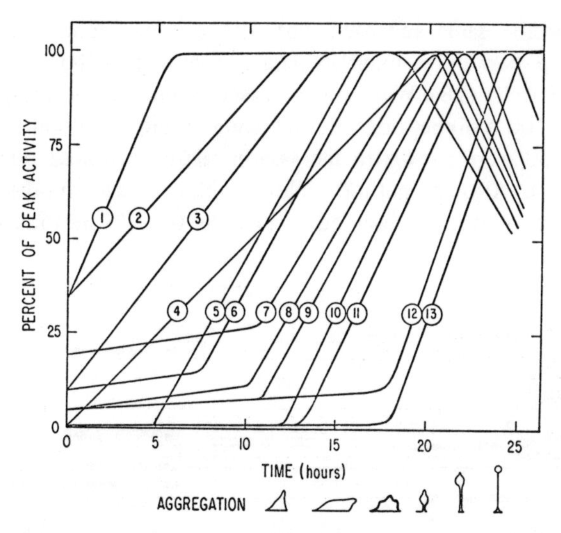

Figure 2. Stage-specific enzymes of *Dictyostelium discoideum*. Vegetative *Dictyostelium* cells are plated on an air buffer interphase for the initiation of development. Samples are then taken at the appropriate times, and specific activities of the enzymes are determined. 1, alanine transaminase; 2, leucine aminopeptidase; 3, *N*-acetylglucosaminidase; 4, α-mannosidase; 5, trehalose phosphate synthetase; 6, threonine deaminase-2; 7, tyrosine transaminase; 8, UDPG-pyrophosphorylase; 9, UDPgal polysaccharide transferase; 10, UDPG-epimerase; 11, glycogen phosphorylase; 12, alkaline phosphatase; and 13, β-glucosidase-2. (See refs. 2, 3, 7, and 9 for details and references.) Reproduced from Loomis (2) with permission of Academic Press.

Different enzymes accumulate at different times of the life cycle. Studies with mutants which are temperature-sensitive for a specific period of development suggest that enzymes characteristic of a given developmental stage cannot appear if enzymes characteristic of an earlier stage have not appeared. Loomis et al. (25) have used such data to construct a map of dependent stages of slime mold development.

Another approach to assessing developmental changes in protein synthesis has been to analyze changes in the differential pulse labeling of total cellular polypeptides. In the first such experiments, Tuchman et al. (26) pulse labeled cells every 2 hr during development with [^{35}S]methionine and then analyzed total pulse-labeled polypeptides on sodium dodecyl sulfate (SDS) polyacrylamide gels. These experiments showed that at least 20 polypeptides dramatically increased or decreased their differential synthesis during development. All of these remain unidentified, with the exception of the major observed change—actin. Tuchman et al. (26) showed that in the first 2 hr of development actin comprised almost 25% of the methionine-labeled polypeptides, whereas in amoebae it was only 4–6% of the labeled protein. Moreover, they showed that at later developmental stages actin synthesis progressively decreased to the level found in vegetative cells. These studies have been followed by Alton (27), who used essentially the same approach, but with a more sophisticated analytical system (i.e., the two-dimensional isoelectric focusing SDS-polyacrylamide gel method of O'Farrell) (28). With this increased resolution, he found changes in approximately 100 polypeptides during slime mold development, of which approximately 50 were shown to increase in relative synthesis during development (Figure 3). Alton's methods allowed him to see approximately 400 proteins, some of which represented as little as 0.007% of total cellular syn-

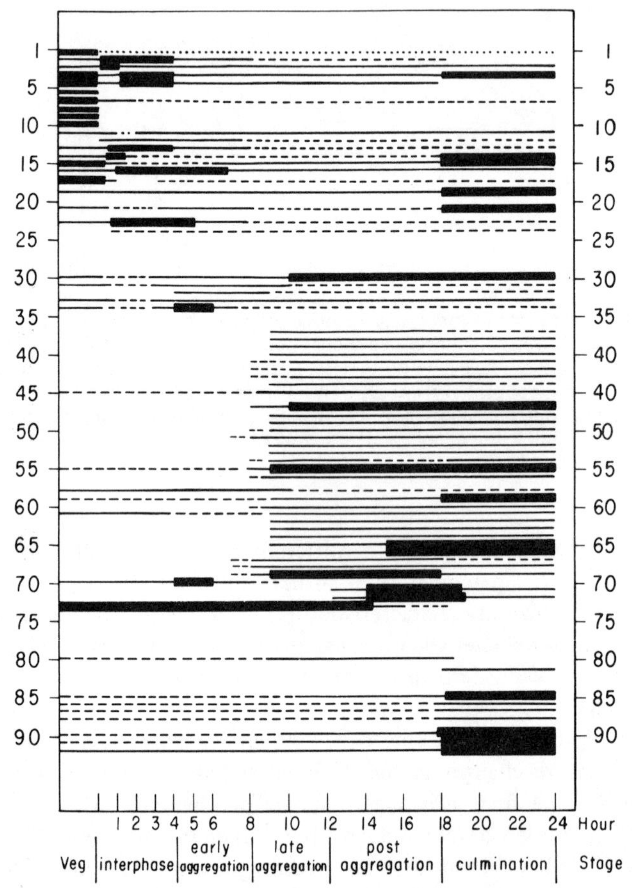

Figure 3. Schematic summary of the program of protein synthesis during *Dictyostelium* development. This summary is based on the examination of gels containing pulse-labeled proteins synthesized during *Dictyostelium* development on an hour-by-hour basis. During the first few hours of development, cells were labeled for 15 min with [^{35}S]methionine and for 1 hr at later times in development. Newly synthesized proteins were examined on two-dimensional gels, according to the method of O'Farrell (28). The numbers on the right and left of the diagram correspond to the polypeptide spots which show changes during development. The base of a line indicates synthesis of the protein could not be detected. – – –, barely detectable synthesis; ——, synthesis; ▬, maximal relative rate of synthesis. (See ref. 27 for details.) Reproduced from Alton (27) with permission of the author.

thesis. Most of the changes observed by Alton occurred at 8–10 hr of development, and maintenance of cells in liquid suspension precluded these changes. Most, but not all, of the observed changes can be mimicked by cell-free protein-synthesizing systems programmed with stage-specific messenger RNA (27). Those that cannot be mimicked may well represent proteins whose synthesis is subject to translational control or problems in efficient in vitro translation of specific mRNAs.

Other approaches to the question of changes in protein synthesis have been reported (29–38). In general, these studies describe changes in the accessibility of certain polypeptides to specific labeling reagents or changes in the binding of certain substrates. Very few of these studies actually determine the appearance or disappearance of certain proteins.

Developmental Changes in RNA Synthesis

As noted above, Alton (27) has shown that cell-free systems programmed with RNA from different developmental stages synthesize in vitro the polypeptides characteristic of that developmental stage. Moreover, the synthesis directed by these RNAs is nonoverlapping; that is, extreme stage specificity is observed. Thus, one can readily conclude that the appearance and disappearance of approximately 100 polypeptides during slime mold development can be explained by the specific availability of active mRNAs for those polypeptides. These results do not address themselves to the periods of synthesis or degradation of these RNAs.

Qualitatively similar results have been obtained in two kinds of DNA-RNA hybridization experiments. In the first, Firtel (39) showed that the fraction of single copy DNA (see below) hybridized to RNA varied if the RNA was extracted from different developmental stages. These workers also showed that hybridization of RNAs of different developmental stages was partially additive and that there was clearly a small, but significant, stage-specific component in all RNA samples. In total, 16,000 different sequences (1,000 nucleotides each) were estimated to be present at all developmental stages, whereas 6,000 new sequences were shown to appear after vegetative growth at specific developmental stages and approximately 6,000 sequences were shown to disappear.

In the set of experiments described above, total cell RNA was used. To examine the changes in the number of polysomal mRNA sequences, a second type of RNA-DNA hybridization experiment has been done (40). In this, DNA complementary to mRNA of amoebae or 18-hr developing cells was prepared by the use of AMV reverse transcriptase (41). Subsequently, the ability of the respective mRNAs to hybridize to the two DNA preparations was evaluated by measuring resistance of labeled DNA to attack by S1 single strand-specific endonuclease (41). These experiments showed that the homologous mRNA-complementary DNA (cDNA) hybridizations rendered the DNA essentially completely nuclease-resistant. The cross-hybridizations (cDNA made from RNA of one stage and mRNA from another) rendered incomplete protection. The results showed that *Dictyostelium* contains approximately 4,000–5,000 different mRNA species during its life cycle, and, of these, approximately 300–400 represent new species which are specific for developing cells (see under "Complexity and Abundancy Classes of Heterogeneous Nuclear RNA and Messenger RNA" for more detailed analysis). This analysis of the number of developmental changes in mRNAs during differentiation is close to the estimated number of genes required for aggregation and later development, determined by genetic mutation frequency analysis (42, 43).

Analysis of hybridization kinetics showed that approximately one-third of these changes represent mRNA present at 100 or more copies per cell, whereas the others are present at only 5–8 copies per cell. The translation of the latter would be undetectable in Alton's experiments (27). These results with the hybridization of cDNA to mRNA are reinforced by experiments in which polysomal, poly(A)-containing mRNA was hybridized to single copy DNA. The number of total mRNAs and the changes observed during development are in agreement with the cDNA results. This analysis also concludes that most of the changes observed in the original experiments (37) are not due to mRNAs but to nontranslated nuclear transcripts. This has been confirmed by direct analysis of the nuclear sequences. It should be pointed out that this type of analysis identifies total changes, whereas the genetics analysis of total genes involved in development estimates "required" or developmentally essential genes (42, 43).

Changes in the synthesis (and processing) of other RNAs have also been reported. Cocucci and Sussman (44) reported that bulk ribosomal RNA of amoebae is turned over with a half-life (first order) of approximately 12 hr during development. More recent results by Jacobson (45) yielded a qualitatively similar conclusion with regard to rRNA turnover, but showed that the decay process was more complex and could not be accounted for by a single first order component. Cocucci and Sussman also showed that this rRNA was replaced by a new, stable component synthesized during development. Experiments by Batts-Young et al. (46) confirmed the results of Cocucci and Sussman and further showed that all species (26 S, 17 S, and 5 S) of rRNA were equally turned over. (The larger mature cytoplasmic rRNA of *Dictyostelium* has been called both 25 S and 26 S in the literature. Its length, as defined by direct sizing and EM analysis, is 4.1 kb (47).) Batts-Young et al. (46) also showed that the primary sequences of the 5 S rRNAs labeled in growth or development were identical and that the primary sequences of all of the major oligonucleotides of the appropriate pairs of 26 S and 17 S rRNAs were also identical. Their results strongly suggest that the turnover of rRNAs is not a mechanism for the introduction of a new species of rRNA into the ribosome.

Developmental changes in tRNA have been assessed by Palatnik et al. (48), who have measured the amino acid acceptance of the transfer RNAs (tRNAs) for 17 different amino acids in amoebae and in developing cells. Furthermore, they have compared the chromatography profiles of isoaccepting species for all of these tRNAs. Their results indicate that it is unlikely that transcription of tRNAs has changed during development. However, they present clear evidence that the post-transcriptional modification of a number of tRNA species is dramatically different in different developmental stages.

All of the above results make it quite clear that, in *D. discoideum*, there are dramatic changes in gene expression during development. Such results have naturally provided the impetus for experiments designed to ultimately determine how specific gene expression is regulated. However, before regulation can be understood, we must first know what a gene looks like, how it is transcribed, and how that transcript is processed. In the following pages, the current answers to these questions are reviewed.

GENERAL PROPERTIES OF NUCLEAR AND MITOCHONDRIAL GENOMES

Characterization of the nuclear and mitochondrial DNAs has been pursued by a variety of methods including fractionation by buoyant density on CsCl, analysis of renaturation kinetics and restriction enzyme digests, isolation of recombinant DNA molecules, and studies of homologous transcripts. Most of this work has been directed at an understanding of the nuclear genome. In this section the general properties of the nuclear and mitochondrial genome are reviewed. Specific genes are considered in subsequent sections.

Components Within Total Cellular DNA; Buoyant Densities and Renaturation Kinetics

The haploid genome size of *D. discoideum* has been determined by renaturation kinetics to be 3.0×10^{10} daltons or approximately 11–12 times the size of the genome of *Escherichia coli* (13). A buoyant density analysis of total cell DNA shows three density peaks in CsCl gradients (12, 49) (Figure 4). The first (p = 1.676 g/cc, "main band" DNA) represents approximately two-thirds of the cellular DNA. The density of the main band DNA corresponds to a GC content of 22–23 mole per cent. A second density peak of 1.682 contains approximately 27% of the cellular DNA and comprises both the mitochondrial DNA (approximately 24% of the cellular DNA) and a nuclear satellite. A third peak, comprising 6–7% of cellular DNA, can be observed at the density of 1.687 (where an *E. coli* standard DNA is taken as 1.704 g/cc). Nuclear DNA also shows three peaks in similar CsCl gradients: the main band peak, comprising approximately 80% of the nuclear DNA; a nuclear satellite at p = 1.682; and an additional satellite at p = 1.687. The relative amounts of the nuclear DNA present in the two satellites are a function of the molecular weight of the DNA (50). At an average molecular weight of approximately 2×10^7 daltons, 15–20% of the nuclear DNA is divided equally between the two satellites. Both satellite bands on CsCl show a relatively broad half-width, suggesting that they contain a variety of components with varying densities. As will be described below, the majority of the two nuclear satellites contains the ribosomal RNA cistrons, and the heterogeneity of the density of the peaks is a function of random shearing of the

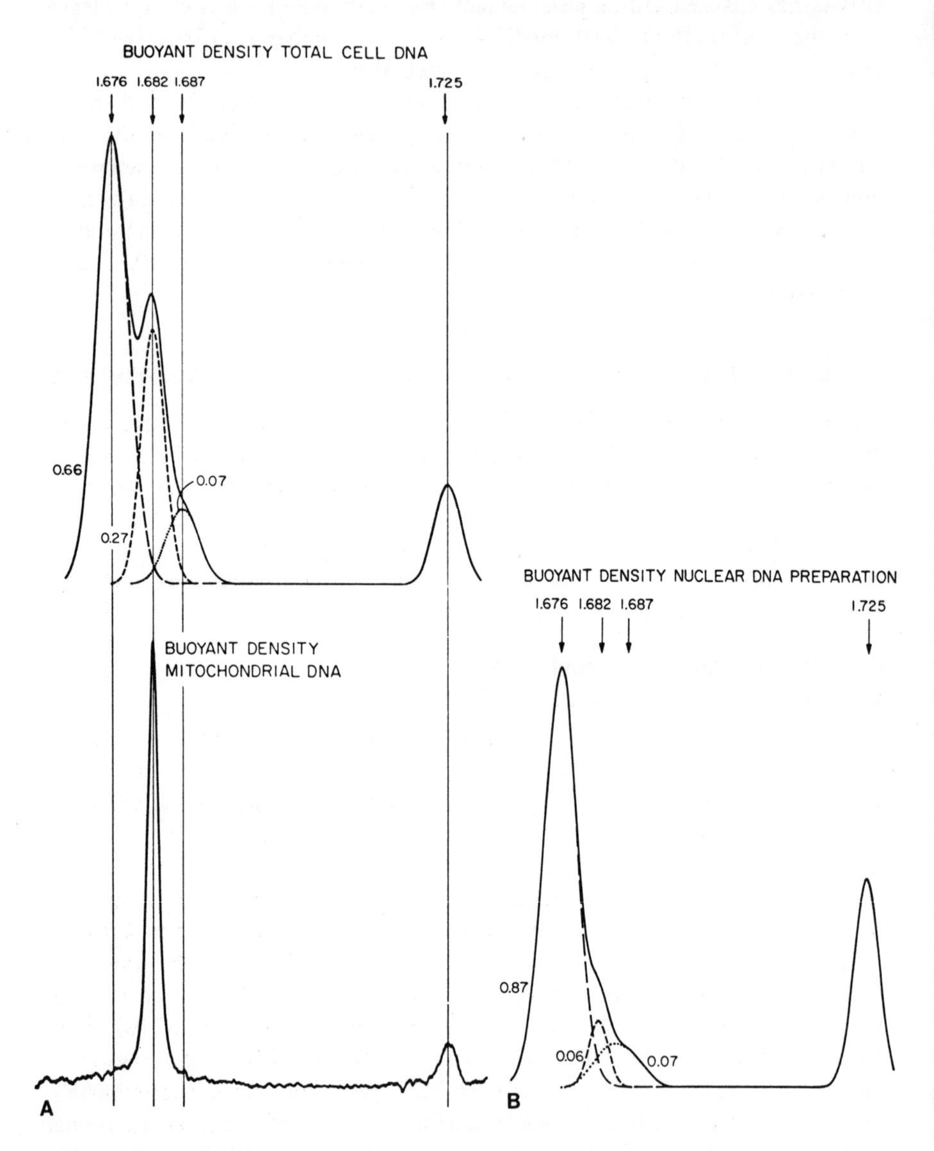

Figure 4. Analytical buoyant density analysis of whole cell, mitochondrial, and nuclear DNA. *A*, *Upper*, computer analysis of X-Y plotter scan of analytical centrifugation run of total cell DNA. *Solid line* represents optical density profile; *dashed lines* represent location sites of individual components suggested by computer analysis. *Lower*, X-Y plotter scan of analytical run of mitochondrial DNA. The band at $\rho = 1.725$ g/cc is marker DNA, *Micrococcus lysodeikeious*. *B*, X-Y plotter scan of analytical run of purified nuclear DNA. (See ref. 13 for details.) Reproduced from Firtel et al. (13) with permission of *J. Mol. Biol.*

ribosomal cistrons which contain both relatively high GC (42% GC) and relatively low GC (22% GC) components. (The calculated average density of ribosomal cistrons is approximately 1.685 g/cc.) Additional data also suggest that other components are present in the two nuclear satellite peaks (49).

Purified mitochondrial DNA from *D. discoideum* has a density of 1.682 (28% GC). Both the band width of the DNA in CsCl gradients and its melting profile suggest that the mitochondrial DNA is probably relatively homogeneous with regard to GC content (13). Renaturation kinetics of the mitochondrial DNA gives a complexity of approximately 35×10^6 daltons. EM analysis of gently lysed *D. discoideum* cells shows that the mitochondrial genome is circular, with a molecular weight agreeing with the physical analysis of the DNA (13, 49).

By analysis of its renaturation kinetics, the nuclear DNA can be separated into two main components: a single copy component, comprising 70–75% of the nuclear DNA, which has a Cot $\frac{1}{2}$ (pure) of 68 (by optical measurements), and a repetitive fraction which has a Cot $\frac{1}{2}$ (pure) of 0.28 (Figure 5). Under the same conditions, the Cot $\frac{1}{2}$ of *E. coli* DNA is 6.5. The complexity of the single copy portion of the genome, after correcting for GC effects, is 7 times greater than that of the *E. coli* genome, or approximately 2×10^{10} daltons. The reiterated fraction has an average complexity of approximately 8×10^7 daltons and an average repetition number of approximately 20 (13, 17, 51). The breadth of the renatured reiterated sequences suggests that they contain multiple families having different reiteration frequencies. Melting profiles of the renatured single copy DNA indicate that these sequences show little base mismatch, as is expected for unique DNA, whereas melting profiles of the repeated sequences show a component having a high degree of homology and a smaller component showing some base mismatch (13). The fraction showing a high degree of homology consists mainly of the ribosomal RNA cistrons (49).

General Organization of Nuclear DNA

As noted above, renaturation profiles of *Dictyostelium* nuclear DNA (sheared to fragments 400 nucleotides long) indicate that approximately 25–30% of the genome contains reiterated sequences and the remainder contains unique sequences. "Linkage" experiments in which tracer DNA of varying lengths is hybridized to an excess of 400-nucleotide "driver" suggest that approximately 40% of the single copy DNA sequences is linked to reiterated DNA with an average single copy sequence length of 1.2 kb. Approximately 20% of the single copy DNA is also linked to repeat sequences, but with a longer interspersion frequency. The remainder of the single copy sequences is not interspersed at fragment lengths of approximately 3–4 kb (17, 51, 52). Repetitive DNA falls into two size classes. Approximately 70% of the repetition DNA is longer than 2 kb. The remaining 30% shows a smaller size distribution, ranging from 0.2 kb to 1 kb or more in length. Sizing in polyacrylamide gels suggests rela-

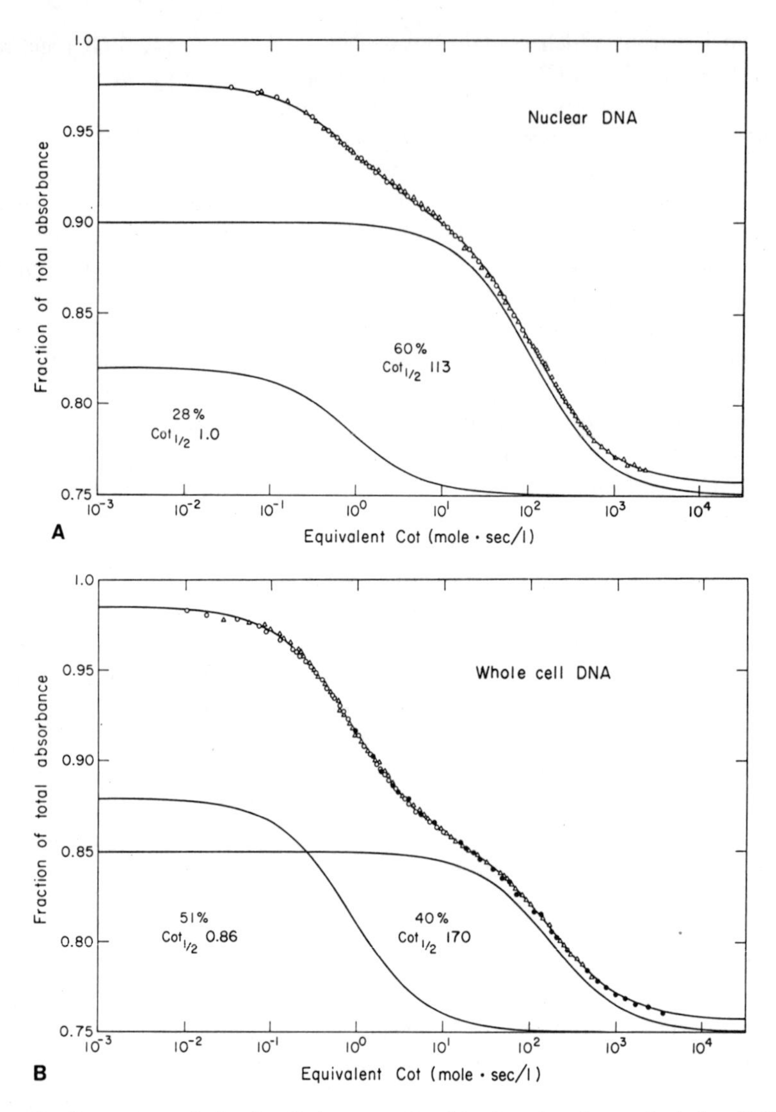

Figure 5. Computer analysis of optical renaturation kinetics of nuclear and whole cell DNA. *A*, nuclear DNA. *B*, whole cell DNA. (See ref. 13 for details.) Reproduced from Firtel et al. (13) with permission of *J. Mol. Biol.*

tively broad size distribution; sizing on Agarose A-15 columns also suggests a similar broad distribution, but does indicate a discrete peak at approximately 2–3 $\times$ 10^5 daltons (17, 51).

The distribution of single copy and repetitive sequences in *D. discoideum* appears to be similar to but not identical with that observed in higher eukaryotes (53, 54). A large fraction of the single copy DNA is interspersed with repetitive sequence elements, and these repetitive sequence elements are distributed in both long and short sequences. In contrast to the situation in toads (*Xenopus laevis*) (55) or sea urchins (56), a much smaller fraction of repetitive sequences in *D. discoideum* appears to be in short interspersed elements, whereas a large fraction is present in long repetitive sequences. From the analysis of the ribosomal RNA genome (see under "Structure and Transcription of Ribosomal RNA Genes"), it is known that approximately 60% of the repetitive sequences present is contained within the ribosomal genes, a function of the long length of the rDNA repeats and the small size of the *D. discoideum* genome. In addition, the size distribution of the "interspersed" repeat sequence in *Dictyostelium* is not as homogeneous as that observed in most metazoans (54–57). At the present time, it is not clear what the functional relationship is of the interspersed repetitive sequences with the single copy genes. Several recombinant plasmids (see under "Structure and Transcription of Messenger DNA") carrying *Dictyostelium* nuclear DNA which contain unique DNA regions complementary to messenger RNA have been examined. In these plasmids, small sequences have been detected which appear to be complementary to multiple mRNA species of mRNA and to renature to reiterated DNA. A more complete analysis of these sequences is not known at the present time.

Poly(dT)$_{25}$ Sequences

Analysis of poly(A) sequences in slime mold nuclear and cytoplasmic heterodisperse RNA (messenger and premessenger RNA) reveals two classes of poly(A) sequences (see Figure 2 and under "Poly(A) on *Dictyostelium* Messenger RNA"): shorter sequences which average 25 nucleotides in length and longer sequences which average 125 nucleotides in length (15, 17, 51, 57, 58). Results described below suggest that the sequence 125 nucleotides in length is located at the 3′ end of the RNA and is added post-transcriptionally. This sequence is analogous to the poly(A) sequences found on a majority of eukaryotic messenger RNAs. The poly(A)$_{25}$ sequence is localized near the 3′ end of the messenger RNA, within approximately 25 nucleotides (average) of the post-transcriptionally added poly(A) (15, 57). Experiments on transcription in isolated nuclei suggest that the poly(A)$_{25}$ sequence is transcribed from the genome rather than added post-transcriptionally (58). Hybridization of [^{3}H] poly(A) and [^{3}H] poly(U) to *D. discoideum* nuclear DNA indicates that it contains poly(dT) and poly(dA) sequences, respectively, totaling approximately 0.3% of the DNA. Melting profiles of the hybrids suggest that they are very short. Depurination of ^{32}P uniformly labeled or [^{3}H] thymidine-labeled nuclear DNA yields a class of poly(dT) sequences with an average size of approximately 25 nucleotides in length. The mass ratio of these sequences indicates that they comprise 0.3% of the nuclear DNA. If one assumes an average size of 25 nucleotides, obtained by sizing on polyacrylamide gels and by oligonucleotide fingerprint analysis, there are approximately 14,000–15,000 (dT)$_{25}$ sequences per genome. Further analysis suggests that these sequences are randomly spaced in the genome rather than

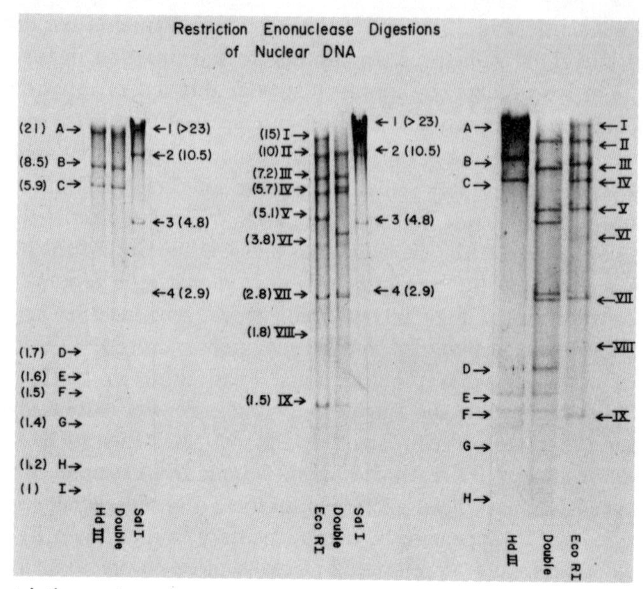

Figure 6. Restriction endonuclease digests of nuclear DNA. Nuclear DNA purified from *D. discoideum* was digested with the restriction endonucleases Eco RI, Hind III, and Sal I, and the three combinations of the enzymes (double digests). The digested DNA was then analyzed by electrophoresis in 0.8% Agarose gels and visualized by ultraviolet fluorescence after staining with ethidium bromide. The major observed restriction bands are labeled as described in refs. 47, 49, 60. The numbers in parenthesis represent molecular weights of the major restriction bands in kilobases (kb). Reproduced from Cockburn et al. (60) with permission of *Cell.*

clustered in a manner similar to that in simple sequence satellites found in many metazoans (17, 57).

The fact that poly$(A)_{25}$ sequences are localized in the majority of *Dictyostelium* messenger RNAs indicates that the poly$(dT)_{25}$ sequences should be associated with gene sequences. Nuclear RNA fragments containing poly$(dT)_{25}$ show appreciable hybridization to poly(A)-containing RNA (poly$(A)^+$ RNA) (17). These results suggest that the majority of gene sequences in *D. discoideum* contain the poly$(dT)_{25}$. Recombinant plasmids carrying nuclear DNA which codes for specific proteins have been shown to contain $(dT)_{25}$ sequences in the plasmids; the size appears to be heterogeneous within the range of approximately 20–30 nucleotides in length.

Figure 7. Mapping of DNA fractionated by buoyant density on CsCl gradients with restriction endonuclease Eco RI. Preparative CsCl gradients containing ^{32}P nuclear DNA and ^{3}H total cell DNA were fractionated into five regions (*A–E*), *A* being from the most GC-rich region and *E* from the most AT-rich region. The individual fractions were digested with Eco RI and analyzed on 0.8% cylindrical Agarose gels. The gels were sliced and counted. The Roman numeral markers represent the position of the major Eco RI restriction bands as shown in Figure 6. M1 to M4 represents the positions of Eco RI digests of *Dictyostelium* mitochondrial DNA. As can be seen, mitochondrial bands can be visualized in ^{3}H total cell DNA and not in the nuclear DNA.

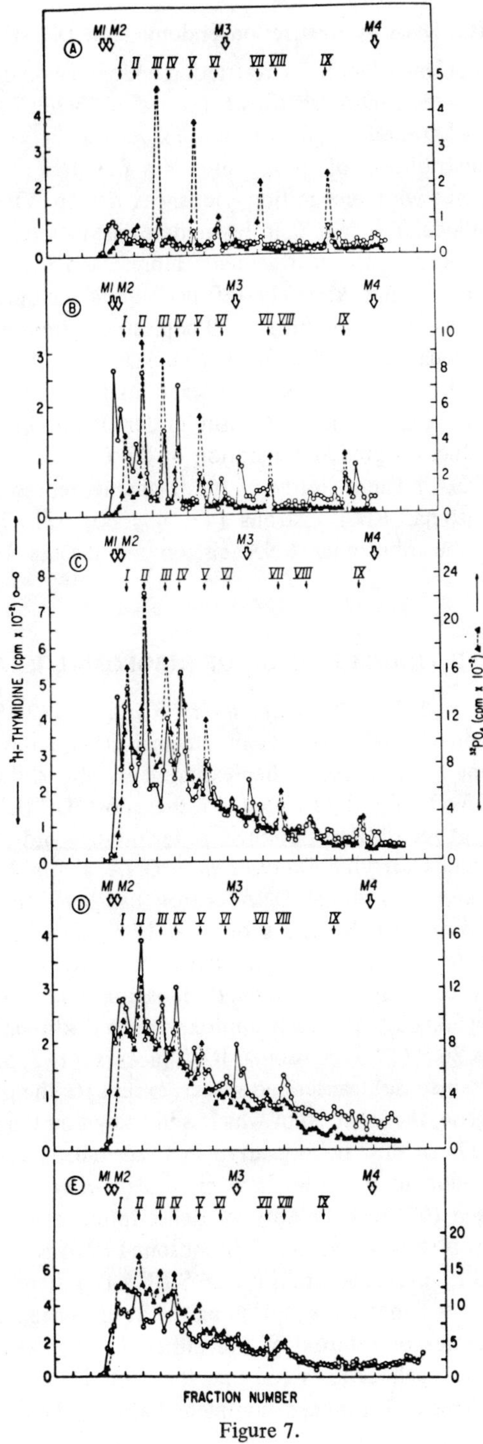
A
B
C
D
E
M1 M2
M3
M4
I II III IV V VI VII VIII IX
^{3}H-THYMIDINE (cpm x 10^{-2})
^{32}PO$_4$ (cpm x 10^{-2})
FRACTION NUMBER

Figure 7.

Analysis of Nuclear DNA by Restriction Endonuclease Digestion

Analysis of the nuclear genome by restriction endonuclease digestion and subsequent Agarose gel electrophoresis shows a series of "bands" superimposed on a heterogeneous background (Figures 6 and 7). As can be seen in Figure 6, the molecular size distribution of these bands is a function of the enzyme used. Restriction enzymes with recognition sequences rich in AT yield fragments in lower average molecular weight than restriction enzymes recognizing sequences rich in GC. A majority of the fragments represented by bands is present in equimolar amounts of approximately 180 per haploid genome, whereas some of the fragments in bands are present in 100 copies per genome or less (49). Most of the bands are localized in the GC-rich nuclear satellites, as determined by analysis of restriction enzyme digests of fractions from CsCl or CsCl-netropsin gradients (49) (see Figure 7). (Netropsin preferentially binds to AT-rich sequences and produces a greater separation of the GC-rich satellite DNAs from the main band DNA.) The majority of the fragments represented by bands are part of the ribosomal RNA cistrons (47, 49, 60, 62). The structure and transcription of the ribosomal RNA cistron are discussed in the following section.

STRUCTURE AND TRANSCRIPTION OF RIBOSOMAL RNA GENES

Earlier studies showed that *Dictyostelium* 17 S and 26 S (25 S) ribosomal RNA hybridized to approximately 1.1% of the nuclear DNA, equivalent to 120–180 copies per genome (13). These earlier experiments also showed that the ribosomal RNA preferentially hybridized to the most GC-rich of the nuclear satellites. By use of restriction endonuclease technology and the construction of recombinant plasmids carrying *Dictyostelium* DNA, a complete structural map of the *Dictyostelium* ribosomal DNA cistron has been formulated (47, 49, 58–63) (Figures 8 and 9). Results show that 17 S, 26 S, and 5.8 S ribosomal RNA hybridized to a discrete set of restriction bands which were linked to nontranscribed spacer sequences. A single ribosomal repeat was approximately 44 kb in length (62, 63), of which approximately 8 kb contained the coding sequences for the 36 S (37 S) ribosomal RNA precursor (47, 61). The remainder of the repeat contained nontranscribed spacer, except for the presence of the 5 S ribosomal RNA gene, the location of which is indicated on the map (47, 61, 64).

The ribosomal cistrons are organized in head-to-head dimers in a fashion similar to that found in the true slime mold *Physarum* and the ciliated protozoa *Tetrahymena* (62, 63). Polarity of transcription of these genes has been determined, and it is shown that the 17 S ribosomal RNA is localized near to the 5′ end of the 36 S transcript and that the 26 S RNA is present at the 3′ end (47). These results indicated that transcription on the head-to-head dimer is antipolar, away from the center of rotation of the dimer. Further analysis showed that these dimers are true palindromes and that the central portion denoted by Eco RI restriction fragment I forms a complete hairpin when quick-cooled after

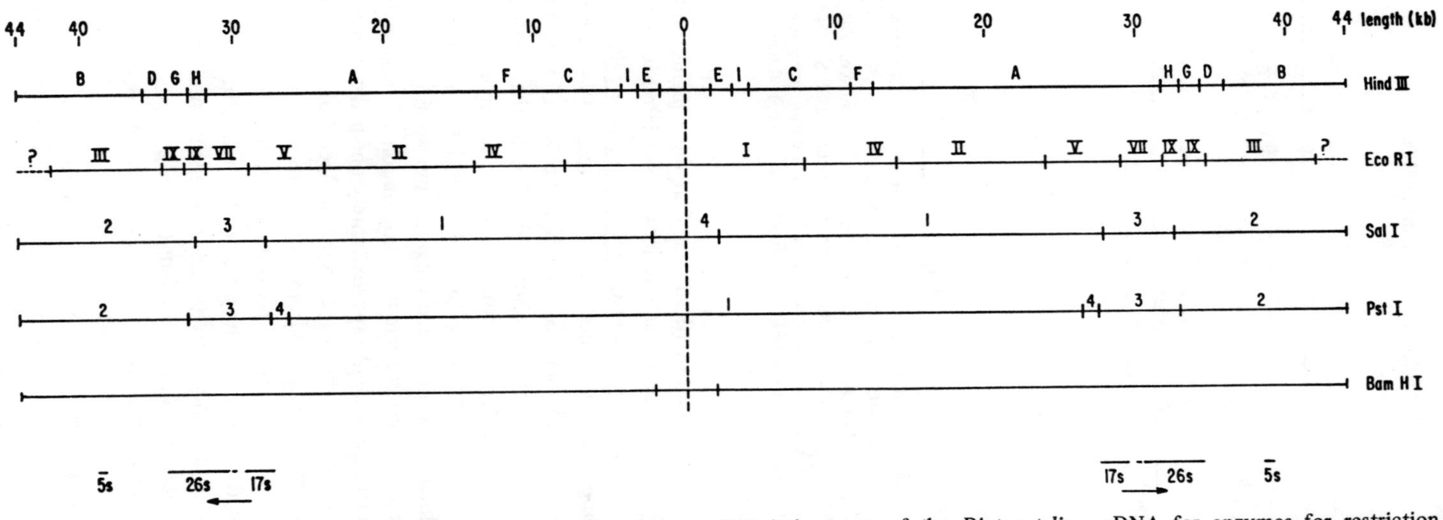

Figure 8. Restriction map of the palindromic rDNA repeat dimer. Restriction map of the *Dictyostelium* rDNA for enzymes for restriction endonucleases Eco RI, Hind III, Sal I, Pst, I, and Bam III is shown. The numbering of the bands for Eco RI, Hind III, and Sal I digests corresponds with that shown in Figure 6. The position and direction of transcription of the ribosomal RNA cistrons are shown. Reproduced from Taylor et al. (62) with permission from Academic Press and Cockburn et al. (63) with permission of *Cell.*

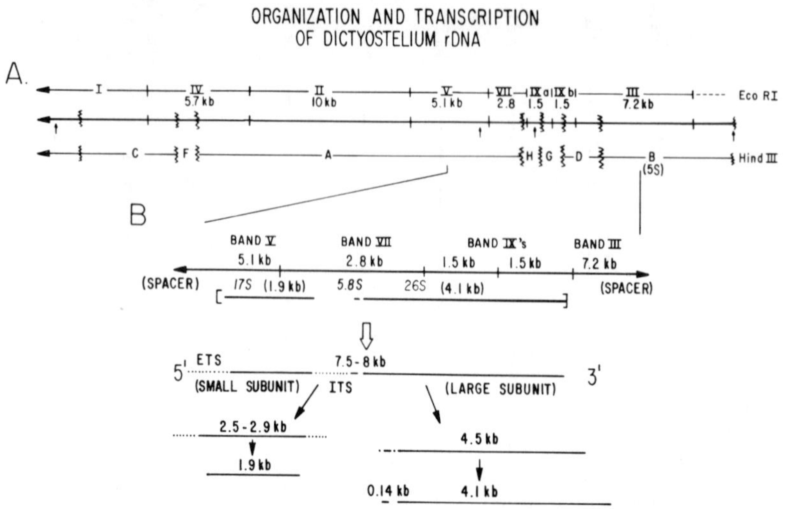

ORGANIZATION AND TRANSCRIPTION OF DICTYOSTELIUM rDNA

Figure 9. Organization of the coding region for *Dictyostelium* 36 S rRNA precursor. The position of the region coding for the 26 S, the 17 S, and 5.8 S rRNA and 36 S rRNA precursor is shown. A processing scheme for ribosomal RNA is shown. The location of the 5′ end of the 36 S precursor, the specific steps of processing at the 28 S rRNA precursor and mature 17 S and 26 S rRNA, and a complete derivation of the map are given in ref. 47. ITS, internal transcribed spacer; ETS, external transcribed spacer. Reproduced from Frankel et al. (47) with permission of *J. Mol. Biol.*

denaturation. Additional experiments suggest that these palindromic dimers are isolated as free DNA molecules not covalently associated with chromosomal DNA. Whole cell lysates analyzed on 0.15–0.2% Agarose gels show a peak of DNA containing approximately one-third of the cell DNA, with a mobility between λ-DNA (31×10^6) and SPOI DNA (85×10^6). Hybridization of ribosomal RNA and cRNA of mitochondrial DNA to the fractions from the Agarose gels indicates that the mitochondrial DNA, as expected, has a mobility in agreement with its complexity of 35×10^6 daltons. The ribosomal cistrons have a mobility which has a mean size ranging from $40–65 \times 10^6$ daltons. EM analysis of similar gentle lysates shows a population of molecules of the expected size of mitochondrial DNA (approximately one-half are open circles) and the rDNA dimer. If no shearing or enzymatic degradation of nuclear DNA existed, one would expect the palindromic dimer to have a size of 97–98 kb or approximately 65×10^6 daltons. It has been concluded from these results that the rDNA as isolated is not covalently associated with the bulk of the chromosomal DNA and is probably present as free elements similar to those found in *Physarum* and *Tetrahymena* in the nucleus (61, 63).

The total ribosomal repeat represents approximately one-sixth of the total nuclear DNA, and each repeat is present approximately 180 times or 90 dimers per genome. The spacer region appears to be heterogeneous with regard to GC content (49). The portion closer to the center of the dimer (Eco RI band I) has a higher GC content (approximately 28% GC) than the spacer region surrounding the 36 S gene (22–23% GC). The 5 S gene has been localized and is present at approximately 3 kb to the 3′ end of the 36 S transcript (64). There appears to be a single 5 S gene per ribosomal repeat. Those restriction fragments coding for the 36 S rRNA have a GC content of 42 mole per cent (58).

Fine structure mapping and a processing scheme of the transcribed portion of the ribosomal gene have been done by using a combination of mapping with restriction endonucleases (47, 49, 61, 65) and EM techniques (47) and by analysis of the methylation, base composition, and oligonucleotide maps of the mature rRNA and the rRNA precursors (65). As shown in the map (Figure 10), the 5′ end of the 36 S transcript contains unconserved regions of approximately 500–1,000 nucleotides in length (exact size is undetermined) which are present in the 21 S precursor to the mature 17 S rRNA. The next region of the 36 S transcript is the 17 S ribosomal RNA (1.9 kb), followed by an internal transcribed spacer (900 nucleotides). Then the 26 S ribosomal RNA in the ribosome maps within 50 nucleotides of the 5′ end on the 26 S ribosomal RNA and is part of the 30 S nuclear precursor to the 28 S species (47).

The processing of the ribosomal RNA in *D. discoideum* is similar to that observed in other eukaryotes, and the organization of the ribosomal genes appears to be similar to that of other nonmetazoan protists, including *Physarum* (66) and *Tetrahymena* (67, 68). In all three cases, the ribosomal cistrons appear to be localized on palindromic extrachromosomal dimers. In the case of *Tetrahymena*, it is known that the 5 S cistrons are not localized on these dimers (69). No information is presently available on the location of the 5 S cistrons in *Physarum*. In *Dictyostelium* and in yeast (70), the 5 S cistron maps toward the 3′ end of the 26 S ribosomal RNA gene sequence; however, the yeast rDNA are chromosomal rather than free and are localized in tandem repeats similar to those in *Xenopus*. In yeast, the 5 S sequence is present on the opposite strand as the coding sequence for the main ribosomal precursor (71). The polarity of the 5 S gene has not yet been determined for *Dictyostelium*. The most unusual property of the *Dictyostelium* ribosomal cistron is the unusual length of the spacer region (approximately 36 kb) which is appreciably longer than any other organism examined. Examination of the ribosomal cistrons in two other cellular slime molds, *Dictyostelium muccuroides* and *Polyspondelium violaceum*, also suggests that these contain very long nontranscribed spacer regions (72). In addition, all the copies of the ribosomal cistrons appear to be homologous as defined by restriction endonuclease mapping. In contrast, the ribosomal cistrons of several metazoans (*Drosophila* and *Xenopus*) show some spacer heterogeneity in different regions of the ribosomal RNA spacers (73–76).

STRUCTURE AND TRANSCRIPTION OF MESSENGER DNA

Poly(A) on *Dictyostelium* Messenger RNA

Early results show that over 90% of *Dictyostelium* cytoplasmic or polysomal heterodisperse RNA could be purified from nonpoly RNA (mostly ribosomal

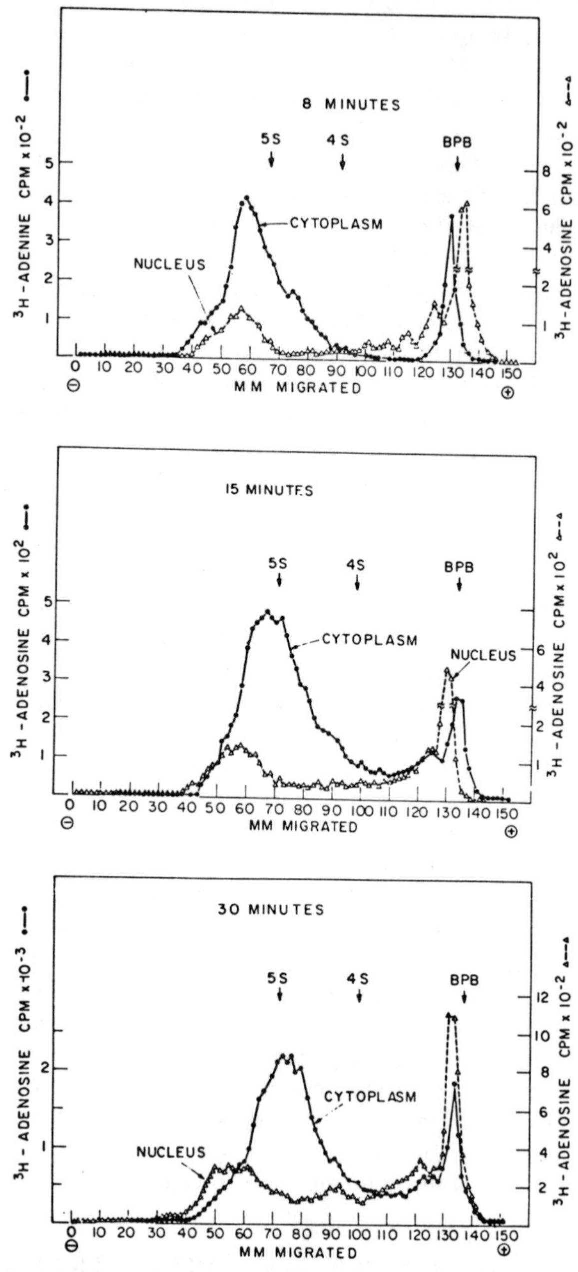

Figure 10. Size of nuclear cytoplasmic poly(A) on *Dictyostelium* heterodisperse RNA. Vegetative cells labeled for the indicated times with [^{3}H]adenosine were fractionated into nuclear and cytoplasmic poly(A)$^+$ RNA (14). Poly(A) was purified from the fractions (77) and analyzed on 10% polyacrylamide gels. [^{32}P]*Dictyostelium* 4 S and 5 S RNA and bromphenol blue were the markers. The nuclear and the cytoplasmic gels were superimposed and plotted on the same figure.

RNA) by preferentially binding to either poly(U)-Sepharose or oligo(dT)-cellulose and, therefore, contained poly(A) sequence (14, 77). More detailed analysis showed that two classes of poly(A) sequences are found in *Dictyostelium* mRNA and are present in equimolar amounts: a short transcribed $poly(A)_{25}$ sequence and a large post-transcriptionally added poly(A) stretch which is 125 nucleotides long in pulse-labeled RNA (14, 17, 56) (see Figure 10). Most mRNAs contain one of each class of poly(A) sequence. Further results showed that both poly(A) sequences are localized near the 3′ end of the messenger RNA and are closely linked, probably separated by only several nucleotides.

Labeling kinetics have indicated that $poly(A)_{125}$ sequence is added at a relatively late step in the transcription-processing steps of the nuclear mRNA precursors (17) (Figure 11). After brief labeling periods of 2 min, over 99% of the RNA label is present in the nucleus. Of the small amount of cytoplasmic label in $poly(A)^+$ RNA, almost all is present in $poly(A)_{125}$ stretches. These experiments also showed that mRNA is synthesized and transported in approximately 5 min and that the $poly(A)_{125}$ label is present in the cytoplasm approximately 3 min prior to the cytoplasmic appearance of the nonpoly(A) sequences of mRNA. These results have led to the conclusion that $poly(A)_{125}$ is added during the last 2 min of the nuclear lifetime of mRNA. Analysis of the $poly(A)_{25}$ shows that it has the same labeling kinetics as the bulk of the mRNA

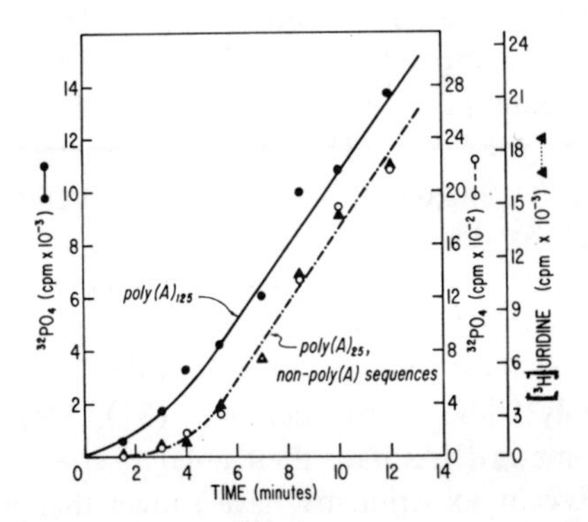

Figure 11. Kinetics of appearance in the cytoplasm of $poly(A)_{25}$ and nonpoly(A) mRNA sequences. Log phase vegetative cells were concentrated and labeled with $[^{32}P]O_4$ and $[^3H]$uracil. Aliquots were taken at the times indicated, and the cells were fractionated into nuclear and cytoplasmic RNA. Sequences containing poly(A) were purified on poly(U)-Sepharose, and total poly(A) was determined as a fraction of ^{32}P labeled in ribonuclease A- and T1-resistant counts from this fraction. Total poly(A) was fractionated by electrophoresis on 10% gel, and the fraction of poly(A) as $poly(A)_{25}$ or $poly(A)_{125}$ was calculated. The amounts of label in the two types of poly(A) were obtained by normalizing the counting in poly(A) to the total poly(A) RNA containing RNA in the sample. Nonpoly(A) RNA sequences shown in this figure were obtained with the use of $[^3H]$uracil as the label. The appearance of $[^3H]$uracil counts agreed with the appearance of $[^{32}P]$ribonuclease-sensitive counts from ^{32}P-labeled RNA bound to poly(U)-Sepharose. ●——●, $poly(A)_{125}$; ○– – –○, $poly(A)_{25}$; ▲– – –▲, nonpoly(A) sequences. (See ref. 17 for details.) Reproduced from Firtel et al. (17) with permission of *Fed. Proc.*

sequences, which is consistent with the fact that it is transcribed from the DNA as part of the primary nuclear transcript.

The size of the post-transcriptionally added poly(A) sequence which is in the cytoplasm in RNA post-label for a short time is indistinguishable in size from the nuclear poly(A) sequence (125 nucleotides). With time, size of the cytoplasmic poly(A) decreases in length and is approximately 100 nucleotides long after 1 hr (see Figure 10). The size continues to decrease and averages only approximately 25–50 nucleotides in messenger RNA which is labeled for at least 24 hr (78). This metabolism of *Dictyostelium* post-transcriptionally added poly(A) is similar to that which was reported for mammalian poly(A) sequences (79). Recent experiments have indicated that a small fraction (5–10%) of the mRNA has only short poly(A) (25–50 nucleotides) and, furthermore, that this mRNA is considerably less stable than the bulk of the mRNA (80).

mRNA labeled in early development (0–2 hr) shows poly(A) which is different from poly(A) labeled for an equivalent time in amoebae (80). In early development, most of the mRNA is not located on polysomes (27). The poly(A) on this mRNA is predominantly short poly(A) (25–50 nucleotides) (80). The RNA which remains on polysomes, however, does have full length poly(A). Furthermore, it is interesting that the full length poly(A) of vegetative mRNA is "metabolized" to very short poly(A) early in development.

Messenger RNA which does not bind to oligo(dT)-cellulose or poly(U)-Sepharose has been detected by assaying for stimulation of in vitro protein synthesis and characterization of the polypeptide products (81–83). Initial experiments with the use of oligo(dT)-cellulose chromatography suggested that this mRNA in the column flow-through (poly(A)-deficient RNA) was representative of all cellular mRNAs (83). Recent experiments, employing poly(U)-Sepharose, suggest that histone mRNA activity is preferentially enriched in the flow-through and that this nonbinding mRNA cannot be detected in the RNA which binds to the column (81–82). Such poly(A)-deficient RNA is frequently referred to as poly(A) RNA; however, it is important to note that RNA with very short poly(A) is most probably present in the flow-through fractions of the oligo(dT)- and poly(U)-Sepharose columns (81). Moreover, there exists the possibility that some mRNAs have their poly(A) stretches masked by poly(U)-rich sequences. Recent experiments have shown that different *Dictyostelium* mRNAs have considerably different content of poly(U)-rich sequences (81).

General Properties of Poly(A)$^+$ RNA

The labeling kinetics of *Dictyostelium* mRNA appears to be similar to that found in higher eukaryotes. The examination of cytoplasmic RNA after short pulse (15 min) labeling shows a heterodisperse pattern of RNA with a peak of sedimentation at approximately 13–15 S superimposed over the peaks of 17 S and 26 S ribosomal RNA (14) (Figure 12). With the use of immobilized poly(U) as a probe for poly(A)-containing RNA, it was shown that the vast majority of this heterodisperse RNA binds to poly(U) and contains poly(A) sequences (see

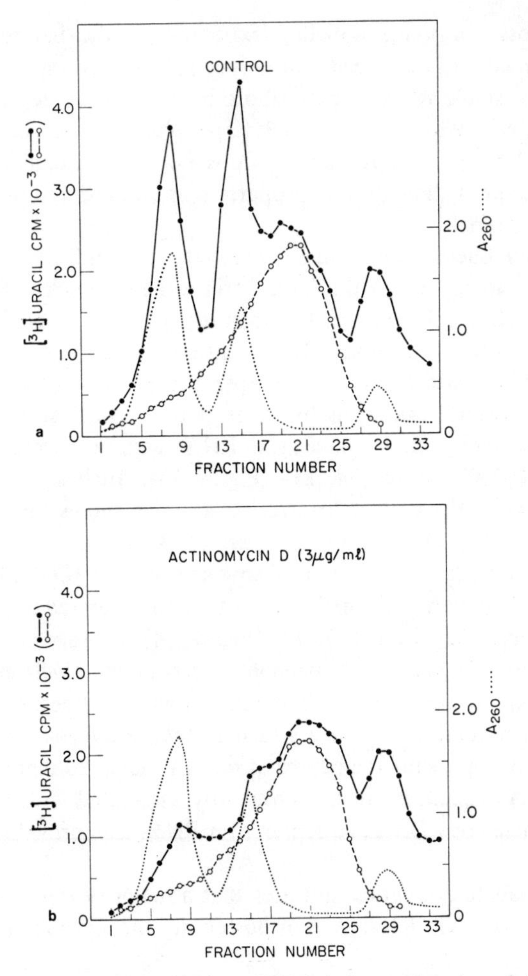

Figure 12. Sedimentation properties of pulse-labeled cytoplasmic poly(A)$^+$ RNA in the presence and absence of 3 μg/ml of actinomycin D. Log-phase cells were labeled for 18 min in the absence (*A*) and presence (*B*) of 3 μg/ml of actinomycin D. Cytoplasmic RNA was extracted and analyzed by centrifugation in SDS-containing sucrose gradients. (See ref. 14 for details.) Optical density profile (A_{260}) shows the ribosomal and tRNA markers. ●——●, total radioactivity; ○– – –○, radioactivity bound to poly(U) filters. (See ref. 21 for details.) Reproduced from Firtel et al. (21) with permission of *J. Mol. Biol.*

above). Since similar kinetics is observed for polysomal RNA as is observed for total cytoplasmic RNA, it is assumed that the majority of these species represents *Dictyostelium* messenger RNA. Other experiments indicate that 95% of pulse-labeled poly(A)$^+$ in vegetative cells is present on polysomes (82). Low levels of actinomycin D (3 mg/ml) were found to preferentially inhibit ribosomal RNA synthesis in *Dictyostelium* by approximately 90% (14). With the use of RNA labeled in the presence of the drug, it was shown that over 90% of the heterogeneous RNAs sedimenting in the region between 5 S and 20 S binds to poly(U)-Sepharose. In longer labeling experiments, the heterodisperse RNA cannot be analyzed without the use of poly(U) affinity chromatography. Under these conditions, stable RNA species, ribosomal and tRNA, represent over 95% of the labeled RNA. With the use of poly(U)-Sepharose, one can determine that approximately 1.5–3% of polysomal RNA is poly(A)-containing mRNA. This RNA has the same sedimentation properties as does the short pulse-labeled heterodisperse material.

Dictyostelium nuclei labeled under a variety of conditions in vivo do not contain material analogous to the large heterogeneous nuclear RNA (hnRNA) found in mammalian cells; rather, the majority of pulse-labeled nuclear RNA which is not precursor to ribosomal RNA contains at least one poly(A) sequence (14). Analysis of the size of the heterodisperse hnRNA on denaturing gradients or denaturing polyacrylamide gels indicates that the nuclear RNA is approximately 20–25% larger than messenger RNA with an average size of approximately 500,000 daltons (14, 81) (Figure 13). Analysis of cytoplasmic or polysomal poly(A)$^+$ RNA by hybridization kinetics shows that over 90% hybridizes to single copy DNA in a vast excess of DNA and 8–10% hybridizes to repeat sequences with a reiteration frequency of 10–100 (14, 17, 77). In contrast, 20–25% of poly(A)$^+$ hnRNA renatures to reiterated sequences and the remainder to single copy DNA (17, 77) (Figure 14). Linkage analysis similar to that described for analyzing interspersion of repeat and single copy sequences gave results which are consistent with a model in which a fraction of the hnRNA molecules has a reiterated sequence which is covalently associated with single copy DNA. In contrast, similar analyses of mRNA indicated that the majority of reiterated mRNA sequences are not covalently associated with sequences transcribed from single copy DNA and represent mRNAs transcribed from reiterated genes (17).

Pulse-chase labeling kinetics indicates that a majority (approximately 70%) of the nuclear poly(A)$^+$ RNA is transported to the cytoplasm, whereas the

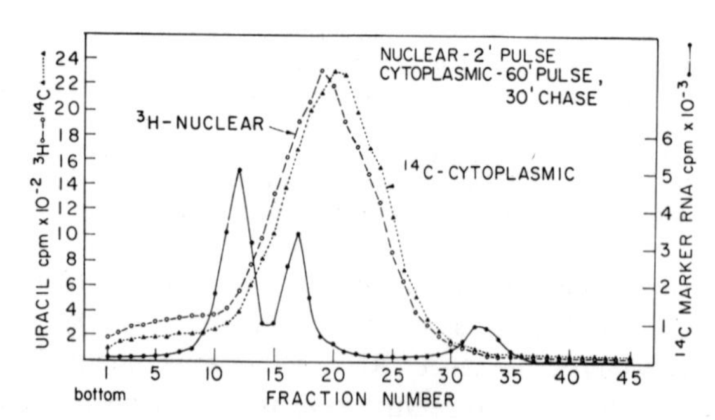

Figure 13. Dimethylsulfoxide gradient of nuclear and cytoplasmic heterogeneous RNA. Poly(A)$^+$ nuclear ^{3}H and cytoplasmic ^{14}C pulse-labeled RNA fractionated on poly(U)-Sepharose were analyzed on 2–8% sucrose gradients in 99% dimethylsulfoxide (14). ^{14}C-labeled ribosomal RNA marker DNA was run in a separate gradient. Reproduced from Firtel and Lodish (14) with permission of *J. Mol. Biol.*

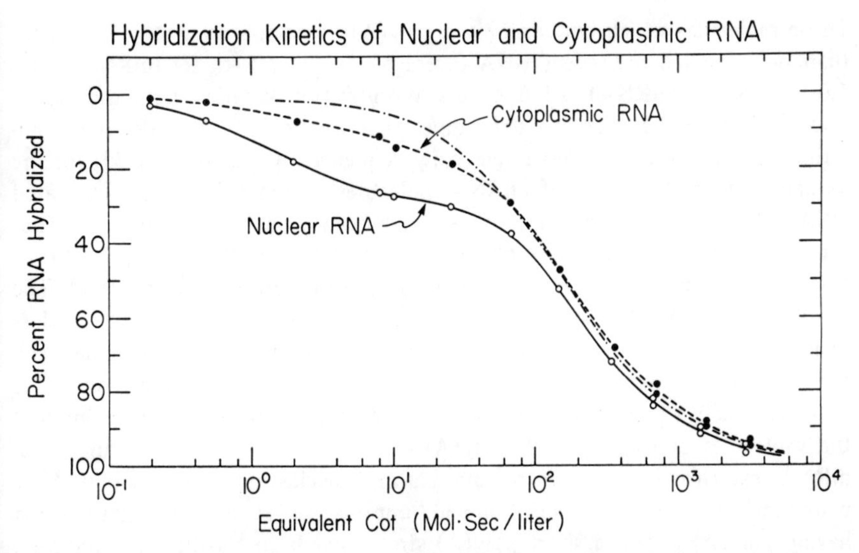

Figure 14. Hybridization kinetics of cytoplasmic and nuclear poly(A)$^+$ RNA with a vast excess of *Dictyostelium* DNA. Purified pulse-labeled cytoplasmic and nuclear poly(A)$^+$ RNA was hybridized to a vast excess of nuclear DNA. ●– – –●, ^{3}H-labeled cytoplasmic poly(A)$^+$ RNA; ○——○, ^{32}P-labeled nuclear poly(A)$^+$ RNA; –•–•, theoretical single component curve with a same $C_0t_{1/2}$ value as RNA hybridizing to single copy DNA. (See ref. 14 for details.) Reproduced from Firtel and Lodish (14) with permission of *J. Mol. Biol.*

remainder is degraded in the nucleus. These pulse-chase experiments, however, contain some degree of error because of the uncertainty of exactly how much newly labeled RNA is synthesized during the early time point in the actinomycin D and the daunomycin chase; however, they do suggest that unlike mammalian hnRNA a major, although not a precisely known, fraction of the hnRNA is transported to the cytoplasm and is the material precursor to mRNA (14, 15).

The vast majority of hnRNA molecules pulse-labeled in the presence of low actinomycin D binds to poly(U)-Sepharose and thus contains at least one poly(A) sequence. In contrast to mRNA, which contains equimolar amounts of poly$(A)_{25}$ and poly$(A)_{125}$, hnRNA contains 4–5 mol of poly$(A)_{25}$ and poly$(A)_{125}$. This stoichiometry suggests that 20–25% of the hnRNA molecules has both a poly$(A)_{25}$ and a poly$(A)_{125}$ and that three-fourths of the molecules contain only poly$(A)_{25}$. These hnRNA molecules which contain only a long poly$(A)_{125}$ and a short poly$(A)_{25}$ and those containing only a poly$(A)_{25}$ have been separated and their properties examined (17). The fractionation was accomplished by using poly(U) affinity chromatography and differential elution of molecules containing increasing amounts of poly(A) with increasing denaturing conditions (formamide gradient or heat elution). Three such classes of poly(A)-containing hnRNA molecules were found: 1) those containing only poly$(A)_{25}$; 2) those containing equimolar ratios of poly$(A)_{25}$ to poly$(A)_{125}$; and 3) those containing only poly$(A)_{25}$ and variable lengths of shorter poly(A).

Those molecules which contain only a poly(A)$_{25}$ are approximately 25% larger than pulse-labeled messenger RNA (average size 1.2–1.5 kb for hnRNA versus 1.0–1.2 kb for mRNA). DNA excess hybridization kinetics, with the use of ribonuclease as the assay to determine what fraction of the hnRNA is transcribed from repeat or from single copy sequences, indicates that 25% of the sequences from this class of hnRNA molecules is transcribed from reiterated DNA, whereas the remainder is transcribed from single copy sequences. That fraction of hnRNA molecules which contains both poly(A)$_{25}$ and poly(A)$_{125}$ has the same size distribution and hybridization properties as does mRNA. The third class of hnRNA, containing poly(A)$_{25}$ and a short heterogeneous size class of larger poly(A), shows size and hybridization properties intermediate between the class of hnRNA containing only poly(A)$_{25}$ and that class containing poly(A)$_{25}$ and a poly(A)$_{125}$. These data have been interpreted as meaning that the molecules containing both poly(A)$_{25}$ and poly(A)$_{125}$ are "mature" pre-mRNA ready for transport to the cytoplasm. The class that has only poly(A)$_{25}$ would include primary transcription and "unprocessed" molecules. That fraction having poly(A)$_{25}$ and a short poly(A) stretch has been classified as molecules being processed and onto which the post-transcriptionally added poly(A) is being polymerized. Since these interpretations are based on the behavior of large populations of molecules, alternative explanations are certainly possible.

Messenger RNA Half-life

The half-life of cytoplasmic and polysomal poly(A)$^+$ RNA (mRNA) has been examined in vegetative and developing cells (82). Steady state labeling kinetics of early log phase *Dictyostelium* vegetative cells labeled with [^{3}H]uracil or [^{3}H]uridine shows a messenger half-life of approximately 3.8 hr for a cell generation time of 9 hr. In the steady state kinetic experiments, labeling kinetics of mRNA and rRNA suggests that the RNA precursor pools reach constant specific activity very rapidly and maintain this for at least 16 hr. Analysis of the results suggested that a single first decay curve fit the data and that the majority of the messenger RNA had the same half-life. Pulse-chase experiments done to examine the stability of cytoplasmic poly(A)$^+$ RNA yielded similar results. Cells pulse-labeled with [^{3}H]uracil or [^{32}P] for varying periods of time were chased by using actinomycin D and daunomycin at concentrations which inhibit messenger RNA synthesis by 99%. These experiments showed a single first order decay component with cytoplasmic or polysomal poly(A)$^+$ RNA with a $t_{1/2}$ of 3.5–4.0 hr. Additional experiments in which the half-life of polysomes was examined showed a similar half-life in vegetative cells (82). It should be noted, however, that a small fraction (10%) of mRNA has only short poly(A). This mRNA codes for a specific family of proteins and is much less stable than bulk mRNA (81).

The half-life of messenger RNA in developing cells has been examined by a combination of pulse-chase labeling experiments (82) and by assays of the stability of translatable messenger RNA after inhibition of mRNA synthesis by drugs (83, 84). (Steady state labeling kinetics is not possible in developing cells under the conditions which have been tried. Kinetics of labeling of RNA in developing cells indirectly suggests that the specific activity of the pools continues to increase in the presence of label instead of reaching a steady state level.) Pulse-chase experiments with the use of actinomycin D and daunomycin of RNA synthesized during the first few hours of development and during the pseudoplasmodium stage show a half-life of approximately 3.5–4.0 hr, similar to that observed in vegetative mRNA (82). The half-life of polysomes during late development also shows a similar half-life. In addition, chemical stability of mRNA on polysomes or on total cytoplasm, as determined by decay of labeled poly(A)$^+$ RNA, is also similar. mRNA half-life as measured by chase and polysome half-life experiments is approximately 2.5–3.0 hr, reproducibly lower than in vegetative or pseudoplasmodial cells. These results suggest that there may be some regulation of gene expression during this stage in development by the change in the rate of degradation of mRNA.

The stability of mRNA for specific proteins has been analyzed by examining the stability of translatable activity of the messenger RNA in developing cells that have been treated by RNA synthesis inhibitors. Lodish et al. (87) have found that actin messenger RNA has a half-life of approximately 4 hr in early developing cells. Cells were allowed to develop normally and were then treated with RNA synthesis inhibitors. RNA was extracted at various times, and the relative amount of the specific mRNA was assayed in an in vitro protein-synthesizing system and quantitated on polyacrylamide gels. Similar experiments have been done for the carbohydrate-binding protein, and similar results have been obtained (85). Although the results of pulse-chase labeling experiments with RNA inhibitors and steady state labeling kinetics show similar results in vegetative cells, one must be cautious in the interpretation of experiments which use drugs during development since the drugs not only inhibit RNA synthesis but also inhibit further differentiation. It is possible that a specific messenger RNA may have a different half-life depending upon the particular developmental stage and that the inhibition of development with the drugs might alter the normal decay kinetics.

The decay of vegetative mRNA during early development has been examined with the use of two different pulse-chase techniques (82). In one case, further labeling of mRNA was inhibited by using actinomycin D and dauomycin. In other experiments, cells were labeled with $[^{32}P]O_4$ and chased for a short time during vegetative growth in the presence of excess phosphate buffer. The cells were then plated for development in the presence of a high concentration of phosphate, under conditions in which no new label could be detected in newly synthesized mRNA. In both cases the chemical stability of cytoplasmic or polysomal poly(A)$^+$ RNA was 3.5–4.0 hr, indistinguishable from that of poly(A)$^+$ RNA in logarithmically growing vegetative cells.

The biological reason for the increase in the rate of mRNA decay between 6 and 10 hr of development is unclear. It is possible that there may be some major developmental switches which require or result in an increase in the rate of loss of specific populations of mRNAs during this time period. The observed chemical and biological half-life of mRNA of vegetative cells and cells from both early and late developmental stages is 3.5–4.0 hr, approximately one-half to one-third of the general time of normal axenic cells. In axenic cells with a 22-hr generation time, the observed mRNA half-life was 5.4 hr, approximately 50% longer than faster growing cells (82). In nature, *D. discoideum* feeds on bacteria and has a generation time of approximately 3.5–4.0 hr. It is not known what the functional half-life of mRNA is under these conditions and whether the observed similarities of the half-life of the mRNA in developing and vegetative axenically grown cells would be the same for cells grown on bacteria; however, any changes in the half-life which may occur do not cause any observable change in either morphological or biochemical patterns of differentiation of *Dictyostelium* grown either axenically or on bacteria which are observed in the laboratory.

Messenger Labeling Kinetics in Developing Cells

Vegetative *Dictyostelium* cells show ideal labeling kinetics in the presence of high specific activity [^{3}H]uridine or [^{3}H]uracil. Indirect experiments suggest that the pools equilibrate rapidly and give steady state labeling kinetics very shortly after introduction of the label. The mRNA and the ribosomal RNA enter the cytoplasm rapidly, with transit times of 5 and 15 min, respectively. In contrast, developing cells label with kinetics which suggests that the pools are not equilibrating, and mRNA or rRNA label with kinetics approaching an exponentially increasing curve rather than a linear curve during the first 1–2 hr of labeling. In addition, the average transit time for the mRNA and ribosomal mRNA appears longer, and there is a preferential transfer of mature 26 S rRNA relative to 17 S rRNA (81, 82, 86). The longer transit times with the messenger RNA in developing cells could be an artifact due to changes in the pool specific activities during the labeling. Other results, however, suggest that the nuclear heterodisperse RNA has a much longer half-life than that in vegetative cells. These observations also showed that the decay of nuclear poly(A)$^+$ RNA in developing cells in the presence of drugs does not follow first order kinetics (see Figure 13). Contrary to the situation in vegetative cells, in which all nuclear poly(A)$^+$ RNA is either transported or decays in 30–45 min, nuclear poly(A)$^+$ during development exhibits a gradual rather than a simple first order decay kinetics. Thus, decay of hnRNA in developing cells may not be due to transport to the cytoplasm but to turnover in the nucleus.

Capping of Poly(A)$^+$ RNA

Most eukaryotic poly(A)$^+$ RNA studied to date has been shown to contain blocked 5′ end "cap" structures (87). In *Dictyostelium*, such structures have been detected in both nuclear and cytoplasmic poly(A)$^+$ RNA (88). The caps found on mRNA are 1) M^{7}GpppApU, 2) m^{7}GpppGpY, 3) m^{7}GpppAmpAp, and 4) M^{7}GpppAmpUp. The mole fractions represented by these structures are 65%, 10%, 10%, and 10%, respectively. These structures are considerably simpler than the large numbers of sequences found on mammalian mRNAs in that methylation of the ribose at the 2′ OH in the last position does not take place (87). Furthermore, in *Dictyostelium*, unlike in mammals, 6-Me-adenosine is absent, and the nucleotide adjacent to the m^7G is always a purine (88).

An examination of the cap structures on nuclear poly(A)$^+$ RNA showed the presence of only M^7GppAp and M^7GpppGp (89). Moreover, the distribution of the nucleotide sequences (3–5 nucleotides) immediately adjacent to the nuclear M^7GpppAp is identical with the sequence adjacent to it in mRNA. These results have led Weiner et al. (89) to suggest that there is no processing (other than capping) at the 5′ end of the primary nuclear transcript. Alternative explanations are that processing precedes capping of nuclear RNA or the less likely possibility that identical cap structures are put on poly(A)$^+$ RNA in the nucleus and the cytoplasm.

Models of Messenger RNA Synthesis

Several models have been prepared for the processing scheme of *Dictyostelium* nuclear mRNA precursors to mRNA (14–17, 19, 89). The data on the differences between the size and renaturation properties of hnRNA and mRNA suggested a model in which *Dictyostelium* mRNA is excised from nuclear precursors which are approximately 25% larger than the message (14–17, 19). These precursors are complete in having poly(A)$_{25}$, but do not have post-transcriptionally added poly(A)$_{125}$. Differences in the repeat and single copy sequences between hnRNA and mRNA represent sequences which are lost during processing. Data obtained from the analysis of the 5′ cap structures give a different interpretation of pre-mRNA processing (89). In this model, *Dictyostelium* mRNA is synthesized in the nucleus and has the same molecular weight as that of the cytoplasmic RNA, or the putative nonconserved sequences are localized at the 3′ end of the hnRNA. It is difficult to resolve this controversy without examining specific cases. It is possible that some hnRNA molecules are processed from a higher molecular weight precursor, whereas others are not. A full understanding of the processing and synthesis of *Dictyostelium* hnRNA must wait until further analysis can be made of specific messages. It is hopeful that recombinant plasmids carrying specific gene sequences will represent a messenger-specific probe which can be used to analyze individual genes. Broad generalization made by examining total populations of RNA may not be relevant to specific cases and may lead to an incorrect conclusion about the regulation of transcription.

Messenger RNA Fractionation

In order to extend the studies of total mRNA to individual molecular species of mRNA, mRNA fractionation procedures have been undertaken. Several different approaches for the direct fractionation of mRNA are underway. mRNA has been successfully fractionated by virtue of its size on polyacrylamide gels containing 99% formamide (90). In these experiments, it was shown that the major slime mold mRNAs could be recognized as "bands" in autoradiograms of polyacrylamide gels containing ^{32}P-labeled mRNA (Figure 15). The RNA in bands is presumed to be mRNA because it 1) is present on polysomes, 2) contains poly(A), 3) can be labeled in the presence of actinomycin D, and 4) does not hybridize to rDNA. Some of the proteins coded for by these mRNAs have been identified by eluting the RNAs from gels and translating them in cell-free systems. From these experiments, the actin mRNA has been identified and can be recovered from gels approximately 50% pure; in general, the major mRNAs code for the major cellular polypeptides, and the spread of translation activity around a band is broader than any band. The latter phenomenon is most likely a consequence of heterogeneity in the length of poly(A) hmdd in RNA, a result of poly(A) metabolism.

This fractionation procedure is suitable for only small quantities of mRNA. Large scale preparative procedures are presently being worked out with the use of lysine-Sepharose, RPC-5, and RPC-6 chromatography (81). Each of these resins appears to fractionate a given mRNA by different criteria. Lysine-Sepharose is a weak ion exchanger and gives mostly separation by size. RPC-5 and RPC-6 resins separate by secondary structure and base composition. The two reverse phase chemotography resins give different types of separation for the same mRNA. From two different resins (two-dimensional fractionation), a large scale purification of a particular mRNA species can be obtained. Preliminary experiments with these methods indicate that mRNAs are differentially fractionated by using a combination.

Another approach has been to fractionate according to content of specific polynucleotide-binding sites. Thus, it has recently been shown that a small but specific fraction of mRNA has only short poly(A) and another fraction of

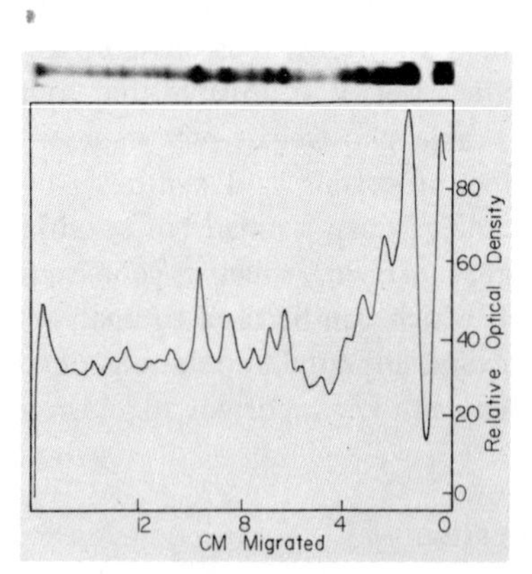

Figure 15. Fractionation of mRNA bands on formamide gels. Log phase amoebae were pulse-labeled with [^{32}P]O_4, and oligo(dT)-binding RNA was isolated from polyribosomes. RNA was fractionated on 4.4% polyacrylamide gels containing 99% formamide. Localization of the RNA was determined by autoradiography of the "wet" gel. The figure shows a photo of the autoradiograph (*top*) and a densitometer tracing of that autoradiograph (*below*). Reproduced from Jacobson et al. (90) with permission of A. S. M. Press.

mRNA which is presumably rich in poly(U) has been shown to preferentially bind to poly(A)-Sepharose (81).

Proteins on Messenger RNA and Heterogeneous Nuclear RNA

Dictyostelium mRNA and hnRNA are associated with a specific subclass of proteins in the form of ribonuclear protein (RNP) particles in the cells (91). The proteins associated with the heterogeneous nuclear RNP (hnRNP) are unique and do not show homology to the bulk nonhistone chromosomal proteins or to total cell protein. A specific protein having a molecular weight of 72,000–74,000 is found specifically associated with poly(A) in hnRNP particles. The hnRNP particles have an average sedimentation coefficient of approximately 55 S, which is expected for hnRNA with an average size of 500,000.

Complexity and Abundancy Classes of Heterogeneous Nuclear RNA and Messenger RNA

Initial work analyzing the complexity and changes in the single copy DNA sequences present in total cell RNA at different stages in *Dictyostelium* development showed that 1) a large fraction (56%, equivalent to 16,000 sequences of 1,000 nucleotides each) of the single copy DNA was transcribed, assuming asymmetric transcription, and 2) there were appreciable changes in the single copy DNA transcripts present at different developmental stages. Approximately 19% of the single copy DNA represented transcripts present in both vegetative and culminating cells. Sequences present in vegetative but not culminating cells numbered 10.5%, and 18% were sequences present in culminating but not vegetative cells. Differences were also noted between aggregating cells, cells at the pseudoplasmodium stage, and vegetative and culminating cells. A more recent analysis has asked questions about the subcellular distribution of these sequences, their complexity, and their relative abundance (Figures 16 and 17). The complexity of polysomal, total cytoplasmic, and nuclear poly(A)$^+$ RNA from vegetative cells and cells from various developmental stages has been determined by RNA excess hybridization to high specific activity, single copy DNA (40). Vegetative nuclear poly(A)$^+$ renders approximately 19% of purified single copy DNA S_1 nuclease resistant and cytoplasmic poly(A)$^+$ RNA renders the DNA 14–15% S_1 resistant. In contrast, polysomal poly(A)$^+$RNA saturates 7–8% of the single copy DNA. Assuming asymmetric transcription, the latter is equivalent to approximately 4,000–4,500 poly(A)-containing mRNAs. The saturation value of total cell, cytoplasmic, or polysomal RNA yields the same values as those obtained with poly(A)$^+$ RNA, indicating that there are very few poly(A)$^+$ sequences. Total cytoplasmic poly(A)$^+$ RNA isolated from 20-hr cells has a saturation value of 17.5%, whereas the nuclear 20-hr RNA has a saturation value of 15%. Twenty-hour polysomal RNA saturates approximately 9% of single copy DNA.

Examination of the hybridization kinetics shows the presence of three kinetic components of abundancy classes of RNA driving the reaction for the

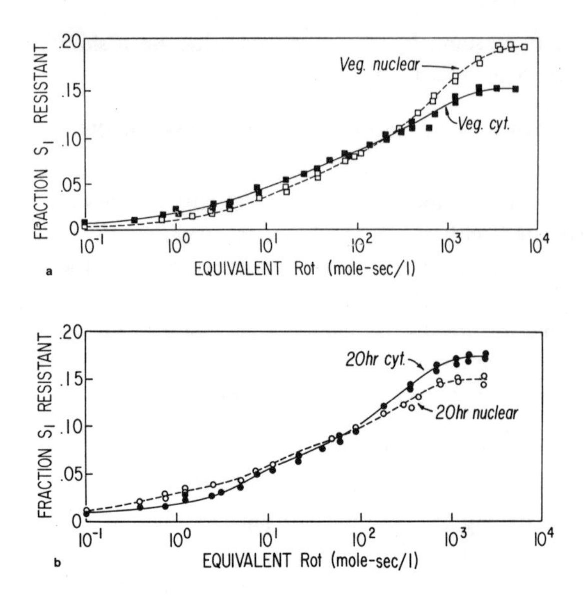

Figure 16. RNA excess hybridization of poly(A)$^+$ nuclear and cytoplasmic RNA from vegetative and 20-hr developing cells to purified single copy DNA. Purified cytoplasmic (*A*) and nuclear (*B*) poly(A)$^+$ RNA was hybridized in vast excess to ^{32}P nick-translated single copy DNA. The fraction of single copy DNA hybridized was determined by resistance to single strand-specific nuclease S1. This fraction is plotted versus equivalent Rot. Reproduced from Firtel et al. (40).

nuclear and cytoplasmic RNA and two components for the polysomal RNA (see Figures 13, 16, and 17). Hybridization kinetics of cellular RNA classes to cDNA complementary to hnRNA or mRNA has yielded additional results. In all, four kinetic or abundancy components of vegetative poly(A)$^+$ RNA can be discriminated. Component 1 contains approximately 10% of the messenger RNA and has a complexity equivalent to 10 sequences of 1,000 nucleotides in length. Each of these messages is present at approximately 1,200 copies per cell. This component is complementary to only 0.02% of the single copy DNA and can be only observed in RNA excess to cDNA. The second transition, which can be distinguished by both RNA excess to cDNA or to single copy DNA, comprises approximately 35–40% of the mass of RNA with a complexity of approximately 300 sequences, each present in approximately 100 copies per cell. The third component comprises approximately 50% of the RNA mass and has a complexity equivalent to approximately 3,600 sequences (cDNA hybridization values), each present at approximately 5 copies per cell. A fourth component, which can be distinguished only by RNA excess hybridization to single copy DNA, contains only 2% of the mass of the RNA with a complexity of approximately 4,000–4,500 sequences in vegetative cells and approximately 5,000–

6,000 sequences in 20-hr cells. Since this component comprises only 2% of the poly(A)$^+$ RNA, it cannot be observed as a component in the cDNA population. This component is found both in the nucleus and the cytoplasm, but is not found on polysomal RNA. The nuclear complexity of this component appears to be slightly higher than that of the cytoplasmic component. The total complexity of polysomal message is approximately 4,000–5,000 in 20-hr cells (40).

The unusual property of this last abundancy class is that it comprises a very small fraction of the total mass of RNA but contains over 50% of the total complexity. Calculations show that these sequences are present in less than one copy for every five cells. The fact that they are not present on polysomes suggests that they are not functional mRNA and do not code for proteins. It is unclear whether their presence in the cytoplasm is due to nuclear leakage or transport in vivo. The fact that this component of the hnRNA has a slightly

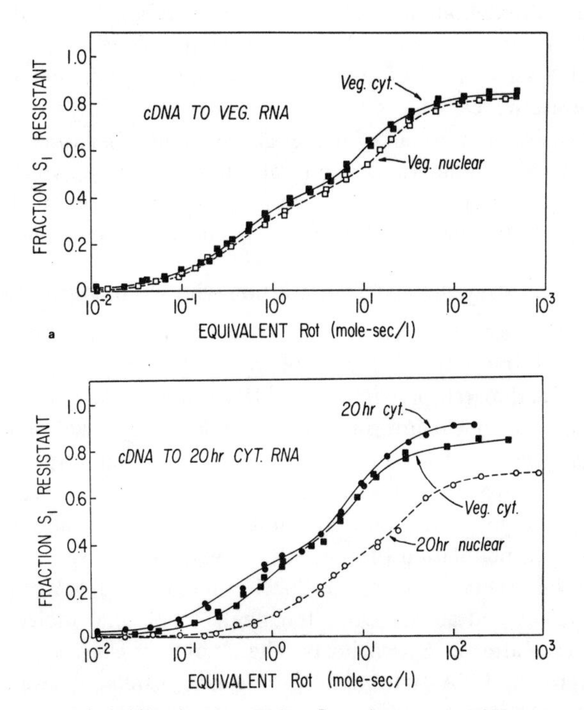

Figure 17. RNA excess hybridization to cDNA. Complementary DNA was made to vegetative or 20-hr poly(A)$^+$ cytoplasmic RNA with reverse transcriptase as described in ref. 41. Appropriate nuclear or cytoplasmic RNA from vegetative 20-hr developing cells was then hybridized in vast excess to the labeled cDNA. The fraction of the cDNA hybridized was determined by resistance to nuclease S1. *A*, cDNA made to vegetative RNA hybridized to a vast excess of vegetative nuclear (□) or cytoplasmic (■) poly(A)$^+$-RNA. *B*, cDNA made to 20-hr poly(A)$^+$ cytoplasmic RNA hybridized to a vast excess of 20-hr poly(A)$^+$ cytoplasmic RNA (●), vegetative poly(A)$^+$ cytoplasmic RNA (■), and 20-hr poly(A)$^+$ nuclear RNA (○). Reproduced from Firtel et al. (40).

greater complexity than that found in the cytoplasm suggests that it may not be due to an artifact resulting from nuclear leakage.

Further analysis of the hybridization kinetics shows that the large fraction of the developmental changes in RNA sequences observed in earlier work using total cell RNA is due to the fourth abundancy class. RNA excess hybridization to cDNA and single copy DNA with the use of polysomal RNA from different developmental stages shows both qualitative and quantitative changes in messenger species; however, these changes are much more limited than those observed with total cell RNA. Approximately 400 new sequences are found associated with polysomes at 20-hr development relative to that of vegetative cells, although over 50% of these sequences is found at a relatively low concentration. In addition, certain sequences found in vegetative cells appear to be more abundant in developing cells and vice versa. These results suggest that not only are new genes being transcribed but certain genes are being regulated quantitatively throughout development. Moreover, the fact that 20-hr nuclear poly(A)$^+$ RNA can only saturate 20-hr cytoplasm cDNA to a level of 75% and vegetative cDNA to a level of approximately 60% suggests that some sequences found in vegetative cells are either not found or found at very low concentrations at 20 hr, and that not all of the sequences transcribed in vegetative cells are transcribed during culmination. The total number of genes which are turned on during development and are transcribed into high or moderate abundancy class RNA is similar to the number of new protein changes observed by Alton (27).

Analysis of Gene Structure and Transcription with Recombinant DNA

The bulk of the work that has been done regarding genome organization, RNA processing, and transcriptional control in *Dictyostelium* has been done with relatively unfractionated preparations of DNA and RNA. These approaches are valuable since they yield information on the gross organization of the genome and on the general trends of RNA processing in transcriptional regulation. However, the drawback of making models based on data observed in heterogeneous populations is that one does not know whether these models may be biased to a given population of molecules or whether any specific gene products conform to the general observation. Specific regions of DNA can be isolated in bulk by covalently attaching DNA fragments to plasmid molecules which are propagated thereafter in *E. coli* hosts. The cloning of DNA sequences which are complementary to RNA transcribed during *Dictyostelium* development allows the study of specific properties of gene structure and genome organization and its use as a probe to ask questions about the processing of the RNA and the transcription of that particular gene throughout the developmental cycle.

Generation of Clones The techniques of molecular cloning of DNA have been described elsewhere in detail and are only mentioned briefly here (92, 93). Recombinant plasmids carrying *Dictyostelium* nuclear and mitochondrial DNA have been made with the use of the restriction-ligation method and the homopolymer extension method (poly(dA)-poly(dT) tailing) (94). In the first, nuclear

DNA is digested with restriction enzyme Eco RI and ligated into the tetracycline-resistant plasmids PSC101 or pMB9, which have been linearized with Eco RI. In the second, poly(dA) tails are synthesized onto randomly sheared *Dictyostelium* DNA (average 10 kb) by using terminal transferase. These fragments are then annealed to linearized vehicle DNA tailed with poly(dT). In both methods, the linked DNA is transformed into *E. coli* and selected for tetracycline resistance. The selection of recombinant plasmids which carry mRNA, tRNA, rRNA genes, or other specific sequences can be accomplished with the use of now well described procedures. Recombinant plasmids carrying regions complementary to poly(A)$^+$ mRNA have been isolated, and the analysis has been initiated (94, 95). All nuclear DNA-mRNA complementary plasmids which have been analyzed at the time of this review were made by using the insertion of randomly sheared *Dictyostelium* DNA into pMB9.

The second type of recombinant plasmids which have been analyzed has been generated by synthesizing complementary DNA (cDNA) to poly(A)$^+$ RNA and cloning it with the use of the homopolymer extension method (96–99). This method generates recombinant plasmids which contain an insert complementary to this specific mRNA, but, unlike the plasmids made with nuclear DNA, the cDNA-pMB9 clones do not contain the entire gene and, more importantly, do not contain any potential coastal regions. By screening with semipurified mRNA preparation, a plasmid carrying a cDNA or a specific mRNA can then be isolated. This cDNA clone can be used as a pure specific probe for the study of the transcription of that specific message or for the isolation of nuclear DNA clones which would contain both a total gene and any surrounding potential regulatory sequences.

Analysis of Recombinant Plasmids Carrying Nuclear DNA The majority of the work focused on the plasmids M4, M6, and KH10 (94, 95). Hybridization of poly(A)$^+$ RNA to immobilized DNA isolated from three plasmids showed that they hybridized substantial quantities of mRNA (M4, 4.5%; M6, 1–1.5%; KH10, 10–12%) and that the relative hybridization changed during *Dictyostelium* development. Several approaches have been used to delineate the type of mRNA complementary sequences present in the plasmid and to determine which genes are contained in the clone sequences.

These experiments, described in Figure 18, show that plasmid M6 hybridizes to a relatively homogeneous size mRNA and that the hybrid is relatively ribonuclease-resistant. In contrast, plasmid KH10 shows hybridization across a broad size range of the polyacrylamide gel. This peak is largely ribonuclease-sensitive, although there is a residual amount of ribonuclease-resistant material. M4 exhibits hybridization properties similar to those observed for both M6 and KH10. It hybridizes to a broad band of RNA which is largely ribonuclease-sensitive and to a sharp peak of a lower molecular weight RNA. This second peak of hybridization is ribonuclease-resistant. Melting profiles of the various types of RNA-DNA hybrids were also analyzed, and they indicate that the RNA hybrids formed between KH10 and M4 and the broad peak of hybridization have

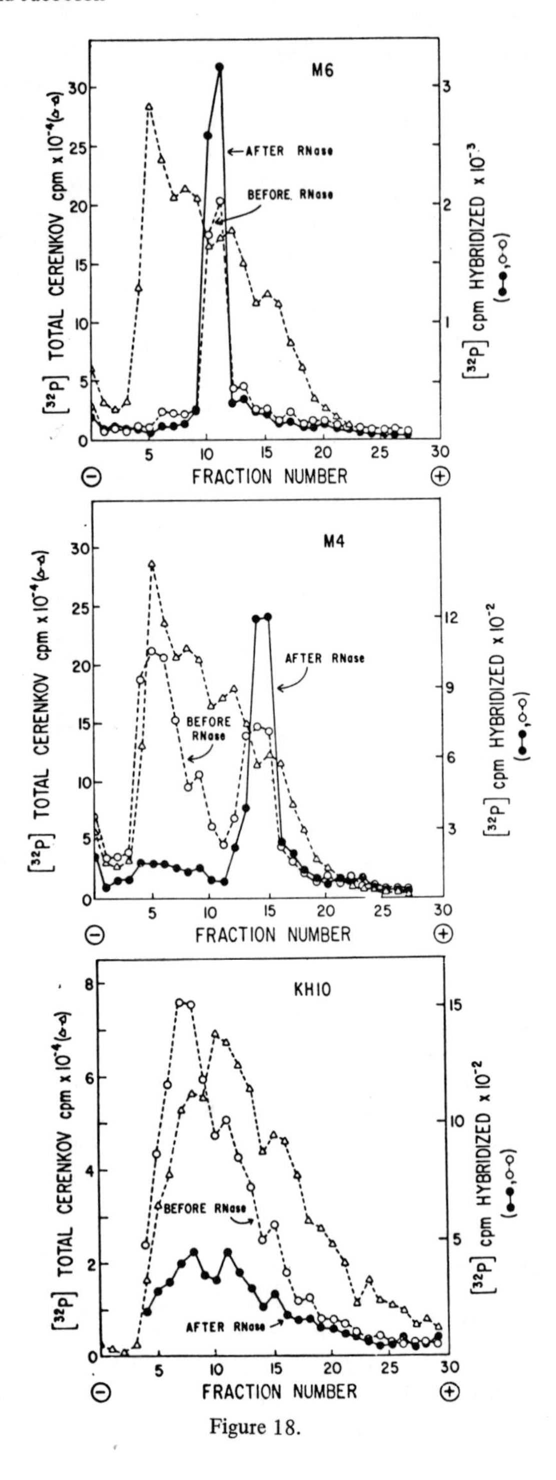
M6
AFTER RNase
BEFORE RNase
[32P] TOTAL CERENKOV cpm x 10^-4 (△-△)
[32P] cpm HYBRIDIZED x 10^-3 (●-●, ○-○)
FRACTION NUMBER
M4
AFTER RNase
BEFORE RNase
[32P] TOTAL CERENKOV cpm x 10^-4 (△-△)
[32P] cpm HYBRIDIZED x 10^-2 (●-●, ○-○)
FRACTION NUMBER
KH10
BEFORE RNase
AFTER RNase
[32P] TOTAL CERENKOV cpm x 10^-4 (△-△)
[32P] cpm HYBRIDIZED x 10^-2 (●-●, ○-○)
FRACTION NUMBER

Figure 18.

relatively low T_m, whereas the RNA hybrid formed with M6 DNA and the sharp peak of hybridization observed for M4 DNA have a much higher T_m, with a sharp transition indicative of a well matched RNA-DNA hybrid. From these results it has been concluded that plasmid M6 and M4 are complementary to a large fraction of *Dictyostelium* mRNA. Additional experiments by these workers suggest that the region of complement between the broad scale of hybridization of mRNA to M4 and KH10 is relatively short, approximately 75–100 nucleotides in length. These results suggest that there may be some sequence in M4 and KH10 which is complementary to a large population of gene sequences. It is assumed that such a sequence would be present in the nontranslated part of the mRNA at the 3′ or 5′ end or both.

The mRNA which specifically hybridizes to plasmid M4 and M6 can be purified by hybridization of poly(A)$^+$ RNA to immobilized DNA, followed by differential elution of the hybridized RNA and analysis of this RNA on agarose or polyacrylamide gels. This analysis shows that specific mRNA hybridizing to M4 has a length of approximately 900 nucleotides, whereas that hybridizing to M6 runs as a dimer of 1,300 and 1,400 nucleotides in denaturing conditions (agarose gels containing methylmercury or polyacrylamide gels containing urea).

Analysis of the structure of the recombinant plasmids has been initiated in order to obtain a more detailed picture of the location of the particular types of sequences which are complementary to mRNA and for future analysis of the location of potential regulatory sequences. The location of specific restriction enzyme sites is mapped within the DNA of all three clones. In addition, the location of the coding region in M4 and M6 has also been localized and has been shown to be totally contained within the plasmid (Figure 19). The length of the coding region complementary to the specific RNA as determined by the mapping data is in agreement with the determined size of the message as analyzed on polyacrylamide or agarose gel electrophoresis.

The organization of the cloned DNA has been further pursued by determining whether the inserted sequences contain reiterated or single copy DNA. Such analysis has been approached by both renaturation kinetics and by hybridization of nick-translated plasmid DNA back to Southern DNA blot filters (99). The hybridization data shown in Figure 20 indicate that M4 is mostly single copy with approximately 5% reiterated DNA sequences. Hybridization kinetics with plasmid M6 indicates that approximately 25–30% of the insert is reiterated 15–20 times, and the remaining sequences renature with single copy

---

Figure 18. Hybridization of size fractionated poly(A)$^+$ RNA to M4, M6, and KH10 DNA filters. In vivo ^{32}P-labeled poly(A)$^+$ RNA was size fractionated on 3.6% polyacrylamide gels in 99% formamide. The gel was sliced, and the RNA eluted by sonication. Each fraction was then hybridized to DNA filters, which were extensively washed, counted by Cerenkov's radiation counter, treated with ribonuclease A (20 μg/ml), and then recounted. △– – –△, total Cerenkov cpm; ○– – –○, Cerenkov hybridization cpm before ribonuclease; ●——●, hybridization cpm after ribonuclease. Reproduced from Kindle and Firtel (94) and Kindle and Firtel (95) with permission of Academic Press.

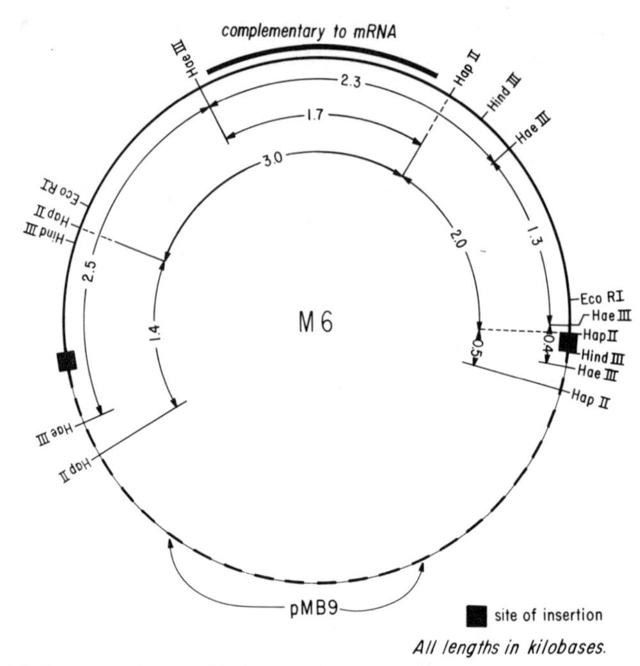

Figure 19. Restriction map of plasmid M6. Restriction map was constructed with the use of standard techniques for restriction enzyme digestion of plasmid DNA, running of Agarose gels, and location of coding region by hybridization of [^{32}P]RNA to Southern DNA blot filters. Reproduced from Kindle and Firtel (94) and Kindle and Firtel (95) with permission of Academic Press.

kinetics. Further analysis indicates that the 1.7-kb Hae III-Hap II fragment which is complementary to mRNA has hybridization kinetics which indicates that 80% of the sequences is reiterated 15–20-fold and the remaining sequences are single-copy 20 kb. Similar analysis of the outside restriction fragments indicates that they are totally single copy. Similar conclusions were obtained by hybridization of ^{32}P nick-translated DNA (92, 101) to Southern DNA blot filters containing *Distyostelium* nuclear DNA (101). In this approach, *Dictyostelium* DNA digested with various restriction enzymes is size-fractionated on Agarose gels. The DNA is denatured in the gel and then eluted and immobilized onto nitrocellulose filters without disturbing the relative position of the fragments. Specific restriction fragments from M6 are isolated on gels, labeled by nick translation, and then hybridized to the filter. In this way, one can determine the molecular weight and size distribution of any sequence complementary to a given RNA or DNA probe. These experiments show that the 1.7-kb Hae-Hap fragment hybridizes to approximately 15–20 bands (Figure 21), whereas the outside sequences hybridize to 1 or 2 bands, again as expected for single copy sequences.

The conclusion that the mRNA complementary gene sequence of M6 is reiterated is reinforced by an additional series of experiments. cRNA made from the 1.7-kb Hae-Hap fragment was used to select a cDNA-containing recombinant plasmid containing the same mRNA complementary sequences. A clone which contained an insert of 1.2 kb, 90% of the length of the M6 complementary message, was isolated. Renaturation kinetics, as well as hybridization of labeled DNA to numbered DNA and to Southern DNA blots, again indicated that the mRNA complementary sequences are reiterated 15–20 fold in the genome (100). Renaturation kinetics also concluded that the entire 1.2-kb fragment is reiterated.

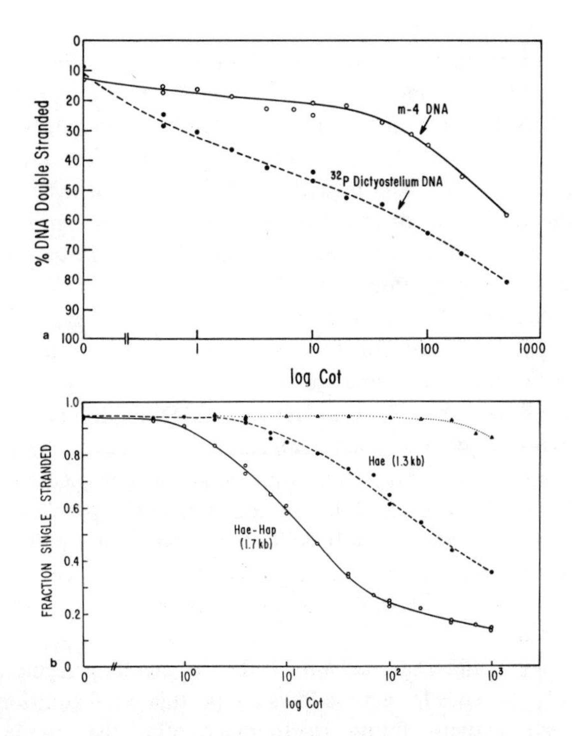

Figure 20. Hybridization of M4 and M6 plasmid DNA to a vast excess of *Dictyostelium* DNA. *A*, ^{3}H nick-translated M4 DNA was hybridized to a vast excess of ^{32}P-*Dictyostelium* DNA. The fraction of the DNA hybridizing was determined by resistance to single strand-specific S11 nuclease. The percentage of the DNA made double-stranded was plotted versus the log Cot. *B*, the 1.3-kb Hae III and 1.7-kb Hae III-Hap II restriction fragments were preparatively purified on Agarose gels. The fragments were nick-translated and hybridized to a vast excess of *Dictyostelium* DNA. Hybridization kinetics was analyzed as described above. △····△, 1.7-kb nick-translated Hae III-Hap II fragment renatured without *Dictyostelium* DNA (control); ○—○, 1.7-kb ^{32}P nick-translated Hae III-Hap II fragment hybridized to excess *Dictyostelium* DNA; ●- - -●, 1.3-kb Hae III fragment hybridized to excess *Dictyostelium* DNA. Reproduced from Kindle and Firtel (94).

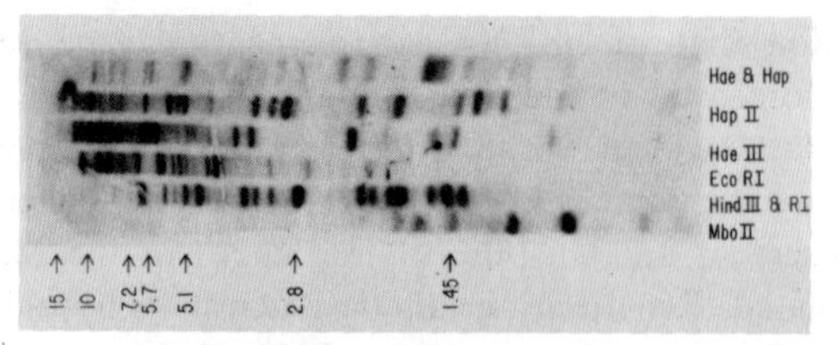

Figure 21. Hybridization of nick-translated M6 coding DNA to total *Dictyostelium* DNA. *Dictyostelium* DNA was digested with various restriction enzymes and run on 0.8% Agarose gel. Nitrocellulose filter blots were made by the method of Southern (101) and hybridized in nick-translated probes made from the 1.7-kb Hae III-Hap II M6 restriction fragment carrying the coding region for actin mRNA. Reproduced from Kindle and Firtel (94) and Kindle and Firtel (95) with permission of Academic Press.

In order to examine and compare regulation of gene activity at the level of transcription, translation, and protein accumulation turnover, it is necessary to be able to identify the protein encoded by a specific clone sequence. Functional identification of the protein (actin, UDPG, parphosphorylase, etc.) is not necessary for the examination of the regulation of the gene but is necessary if one is to correlate the regulation of the mRNA with the physiological role in differentiation of the organism. In order to identify the protein by the clone sequences, mRNA complementary to M4 and M6 is isolated by hybridization to immobilized DNA and translated in in vitro synthesizing systems. M4-specific complementary RNA codes for low molecular weight protein, whereas M6 appears to code for *Dictyostelium* actin.

Present work on the analysis of recombinant plasmids carrying *Dictyostelium* genes has yielded several interesting conclusions. Two clones were isolated which appear to contain sequences which are complementary to a large population of mRNAs. The function and location of this particular sequence or sequences of mRNAs have not been identified. The second point is the identification of a gene found for actin in *Dictyostelium*. Sequences complementary to actin mRNA have been shown to be reiterated in the genome and not tandemly repeated.

Synthesis of actin in *Dictyostelium* is developmentally regulated (26). The presence of multiple genes for actin allows the postulate ion additional models in regulation of gene activity during development, other than the classic ones in which a specific gene is turned on or turned off at various developmental times. Although it is not known how many of the actin genes are transcribed during any particular part of *Dictyostelium* differentiation, it is possible that one set of genes may be turned on at one developmental stage, whereas another set of genes may be turned on at another developmental stage. Kindle and Firtel have found microheterogeneity of the gene regions which would suggest some microheterogeneity in actin protein in *Dictyostelium*. Presently, no such microheterogeneity has been observed, and thus no conclusion can be made for either the function or the regulation of this specific gene family. This, however, represents a new area of interest in the examination of the regulation of differentiation in *Dictyostelium.*

The work on the clone DNA today indicates its value as a powerful tool for the study of the fine structural organization of the genome and as a probe for examination of the differential transcription and processing of mRNA precursors. The surface of the large mass of potential information that can be obtained by using cloned DNA fragments has only been scratched. The most interesting analysis will probably come when sequences and factors which regulate the specific initiation of transcription and translation can be localized. The availability of specific cloned sequences will then allow further analysis which could not have been possible without their isolation. In addition, work on cloned DNA sequences can be linked to in vitro transcription studies to examine the role of chromosomal proteins in the regulation of transcription. It is also possible to perceive the use of recombinant plasmids to make specific mutations and specific sites in the gene and to ask what effect these have on in vitro transcription translation. The future possibilities of this type of work may bypass the present inability to use genetic studies to examine the phenotypic expressions of mutations in regulatory sequences in *Dictyostelium.*

Analysis of Recombinant Plasmids Carrying Complementary DNA

A set of experiments similar to those used to analyze nuclear DNA containing recombinant plasmids has been used to examine plasmids carrying cDNA (Figure 22). With the use of hybridization selection procedures, recombinant plasmids containing cDNAs to moderately abundant mRNAs have been obtained (98, 99). The molecular weight of the RNA complementary to four such clones has been

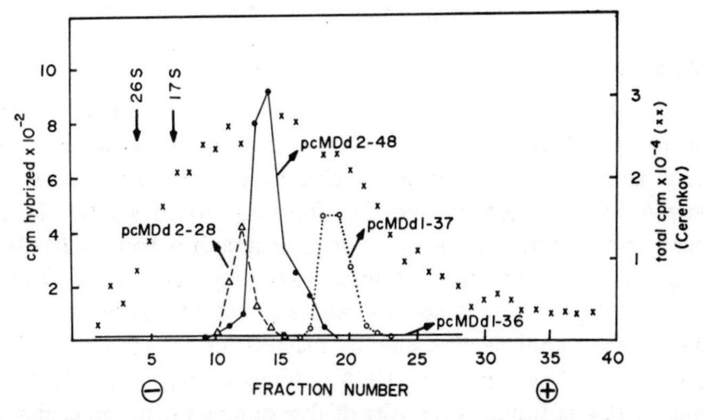

Figure 22. Hybridization of cDNA plasmid DNA to poly(A)$^+$ mRNA size fractionated on formamide gels. Size fractionated in vivo ^{32}P-labeled poly(A)$^+$ RNA gels were hybridized to DNA filters carrying the specific cDNA plasmid DNAs. (See text and Figure 19 for details.) The locations of the 17 S and 26 S ribosomal marker were determined by hybridization to plasmid DNA carrying the coding region for the ribosomal RNA. Sufficient ribosomal RNA contamination is present in the messenger RNA preparation to allow this identification. Reproduced from Roewekamp et al. (98).

examined as described in the legend to Figure 22. Three clones–pcMDd2-28, pcndd2-48, and pcMDd1-37–show hybridizations to three individual, specific peaks of mRNA. In addition, the results show the amount of label which hybridizes is different for the different mRNAs, suggesting that they are transcribed at different levels in the cell. This analysis has been confirmed by RNA excess hybridization kinetics to labeled plasmid DNA. One plasmid, pcMDd1-36, shows no hybridization; however, by restriction digest analysis and by hybridization of pcDd1-36 DNA to *Dictyostelium* DNA, it has been shown that the plasmid does contain an inserted sequence (102). It is assumed that mRNA complementary to this particular cDNA is present at too low an abundance to be detected by the techniques being used. These workers have shown that the cDNA clones appear to be most useful for initial selection of a probe which is specific for a given mRNA (102). The ability to isolate semipurified mRNA fractions, combined with the cDNA cloning technology, allows the use of semi-purified mRNA to obtain a specific probe for a specific gene sequence. Such an approach will yield probes which are useful for both analyzing the transcription of a particular mRNA directly and selecting nuclear DNA-containing clones with the gene and flanking regulatory sequences. These approaches are presently being used in several laboratories analyzing *Dictyostelium* molecular biology.

TRANSCRIBING ENZYMES AND OTHER NUCLEAR PROTEINS

Analysis of the proteins important in genome structure, transcription, and replication has begun. Such studies include the isolation and characterization of RNA polymerases, DNA polymerase, and histones, as well as studies on the properties of total chromatin and the transcriptional activity of whole nuclei.

RNA Polymerase

The isolation of DNA-dependent RNA polymerases from slime mold nuclei was first reported by Pong and Loomis (104). They resolved three peaks of RNA polymerase activity by conventional chromatography procedures (105) and characterized two of these (I and II). The two enzymes differ in their ionic preferences and sensitivities to the drug α-amanitin (one α-AMA^{r} and one α-AMA^{s} species). The activities of these two enzymes isolated from different developmental stages showed no differences in activity, although the nature of the assays and the templates was such that significant differences could well have been missed. One of the enzymes (RNA polymerase II) was purified to apparent homogeneity. It was found to consist of five subunits with molecular weights of 170,000, 150,000, 28,000, 21,000, and 15,000.

Experiments on transcription in isolated nuclei (58) indicated that there were probably three species of RNA polymerase in *D. discoideum,* two of which were resistant to low levels of α-amanitin and one which was α-amanitin-resistant enzyme. Messenger RNA precursors were synthesized in vitro by two enzymes, the α-amanitin-sensitive species and one of the α-amanitin-resistant species. The α-amanitin-resistant species accounted for 10–20% of the total in vitro synthesized mRNA precursor.

Experiments by Yagura et al. (106) confirmed the observations. They showed that, by the use of phosphocellulose chromatography, three species of RNA polymerase from amoebae could be resolved and characterized. (No experiments have been reported which test whether one of the α-AMA^r species is sensitive to high levels of the drug as has been found for polymerase III (105).) Moreover, Yagura et al. (106) reported that they are unable to detect the second α-amanitin-resistant species in nuclei of cells from the culmination stage of development. The significance of this result is not clear at present.

Transcription in Isolated Nuclei

Several labs have reported that isolated slime mold nuclei, when incubated with all four ribonucleoside triphosphates, will support the in vitro synthesis of RNA (58, 107–109). In one study it was shown that the products of in vitro synthesis were indistinguishable from cellular precursors to mRNA and rRNA by virtue of size, base composition, sensitivity of synthesis to actinomycin D, poly(A) content, and hybridization to nuclear DNA (58).

Studies of transcription in isolated nuclei have led to several interesting observations. These include the following:

1. The poly(A) found on in vitro synthesized mRNA precursor was shown to be only 25 nucleotides long (48). This led to the demonstration of two classes of poly(A) on slime mold mRNA and the demonstration that one of these ($poly(A)_{25}$) could be accounted for as a transcript of $poly(dT)_{25}$ sequences in DNA (see under "General Properties of Nuclear and Mitochondrial Genomes" and "Structure and Transcription of Messenger DNA").
2. Two RNA polymerase activities (one α-amanitin-resistant and one α-amanitin-sensitive) were shown to synthesize poly(A)-containing RNA in isolated nuclei (48) discussed above).
3. The size of the in vitro synthesized poly(A)-containing RNA was found to be identical with that of its in vivo (nuclear) counterpart (15, 58). This result reinforced the in vivo observations on the upper limits of the size of the mRNA precursors (see under "Structure and Transcription of Messenger DNA").
4. A growth inhibitor which accumulates in stationary phase cells apparently inhibits transcription in isolated nuclei (108). This inhibitory effect may be specific for the in vitro initiation of ribosomal RNA synthesis.
5. Addition of cyclic AMP to cells prior to isolation of nuclei yields nuclei which are apparently more active for in vitro RNA synthesis (108).

The experiments cited above have been done under conditions in which most synthesis represents completion of nascent RNA chains. For mRNA precursor synthesis, it was shown that standard conditions led to the addition of 200–300 nucleotides to the 3′ ends of existing RNA chains (48).

Recently, conditions have been established for efficient in vitro initiation of RNA chains (110). γ-^{32}P-labeled triphosphates have been used in these studies to selectively monitor initiations. Figure 23 shows that efficient in vitro initiation with [γ-^{32}P]GTP is very dependent upon both salt and GTP concentrations; elongations of RNA chains ([^{3}H]UTP incorporation) are equally efficient

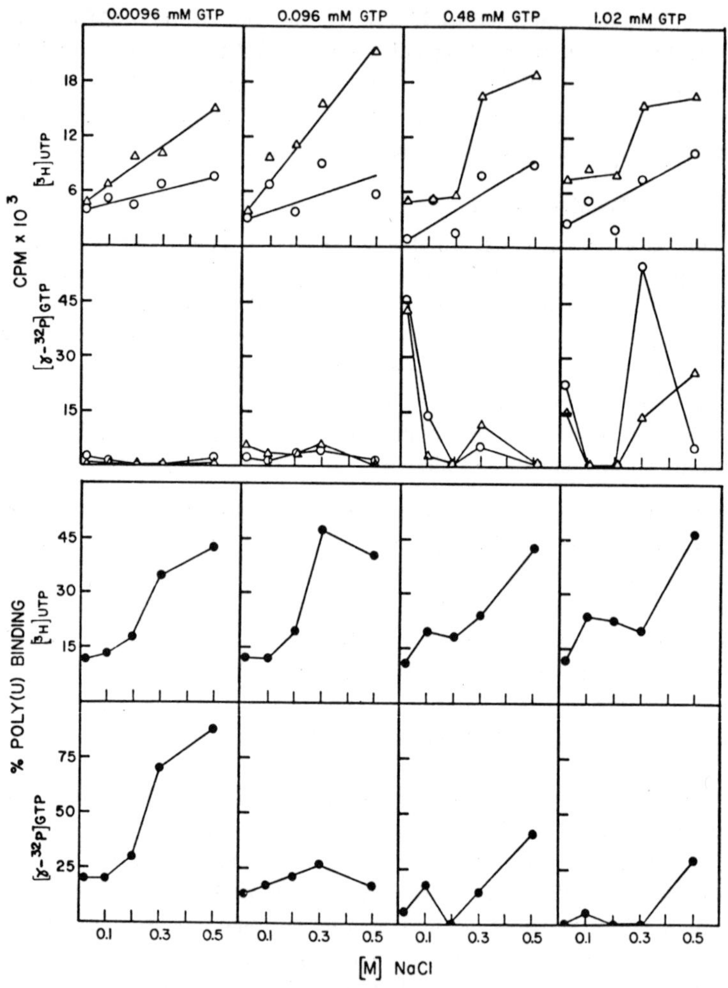

Figure 23. Initiation of RNA synthesis in isolated nuclei. RNA was synthesized in isolated slime mold nuclei by the general procedures outlined previously (58, 110). [γ-^{32}P]GTP and [^{3}H]UTP were added to monitor initiation and elongation of RNA chains, respectively. [^{3}H]UTP, ATP, and CTP were present at 0.16 mM. [γ-^{32}P]GTP was present at the concentrations indicated on the figure. NaCl replaced KCl in the reaction mix and was present at the indicated concentrations. The *upper half* of the figure depicts acid-precipitable radioactivity accumulated in 5′ (○) or 15′ (△) of incubation. The *lower half* of the figure depicts poly(U)-binding radioactivity in RNA purified from a 15′ reaction. Poly(U) binding is corrected for nonspecific binding to blank filters. (Data taken from ref. 111.)

at all GTP concentrations tested, but very dependent upon salt concentration. Figure 23 also shows that high GTP concentrations depress the in vitro initiation of poly(A)-containing RNA labeled with [γ-^{32}P]GTP but do not affect elongation of poly(A)-containing RNA.

Further experiments show that, under optimal conditions, at least 50% of the poly(A)-containing RNA can be initiated in vitro with either ATP or GTP. The ratio of full length poly(A)-containing RNAs initiated with ATP and GTP is 4:1, respectively. The size distribution of total [γ-^{32}P]ATP- and [γ-^{32}P]GTP-labeled molecules is skewed toward low molecular weight RNA (Figure 24); however, the size of the poly(A)-containing γ-labeled RNA is essentially the same as in vivo ^{3}H-labeled poly(A)$^+$ hnRNA (Figure 25).

The use of both γ^{32}P-labeled and ^{3}H-labeled nucleotides in the same reaction has allowed the determination of the in vitro rate of nucleotide polymerization. Figures 26 and 27 depict an experiment in which isolated nuclei were incubated with both [^{3}H]ATP and [γ-^{32}P]GTP. Several time points were taken during the course of in vitro synthesis and assayed for both acid precipitable radioactivity and poly(U)-binding radioactivity. Since the nuclei, as isolated, are loaded with RNA polymerase molecules already on DNA, the poly(U)-binding ^{3}H increases linearly from zero time. However, since γ-labeled molecules must traverse an entire gene before terminating at a poly(dT)$_{25}$ site, the poly(U)-binding [^{32}P] curve shows a lag, followed by a linear increase. The time lag between the first appearance of [^{3}H] and [^{32}P] poly(U)-binding RNAs is the "transit time" for RNA polymerase traversing the shortest genes. Appropriate calculations indicate that such RNA polymerases are incorporating 12–15 nucleotides/sec at 22°C. This represents approximately one-third the rate of *E. coli* RNA polymerase operating at maximal efficiency.

Basic Nuclear Proteins

Three different laboratories have described the isolation and characterization of the basic nuclear protein of *D. discoideum* amoebae (110–114). There are a large number of discrepancies in these results, although the work by Charlesworth and Parish is the most detailed and convincing.

The ratio of DNA to RNA protein in purified nuclei is reported to be 1:2.2:37.6 by Charlesworth and Parish (113), 1:8:29 by Osborn and Ashworth (114), and 1:3.4:22 by Coukell and Walker (112). Coukell and Walker observed 10 predominant basic nuclear protein bands on urea-polyacrylamide gels. These could be broken down into 6 "fractions," 5 of which were considered to roughly comigrate with the histones of calf thymus. Charlesworth and Parish observed 6-7 prominent basic protein bands which could be repeated into 5 fractions. Four of the latter fractions comigrated with calf thymus histones. Both groups show that the slowest migrating of these bands comigrates with calf thymus histone F1 (on urea gels). However, Coukell and Walker showed that this comigration did not hold on gels containing SDS and, further, that only 2 of 5 slime mold bands comigrated with calf thymus histones.

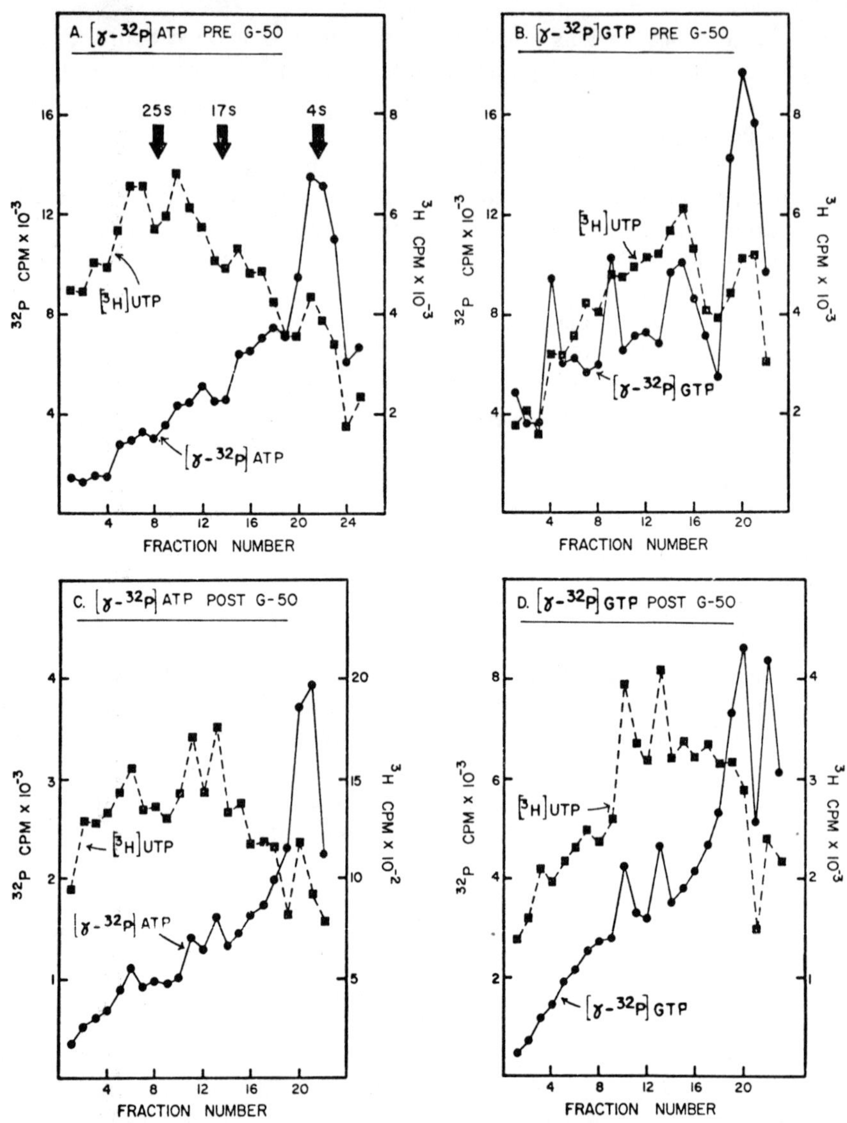

Figure 24. Sucrose gradient fractionation of RNA labeled in isolated nuclei with [γ-^{32}P]ATP, [γ-^{32}P]GTP, or [^{3}H]UTP. RNA was synthesized in isolated nuclei as described previously (58, 110, 111). Two separate reactions were prepared, one with 0.48 mM [γ-^{32}P]ATP, 0.16 mM GTP, and 0.2 M NaCl, and one with 0.48 mM [γ-^{32}P]GTP, 0.16 mM ATP, and 0.1 M NaCl. All reactions included [^{3}H]UTP and CTP at 0.16 mM. After 20 min, reactions were terminated with SDS and diethylpyrocarbonate and split into two parts. Half of each reaction was layered directly onto 15–30% sucrose-SDS gradients and centrifuged as described (58). The other half of each reaction was prefractionated on Sephadex G-50 to separate macromolecules from labeled nucleotides. The macromolecular fractions were then concentrated by ethanol precipitation and layered and run on similar sucrose gradients. Fraction 1 represents the bottom of the respective tubes. Reproduced from Mabie et al. (111) with permission of the authors.

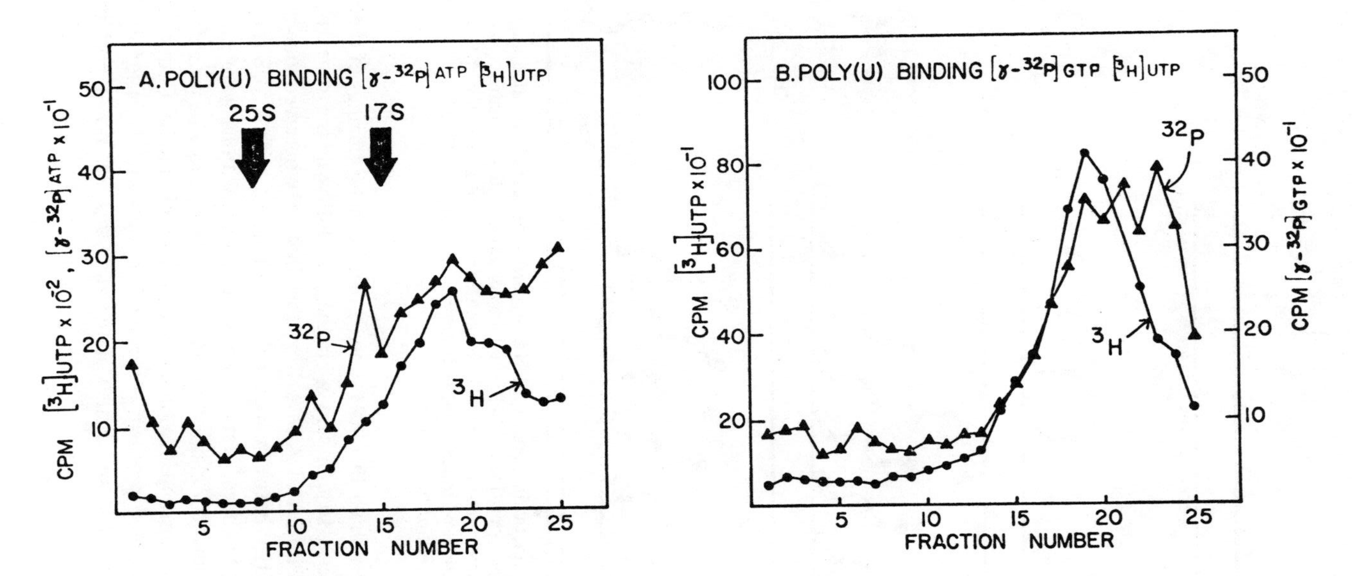

Figure 25. Sucrose gradient fractionation of in vitro labeled poly (U)-binding RNA. As described in Figure 24, RNA was synthesized in vitro (in isolated nuclei), fractionated on Sephadex G-50, and then centrifuged on sucrose gradients. Subsequently, each fraction was assayed for binding to poly(U) filters. Reproduced from Mabie et al. (111) with permission of the authors.

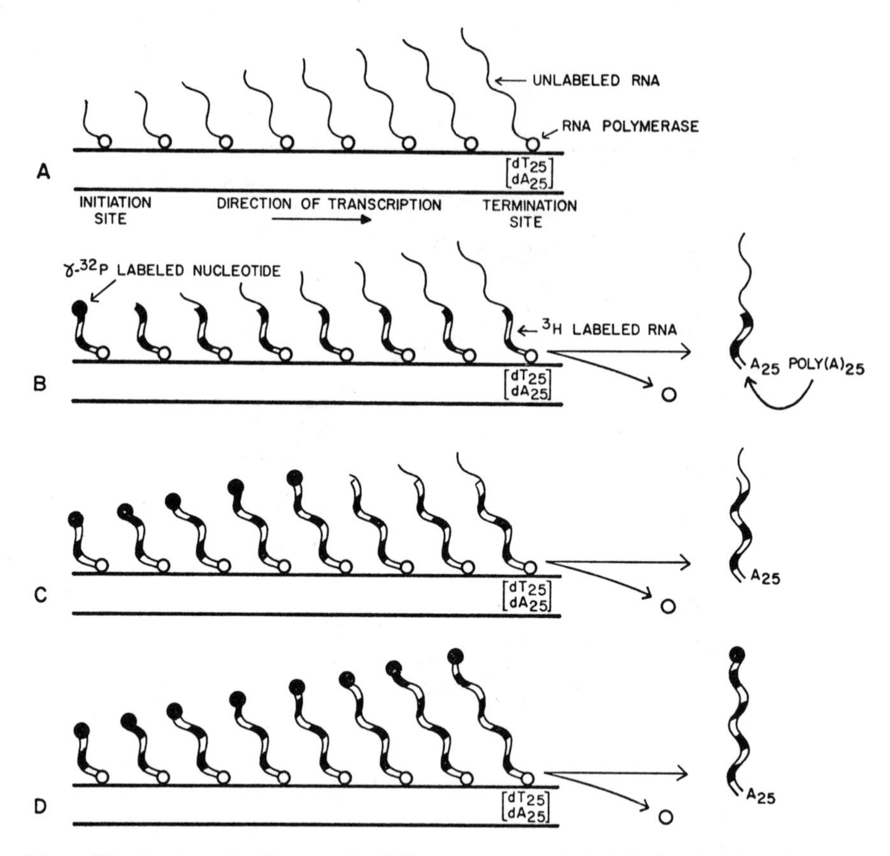

Figure 26. A schematic diagram of mRNA precursor synthesis in isolated nuclei. The figure depicts an average gene loaded with unlabeled nascent RNA chains attached to RNA polymerase molecules at zero time (*A*); the consequences of a very short period of synthesis (*B*), the consequences of longer synthesis periods (*C* and *D*). Reproduced from Mabie et al. (111) with permission of the authors.

Osborn and Ashworth (114) showed densitometer tracings of their urea gels. These are difficult to compare with the gel photos in the other two reports. Nevertheless, Osborn and Ashworth report that they observed nine major bands and little similarity between their gel patterns (urea gels or SDS-gels) and those of calf thymus histones. Data for the latter observations were not presented. Osborne and Ashworth claim that one of the basic proteins present in amoebae nuclei is absent from nuclei isolated from pseudoplasmodia. Coukell and Walker observed a substantial problem of proteolysis during the isolation of the nuclear protein. Charlesworth and Parish stated that this was not a problem in their experiments.

Chromatin

Chromatin has been isolated from nuclei of slime mold amoebae by two procedures (115, 116). The first involves detergent lysis procedures, sonic

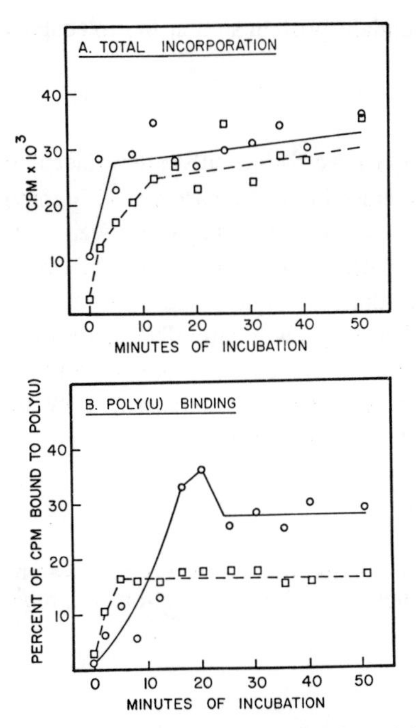

Figure 27. Transit time for RNA polymerase in isolated nuclei. RNA was synthesized in isolated nuclei with [γ-^{32}P]GTP and [^{3}H]ATP. All nucleotides were at 0.16 mM. At the indicated times, samples were withdrawn for (*A*) an assay of acid-precipitable radioactivity or (*B*) an assay of poly(U)-binding RNA. □, ^{3}H cpm; ○, ^{32}P cpm. Reproduced from Mabie et al. (111) with permission of the authors.

disruptions of nuclei, and isolation of chromatin on differential sucrose gradients (115). Such preparations show a protein to DNA ratio of 3.0. Less than 0.3% of the chromatin protein could be attributable to cytoplasmic contamination. The most significant nonhistone chromosomal proteins were found to comigrate on SDS-gels with actin and myosin. The total complexity of the nonhistone chromosomal proteins was less than that seen in higher eukaryotes. These chromatin preparations do not have a significant amount of endogenous RNA polymerase activity. Transcription of this chromatin made from log phase cells by exogenous *E. coli* RNA polymerase yields a product which, in preliminary experiments, has been shown to anneal almost completely from RNA by cDNA made to vegetative mRNA (115).

A second procedure for chromatin isolation involves lysis of the cells with amphotericin B (115), followed by differential sucrose gradient centrifugation. This group has paid considerable attention to the endogenous RNA polymerase activity of this preparation. It has been shown that at least one α-amanitin-resistant and one α-amanitin-sensitive enzyme are present in this chromatin and that these enzymes are active in vitro. The nonhistone chromosomal proteins in these chromatin preparations have also been characterized, and it is interesting to note that they include an abundant protein similar in molecular weight to actin (115, 117).

DNA Polymerase

L. D. Loomis et al. (117) have found only one molecular weight species (MW 127,000) of DNA polymerase in *D. discoideum.* This enzyme was shown to be present in cells of all developmental stages, as well as in spores. Its general properties are consistent with those of the α form of DNA polymerase found in mammalian cells, except that, in slime molds, no variation in enzyme activity was observed as a function of growth rate. These authors strongly suggest that the B form of DNA polymerase is absent from slime molds, as opposed to the possibility that it was present and not detected. They note that other lower eukaryotes also appear to lack the B form.

ACKNOWLEDGMENTS

The authors would like to thank William F. Loomis, Karen L. Kindle, and Mary Jane Newkirk for assistance and helpful suggestions in preparing this manuscript.

REFERENCES

1. Sussman, M. (1956). Annu. Rev. Microbiol. 10:21.
2. Loomis, W. F. (1975). *Dictyostelium discoideum:* A Developmental System. Academic Press, New York.
3. Jacobson, A., and Lodish, H. F. (1975). Annu. Rev. Genet. 9:145.
4. Cotter, D. A. (1975). *In* P. Gerhardt, A. L. Sadoff, and R. N. Costilow (eds.), Spores, Vol. VII. ASM Press, Washington, D. C.
5. Killick, K. A., and Wright, B. C. (1974). Annu. Rev. Microbiol. 28:139.
6. Robertson, A., and Grutsch, J. (1974). Life Sci. 15:1031.
7. Sussman, M., and Brackenbury, R. (1976). Annu. Rev. Plant Physiol. 27:229.
8. Bonner, J. T. (1967). The Cellular Slime Molds, Ed. 2. Princeton University Press, Princeton, New Jersey.
9. Newell, P. C. (1971). Essays Biochem. 7:87.
10. Franke, J., and Kessin, R. Proc. Natl. Acad. Sci. U.S.A., in press.
11. Katz, E. R., and Sussman, M. (1973). Proc. Natl. Acad. Sci. U.S.A. 69:495.
12. Williams, K. L., Kessin, R. H., and Newell, P. C. (1974). J. Gen. Microbiol. 84:59.
13. Firtel, R. A., and Bonner, J. (1972). J. Mol. Biol. 66:339.
14. Firtel, R. A., and Lodish, H. F. (1973). J. Mol. Biol. 79:295.
15. Lodish, H., Firtel, R. and Jacobson, A. (1973). Cold Spring Harbor Symp. Quant. Biol. 38:899.
16. Jacobson, A., Firtel, R., and Lodish, H. F. (1974). Brookhaven Symp. Biol. 26:307.
17. Firtel, R. A., Kindle, K., and Huxley, M. P. (1976). Fed. Proc. 35:13.
18. Konijn, T. M., Van de Meene, J. G. C., Bonner, J. T., and Barkley, D. S. (1967). Proc. Natl. Acad. Sci. U.S.A. 58:1152.

19. Firtel, R. A., Jacobson, A., Tuchman, J., and Lodish, H. F. (1974). 8th Intl. Cong. Gen. Genet. 78:355.
20. Roth, R., Ashworth, J. M., and Sussman, M. (1968). Proc. Natl. Acad. Sci. U.S.A. 59:1235.
21. Firtel, R. A., Baxter, L., and Lodish, H. F. (1973). J. Mol. Biol. 79:315.
22. Franke, J., and Sussman, M. (1973). J. Mol. Biol. 81:173.
23. Wright, B. E. (1973). Critical Variables in Differentiation. Prentice-Hall, Englewood Cliffs, New Jersey.
24. Ma, G., and Firtel, R. A., in press.
25. Loomis, W. F., White, S., and Dimond, R. L. (1976). Dev. Biol. 53:177.
26. Tuchman, J., Alton, T., and Lodish, H. F. (1974). Dev. Biol. 40:116.
27. Alton, T. (1977). Ph.D. thesis, Massachusetts Institute of Technology.
28. O'Farrell, P. H. (1975). J. Biol. Chem. 250:4007.
29. Smart, J. E., and Hynes, R. O. (1974). Nature 251:319.
30. Siu, C.-H., Lerner, R. A., Ma, G., Firtel, R. A., and Loomis, W. F. (1976). J. Mol. Biol. 100:157.
31. Rosen, S. D., Kafka, J. A., Simpson, D. L., and Barondes, S. H. (1973). Proc. Natl. Acad. Sci. U.S.A. 70:2554.
32. Wilhelms, O. H., Luderitz, O., Westphal, O., and Gerisch, G. (1974). Eur. J. Biochem. 48:89.
33. Hohl, H. R., and Jehli, J. (1973). Arch. Microbiol. 92:179.
34. Gregg, J. H., and Badman, S. (1970). Dev. Biol. 22:96.
35. Killick, K. A., and Wright, B. E. (1974). Annu. Rev. Microbiol. 28:139.
36. Hemmes, D. E., Kojima-Buddenhagen, E. S., and Hohl, H. R. (1972). J. Ultrastruct. Res. 41:406.
37. Malchow, D., and Gerisch, G. (1974). Proc. Natl. Acad. Sci. U.S.A. 71:2423.
38. Sussman, M., and Boschwitz, C. (1975). Exp. Cell Res. 95:63.
39. Firtel, R. A. (1972). J. Mol. Biol. 66:363.
40. Firtel, R. A., Johnson, P., Wright, C., Kindle, K. L., and Huxley, M. P., manuscript in preparation.
41. Verma, I., Firtel, R. A., Lodish, H. F., and Baltimore, D. (1974). Biochemistry 13:3917.
42. Loomis, W. F., in press.
43. Williams, K. L., and Newell, J. C. (1976). Genetics 82:287.
44. Cocucci, S. M., and Sussman, M. (1970). J. Cell. Biol. 45:399.
45. Jacobson, A., unpublished observations.
46. Batts-Young, B., Lodish, H. F., and Jacobson, A., manuscript in preparation.
47. Frankel, G., Cockburn, A. F., Kindle, K. L., and Firtel, R. A. (1977). J. Mol. Biol. 129:539.
48. Palatnik, C. M., Katz, E. R., and Brenner, M. (1976). J. Biol. Chem. 252:694.
49. Firtel, R. A., Cockburn, A., Frankel, G., and Hershfield, V. (1976). J. Mol. Biol. 102:831.
50. Firtel, R. A., unpublished observations.
51. Firtel, R. A., and Kindle, K. (1975). Cell 5:401.
52. Firtel, R. A., unpublished observations.
53. Davidson, E. H., Galau, G. A., Angerer, R. C., and Britten, R. J. (1975). Chromosoma 51:253.
54. Goldberg, R. A., Gain, W. R., Ruderman, J. V., Moore, G. P., Varnett, T.

R., Higgins, R. C., Gelfand, R. A., Galan, G. H., Britten, R. J., and Davidson, E. H. (1975). Chromosoma 51:225.
55. Davidson, E. H., Hough, B. R., Amenson, C. J., and Britten, R. J. (1973). J. Mol. Biol. 77:1.
56. Graham, D. E., Neufeld, B. R., Davidson, E. H., and Britten, R. J. (1974). Cell 4:127.
57. Jacobson, A., Firtel, F. A., and Lodish, H. F. (1974). Proc. Natl. Acad. Sci. U.S.A. 71:1607.
58. Jacobson, A., Firtel, R. A., and Lodish, H. F. (1974). J. Mol. Biol. 82:213.
59. Bender, W., McKeown, M., Davidson, N., and Firtel, R. A., unpublished observations.
60. Cockburn, A. F., Newkirk, M. J., and Firtel, R. A. (1976). Cell 9:605.
61. Maizels, N. (1976). Cell 9:431.
62. Taylor, W. C., Cockburn, A. F., Frankel, G. A., Newkirk, M. J., and Firtel, R. A. UCLA Symposium on Molecular and Cell Biology, Vol. VI. Academic Press, New York, in press.
63. Cockburn, A. F., Taylor, W. C., Newkirk, M. J., and Firtel, R. A., in press.
64. Taylor, W. C., Cockburn, A. C., Frankel, G., Newkirk, M. J., and Firtel, R. A., manuscript in preparation.
65. Batts-Young, B., Maizels, N., and Lodish, H. F., in press.
66. Vogt, V. M., and Braun, R. (1976). J. Mol. Biol. 106:567.
67. Karrer, K. M., and Gall, J. F. (1976). J. Mol. Biol. 104:421.
68. Engberg, J., Anderson, P., and Leick, V. (1976). J. Mol. Biol. 104:455.
69. Kimmel, A. (1977). Ph.D. thesis, University of Rochester.
70. Rubin, G. M., and Sulston, J. E. (1973). J. Mol. Biol. 79:521.
71. Davis, R., personal communication.
72. Newkirk, M. J., and Firtel, R., unpublished observations.
73. Tartof, K. D., and Dawid, I. B. (1976). Nature 263:27.
74. Pellegrini, M., Manning, J., and Davidson, N., in press.
75. Wellauer, P. K., Reeder, R. H., Carroll, D., Brown, D. D., Deutch, A., Higashinakagawa, T., and Dawid, I. B. (1974). Proc. Natl. Acad. Sci. U.S.A. 71:2823.
76. Wellauer, P. K., Reeder, R. H., Dawid, I. B., and Brown, D. D. (1976). J. Mol. Biol. 105:487.
77. Firtel, R. A., Jacobson, A., and Lodish, H. F. (1972). Nature (New Biol.) 239:225.
78. Firtel, R. A., and Jacobson, A., unpublished observations.
79. Sheiness, D., and Darnell, J. E. (1973). Nature (New Biol.) 241:205.
80. Jacobson, A., and Palatnik, C., unpublished observations.
81. Jacobson, A., unpublished observations.
82. Firtel, R. A., unpublished observations.
83. Lodish, H. F., Jacobson, A., Firtel, R., Alton, T., and Tuchman, J. (1974). Proc. Natl. Acad. Sci. U.S.A. 71:5103.
84. Lodish, H. F., Alton, T., Margolski, S., Dottin, R., and Weiner, A. *In* C. F. Fox (ed), Developmental Biology. Benjamin, New York, in press.
85. Ma, G., and Firtel, R. A., unpublished observations.
86. Kessin, R. (1971). Ph.D. thesis, Brandeis University.
87. Shatkin, A. (1977). Cell 9:645.
88. Dottin, R. P., Weiner, A. M., and Lodish, H. L. (1976). Cell 8:233.
89. Weiner, A., Dottin, R., and Lodish, H. F., personal communication.

90. Jacobson, A., Lane, C. D., and Alton, J. (1975). *In* D. Schlessinger (ed.), Microbiology 1975, p. 490. ASM Press, Washington, D.C.
91. Firtel, R. A., and Pederson, T. (1975). Proc. Natl. Acad. Sci. U.S.A. 72:301.
92. Wensink, P. C., Finnegan, D. J., Donelson, J. E., and Hogness, D. S. (1974). Cell 3:315.
93. Glover, D. M., White, R. L., Finnegan, D. J., and Hogness, D. S. (1975). Cell 5:149.
94. Kindle, K., and Firtel, R. A., manuscript in preparation.
95. Kindle, K. L., and Firtel, R. A. UCLA Symposium on Molecular and Cell Biology, Vol. VI. Academic Press, New York, in press.
96. Higuchi, R., Paddock, G. V., Wall, R., and Salser, W. (1976). Proc. Natl. Acad. Sci. U.S.A. 73:3671.
97. Maniatis, T., Kee, S. G., Elstradiadis, A., and Kafatos, F. C. (1976). Cell 8:163.
98. Roewekamp, W., Firtel, R. A., Higuchi, R., and Salser, W., manuscript in preparation.
99. Roewekamp, W., and Firtel, R. A., manuscript in preparation.
100. Schachat, F. H., and Hogness, D. S. (1974). Cold Spring Harbor Symp. Quant. Biol. 38:371.
101. Southern, E. M. (1975). J. Mol. Biol. 98:503.
102. Maniatis, T., Jeffrey, A., and Kleid, D. G. (1975). Proc. Natl. Acad. Sci. U.S.A. 72:1184.
103. Firtel, R. A., unpublished observations.
104. Pong, S. S., and Loomis, W. F. (1973). J. Biol. Chem. 248:3933.
105. Roeder, R. G. (1976). *In* R. Losick and M. Chamberlin (eds.), RNA Polymerase, p. 285. Cold Spring Harbor Laboratory Press, Cold Spring Harbor.
106. Yagura, T., Yanazisawa, M., and Iwabuchi, M. (1976). Biochem. Biophys. Res. Commun. 68:183.
107. Soll, D., and Sussman, M. (1973). Biochem. Biophys. Acta 319:312.
108. Yager, J., and Soll, D. R. (1975). Biochem. Biophys. Acta 390:46.
109. Farrell, C. A., and DeToma, F. J. (1973). Biochem. Biophys. Res. Commun. 54:1504.
110. Jacobson, A. (1976). *In* J. Last and J. Laskin (eds.), Eukaryotes at the Subcellular Level, p. 161. Marcel Dekker.
111. Mabie, C., Lowney, K., and Jacobson, A., manuscript in preparation.
112. Coukell, M. B., and Walker, I. O. (1973). Cell Differ. 2:87.
113. Charlesworth, M. C., and Parish, R. W. (1975). Eur. J. Biochem. 54:307.
114. Osborn, P. J., and Ashworth, J. M. (1975). Cell Differ. 4:237.
115. Pederson, T. Biochemistry, in press.
116. Kestler, D. P., and Rossomosido, E. F., personal communication.
117. Loomis, L. D., Rossomondo, E. F., and Chang, L. M. S. (1976). Biochem. Biophys. Acta 425:469.

Index